IEEE Recommended Practice for Electric Power Systems in Commercial Buildings

Published by
The Institute of Electrical and Electronics Engineers, Inc

Distributed in cooperation with
Wiley-Interscience, a division of John Wiley & Sons, Inc

ANSI/IEEE
Std 241-1983
(Revision of IEEE
Std 241-1974)

An American National Standard

IEEE Recommended Practice for Electric Power Systems in Commercial Buildings

Sponsor

**Commercial Buildings Power Systems Committee
of the
IEEE Industry Applications Society**

Approved September 17, 1981

IEEE Standards Board

Approved July 8, 1983

American National Standards Institute

Second Printing
September 1984

ISBN 0-471-89357-9

Library of Congress Catalog Number 82-83449

© Copyright 1983 by

The Institute of Electrical and Electronics Engineers, Inc

January 16, 1983   *SH08805*

Foreword

(This Foreword is not a part of ANSI/IEEE Std 241-1983, Recommended Practice for Electric Power Systems in Commercial Buildings.)

Over 30 years ago AIEE published a manual entitled "Interior Wiring Design for Commercial Buildings." This was followed in 1964 by the first edition of IEEE Std 241, Electrical Systems for Commercial Buildings, known as the Gray Book. In 1974 a first revision to this standard was published. This publication is the second revision.

The purpose of the IEEE Gray Book is to promote the use of sound engineering principles in the design of commercial buildings. It is hoped that it will alert the electrical engineer or designer to the many problems that are encountered in designing electrical systems for commercial buildings and to develop a concern for the professional aspects of commercial building engineering. The IEEE Gray Book is not intended to be a complete handbook; however, it can direct the engineer to texts, periodicals, and references, for commercial buildings and in particular act as a guide through the myriad of codes, standards, and practices published by the IEEE and other professional associations and governmental bodies.

The standard has been revised by updating all sections and by rewriting many paragraphs in most of the sections. Section 14, Facility Automation has been completely rewritten. Section, 17, Energy Conservation has been added.

This revised standard is published under the auspices of the IEEE Standards Board which assured a complete review and ballot throughout the IAS. The standard was developed on a cencensus basis; differences of opinion, which in most cases were over the approach to a particular subject, were resolved through discussions with the parties concerned. Seventy contributors, who are employed by users, consultants, engineering companies, utilities, governmental and similar organizations, or who are engineers who have retired but remained professionally active, have made this edition possible. Contributions by individuals ranged from changing a few words to the writing of entire sections.

The Commercial Buildings Committee of the IAS is eager to receive comments and suggestions for future revisions. These should be directed to the

Secretary
IEEE Standards Board
345 East 47 Street
New York, NY 10017

When the IEEE Standards Board approved this standard on September 17, 1981, it had the following membership:

I. N. Howell, Jr, *Chairman*   **Irving Kolodny,** *Vice Chairman*

Sava I. Sherr, *Secretary*

G. Y. R. Allen	Jay Forster	F. Rosa
J. J. Archambault	Kurt Greene	R. W. Seelbach
J. H. Beall	Loering M. Johnson	J. A. Stewart
J. T. Boettger	Joseph L. Koepfinger	W. E. Vannah
Edward Chelotti	J. E. May	Virginius N. Vaughan, Jr
Edward J. Cohen	Donald T. Michael*	Art Wall
Len S. Corey	J. P. Riganati	Robert E. Weiler

*Member emeritus

Electric Power Systems in Commercial Buildings

3rd Edition

Electric Power Systems in Commercial Buildings

Working Group Members and Contributors

Thomas E. Sparling, *Chairman*

Richard H. McFadden, *Chairman Publications*

Industry Power Systems Department

Chapter 1—Introduction: Daniel L. Goldberg, *Chairman;* Samuel Bogen, Donald T. Michael, Walter E. Thomas, R. Gerald Irvine

Chapter 2—Load Characteristics: W. E. Thomas, *Chairman;* Louis F. Flagg, Daniel L. Goldberg, Charles R. Heising, George Hutchinson, R. Gerald Irvine, George J. Lavoie, Alfred B. Marden, Thomas E. Sparling, William Tao, Hugh Wayne

Chapter 3—Voltage Considerations: Donald T. Michael, *Chairman;* Carl E. Becker, Donald S. Brereton, A. B. Gospel, C. T. Hutchinson, L. Ilgen, R. Gerald Irvine, D. R. Kinitz, George J. Lavoie, R. Miller, James W. Patterson, R. C. Seebald

Chapter 4—Power Sources and Distribution Systems: Alfred B. Marden, *Chairman;* René Castenschiold, John Erb, Andrew J. Sebold, Hugh Wayne

Chapter 5—Power-Distribution Apparatus: John Erb, *Chairman;* René Castenschiold, John Erb, J. Ferguson, J. Frank, Arthur Freund, H. Greene, A. A. Regotti, D. Stirling, K. W. Swain, J. C. Wilson

Chapter 6—Controllers: René Castenschiold, *Chairman;* Manuel J. DeLerno, Frank W. Kussy, James M. Rice

Chapter 7—Service, Vaults, and Electrical Equipment Rooms: Marvin W. Anderson, *Chairman;* Andrew J. Sebold, Thomas E. Sparling

Chapter 8—Wiring Systems: Frank C. Johnston, *Chairman;* James Daly, Russell Ohlson, Howard Stickley

Chapter 9—Systems Protection and Coordination: Richard Koestner, *Co-Chairman,* Thomas E. Sparling, *Co-Chairman;* René Castenschiold, John Erb, R. Gerald Irvine, George J. Lavoie, Walter E. Thomas

Chapter 10—Lighting: William Fisher, *Chairman;* René Castenschiold, Daniel L. Goldberg, George T. Hutchinson, R. Gerald Irvine, Al Kiener, Laurence J. Maloney, Don Ross

Chapter 11—Electric Space-Conditioning: James R. Pfafflin, *Chairman;* William H. Chadwick, Edgar D. Kauffman, Laurence J. Maloney, Alfred B. Marden, Morris L. Markel, Thomas E. Sparling

Chapter 12—Transportation: Elmer H. Sumka, *Chairman;* René Castenschiold, Edward A. Donoghue, John J. Faup, William S. Lewis, George R. Strakosch, Robert W. Young

Chapter 13—Communication Systems Planning: Robert Reese, *Chairman;* Tony D'Urso, R. M. Johnson, Harry Pfister, Robert Turner, Robert Greenquist, Ahmed Nagia, Orv Evinrude

Chapter 14—Facility Automation Systems: Morton Isaacs, *Chairman;* George S. Rye, Robert J. Boles, Robert Wanek, Herman Turk

Chapter 15—Expansion, Modernization, and Rehabilitation: Daniel L. Goldberg, *Chairman;* S. Leo Perlstein, Louis F. Flagg, R. Gerald Irvine

Chapter 16—Special Requirements by Occupany: Thomas E. Sparling, *Chairman;* Jerry L. Dinsdale, Daniel L. Goldberg, George T. Hutchinson, George J. Lavoie, R. Gerald Irvine

Chapter 17—Energy Conservation: Eugene R. Smith, *Chairman;* Daniel L. Goldberg, E. G. Kiener, Carl E. Becker, René Castenschiold

Contents

FIGURES

1. Introduction

1.1 General Discussion. IEEE Std 241-1983, IEEE Recommended Practice for Electric Power Systems in Commercial Buildings commonly known as the IEEE Gray Book, is published by the Institute of Electrical and Electronics Engineers (IEEE) to provide a recommended practice for the electrical design for commercial buildings. It has been prepared on a voluntary basis by engineers and designers functioning as the Gray Book Working Group within the IEEE Commercial Building Power Systems Committee.

This recommended practice will probably be of greatest value to the power-oriented engineer with limited commercial-building experience. It can also be an aid to all engineers responsible for the electrical design of commercial buildings. However, it is not intended as a replacement for the many excellent engineering texts and handbooks commonly in use, nor is it detailed enough to be a design manual. It should be considered a guide and general reference on electrical design for commercial buildings.

Tables, charts, and other information which have been extracted from codes, standards, and other technical literature are included in this recommended practice. Their inclusion is for illustrative purposes; where the correctness of the item is important, the referenced document should be used to assure that the information is complete, up-to-date, and correct. It is not possible to reproduce the full text of these items in this recommended practice.

1.2 Commercial Buildings. The term *commercial, residential,* and *institutional buildings,* as used here, encompasses all buildings other than industrial buildings and private dwellings. It includes office and apartment buildings, hotels, schools, and churches, steamship piers, air, railway and bus terminals, department stores, retail shops, governmental buildings, hospitals, nursing homes, mental and correctional institutions, theatres, sports arenas, and other buildings serving the public directly. Buildings, or parts of buildings, within industrial complexes, which are used as offices or medical facilities or for similar nonindustrial purposes, logically fall within the scope of this recommended practice. Thus the specific use of the building in question, rather than the nature of the overall development of which it is part, determines its electrical design category.

While industrial plants are primarily machine- and production-oriented, commercial, residential, and institutional buildings are primarily people- and public-oriented. The fundamental objective of commercial building design is to provide a safe, comfortable, energy-efficient, and attractive environment for living, working, and enjoyment. The electrical design must satisfy these criteria if it is to be successful.

Today's commercial buildings, because of their increasing size and complexity, have become more and more dependent upon adequate and reliable electrical systems. One can better understand the complex nature of modern commercial buildings by examining the systems, equipment, and facilities listed in 1.2.1.

1.2.1 System Requirements for Commercial, Residential, and Institutional Buildings. The systems, equipment, and facilities which must be provided to satisfy functional requirements will vary with the type of building, but will generally include some, or all, of the following.

(1) Lighting: Interior and exterior, both utilitarian and decorative

(2) Communications: Telephone, telegraph, computer link, radio, closed-circuit television, code call, public address, paging, electronic intercommunication, pneumatic tube, doctors' and nurses' call, and a variety of other signal systems

(3) Fire alarm and control: Fire pumps and sprinkler, fire-detection, and alarm systems

(4) Transportation: Elevators, moving stairways, dumbwaiters, and moving sidewalks

(5) Space-conditioning: Heating, ventilation, and air-conditioning

(6) Sanitation: Garbage and rubbish storage and removal, incinerators, and sewage handling

(7) Plumbing: Hot- and cold- water systems and water treatment facilities

(8) Security watchmen and burglar alarms, electronic access systems

(9) Business machines: Typewriters, computers, calculating machines, and duplicating machines

(10) Refrigeration equipment

(11) Food handling and preparation facilities

(12) Building maintenance facilities

(13) Lightning protection

(14) Automated building control systems

(15) Entertainment facilities and specialized audio-visual and lighting systems

(16) Medical facilities

1.2.2 Electrical Design Elements. In spite of the wide variety of commercial, residential, and institutional buildings, some electrical design elements are common to all. These elements, listed below, will be discussed generally in this section and in detail in the remaining sections of this recommended practice. The principal design elements which must be considered in the design of the power, lighting, and auxiliary systems include

(1) Magnitudes, characteristics, demand, and coincidence or diversity of loads and load factors

(2) Service, distribution, and utilization voltages and voltage regulation

(3) Flexibility and provisions for expansion

(4) Reliability

(5) Safety of personnel and property

(6) Initial and maintained cost

(7) Operation and maintenance

(8) Fault current and system coordination

(9) Power sources

(10) Distribution systems

(11) Standby and emergency power

(12) Energy conservation and demand and energy control

(13) Conformance with regulatory requirements

1.2.3 Grouping of Commercial Buildings. Arbitrary principal groupings of commercial buildings are:

(1) Large multistory buildings, office buildings, and apartment buildings

(2) Small public buildings and stores such as retail shops and supermarkets

(3) Institutional buildings such as hospitals, large schools and colleges

(4) Airport, railroad, and other transportation terminals

1.3 IEEE Publications. The IEEE publishes several standards similar to the IEEE Gray Book, prepared by the Industrial Power Systems department of the IEEE Industry Applications Society.[1]

[1] IEEE Std 142-1982, IEEE Recommended Practice for Grounding of Industrial and Commercial Power Systems.

[2] IEEE Std 141-1976, IEEE Recommended Practice for Electric Power Distribution for Industrial Plants.

[3] IEEE Std 242-1975, IEEE Recommended Practice for Protection and Coordination of Industrial and Commercial Power Systems.

[4] ANSI/IEEE Std 399-1980, IEEE Recommended Practice for Industrial and Commercial Power Systems Analysis.

[5] ANSI/IEEE Std 446-1980, IEEE Recommended Practice for Emergency and Standby Power Systems for Industrial and Commercial Applications.

[6] ANSI/IEEE Std 493-1980, IEEE Recommended Practice for the Design of Reliable Industrial and Commercial Power Systems.

1.3.1 Industry Applications Society. The IEEE is divided into 31 groups and societies which specialize in various technical areas of electrical engineering. Each group or society conducts meetings and publishes papers on developments within its specialized area. The Industry Applications Society (IAS) presently encompasses 23 technical committees covering electrical engineering in specific areas (petroleum and chemical industry, cement industry, glass industry, industrial and commercial power systems, and others). Papers of interest to electrical engineers and designers involved in the field covered by the IEEE Gray Book are, for the most part, contained in the *Transactions* of the *IAS*.

IEEE Std 241-1982 is published by the IEEE on behalf of the Commercial Buildings Committee of the Industrial Power Systems Department of the Industrial Applications Society acting through the Gray Book Working Group. Individuals who desire to participate in the activities of the committees, subcommittees, or working groups in the preparation of revision of texts such as this should write or call IEEE Standards Office, 345 East 47 Street, New York, NY 10017.

1.4 Professional Registration. Most regulatory agencies require that design for public and other commercial buildings be prepared under the jurisdiction of state-licensed professional architects or engineers. Information on such registration may be obtained from the appro-

---

[1] These documents can be obtained from the Institute of Electrical and Electronics Engineers, Service Center, 445 Hoes Lane, Piscataway, N J 08854.

priate state agency or from the local chapter of the National Society of Professional Engineers.

To facilitate obtaining registration in different states under the reciprocity rule, a national professional certificate is issued by the National Bureau of Engineering Registration[2] to engineers who obtained their home-state license by examination. All engineering graduates are encouraged to start on the path to full registration by taking the engineer-in-training examination as soon after graduation as possible. The final written examination in the field of specialization is usually conducted after four years of progressive professional experience.

1.5 Codes and Standards

1.5.1 National Electrical Code. The electrical wiring and design recommendations in the National Electrical Code (NEC), ANSI/NFPA 70-1981, are vitally important guidelines for commercial-building engineers. The NEC is revised every three years, and care should be taken to use the edition that is current at the time of construction. The NEC is published by and available from the National Fire Protection Association (NFPA),[3] and the American National Standards Institute (ANSI)[4]. It does not represent a design specification but only identifies minimum requirements for the safe installation and utilization of electricity in the building. It is strongly recommended that the introduction to the NEC, Article 90, covering purpose and scope, be carefully read. The NEC is also available from the American National Standards Institute and from each state's Board of Fire Underwriters, usually located in the State Capitol. The NFPA *Handbook*, of the *National Electrical Code*, sponsored by the NFPA, SPP-6C-1981, contains the complete NEC text plus explanations. This book is edited to correspond with each edition of the NEC.

The NFPA publishes the following related documents containing requirements on electrical equipment and systems:

(1) NFPA No PPH-1476, Fire Protection Handbook (1976) 14th Edition

(2) NFPA No 101-HBK, Life Safety Code Handbook (1978) 1st Edition

(3) NFPA No 20, Centrifugal Fire Pumps 1978

(4) NFPA No 70 B, Electrical Equipment Maintenance 1977

(5) NFPA No 70 E, Electrical Safety Requirements for Employee work places

(6) NFPA No 71, Central Station Signaling Systems 1977

(7) NFPA No 72 A, Local Protective Signaling Systems 1979

(8) NFPA No 72 B, Auxiliary Protective Signaling Systems 1979

(9) NFPA No 72 C, Remote Station Protective Signaling Systems 1975

(10) NFPA No 72 D, Proprietary Protective Signaling Systems 1979

(11) NFPA No 72 E, Automatic Fire Detectors 1978

(12) NFPA No 75, Protection of Electronic Computer/Data Processing Equipment 1976

(13) NFPA No 76 A, Essential Electrical Systems for Health Care Facilities 1977

(14) NFPA No 76 C, The Safe Use of High Frequency Electricity in Health Care Facilities 1975

(15) NFPA No 77, Static Electricity 1977

---

[2] PO Drawer 1404, Columbia, SC

[3] National Fire Protection Association, Batterymarch Park, Quincy, MA 02269.

[4] Copies of the National Electrical Code are available from the Sales department of American National Standards Institute, 1430 Broadway, New York, NY 10018.

(16) NFPA No 78, Lightning Protection Code 1980

(17) NFPA No 101, Life Safety Code 1976

Applicable Documents in Preparation[5]

1.5.2 Local State and Federal Codes and Regulations. While most municipalities, counties, and states use the NEC without change, some have their own codes. In some instances the NEC is adopted by local ordinance as part of the building code, with deviations from the NEC listed as addenda. It is important to note that only the code adopted as of a certain date is official, and that governmental bodies may delay adopting the latest code. Federal rulings may require use of the latest NEC regardless of local rulings, so that reference to the enforcing agencies for interpretation on this point may be necessary.

Some city and state codes are almost as extensive as the NEC. It is generally accepted that in the case of conflict, the more stringent or severe interpretation applies. Generally the entity responsible for enforcing the code has the power to interpret it.

Failure to comply with NEC or local code requirements can affect the owner's ability to obtain a certificate of occupancy and may have a negative effect on insurance.

Legislation by the US Federal Government has had the effect of giving standards, such as those of the American National Standards Institute (ANSI), the impact of law. The Occupational Safety and Health Act, administered by the US Department of Labor, permits federal enforcement of codes and standards. The Occupational Safety and Health Administration (OSHA) has adopted the 1971 (or later) NEC for new electrical installations and equipment within the scope of subpart S—Electrical — of OSHA regulations and also for major replacements, modification, or repair installed after March 5, 1972. A few articles and sections of the NEC have been deemed by OSHA to apply retroactively.

A number of states have enacted legislation embodying various energy conservation standards such as ASHRAE/IES 90-80, Energy Conservation in New Building Design.[6] Such standards establish energy or power budgets that materially affect architectural, mechanical, and electrical designs.

1.5.3 Standards and Recommended Practices. A number of organizations in addition to the NFPA publish documents which affect electrical design. Adherence to these documents can be written into design specifications.

The American National Standards Institute (ANSI) approves the standards of many other organizations. ANSI coordinates the review of proposed standards among all interested affiliated societies and organizations to assure a concensus approval. It is in effect a *clearing house* for technical standards of all types.[4]

Underwriters Laboratories, Inc[7] (UL) is a non-profit organization, operating

---

[5] When the following documents are completed, approved and published, they become a part of this listing.

NFPA No 72 F, Emergency Communications Systems in High Rise and Other Occupied Buildings

NFPA No 100, Emergency Power Supply Systems

[6] American Society of Heating, Refrigerating, and Air-Conditioning Engineers, 1791 Tullie Circle, N.E., Atlanta, GA 30329.

[7] 1285 Walt Whitman Road, Melville, NY 11746

laboratories for investigation with respect to hazards affecting life and property, materials and products, especially electrical appliances and equipment. In so doing, they develop test standards. Equipment which has been tested by UL, and found to conform to their standards, is known as *listed* or *labeled* equipment.

The Edison Electric Institute[8] (EEI), representative of the investor-owned utilities, publishes several informative handbooks such as

(1) Electric Heating and Cooling Handbook

(2) A Planning Guide for Architects and Engineers

(3) A Planning Guide for Hotels and Motels

(4) Electric Space-Conditioning

(5) Industrial and Commercial Power Distribution

(6) Industrial and Commercial Lighting

(7) An energy analysis computer program, AXCESS, for forecasting electricity consumption and costs in existing and new buildings.

The National Electrical Manufacturers Association[9] (NEMA) represents equipment manufacturers. Their publications serve to standardize the manufacture of, and provide testing and operating standards for electrical equipment. The design engineer should be aware of any NEMA standard which might affect the application of any equipment specified by him.

The IEEE publishes several hundred electrical standards relating to safety, measurements, equipment testing, application, and maintenance. The following three publications are general in nature:

(1) ANSI/IEEE Std 100-1977, IEEE Standard Dictionary of Electrical and Electronics Terms.

(2) ANSI/IEEE Std 315-1975, Graphic Symbols for Electrical and Electronics Diagrams.

(3) ANSI Y32.9-1972, American National Standard Graphic Symbols for Electrical Wiring and Layout Diagrams Used in Architecture and Building Construction, is of importance for the preparation of plans.

The Electric Energy Association[10] (EEA) publishes specification sheets for electrically space-conditioned buildings as well as other informative pamphlets.

1.6 Handbooks. The following handbooks have, over the years, established reputations in the electrical field. This list is not intended to be all-inclusive, and other excellent references in addition to those listed in 1.3 are available, but are not listed here because of space limitations.

(1) FINK, D.G., and BEATY, H.W., *Standard Handbook for Electrical Engineers,* 11th ed. New York: McGraw-Hill, 1978. Virtually the entire field of electrical engineering is treated, including equipment and systems design.

(2) CROFT, T., CARR, C.C., WATT, J.H., *American Electricians Handbook,* ninth ed. New York: McGraw-Hill, 1970. The practical aspects of equipment, construction and installation are covered.

(3) *Lighting Handbook,* Illuminating Engineering Society,[11] (IES) 1981. All aspects of lighting, including seeing, recommended lighting levels, lighting calculations, and design are included in extensive detail in this comprehensive text.

---

[8] 1111 19th Street, Washington, DC 20036

[9] 2101 L Street, Suite 300, Washington, D.C. 20037

[10] 437 Madison Avenue, New York, NY 10022

[11] 345 East 47 Street, New York, NY 10017

(4) *Electrical Transmission and Distribution Reference Book*, (Westinghouse Electric Corporation[12], 1964). All aspects of transmission and distribution, performance and protection, are included in detail.

(5) *Applied Protective Relaying* (Westinghouse Electric Corporation, 1976). The application of protective relaying to commercial-utility interconnections, protection of high-voltage motors, transformers, cables, are covered in detail.

(6) *ASHRAE Handbook* (American Society of Heating, Refrigerating, and Air-Conditioning Engineers[13] (ASHRAE). This series of reference books in four volumes, which are periodically updated, detail the electrical and mechanical aspects of space conditioning and refrigeration.

(7) *Motor Applications and Maintenance Handbook*, SMEATON, R.S., ed. McGraw-Hill, 1969. Contains extensive, detailed coverage of motor load data and motor characteristics to better coordinate electric motors with machine mechanical characteristics.

(8) *Industrial Power Systems Handbook*, BEEMAN, D.L., ed. McGraw-Hill, 1955. A text on electrical design with emphasis on equipment, including that applicable to commercial buildings.

(9) *Electrical Maintenance Hints* (Westinghouse Electric Corporation, 1974). The preventative maintenance procedures for all types of electrical equipment and the rehabilitation of damaged apparatus are discussed.

(10) *Lighting Handbook* (Westinghouse Electric Corporation, 1976). The application of various light sources,

fixtures, and ballasts to interior and exterior commercial, industrial, sports, and roadway lighting projects.

(11) *Underground Systems Reference Book* (Edison Electric Institute, 1957). The principles of underground construction and the detailed design of vault installations, cable systems, and related power systems are fully illustrated, cable splicing design parameters are thoroughly covered.

(12) *Switchgear and Control Handbook*, SMEATON, R.S., ed. McGraw-Hill, 1977. Concise, reliable guide to important facets of switchgear and control design, safety, application, and maintenance including high- and low-voltage starters, circuit breakers, and fuses.

(13) *How to Design Electrical Systems*, McPARTLAND, J., and the editors of Electrical Construction and Maintenance magazine. McGraw-Hill, 1968.

McPARTLAND, J., and NOVAK, W. *Electrical Design Details*. McGraw-Hill, 1960. These references offer an insight for the younger engineer into the systems approach to electrical design.

A few of the older texts may not be available for purchase, but are available in most professional offices and libraries.

1.7 Periodicals. Spectrum, the basic monthly publication of the IEEE, covers all aspects of electrical and electronic engineering with limited material on commercial buildings. This publication, however, does contain references to IEEE books and other publications, technical meetings and conferences, IEEE group, society, and committee activities, abstracts of papers and publications of the IEEE and other organizations and other material essential to the professional advancement of the elec-

---

[12] Printers Division, Forbes Road, Trafford, PA 15085

[13] ASHRAE, Publication Sales, 1791 Tullie Circle, N.E., Atlanta, GA 30329

trical engineer. The Transactions of the Industrial Applications Society of the IEEE are directly useful to commercial building electrical engineers.

Following are some other well-known periodicals:

(1) *ASHRAE Journal*, American Society of Heating, Refrigerating and Air-Conditioning Engineers.[13]

(2) *Electrical Construction and Maintenance*, McGraw-Hill, 1221 Avenue of the Americas, New York, NY 10020.

(3) *Electrical Consultant*, One River Road, Cos Cob, CT 06807.

(4) *Fire Journal*, National Fire Protection Association, Batterymarch Park, Quincy, MA 00269.

(5) *IAEI News*, International Association of Electrical Inspectors, 802 Busse Highway, Park Ridge, IL 60068.

(6) *Lighting Design and Application*, Illuminating Engineering Society, 345 East 47th Street, New York 10017.

(7) *Plant Engineering*, 1301 South Grove Avenue, Barrington, IL 60010.

(8) *Professional Engineer*, National Society of Professional Engineers, 2029 K Street Northwest, Washington, DC 20006.

(9) *Specifying Engineer*, Cahners Publishing Company, 5 South Wabash Avenue, Chicago, IL 60603.

1.8 Manufacturers' Data. The electrical industry through its associations and individual manufacturers of electrical equipment issues many technical bulletins, data books, and magazines. While some of this information is difficult to obtain, copies should be available to each major design unit. The advertising sections of electrical magazines contain excellent material, usually well illustrated and presented in a clear and readable form, concerning the construc-tion and application of equipment. Such literature may be promotional; it may present the advertiser's equipment or methods in a *best light* and should be carefully evaluated. Manufacturers' catalogs are a valuable source of equipment information. Some manufacturers complete catalogs are quite extensive covering several volumes. However, these companies may issue condensed catalogs for general use. A few manufacturers publish regularly scheduled magazines containing news of new products and actual applications. Data sheets referring to specific items are almost always available from the sales offices. Some technical files may be kept in microfilm for use either by projection or by printing at larger design offices. Manufacturers' representatives, both sales and technical, can do much to provide complete information on a product.

1.9 Safety. Safety of life and preservation of property are two of the most important factors in the design of the electrical system. This is especially true in commercial buildings because of public occupancy, throughfare, and high occupancy density. In many commercial buildings the systems operating staff have very limited technical capabilities and may not have any specific electrical training.

Various codes provide rules and regulations as minimum safeguards of life and property. The electrical design engineer must often provide greater safeguards than outlined in the codes according to his best judgement, while also giving consideration to utilization and economies.

Personnel safety may be divided into two categories

(1) Safety for maintenance and operating personnel

(2) Safety for the general public

Safety for maintenance and operating personnel is achieved through proper design and selection of equipment with regard to enclosures, key-interlocking, circuit breaker and fuse interrupting capacity, the use of high-speed fault-detection and circuit-opening devices, clearances from structural members, grounding methods, and identification of equipment.

Safety for the general public requires that all circuit-making and-breaking equipment as well as other electrical apparatus be isolated from casual contact. This is achieved by using locked rooms and enclosures, proper grounding, limiting of fault levels, installation of barriers and other isolation, proper clearances, adequate insulation, and other similar provisions outlined in this standard.

The National Electrical Safety Code (NESC), ANSI C2, 1981 edition is available from the IEEE and ANSI. A cloth-bound hardcover edition may be obtained from John Wiley & Sons Inc, New York. It covers outdoor distribution systems, supply and communications systems, overhead lines, high-voltage systems, and other items related to the supply of building power.

Circuit protection is a fundamental safety requirement of all electrical systems. Adequate interrupting capacities are required in services, feeders, and branch circuits. Selective, automatic isolation of faulted circuits represents good engineering. Fault protection should be designed and coordinated throughout the system down to the last branch circuit overcurrent device. Physical protection of wiring by means of approved raceways under all probable conditions of exposure to electrical, chemical, and mechanical damage is necessary. Such raceways should be of sufficient size for future expansion. Additional raceways and spare conductors for future use may be installed within allowable financial constraints. If the raceways are properly constructed and bonded, they can minimize power interruptions. The design engineer should locate equipment where suitable ambient temperatures exist and ventilation is available. The operation of fault-detection and circuit-interruption devices under conditions of abnormal voltage and frequency should be assured.

1.9.1 Appliances and Equipment. Improperly applied or inferior materials can cause electrical failures. The use of appliances and equipment listed by the Underwriters Laboratories, Inc, or other approved laboratories is recommended. The Association of Home Appliance Manufacturers[14] (AHAM) and the Air-Conditioning and Refrigeration Institute[15] (ARI) specify the manufacture, testing, and application of many common appliances and equipment.

High-voltage equipment is manufactured in accordance with NEMA, ANSI, and IEEE standards, and the engineer should make sure that the equipment he specifies conforms to these standards. Properly prepared specifications can prevent the purchase of inferior or unsuitable equipment. The lowest initial purchase price may not result in the lowest cost after taking into consideration operating, maintenance, and owning costs. Value engineering is an organized approach to identification of unnecessary costs utilizing such methods as life cycles, cost analysis, and related techniques.

---

[14] 20 North Wacker Drive, Chicago, IL 60606
[15] 1815 North Fort Myer Drive, Arlington, VA 22209

1.9.2 Operational Considerations. When the design engineer lays out equipment rooms and locates electrical devices, he cannot always avoid having some areas accessible to *unqualified* persons. Dead-front construction should be utilized whenever practical. Where dead-front construction is not available, all exposed electrical equipment should be placed behind locked doors or gates. This will result in a reduction in electrical failures caused by human error, as well as improved safety.

A serious cause of failure, which is attributable to human error, is unintentional grounding or phase-to-phase short circuiting of equipment which is being worked upon. By careful design such as proper spacing and barriers, and by enforcement of published work-safety rules, the engineer can minimize unintentional grounding and phase-to-phase, and ground faults in the distribution equipment. High-quality workmanship is an important factor in the prevention of electrical failures. Therefore, the design should incorporate features that are conducive to good workmanship.

Protective devices such as ground-fault relays and ground-fault detectors (ungrounded systems) will minimize damage from electrical failures. Electrical fire and smoke can cause building staff to disconnect all electric power to a building, even if there is not direct danger to the occupants. Electrical failures which involve smoke and noise, even though occurring in nonpublic areas, may frighten building occupants.

1.10 Maintenance. Maintenance is essential to proper operation. The installation should be so designed that building personnel can perform most of the maintenance with a minimum need for specialized services. Design details should provide proper space and accessibility so that equipment can be maintained without difficulty and excessive cost.

The engineer should consider the effects of a failure in the system supplying the building. Generally, the external systems are operated and maintained by the electrical utility, though at times they are a part of the commercial building distribution system.

Where continuity of service is essential, suitable emergency and standby equipment should be provided. Such equipment is needed to maintain minimum lighting requirements for passageways, stairways, and critical areas as well as to supply power to critical loads. These systems are usually installed within the building, and they include automatic or manual equipment for transferring loads on loss of normal supply power or for putting battery- or generator-fed equipment into service.

1.11 Design Considerations. Electrical equipment usually occupies a relatively small percentage of the total building space, and in design it may be easier to relocate electrical service areas than mechanical areas or structural elements. Allocation of space for electrical areas is often given secondary consideration by architectural and related specialties. In the competing search for space, the electrical engineer is responsible for fulfilling the requirements for a proper electrical installation while at the same time recognizing the flexibility of electrical systems in terms of layout and placement.

Today, architectural considerations and appearances are of paramount importance in determining the salability of a building. Aesthetic considerations may play an important role in the selection of equipment, especially lighting equipment. Provided that the dictates of good practice, code requirements, and environmental considerations are not violated, the electrical engineer may have to compromise in his design to accommodate the desires of other members of the design team.

1.11.1 Coordination of Design. The electrical engineer is concerned with professional associates such as the architect, the mechanical engineer, the structural engineer, and where underground services are involved, the civil engineer. He is also concerned with the builder and the building owner or operator who, as clients, may take an active interest in the design. More often the electrical engineer will work directly with the coordinator of overall design activities, usually the architect. He must cooperate with the safety engineer, fire protection engineer, perhaps the environmental engineer, and a host of other concerned people, such as interior decorators, all of whom have a say in the ultimate design.

The electrical designer must become familiar with local rules and know who are the authorities having jurisdiction over the design and construction. It can be embarrassing to have an electrical project held up at the last moment because proper permits have not been obtained such as, for example, a permit for a street closing to allow installation of utilities to the site. Local contractors are usually familiar with local ordinances and union work rules and can be of great help in avoiding pitfalls. Union work practices may, for reasons of safety or other considerations, discourage the use of certain materials and techniques.

In performing electrical design, it is essential, at the outset, to prepare a checklist of all the design stages that have to be considered. Major items include temporary power, access to the site, and review by others. It is important to note that certain electrical work may appear in nonelectrical sections of the specifications. For example, furnishing and connecting of electric motors may be covered in the mechanical section of the specifications. For administrative control purposes, the electrical work may be divided into a number of contracts, some of which may be under the control of a general contractor and some of which may be awarded to electrical contractors. Among items with which the designer will be concerned are preliminary cost estimates, final cost estimates, plans or drawings, specifications (which are the written presentation of the work), materials, manuals, factory inspections, laboratory tests, and temporary power. He may well be involved in providing information on how electrical considerations affect financial justification of the project in terms of owning and operating costs, amortization, return on investment, and related items.

Many electrical designs follow the concept of competitiveness in the commercial sense. Here cost is a primary consideration, and such designs tend toward minimum code requirements as the standard. There is great pressure on the designer to consider cost above maintainability and long life. However, the experienced designer can usually adopt effective compromises.

Where the owner or builder is the ultimate occupant, and in buildings such as libraries, city halls, and hospitals, considerations of safety, long life, use by the public, and even prestige may dictate a type of construction known as institutional. Such design emphasizes reliability, resistance to wear and use, safety to public, and special aesthetic considerations such as the agelessness of the structure.

Smaller buildings, shops, and stores, may provide more latitude to the designer in that he is, within budget limitations, subject to a minimum of control in selecting lighting fixtures, equipment, and accessories.

1.11.2 Flexibility. Flexibility of the electrical system means the adaptability to development and expansion as well as to changes to meet varied requirements during the life of the building. Often a designer will be faced with providing utilities to a building where the loads may be unknown. For example, many office buildings are constructed with the tenant space designs uncompleted. In some cases the designer will provide only the core utilities available for connection by others to serve the working areas. In other cases, the designer may lay out only the basic systems and, as the tenant requirements are developed, fill in the details. Often the tenant provides all of his own working space designs.

Because it is difficult and costly to increase the capacity of risers and feeders later, it is important that sufficient capacity be provided initially in these circuits. Extra conductors or raceway space should be included in the design stage if additional loads may be added later. This consideration is particularly important for commercial buildings with the increasing use of business machines and air conditioning. The cost and difficulties in obtaining space for new feeders and larger switchgear which would be required when modernizing or expanding a building may well be considered in the initial design. A load growth margin of 50% applied to the installed capacity of the major feeders is often justified where expansion is anticipated. Each project deserves careful consideration of the proper load growth margin to be allowed.

Flexibility in an electrical wiring system is enhanced by the use of oversize or spare raceways, cables, and equipment. The cost of making such provisions is usually relatively small in the initial installation.

Empty shafts and holes through floors may be provided at relatively low cost for future work. Openings through floors should be filled in with fireproof, easily removed materials to prevent the spread of fire and smoke between floors. For computer rooms and the like, flexibility is frequently provided by raised floors made of removable panels, providing access to a wiring space between the raised floor and the slab below. Engineers experienced in electrical rehabilitation realize the almost inevitable worth of such facilities.

1.11.3 Specifications. A contract for installation of electrical systems consists of a written document and drawings. The document is referred to as the *contract* and contains legal and engineering parts. The *legal* part gives the general terms of the agreement between contractor and owner, such as payment, working conditions, and time requirements; and it may include clauses on performance bonds, extra work, and damages for breach. The engineering part consists of the specifications.

The specifications give descriptions of the work to be done and the materials to be used. It is common practice in larger installations to use a standard outline format listing division, section, and subsection titles or subjects of the Contract Standards Institute (CSI). Where several specialities are involved, Divison 16 covers the electrical installation, Division 15 covers the mechanical part of the work and, in some formats, the building control system is included in Division 17. It is important to note that some electrical work will almost always be included in Divisions 15 and 17 and occasionally in other divisions. Each division has a detailed breakdown of various items, such as switchgear, motor starters and lighting equipment, specified by CSI.

In order to assist the engineer in preparing contract specifications, standard technical specifications covering construction, application, technical and installation details are available from technical publishers and manufacturers (which may require revision to avoid proprietary specifications). Large organizations such as the US Government General Services Administration develop their own standard specifications.

1.11.4 Drawings. The designer will usually be given preliminary architectural drawings as a first step. These will permit him to arrive at the preliminary scope of the work, roughly estimate the requirements for and determine in a preliminary way the location of equipment and the methods and types of lighting. In this stage of the design, such items as hung ceilings, recessed or surface mounted fixtures, and general types of distribution will be decided. It is important to discuss the plans with the senior engineer, and with the architect who has the advantage of knowing the type of construction and building finishes. The mechanical engineer will indicate the mechanical loads that will exist. It is during this period that the designer should emphasize his need for room to hang conduits, crawl spaces, special structural reinforcements for heavy equipment, special floor loadings, and many other items that may be required. It is much more difficult to obtain such special requirements once the building design has been committed.

The one-line diagrams should then be prepared in conformity with the utility's service requirements. It may be desirable to furnish the utility with a set of service entrance power requirements, requesting from them a formal service layout. Electrical drawings necessarily are based on architectural drawings and while prepared at the same time as the structural and mechanical drawings, they are usually the last ones completed because of resolution of interferences.

Checking is an essential part of the design process. The checker looks for design deficiencies in the set of plans. It is usually a shock to the young designer or draftsman when he receives his first corrected drawing marked up in red indicating all kinds of corrections that are required. The designer can help the checker by having for ready reference the catalog sections detailing the equipment he has selected. The degree of checking is a matter of design policy.

1.11.5 Manufacturers' or Shop Drawings. After the design has been completed and contracts are awarded, manufacturers and other suppliers will submit manufacturers' or shop drawings for approval or information. It is important to return these shop drawings as quickly as possible, otherwise the contractor may claim that his work was delayed by failure to receive approval or other permission to proceed. Unless drawings are unusable, it is a good idea not to reject them but to stamp the drawings *approved as noted* and mark them to show changes and corrections. The supplier can then make whatever changes are indicated and will not have to wait for a completely approved set of drawings before commencing work. In returning corrected shop drawings, remember that the contract for supplying the equipment is usually with the general contractor and that the official chain of communication is through him. Sometimes he will permit direct communication with a subcontractor or a manufacturer; however, the subject of such communication should always be confirmed in writing with the general contractor.

1.12 Estimating. A preliminary estimate is usually requested. Sometimes the nature of a preliminary estimate makes it nothing more than a good guess. Enough information is usually available however, to perform the estimate on a square-foot or similar basis. The preliminary estimate may become part of the project's economic justification.

A second estimate is often provided after the project has been clearly defined, but before any drawings have been prepared. The electrical designer can determine the type of lighting fixtures and heavy equipment which is to be used from sketches and architectural layouts. Lighting fixtures as well as most items of heavy equipment can be priced directly from the catalogs, using appropriate discounts.

The most accurate estimate is made when drawings have been completed and bids are about to be received or the contract negotiated. The estimating procedure of the designer in this case is similar to that of the contractor's estimator. It involves first the takeoffs, that is, counting the number of receptacles, lighting fixtures, lengths of wire and conduit, determining the number and types of equipment, and then applying unit costs for labor, materials, and equipment.

The use of standard estimating sheets is a big help: *Electrical Estimating* by R. ASHLEY, McGraw-Hill, 1961, contains much information. This book is most helpful in electrical estimating, and it shows various forms including those of the National Electrical Contractors' Association (NECA) which normally are available only to contractors. For the preliminary estimate, a general estimating book is suggested that gives square-foot figures and other general costs: *Building Construction Cost Data and Mechanical and Electrical Cost Data* by R. SNOW MEANS.[16] Estimating data on costs and man hours are presented in *Sherlock's Blue Book of Electrical Estimating* (Estimating Handbook Associates[17]).

A number of different estimating books are available from technical publishers. Two other books of value are:

(1) *Building Cost File*, Van Nostrand Reinhold Company. This is published annually, and in editions for different sections of the US.[18]

(2) *The Richardson Rapid System*, vol 3, *Mechanical and Electrical*, Richardson Engineering Services, Inc P O Box 1055, San Marcos, CA 92069.

The estimator-designer must include special costs such as vehicles, temporary connections, temporary or construction power, rental of special tools, scaffolding, and many other items. Because of interference with local operations, as at a public terminal, work may have to be performed during overtime periods. Electricians generally receive overtime premium pay, usually at a rate of time and a half or double time.

Electrical base pay may represent about half the total cost when considering employee benefits, overhead, and supervision. The designer will typically estimate 15 to 25% for overhead and 10% for profit, with possibly an addi-

[16] 100 Construction Plaza, Kingston, MA 02364.

[17] 5848 North Leonard Avenue, Chicago, IL 60646.

[18] 100 Construction Plaza, Kingston, MA 02364.

tional 5 to 10% markup where the electrical contractor is a subcontractor.

In pricing equipment and materials, manufacturers' catalogs can be used. There is often an appropriate discount to be applied which may be listed in the front of the catalog. The determination of the correctness of this discount and which discount table is to be used must be made through the distributor or manufacturer. Many companies publish a catalog with list prices and simply issue revised discount lists to take care of price changes. Certain items as, for example, copper cable, vary in price from day to day, dependent upon the cost of base materials.

Extra work (extras) is that work performed by the contractor which has to be added to the contract for unforeseen conditions or changes in the scope of work. The contractor is not usually faced with competition in making these changes and extra work is, therefore, expected to be more costly than the same work if included in the original contract.

Extra cost on any project can be minimized by greater attention to design details in the original planning stage. On rehabilitation or modification work, extras are more difficult to avoid; however, with careful field investigation, extras can be held to an absolute minimum.

1.13 Contracts. Contracts for construction may be awarded on either a lump-sum or a unit-price basis, or on a cost-plus (time-and-material) basis. A lump sum involves pricing the entire job as one or several major units of work. It is up to the contractor in this case to take the gamble that he has determined the quantities of work properly.

The unit-price basis simply specifies so much per unit of work, for example, so many dollars per foot of 3-inch conduit. In this case the designer takes the risk that his original estimate was sufficient to cover the actual units of work. The lump-sum contract is usually preferable where the design can be worked out in sufficient detail. The unit-price contract is desirable where it is not possible to indicate exactly the quantities of work to be performed and where a contractor, in order to provide a lump-sum contract, might have to overestimate the job to cover items which he could not determine exactly from the drawings.

If the unit-price basis is used, the estimated quantities should be as accurate as possible, otherwise it may be advantageous for the contractor to quote unit prices of certain items as high as possible and reduce other items to a minimum figure. It could be to his advantage to list those items highest on which he would receive payment first or which would be most likely to increase in quantity.

The time-and-material basis is valuable for emergency or extra work where it would be impractical to use either of the above two methods. It has the disadvantage of requiring a close audit of manpower and material expenditures of the contractor.

1.14 Building Access and Loading. Most designers are aware of the need for checking that the equipment will fit into the area specified, and that the floor load rating is adequate for the weight of the equipment. The designer is usually aware that he must check door openings and the size of elevators to get equipment into a building. However, it is easy to forget that equipment has to be moved across floors, and that the floor load ratings of the access areas for moving the equipment must be adequate. If floor strengths are not adequate, provision must be made to reinforce the floor or, more simply, to spread the weight so that the floor loading will not exceed load limitations.

It is important to review weights and loadings with the structural engineers. Sometimes it is necessary to remove windows and even to make minor structural changes in order to move heavy pieces of equipment into a building. Provisions must also be made for removal of equipment for replacement purposes.

Clearances must be in accordance with code provisions regarding working space. Clearance must also be provided for installation, maintenance, and such items as cable pulling and switchgear drawout.

It is often essential to phase items of work in order to avoid conflict with other electrical work or work of other trades.

1.15 Contractor Performance. Contractors may be selected on the basis of bid or quoted price or by negotiation. Governmental or corporate requirements may mandate the selection of the lowest qualified bidder.

Where the relative amount of electrical work is large, it may be awarded to an electrical contractor. In other instances, the electrical work may be awarded as a subcontract by the overall or general contractor, except where prohibited by state law, as in New York, for public works.

The performance of the work will usually be inspected by representatives of the owner or architect-engineer. The work is subject to the inspection of governmental and other assigned approval agencies such as insurance carriers.

The designer may communicate with the contractor only to the extent permitted by the agency exercising control over the contract: the architect, builder, or general contractor, as may be appropriate. It is essential that the designer, in attempting to expedite the contract, not place himself in the position of requesting or interpreting into the contract things not clearly required by the specifications or drawings without proper authorization.

The contract may require the contractor to deliver at the end of the work revised contract drawings, known as as-built drawings. These show all changes in the work which may have been authorized or details which were not shown on the original drawings.

1.16 Environmental Considerations. In all branches of engineering, an increasing emphasis is being placed on social, and environmental concerns. Today's engineer must consider air, water, noise, and all other items which have an environmental impact. The limited availability of energy sources and the steadily increasing cost of electric energy require a concern with energy conservation on the part of the engineer.

These items are becoming more than just a matter of conscience or professional ethics. Laws, codes, rules, and standards issued by legislative bodies, governmental agencies, public service commissions, insurance, and professional organizations (including groups whose primary concern is the protection of the environment and conservation of natural resources) increasingly require an assessment of how the project may affect the environment.

The electrical engineer may participate in studies such as total energy compared to utility power, electric heating versus fossil fuel, a comparison of boilers, purchased steam, and the heat-pump. The use of steam turbines compared to absorption units and electric drives for air conditioning may be evaluated. In these studies the effects of noise, vibration, exhaust gases, cooling methods, and energy requirements must be considered in relationship to the immediate and sometimes general environment.

Following are some examples indicating a concern for energy conservation and the environment. Methods of reducing energy usage are also covered in 2.1.1.

(1) Use of efficient lighting fixtures and a concern with the quality of lighting as well as the quantity of lighting.

(2) Dividing lighting circuits so that half or partial lighting is available during periods of building cleaning or low activity, or even during utility brownouts, and so that areas not in use can be separately darkened.

(3) Use of high power-factor ballasts, motors, and other devices which will minimize line and equipment losses.

(4) Use of automatic devices for shutting down or reducing the level of operation of nonessential equipment. Multi- or variable-speed equipment with regulator or feedback control can materially reduce energy requirements. This includes such simple devices as photocell control of lighting or even integrated and computerized control of the heating and ventilating system.

(5) Use of *waste* heat, including that of lighting fixtures, as part of the space-conditioning system.

(6) Use of materials of high availability in construction specifications. Aluminum can in a number of instances replace copper; glass or plastics may replace metals in many applications.

(7) Use of high temperature rated insulation which may result in savings of insulating, conductor, and enclosure materials, although resulting in an increase of energy loss in the conductors which should be considered.

(8) Use of low-noise equipment. Noise ratings have been developed for many types of lighting ballasts and transformers. Sound levels for transformers and rotating equipment must be considered in most commercial areas. Sound-deadening techniques for equipment rooms and enclosures may be utilized.

(9) Use of efficiency and losses to determine the acceptability of equipment. For example, the cost of rated transformer losses for various loads, calculated for given periods of time such as ten or twenty years, have been added on a weighted basis to the transformer first cost in evaluating the low bid.

(10) The use of aesthetically pleasing housing and enclosures to minimize space and construction requirements. Pad-type transformers can eliminate unsightly fences and walls. Landscaping architects can provide pleasing designs of trees and shrubbery to completely conceal outdoor substations, and, of course, overhead lines may be replaced by underground systems.

(11) Consideration of local area usage in selecting locations for electrical equipment. Substations situated in residential areas must be carefully located so as not to create a local nuisance. Floodlighting must not spill onto adjacent areas where it may provide undesirable glare or lighting levels.

Many of the excellent currently available technical texts and references were written before the present era of environmental concern. It is essential to refer to current technical and related literature to keep up to date on developments in the areas of environmental protection and energy conservation, that is, federal and local Environmental Protection Agency guidelines and judicial rulings on environmental litigation.

2. Load Characteristics

2.1 General Discussion. The electrical power distribution system in a building exists solely to serve the loads — the electrical utilization devices. The power distribution system must accomplish that assignment safely and economically, provide sufficient reliability to adequately satisfy the requirements of the building (and its users), and incorporate sufficient flexibility to accommodate changing loads during the life of the building.

This section is intended to provide typical load data and a suggested methodology for determining individual and total connected and total demand load characteristics of a commercial building. The engineer should make provisions for load growth as well as building expansion in order to provide adequate electrical capacity or provision for electrical equipment expansion during the expected life of the building.

The steadily increasing sophistication of some of the load devices — complex communication systems, electronic-data processing equipment, closed-circuit TV security systems, heating-ventilating and air-conditioning systems, and centralized automated building control systems, etc increases the difficulty of determining initial load, forecasting future loads and establishing realistic demand factors.

The electrical engineer must determine the building electrical load characteristics early in the preliminary design stage of the building in order to select the proper power distribution system and equipment having adequate power capacity with proper voltage levels, and sufficient space and ventilation to maintain proper ambients. Once the power system is determined, it often is difficult to make major changes because of coordination required with other disciplines. Architects, mechanical and structural engineers simultaneously will be developing their designs and making space and ventilation allocations. It is imperative therefore, from the start, that the electrical systems be correctly selected based on realistic load data or *best possible* typical-load estimates, or both, (because all final, finite load data is not available during the preliminary design stage of the project). When using estimated data, it should be remembered that the *typical data* applies only to the condition from which the data was taken and most likely an adjustment to fit the particular application will be required.

While much of the electrical require-

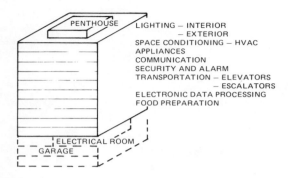

LIGHTING — INTERIOR
— EXTERIOR
SPACE CONDITIONING — HVAC
APPLIANCES
COMMUNICATION
SECURITY AND ALARM
TRANSPORTATION — ELEVATORS
— ESCALATORS
ELECTRONIC DATA PROCESSING
FOOD PREPARATION

**Fig 1
Groups of Loads in a
Typical 10-Story Commercial Building**

ments of building equipment such as ventilating, heating/cooling, lighting, etc are furnished by other disciplines, the electrical engineer must also furnish to the other disciplines such data as space, accessibility, weight and heat dissipation requirements for the electric power distribution apparatus. This involves a continuing exchange of information that starts as preliminary data and is upgraded to be increasingly accurate as the design progresses. Documentation and coordination throughout the design process is imperative.

At the beginning of the project, the electrical engineer should review the utility's rate structure and the classes of service available. Information pertaining to demand, energy, and power factor should be developed to aid in evaluating, selecting, and specifying the most advantageous utility connection. As energy resources become more costly and scarce, items such as energy efficiency, power demand minimization, and energy conservation must be closely considered to reduce both energy consumption and utility cost.

System power (that is, energy) losses

must be considered as part of the total load in sizing mains and service equipment. ANSI/NFPA 70-1981, [3][19] recommends that the total voltage drop from electrical service entrance to the load terminals of the furthest piece of equipment served should not exceed 5% of the system voltage and thus the energy loss, $I^2 R$, will correspondingly be limited.

Listed below are typical load groups and examples of classes of electrical equipment which should be considered when estimating initial and future loads.

(1) *Lighting.* Interior (general, task, exits and stairwells), exterior (decorative, parking lot, security) normal and emergency

(2) *Appliances.* Business and copying machines, receptacles for vending machines, and general use

(3) *Space Conditioning.* Heating, ventilating, cooling, cleaning, pumping, and air handling units

(4) *Plumbing and Sanitation.* Water pumps, hot water heaters, sump and sewage pumps, incinerators, and waste handling

(5) *Fire Protection.* Fire detection, alarms, and pumps

(6) *Transportation.* Elevators, dumbwaiters, conveyors, escalators, and moving walkways

(7) *Data Processing.* Central processing and peripheral equipment, Uninterruptible Power Supply (UPS) systems including related cooling

(8) *Food Preparation.* Cooling, cooking, special exhausts, dishwashing, disposing, etc

(9) *Special Loads.* For equipment and facilities in mercantile buildings, restaurants, theaters, recreation and

---

[19] The numbers in brackets correspond to those in the references at the end of this Section.

Table 1
Load Survey of Equipment Utility Requirements

Room Number	Equipment Load Description	Electrical Requirements						Load Duration Continuous or Cyclical	Mechanical Requirements							
		volts	amperes	kW	hp	phase	hertz		Steam	Hot Water	Cold Water	Waste	Gas	Air	Exhaust	Others

sports complexes, religious buildings, terminals and airports, health care facilities, laboratories, broadcasting stations, etc.

(10) *Miscellaneous Loads.* Security, central control systems, communications, audio-visual, snow melting, recreational or fitness equipment, incinerators, shredding devices and waste compactors, shop or maintenance equipment, etc

2.1.1 Load Survey. Power systems for different buildings are seldom the same because load requirements differ from building to building. Therefore, design of an electrical power distribution system should begin with a load survey to identify the size, location, and nature of the various loads. In assembling this information, Table 1, Load Survey of Equipment Utility Requirements, may be helpful. This is not an easy task; it should not be undertaken lightly.

Most of the data for making the load survey is usually obtained from those involved in designing the building and its integral systems (for example, lighting, heating-ventilating and air conditioning, and transportation). Useful information may be obtained from meter readings or measurements for similar buildings, from electric utility companies, from equipment manufacturers, associations, or from some governmental agencies.

The load survey provides an opportunity to identify the load of the utilization equipment and the voltages at which it can be served. The lighting load may be 30% to 50% of the electrical load in office-type buildings; in contrast, it may be only incidental in luxurious restaurants or hotels. Ultimately the power system must serve all the loads. The load survey allows definition of the continuity of operation that is required (for example, for safety or security of occupants such as for stairway or exit lighting, certain ventilating fans, fire pumps, availability of certain elevators for fire-fighting and rescue personnel, etc).

In addition, the load survey may identify those loads that can be considered for load-shedding during emergency operation or to minimize energy consumption or peak energy demands. The load survey can be used to identify utilization equipment having special requirements (for example, computers or certain lighting circuits, etc may impose special requirements such as extreme reliability or continuity of supply, low *noise* levels, or ungrounded operation, etc). These load characteristics or requirements need to be identified as early in the project as possible because they may necessitate special power distribution apparatus.

The location and magnitude of major loads must be carefully noted since such information may have considerable influence on the economic justification for the location and reliability aspects of the power service selected.

2.1.2 Relation to Power Company. Just as the individual and collective load requirements of one building differ from all other buildings, each electric utility differs to some degree from every other utility in its rates structure, service policies, and requirements making it important for the electrical engineer to contact the utility company early in the design phase. But, before beginning to discuss rate structure and availability of service, the engineer should develop the load survey to estimate initial and future loads and their electrical characteristics, to permit conveying to the electric utility the following data:

(1) Initial demand and connected load, and possible expansion

(2) Average usage or load factor

(3) Seasonal and time-of-day variations

(4) Power factor of total load

(5) Ratings of largest loads and associated switching (that is, starting) requirements

(6) Required reliability and expected continuity of service

(7) Identification of interruptible loads, to permit consideration of demand limiting

(8) Identification of loads sensitive to voltage and frequency transients

A detailed discussion on the various aspects of planning for utility service and the many factors affecting electric utility rates is presented in Section 4.

The electrical engineer should establish, in consultation with the electric utility, the special service classifications and incentive tariffs that are available to customers employing heat-recovery space-conditioning systems, thermal-storage designs, solar energy, off-peak space-conditioning systems, or similar special systems to minimize electrical energy consumption.

The electrical engineer should analyze the features of rate structures that serve to penalize poor loads. Ratchet clauses cause utility customers to pay a demand charge on the highest demand established during a number of preceding months and is an incentive to control demands. Increased seasonal and time of day rates may result in higher electric rates during the high rate periods.

Several techniques are available to the electrical engineer to reduce the cost of electric power. These techniques include the following:

(1) *Load Limiters.* Load limiters, demand limiters, programmable energy controllers, or load-shedding controllers are devices programmed to control building loads in such a sequence or manner

that the billing demand remains at an optimized value.

(2) *Power-Factor-Correcting Equipment.* Many utilities have the authority to levy *power factor penalties* or surcharges on those users whose power factor is below some specified level, often 85% (but sometimes as high as 95%). Whenever economically feasible, synchronous motors should be selected or capacitors used to compensate for the *lagging* power factor, particularly caused by induction motors and lamp ballasts, to improve the overall power factor of the system.

(3) *Power-Factor-Improvement Techniques.* Because the power factor of an induction motor is lowered considerably when the motor is loaded to less than 75%—80% of rated load (even though motor efficiency remains relatively high and constant down to about 25% load), proper sizing of induction motors for the respective application serves to improve the load power factor and minimize the investment in power-factor-correcting equipment. Power-factor-correcting capacitors are commonly installed to be switched with the respective motor starter, that is, connected at the motor terminals or at the motor control center. The use of high-power-factor ballasts in lighting equipment can improve power factor significantly in buildings where lighting is an appreciable part of the total load.

(4) *High-efficiency Motors.* Use of higher-efficiency motors, which utilize improved materials and modified (from standard motor) construction, may result in a considerable reduction in energy consumption. Due to the lack of uniform testing procedures among the various suppliers, the electrical engineer should exercise caution when evaluating various motor sources on the basis solely of published values of efficiency.

(5) *Motor Speed Control.* For certain motor applications such as pumps and blowers, where energy can be saved by reduced-speed operation when rated output is not needed, ac induction motors with solid-state adjustable-frequency controllers or multi-speed motors may be economically justified.

(6) *Regenerative Systems.* Energy can be saved in some motor-driven applications, where under certain operating conditions the load is capable of driving the motor, by utilizing regenerative systems. A loaded descending elevator or an empty ascending elevator, for example, can return energy to the building power system. The designer must analyze system performance during abnormal conditions to prevent equipment malfunction and damage.

(7) *Programmed Loads.* Certain loads can be programmed to save energy by being switched off during the hours when the space is unoccupied, or the systems are not required.

(8) *Switched Loads.* The need to provide flexible lighting systems should be satisfied by the choice of luminaire systems and lighting circuitry design. Engineering analyses of lighting systems should consider the following:

(a) Luminaires with the capability of having individual lamps or pairs of lamps switched so that the illumination levels can be set to match the task.

(b) Ceiling and luminaire systems that allow the individual liminaires to be removed or installed as the illumination levels vary for the task being performed.

(c) Use of photosensitive controls for exterior lighting.

(d) Use of separate circuits for lighting along the interior perimeter of the building so that as more light is supplied by sunlight during the day the interior perimeter lighting can be reduced either manually or by automatic controls.

(9) *Medium-Voltage Service* (2.4 to 72.5 kV). It may be possible to reduce billing costs by connecting the building loads through a transformer to the utility's primary service lines.

(10) *Medium-Voltage Distribution* (2.4 to 15 kV). Energy losses within the building may be reduced through designing the distribution system for some voltage above the utilization level of the smaller loads.

(11) *Redistribution*. Building owners may redistribute electricity through meters to office and apartment tenants, as the utility's regulations allow. Energy consumption is usually less in buildings where the tenants are paying electricity costs directly than in master-metered buildings.

2.1.3 Relation to National Electrical Code.

The first section of ANSI/NFPA 70-1981 [3], calls for the practical safeguarding of persons and property from hazards arising from the use of electricity. It further states that compliance will not necessarily result in a load-serving electrical system that is *efficient*, *convenient*, *adequate*, *or expandable*. Elsewhere the NEC (see 1.5) states that it does not represent a design specification but only identifies minimum requirements. For example, the NEC establishes certain minimum electrical system capacity requirements for general lighting receptacles, etc based on type of occupancy and demand factors. It is essential that the electrical power system designer therefore, be very knowledgeable of the contents of the latest edition of ANSI/NFPA 70-1981, along with any local electrical codes.

2.2 Load Characteristics. During the process of determining the total capacity of the electrical power distribution system for the building, in addition to noting the size and location of each load, much consideration must be given to the various operating or load characteristics, for example, repetitive starting or cycling of a load from lightly loaded, to full load, etc. The possibility of non-coincidence of many of the loads often invites consideration of diversity or demand factors. A method for using typical data for load estimation and system sizing for power systems in commercial buildings is introduced in 2.5.

2.2.1 Lighting. Electric lighting can be a high technology load. As a result of research and development by manufacturers, many highly efficient light sources, luminaires, and auxiliary equipment have been introduced. Research in basic seeing factors has provided greater knowledge of many of the fundamental aspects of quality and quantity of lighting. Consequently it is now possible to utilize considerably less lighting energy than in the past. Section 10 concentrates on these factors, with considerable attention to ways to reduce energy consumption of lighting while providing adequately for the seeing requirements and the well-being of the occupants and the objectives of the owners. Additional information which regards system design to properly serve the lighting loads can be found in 4.9.1.

Traditionally, lighting loads have accounted for 30% to 50% of the load in air-conditioned commercial buildings. The total lighting load for various buildings has commonly ranged from 3 to 10 VA/ft^2. The individual area lighting loads (either in watts or voltamperes per square foot) varies directly with the required illumination level and inversely with the efficiency of the lighting fixtures (including incandescent versus fluorescent lamps with their rule-of-thumb 2 to 1 ratio of VA/ft^2 for the same value of lighting). While stressing that ANSI/ NFPA 70-1981 [3] is not a design manual, the electrical engineer must be aware that the NEC does include, for example, Article 220 on Branch Circuit and Feeder Calculations, for various types of occupancies in commercial buildings. Minimum lighting loads for each square foot of floor area which help to identify the minimum capacities for the associated branch and feeder circuits of the power distribution system are specified. While some adjustment of the lighting-feeder loads may be possible, or justified, by utilization of a demand factor, in general the use of a 100% demand factor is highly recommended, particularly while the loads are still being estimated.

The engineer should recognize a consistently increasing trend in exterior lighting for security and decorative effect and then provide service and feeder capacity for resulting future increases in loads. Not only should added circuit capacity be provided, but consideration should also be given to space in distribution equipment for the added branch circuits.

Criteria for controlling the energy consumption of lighting systems in, and connected with, building facilities have been prepared by the American Society of Heating, Refrigerating and Air Conditioning Engineers (ASHRAE) in concert with the Illuminating Engineering Society (IES). They are identified in Section 9 of Design Standard ASHRAE/IES 90-80 [5] which establish an upper limit of power to be allowed for lighting systems plus guidelines for designing and managing those systems. A simplified method based on the above standard for

Table 2
Typical Unit Lighting Power Density (UPD) for Commercial Task/Areas

Task/Areas	Unit Power Density (W/ft^2)*	
	Large Rooms	Small Rooms
Classrooms	2.2	2.7
Computer equipment rooms	1.7	2.1
Conference rooms	1.3	2.6
Corridors	0.6	0.9
Dining rooms	1.4	2.1
Drafting rooms	4.7	7.0
Hospital, surgical suites	7.6	11.6
Hospital, patient rooms	1.4	
Kitchens	1.7	2.5
Laboratories	3.2	6.8
Library, reading area	2.2	3.3
Library, active book stack, area	0.7	1.1
Offices, accounting	3.2	4.8
Offices, general	2.2	3.3
Machine shop	1.3	2.5
Storage rooms, inactive	0.2	
Storage rooms, active	0.4	0.6
Stores, circulation areas	0.9	1.8
Stores, merchandising areas	3.8	5.7
Tennis, indoor clubs	1.9	
Basketball courts, college	1.4	

*Watts per square foot

determining the lighting power budget is known as Unit Power Density (UPD) Procedure. Selected UPD values for various tasks/areas are given in Table 2. Lighting performance depends very much on the size and proportions of the room, and the reflectances of the surfaces. Table 2 gives the extreme UPD values for typical rooms. Where spaces are of fixed dimensions, such as tennis and basketball courts, a single value is given. Exact use of the procedure is detailed in the referenced publication.

2.2.2 General Purpose Receptacles for Appliance Loads. Power required for appliances depends largely on type of space usage. Commercial building appliances include such loads as typewriters, copiers, communication equipment, and office automation equipment. Loads for large computers, plug-in type air conditioners, cooking and laundry equipment, etc should be considered separately. In contrast to lighting, the overall demand factor for appliances is very low. Article 220 of ANSI/NFPA 70-1981 [3] on Branch Circuit and Feeder Calculations provides information on allowable (for minimum safety) use of *Demand Factors for Nondwelling Receptacle Loads.* In general, 1 W/ft² of net demand is adequate for most commercial buildings; however, the wiring (feeders and branch circuits) to serve the connected load is often installed with capacity for 1.5 A per duplex outlet, or 180 VA/100 ft² of office area. Typical unit load data for various occupancies is given in Table 3 and for apartments in Table 4.

2.2.3 Space Conditioning and Associated Auxiliary Equipment. Building design engineers are increasingly using the concept of controlled environment. Space conditioning generally refers to heating, ventilating, cleaning, and cooling systems. The connected and demand power required for space-conditioning systems depends largely on the climatic conditions (that is, building geographical location) and building envelope design, interior load such as lighting and number of occupants, appliances and special process loads (for example, data processing). All of the above factors can have major influence on a space-conditioning load. The actual electrical requirements can best be obtained from those responsible for the design of the space-conditioning system. Detailed discussions of definitions, equipment ratings,

Table 3
Typical Appliance/General Purpose Receptacles Loads
(Excluding Plug-in Type A/C and Heating Equipment)

Type of Occupancy	Unit Load (VA/ft²)		
	Low	High	Average
Auditoriums	0.1	0.3	0.2
Cafeterias	0.1	0.3	0.2
Churches	0.1	0.3	0.2
Drafting rooms	0.4	1.0	0.7
Gymnasiums	0.1	0.2	0.15
Hospitals	0.5	1.5	1.0
Hospitals, large	0.4	1.0	0.7
Machine shops	0.5	2.5	1.5
Office buildings	0.5	1.5	1.0
Schools, large	0.2	1.0	0.6
Schools, medium	0.25	1.2	0.7
Schools, small	0.3	1.5	0.9

Other Unit Loads:
  Specific appliances — ampere rating of appliance
  Supplying heavy-duty lampholders — 5 A/outlet

Table 4
Typical Apartment Loads

Type	Load
Lighting and convenience outlets (except appliance)	3 VA/ft²
Appliance circuits	3 kW
Range	8 to 12 kW
Microwave Oven	1.5 kW
Refrigerator	0.3 to 0.5 kW
Freezer	0.3 to 0.5 kW
Dishwasher	1.0 to 2.0 kW
Garbage disposal	0.33 to 0.5 hp
Clothes washer	0.33 to 0.5 hp
Clothes dryer	1.5 to 6.5 kW
Water heater	1.5 to 9.0 kW
Air conditioner (0.5 hp/ room	0.8 to 4.6 kW

selection factors, system operation, calculation methods, etc are included in Section 11.

Table 5
Total Connected Electrical Load for Air Conditioning Only

Type of Building	Conditioned Area (VA/ft^2)
Bank	7
Department store	3 to 5
Hotel	6
Office building	6
Telephone equipment building	7 to 8
Small store (shoe, dress, etc)	4 to 12
Restaurant (not including kitchen)	8

When exact loads are not known or cannot be determined, an approximate preliminary load can be determined as outlined in 2.2.3.1, 2.2.3.2 and 2.2.3.3.

2.2.3.1 Air Conditioning. The air-conditioning load will consist of the motor drives for compressors, chilled-water pumps, condensate pumps, evaporative condensers or cooling towers, air-distribution fans or blowers, motorized dampers and valves, and associated control circuits. For rough estimating purposes it may be assumed that 1 ton of refrigeration will require 1 hp of motor drive for refrigeration units only, or approximately 1 kVA of load. The refrigeration unit or compressor will usually constitute about 55% to 70% of the total connected air-conditioning load. The remaining load may consist of pumps, fans, and other auxiliaries. It is customary, therefore, to apply a factor of from 1.6 to 2.0 to the total tonnage involved, and the result will be a fair estimate of the total connected load to be expected. The above factors would apply in most cases for systems of 100 t and larger. On systems below this figure, a factor of about 2.3 may be used for preliminary estimates. Where many small-unit air-conditioning machines are used, a factor of 2.8 is suggested.

In air-conditioning systems utilizing refrigeration machines that operate on the absorption principle, the heavy compressor load is eliminated but the auxiliary equipment load is still present. This type of system will usually reduce the electrical load to about 40% to 50% of that required for a full electric drive system, or to about 0.7 to 1.0 kVA per ton.

Table 5 gives the approximate air-conditioning load which might occur in the average commercial building. Loads include compressors and all auxiliary equipment involved in the cooling and ventilating system.

2.2.3.2 Auxiliary Equipment. The electrical load for boiler-room and mechanical auxiliary equipment will not normally constitute a large portion of the building load. Usually it will not exceed 5% of the total load (not including air conditioning) but it may be as high as 10% in schools. In small commercial buildings the auxiliary equipment load will consist of small units, many of which may be served by fractional-horsepower motors.

While larger buildings will have some fractional-horsepower equipment, some of the fans and pumps required may be relatively large, 10 to 20 hp being common and 30 to 75 hp or more being quite possible. The electrical engineer should consult the mechanical designers on the possible use of large motors or electric heating loads which might affect the preliminary load estimate.

The major pieces of equipment frequently encountered are listed below:

(1) Induced-draft or forced-draft fans

(2) Ventilation or exhaust fans

(3) Pumps for boiler feed, condensate return, sumps, sewage ejectors, and water circulation

(4) Fire and house-service tank pumps

(5) Air compressors and service equipment

(6) Electric heating and auxiliary heating elements

(7) Control devices and circuits

(8) Electronic air cleaners

The induced-draft or forced-draft fans are normally located in the boiler room and range in size from small fractional-horsepower units to 25 hp or more. Exhaust fans are usually small units scattered throughout the buildings, although in some cases exhausting is handled by a single large fan of 20 hp or more. Where fans are supplied with adjustable sheaves for speed control, horsepower requirements of most centrifugal fans vary as the cube of the speed.

2.2.3.3 **Heating.** Electric heating loads may range in size from many 10 kW or larger units, comprising the building's total heat source and amounting to one third to one half of the total electrical load, down to relatively small loads serving specific areas as supplemental heaters rated 10 kW or less. Other units may provide the building's hot water supply, again ranging from large electric boilers to small (1 to 4 kW) units.

A building surrounded by air colder than the interior air temperature is constantly dissipating heat. The rate of dissipation is controlled by many factors such as outside temperature, wind velocity, area of exposed surfaces, types of construction materials, amounts of insulation used, fresh-air requirements, and the type of usage. The amount of heat required to maintain comfort in a structure may be determined by taking all these factors into consideration. See Section 11, for more information.

With a known heat loss, the electrical load in kilowatts can be obtained by dividing the estimated heat loss (Btu/h) by 3413 since there are 3413 Btu in 1 kWh of electricity. Usually it is necessary to use a demand factor of 100% for electric heating loads.

Loads larger than a few hundred watts should be connected to the power panels in order to prevent excessive voltage drop on the lighting circuits. Installed heating loads should not be supplied from lighting panelboards.

Table 6 is based on a building with the insulation necessary to provide proper comfort and operating economy. These values are used for the all-weather-comfort standard. Other valuable data are available in Section 11.

2.2.4 **Plumbing and Sanitation.** Generally, for a commercial building, the loads of plumbing and sanitation equipment are not large. Typical loads for water-pressure boosting systems and electric hot-water heater heating are identified in Tables 7 and 8. Sump and sewage pumps are usually small, often applied in pairs with an electrical or mechanical alternator control, so that allowing for several 2 hp duplex units is a satisfactory allowance for the basement (that is, boiler room) of most buildings.

Table 6
All-Weather Comfort Standard
Recommended Heat-Loss Values

	Design Heat Loss per Square Foot of Floor Area	
Degree Days	(Btu/h)	(watts)
Over 8000	40	11.7
7001 to 8000	38	11.3
6001 to 7000	35	10.3
5001 to 6000	32	9.4
3001 to 5000	30	8.8
Under 3001	28	8.2

Table 7
Typical Power Requirement (kW) for
High-Rise Building Water-Pressure
Boosting Systems

Building Type	Unit Quantity	Number of Stories			
		5	10	25	50
Apartments	10 apt./ floor	—	15	90	350
Hospitals	30 patients/ floor	10	45	250	—
Hotels/ Motels	40 rooms/ floor	7	35	175	450
Offices	10 000 ft² / floor	—	15	75	250

Table 8
Typical Power Requirement (kW) for
Electric Hot-Water Heating System

Building Type	Unit Quantity	Load
Apartments/ Condominiums	20 apt/condo	30
Dormitories	100 residents	75
Elementary schools	100 students	6
High schools	100 students	12
Restaurant (full service)	100 servings/h	30
Restaurant (fast service)	100 servings/h	15
Nursing homes	100 residents	60
Hospitals	100 patient beds	200
Office buildings	10 000 ft²	5

2.2.5 Fire Protection. The largest load for fire protection will usually be a fire pump, when required to maintain system pressure beyond the capacity of the city water system. (Pertaining to design of the power system rather than load magnitude, the fire pump is one of the few loads ever permitted to be connected to the power source ahead of the service disconnect device.) Typical power load data for fire pumps are given in Table 9.

Fire detection and alarm systems are highly critical loads, but their magnitudes are generally so small that they can

Table 9
Typical Power Requirements (kW) for
Fire Pumps in Commercial Buildings
(Light Hazard)

Area/Floor (ft²)	Number of Stories			
	5	10	25	50
5000	40	65	150	250
10 000	60	100	200	400
25 000	75	150	275	550
50 000	120	200	400	800

usually be neglected when identifying the total load of the building.

2.2.6 Transportation Systems. Transportation equipment for commercial buildings includes elevators, escalators, conveyors, dumbwaiters, and pneumatic conveying systems. There is no simple rule-of-thumb method for determining the number and type of elevators or escalators required in a particular size or type (occupancy) of commercial building. Manufacturers of the equipment or specialized consultants are the best source of load information. For additional information on these loads see Section 12. When determining this total load, typical demand factors might be 0.85 for two elevators, 0.75 for four elevators, and somewhat lower value for additional elevators.

2.2.7 Data Processing. The power requirements of data-processing equipment will vary over a wide range. For smaller installations, consisting of appliance-type loads, single-phase power at 120 V may be adequate. For larger installations, including computer and peripheral (or support, or auxiliary) equipment, it may be necessary to supply 208Y/120 or 480Y/277 V power.

Data-processing installations may also be categorized as requiring continuity of high-quality power supply with the flexibility to facilitate changing the loads or location of the equipment, or both. Continuity is of prime importance to avoid loss of information stored in memory units. Power supply distortions such as voltage dips, spikes, and harmonics are considered noise to the computer since the input voltage or signal is modified in an undesired manner. These installations will also require high-level lighting and air conditioning, plus possibly a raised floor to accommodate air-handling plus power, signal and communications conductors.

Some installations may include a computer central processing unit (CPU) which may utilize high-frequency power distribution. These special requirements include the use of 60/415 Hz motor-generator sets or static inverters. The electrical engineer must be aware of the reduced ampacity of conductors and increased voltage drop at the higher frequency.

It should be stressed that if there is any possibility that electronic computers will be installed, the manufacturer of such equipment must be consulted well in advance to determine the specific electrical requirements.

A few illustrative excerpts taken from one computer manufacturer's specifications are given here.

(1) Air-conditioned space should be provided for the machine room, magnetic-tape storage and engineering areas.

(2) Raceways should be 6 to 10 inches deep and 10 to 12 inches wide and should be provided with removable covers.

(3) It is recommended that a minimum average illumination of 50 fc be maintained 30 inches above the floor in the general machine room and engineering areas; specific local areas should be at 70 to 85 fc.

Typical electronic data-processing-machine power-service requirements are listed below.

(1) 208 V, three-phase, four-wire, 60 Hz services incorporating a fifth wire insulated equipment-grounding conductor.

(2) Voltage variations not greater than ±3%.

(3) Line-to-line voltages balanced to within 2.5% with machine operating.

(4) Frequency variation not greater than ± 0.5 Hz.

(5) Maximum total harmonic content of the power system waveforms on the electric power feeders not in excess of 5% with the equipment not operating.

It may be desirable to provide a separate transformer bank, motor-alternator set, or complete rectifier-battery-inverter assembly for the electronic data-processing machinery. Analysis of filtering equipment and surge-protective equipment on the incoming utility power line may also be required to minimize the likelihood of improper operation due to line transients. If the area is fed by a low-voltage secondary-network power system, the customer should consult the local equipment manufacturer regarding the advisability of a separate transformer bank. Line inductive reactance at the wall box should not exceed 0.0173 Ω per line and can consist of either the overall reactance of the entire power system or the subtransient reactance of the separate alternator. The reactance-to-resistance ratio may be as low as 2 with no upper limit.

Typical loads for a medium to large installation could be as follows:

Central processing unit	35 kVA
Miscellaneous (tape, disks, card readers, printers)	80 kVA
400 Hz motor-generator set	64 kVA
Air conditioning	30 t

Additional comments on electrical

Table 10
Typical Loads in Commercial Kitchens

	Number Served	Connected Load (kW)
Lunch counter (gas ranges, with 40 seats)		30
Cafeteria	800	150
Restaurant (gas cooking)		90
Restaurant (electric cooking)		180
Hospital (electric cooking)	1200	300
Diet kitchen (gas cooking)		200
Hotel (typical)		75
Hotel (modern, gas ranges, three kitchens)		150
Penitentiary (gas cooking)		175

power requirements for these loads can be found in 16.11. An early check with the utility company supplying electric service may provide valuable data on supply reliability.

2.2.8 Food Preparation. The magnitude of the electrical load depends more upon the number of meals served at one time than upon the total size of the space. The load also depends upon whether electricity or gas is used to provide the heat for the main equipment (ovens and ranges). However, the additional devices using electric power, (that is, fryers, microwave ovens, stock kettles, warming tables, meat slicers and saws, coffee pots, toasters, waffle irons, mixers, potato peelers, etc) may present a sizable load and should not be overlooked in the system design. Besides the equipment directly involved in food preparation, there will be additional service equipment including lighting, dishwashing and garbage disposal equipment, exhaust fans, makeup air heaters, hotwater booster heaters, etc. In addition there may be refrigeration equipment,

varying from walk-in type refrigerators to freezer units or deep-freeze lockers. (Commercial freezer or cold-storage plants present different system design problems; they are considered an industrial type of building, and consequently are excluded from this section.) Where the utility power supply is subject to prolonged outages, freezer or refrigerated loads may require transfer to an alternate, or emergency power source. (This may also apply to laboratories where sensitive experimental materials are kept under refrigeration.) Table 10 provides some approximate total-load data for commercial kitchens such as might be located in a commercial building.

The cooling load for the kitchen should not be overlooked since heat gain in the kitchen is often large. This heat can be removed by exhaust fans (for example, range hoods, room exhausts), air conditioning, or a combination of the two. When ventilation alone is used, fan capacity to provide one air change per minute may be necessary. There are so many variables in heat gains for kitchen equipment that a general rule-of-thumb cannot be used for the load required to air condition a commercial kitchen.

Additional comments on electrical power requirements for these loads can be found in 16.29.

2.2.9 Miscellaneous or Special Loads. There are many loads that do not qualify in the preceding schedule of major load groups that possibly might appear in any new commercial building. These additional loads will generally be small (but could be major in size, for example, broadcasting equipment) and occur only occasionally in commercial buildings. Therefore, these loads can best be categorized as miscellaneous or special loads since they vary so widely with regards

Table 11
Types of Electrical Load Equipment

(1) Broadcasting equipment (radio, television)
(2) Control and monitoring systems
  (centralized, local)
  (a) security
  (b) fire alarm
(3) Fire pumps
(4) Health care facilities
  (a) sterilizers
  (b) X-ray machines
  (c) diathermy equipment
  (d) miscellaneous
(5) Incinerators
(6) Intercommunication systems
(7) Kitchen equipment
  (a) cooking
  (b) disposal
  (c) refrigeration and freezing
  (d) ventilation
  (e) washing
(8) Laboratory equipment
  (a) air compressors and vacuum pumps
  (b) centrifuges
  (c) furnaces and refrigerators
  (d) incubation and cold rooms
  (e) sterilizers
  (f) miscellaneous
(9) Lighting
  (a) general
  (b) task
  (c) exits and stairwells
  (d) emergency
  (e) exterior
(10) Office equipment
  (a) Addressing machines
  (b) Calculating machines
  (c) Copying cameras, microfile
    equipment

  (d) Duplicating machines
    (i) blueprinters
    (ii) xerographic
  (e) Dictating and transcribing machines
  (f) Electronic data processing
    equipment
  (g) Typewriters
(11) Public-address system
(12) Radio and television receivers
(13) Recreational and fitness equipment
  (a) swimming pool heaters and pumps
  (b) exercise equipment
(14) Shop equipment
  (a) drill press
  (b) lathe
  (c) saw
  (d) welder
(15) Shredding devices, waste compactors
(16) Snow-melting
(17) Space-conditioning (central or individual)
  (a) heating
  (b) cooling
  (c) ventilating
  (d) cleaning
(18) Television equipment, closed-circuit
(19) Transporting equipment
  (a) dumbwaiters
  (b) conveyors
  (c) elevators
  (d) electric stairways
(20) Visual-aid equipment
  (a) motion picture, slide and overhead
    projectors
  (b) tape recorders
(21) Water heaters
(22) Water pumps
(23) Others

size and frequency of appearance. However, if such load apparatus will eventually be included in the building, they should be considered (even if only in the form of spare feeder or branch circuit or space for future protective device) when the power system is initially being designed. Multi-story office buildings may require approximately $1-2$ VA/ft^2 for such general or micellaneous loads. A partial checklist of such loads is provided in Table 11. Some apparatus, including electric typewriters or visual-aid equipment, can be served (operated) from the usual 15 A or 20 A receptacles, neither creating any appreciable voltage fluctuations nor demanding critical voltage regulation or emergency power source. Others, such as intercommunication, photographic reproduction or X-ray equipment, may amount to small loads yet require a high-quality (for example, stabilized voltage) power source. Some of these special loads, such as, welders, may draw heavy currents for short times in repetitive cycles so that voltage variation (potential light flicker) should be investigated.

2.3 Electromagnetic Hazards, Pollution, and Environmental Quality.

The increasing use of electronic equipment calls for some consideration of the electromagnetic environment created by this equipment and also the effect of external electromagnetic influences on its performance. Specifically, computers, communications equipment, and other low-level electronic systems require special analysis of the grounding system. Inadequate grounding can be both a shock hazard and a source of noise input to the computer.

Since the cost of providing shielding after construction is quite high, the electrical engineer should analyze shielding requirements for sensitive equipment before construction. Some of the following applications may require the degree of control of electromagnetic energy achieved only by a shielded enclosure:

(1) Research and development laboratories for low-noise circuitry work

(2) Research and development laboratories using high-energy radio-frequency devices

(3) Computer facilities

(4) Test and measurement laboratories

(5) Terminal equipment facilities for both line and radio-frequency transmission systems

(6) Hospital and other biomedical research and treatment rooms

(7) Control and monitoring equipment in strong fields of other emitters or strong radio-frequency fields from industrial sources

2.4 Additions to Existing Systems.

Whenever it is contemplated that the occupancy of a commercial-industrial-institutional building is to be changed or if the building is to be expanded or modernized, depending upon the nature and magnitude of the changes in the total and individual loads, an engineering study of the existing electrical power distribution system should be included in the initial planning of the building change. Additional comments pertinent to this subject is offered in 15.2 and 15.9 (that is, discussing need for accurate drawings of details of the existing building and loads).

2.5 Total Load Considerations.

If all the connected loads in the building are arithmetically totalled, (that is, all expressed in hp, or kW, or kVA, or A at some specified voltage) to identify the total building load, the resultant number will in most cases seem to require a larger power system capacity than will be realistically needed to adequately serve the loads. The average load on the power system is usually less than the total connected load; this is termed the demand load. It may vary depending on the time interval over which the load is averaged. Certain loads may at times be turned off or operating at reduced power levels reducing the system power requirements (that is, total load). This effect is termed *diversity* and it may be expressed as a diversity factor.

The value of demand or diversity to be used is highly dependent upon the location in the particular power system being considered. Diversity factors become larger as the loads are totalled nearer to the power source and include more of the diverse building components. The following factors (or definitions) are commonly used, when totalling loads, to facilitate system planning. See ANSI/ IEEE Std 100-1977 [4].

demand (or demand load). The electric load at the receiving terminals averaged over a specified interval of time. Demand is expressed in kilowatts, kilovoltamperes, kilovars, amperes, or other suitable units. The interval of time is generally 15 min, 30 min, or 1 h.

NOTE: If there are two 50 hp motors (which drive 45 hp loads) connected to the electric power system but only one load is operating at any time, the demand load is only 45 hp but the connected load is 100 hp.

coincident demand. Any demand that occurs simultaneously with any other demand; also the sum of any set of coincident demands.

maximum demand. The greatest of all the demands that have occurred during the specified period of time; determined by measurement over a prescribed time interval.

demand factor. The ratio of the maximum demand of a system to the total connected load of the system.

NOTES: (1) Since **demand load** cannot be greater than the **connected load**, demand factor cannot be greater than unity.
(2) Only those demand factors permitted by the National Electrical Code (for example, services and feeders) may be considered in sizing the electrical system; otherwise the circuit must be sized to support the connected load.

connected load. The sum of the continuous ratings of the power-consuming apparatus connected to the system or any part thereof in watts, kilowatts, or horsepower.

gross demand load. The summation of the demands for each of the several group loads.

peak load. The maximum load of a specified unit or group of units in a stated period of time.

branch circuit load. The load on that portion of a wiring system extending beyond the final overcurrent device protecting the circuit. (See ANSI/NFPA 70-1981 [3], Article 220-2 for complete details and exceptions.)

diversity factor. The ratio of the sum of the individual maximum demands of the sub-divisions of the system to the maximum demand of the complete system.

NOTE: Since maximum demand of a system cannot be greater than the sum of the individual demands, diversity factor will always be equal to or greater than unity.

load factor. The ratio of the average load over a designated period of time to the peak load occurring in that period.

load profile. The graphic representation of the demand load, usually on an hourly basis, for a particular day. The demand load for typical groups of loads (for example, heating, cooling, lighting, etc) may be accumulated to determine the demand load of the system; the highest point of the load profile will be the maximum demand load of the system. See Figs 2(a) and 2(b) for typical load-profile representations.

Information on these factors for the various loads and groups of loads is essential in designing the system. For example, the sum of the connected loads on a feeder, multiplied by the demand factor of these loads, will give the maximum demand which the feeder must carry. The sum of the individual maximum demands on the circuits associated with a load center or panelboard, divided by the diversity factor of those circuits, will give the maximum demand

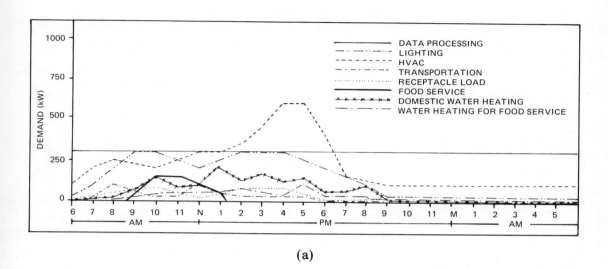

(a)

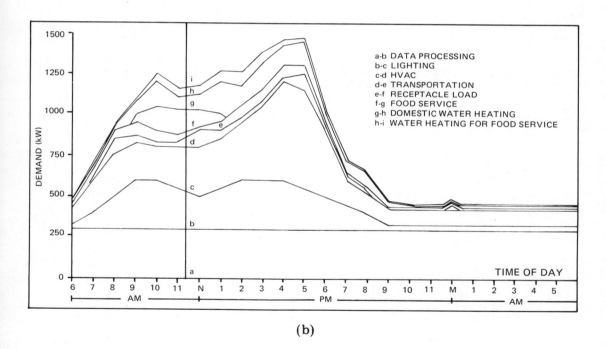

(b)

**Fig 2
(a) Individual Group Load Profile
(b) Cumulative Load Profile**

Table 12
Comparison of Maximum Demand

Type of Store	Shopping Center A, New Jersey No Refrigeration*		Shopping Center B, New Jersey Refrigeration		Shopping Center C, New York Refrigeration	
	Gross Area (ft^2)	(W/ft^2)	Gross Area (ft^2)	(W/ft^2)	Gross Area (ft^2)	(W/ft^2)
Bank					4000	9.0
Book	3700	6.0	2500	6.7		
Candy	1600	6.9			2000	10.8
Department	343 500	4.7	222 000	7.3	226 900	8.0
	84 000	3.1	114 000	5.6		
Drug	7000	6.1	6000	7.7		
Men's wear	17 000	5.5	17 000	9.9	2000	10.8
	28 000	4.9	9100	8.8		
Paint					15 600	8.5
Pet					2000	12.1
Restaurant					4000	9.0
Shoe	11 000	6.3	7000	12.5	3300	15.4
	4000	8.0	4400	12.9	2100	9.0
Supermarket	32 000	5.7	25 000	8.6	37 600	11.5
Variety	31 000	4.6	24 000	6.8	37 400	7.1
	30 000	4.4			30 000	7.0
Women's wear	20 400	4.7	19 300	8.9	1360	13.0
	1000	5.8	4500	9.6	1000	11.7

*Loads include all lighting and power, but no power for air-conditioning refrigeration (chilled water) which is supplied from a central plant.

Table 13
Connected Load and Maximum Demand by Tenant Classification

	Classification	Connected Load (W/ft^2)	Maximum Demand (W/ft^2)	Demand Factor
10	Women's wear	7.7	5.9	0.75
3	Men's wear	7.2	5.6	0.78
6	Shoe store	8.5	6.9	0.79
2	Department store	6.0	4.7	0.74
2	Variety store	10.5	4.5	0.45
2	Drug store	11.7	6.7	0.57
5	Household goods	5.4	3.9	0.76
10	Specialty shop	8.1	6.8	0.79
4	Bakery and candy	17.1	12.1	0.71
3	Food store (supermarkets)	9.9	5.9	0.60
5	Restaurant	15.9	7.1	0.45

NOTE: Connected load includes an allowance for spares.

Table 14
Electrical Load Estimation

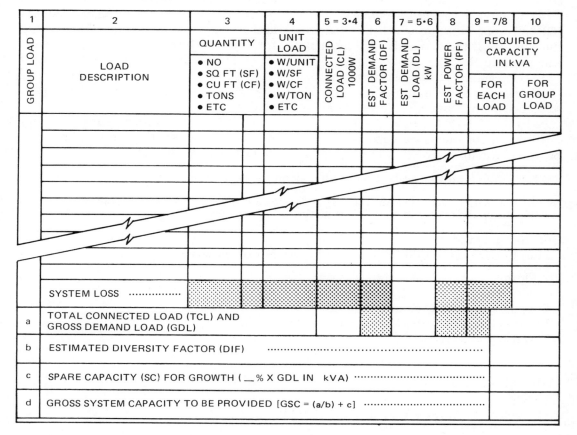

1	2	3	4	5 = 3·4	6	7 = 5·6	8	9 = 7/8	10
GROUP LOAD	LOAD DESCRIPTION	QUANTITY • NO • SQ FT (SF) • CU FT (CF) • TONS • ETC	UNIT LOAD • W/UNIT • W/SF • W/CF • W/TON • ETC	CONNECTED LOAD (CL) 1000W	EST DEMAND FACTOR (DF)	EST DEMAND LOAD (DL) kW	EST POWER FACTOR (PF)	REQUIRED CAPACITY IN kVA — FOR EACH LOAD	FOR GROUP LOAD
	SYSTEM LOSS								
a	TOTAL CONNECTED LOAD (TCL) AND GROSS DEMAND LOAD (GDL)								
b	ESTIMATED DIVERSITY FACTOR (DIF)								
c	SPARE CAPACITY (SC) FOR GROWTH (__ % X GDL IN kVA)								
d	GROSS SYSTEM CAPACITY TO BE PROVIDED [GSC = (a/b) + c]								

at the load center and on the circuit supplying it. By the use of the proper factors, as outlined, the maximum demands on the various parts of the system from the load circuits to the power source can be estimated. Tables 12 and 13 provide typical maximum demand and demand-factor data for various types of occupancies.

2.5.1 Estimation of Building Load. A suggested procedure for determining the demand load of a building is given in the following steps. Calculations can be summarized in tabulated form as shown in Table 14.

(1) Determine the quantity of load units and the power requirement of each load.

(2) Determine the demand factor (DF) of the load or group of loads by the definition given in 2.5, or from Table 13, or from the NEC [3] (see 2.1.3).

(3) (a) Determine the demand load (DL) for present and future operating conditions; it is the product of connected (CL) and demand factor (DF).

(b) Estimate (for column 8) the power factor (decimal value) of the particular load when operating at its intended rated capacity. The various loads divided by their respective power factor (decimal value) will determine the required *source capacity* in kilovoltampere.

(4) Compute the gross demand load (GDL) of the building which is equal to the sum of all the demands of individual and group loads.

(5) Determine the diversity factor (DIF) of the system by estimation or reference from similar projects or reference to ANSI/NFPA 70-1981 [3].

(6) Estimate the spare capacity to be provided for load growth and identified future loads, such as data processing, food service, air conditioning, etc. Use either a *blanket* percent against the gross demand load or apply estimated percent against each load (or group of loads) and for *c* utilize the sum of these increments.

(7) Determine the required capacity from steps (4), (5) and (6). (When load profile is used, step (5) can be eliminated.)

(8) Select a system with capacity which will satisfy the required capacity determined in step (7).

2.6 Example: Sample Partial Load Calculation for an Office Building

NOTE: Calculations according to NEC [3] are accepted practice in some jurisdictions only, thus the example shown is intended to indicate one particular method only.

Outside Dimensions:
100 ft $\cdot$ 160 ft (16 000 ft^2 /floor), 18 stories of office space.

Per Floor
Deduct Areas:

Corridors (including stairways):	1050 ft^2	(0.5 W/ft^2)
Electrical and Janitors Closets:	150 ft^2	(0.5 W/ft^2)
Elevator and Vent Shafts:	800 ft^2	(no floor)
	2000 ft^2	

Code Calculations ANSI/NFPA 70-1981 [3] Table 220-2(b)

Lighting:		kW	
(16 000 − 2000) (3.5 W/ft^2)	=	49.0	
(1200) (0.5) w/ft^2	=	0.6	
	Total		49.6 kW

General Purpose Receptacles:

(16 000 − 800) (1.0 w/ft^2)	=	Connected total = 15.2	
(15.2 − 10) (0.5 + 10 kW)	=	Demand total =	12.6 kW
		Total/Floor	62.2 kW

Actual Load
Useable office area = 13 000 ft^2
Lighting:
1 – 2 $\cdot$ 4 four-lamp lay-in
competitive troffers/64 ft^2 (85 fc)
Conventional 3200 lumen lamps, standard ballasts

13 000/64 = 203 fixtures (93 w/ballast or pair of lamps $\cdot$ 2 = 186 W/fixture

Offices − 37.76 kW

2-Lamp Fixtures

Corridors, stairways, lobby	24
Restrooms	6
Mechanical, electrical	4
Total	34 (93 W/fixtures) = 3.16 kW
Lighting Load Total/Floor	40.9 kW

40.9/16 000 = 2.56 W/ft^2

Use of very low heat ballasts, low-wattage high-efficacy lamps or improved luminaire configuration, or both, can lower wattage per square foot for the same footcandle level by as much as 25%. Solid state ballasts will probably reduce the required input wattage even more.

General Purpose Receptacles:
Office Area — 1/100 ft^2

13 000/100	130
Other Areas	9
Connected Total	139 at 180 W at 100% of duplex receptacle = 25.0 kW

(25.0 − 10) (0.5) + 10 kW = Demand total = 17.5 kW

Actual Load   Total/floor   58.4 kW

Building. Note that the building switchboard capacity for office lighting must be not less than (49.6) (18) = 892.8 kW but each floor panel need have breakers and branch circuits provided for only the load actually installed. The building switchboard capacity for office general purpose receptacles must be not less than (15.2) (18) = 273.6; (273.6 − 10) (0.5) + 10 = 141.8 kW.

NOTE: These calculations are for general purpose lighting and general purpose receptacles only. Additions must be made for all other loads. Calculations are made on the kilowatt basis as described in ANSI/NFPA 70-1981 [3]. When calculating load currents the voltamperes (or kVA values should be used.

2.7 References

[1] McWILLIAMS, D.W., Users' Needs: Lighting, Start-up Power, Transportation, Mechanical Utilities, Heating, Refrigeration and Production, IEEE *Transactions* on Industry and Applications (IAS), vol 1A—10, Mar/Apr, 1974.

[2] BAUER, G.M., Users' Needs: Space Conditioning, Fire Protection, Data Processing, Life Support and Life Safety Systems, Communication Systems and Signal Circuits, IEEE *Transactions* on Industry and Applications (IAS), vol 1A—10, Mar/Apr, 1974.

[3] ANSI/NFPA 70-1981, National Electrical Code.[20]

[4] ANSI/IEEE Std 100-1977, IEEE Standard Dictionary of Electrical and Electronics Terms.

[5] ASHRAE/IES Std 90-80, Energy Conservation in New Building Design.[21]

[20] The National Electrical Code is published by the National Fire Protection Association, Batterymarch Park, Quincy, MA 02269. Copies are also available from the Sales department of American National Standards Institute, 1430 Broadway, New York, NY 10018.

[21] ASHRAE documents are available from Publication Sales, ASHRAE, 1791 Tullie Circle, NE, Atlanta, GA, 30329.

3. Voltage Considerations

3.1 General. An understanding of system voltage nomenclature and the preferred voltage ratings of distribution apparatus and utilization equipment is essential to ensure proper voltage identification throughout a power distribution system. The dynamic characteristics of the system must be recognized and the proper principles of voltage control applied so that satisfactory voltages will be supplied to all utilization equipment under all normal conditions of operation.

3.1.1 Definitions. The following terms and definitions from ANSI C84.1-1977 [2][22] are used to identify voltages and voltage classes used in electric power distribution.

3.1.1.1 System Voltage Terms

3.1.1.1.1 system voltage. The root-mean-square phase-to-phase voltage on an alternating-current electric system.

3.1.1.1.2 nominal system voltage. The root-mean-square phase-to-phase voltage by which the system is designated and to which certain operating characteristics of the system are related.

NOTE: The nominal system voltage is near the voltage level at which the system normally operates. To allow for operating contingencies, systems generally operate at voltage levels about 5 to 10% below the maximum system voltage for which system components are designed.

3.1.1.1.3 maximum system voltage. The highest root-mean-square phase-to-phase voltage which occurs on the system under normal operating conditions, and the highest root-mean-square phase-to-phase voltage for which equipment and other system components are designed for satisfactory continuous operation without derating of any kind.

NOTE: When defining maximum system voltage, voltage transients and temporary overvoltages caused by abnormal system conditions such as faults, load rejection, etc are excluded. However, voltage transients and temporary overvoltages may affect equipment life and operating performance as well as conductor insulation and are considered in equipment application.

3.1.1.1.4 service voltage. The root-mean-square phase-to-phase or phase-to-neutral voltage at the point where the electrical system of the supplier and the user are connected.

3.1.1.1.5 utilization voltage. The root-mean-square phase-to-phase or

---

[22]The numbers in brackets correspond to those in the references at the end of this section.

phase-to-neutral voltage at the line terminals of utilization equipment.

3.1.1.2 System Voltage Classes

3.1.1.2.1 low voltage. A class of nominal system voltages less than 1000 V.

3.1.1.2.2 medium voltage. A class of nominal system voltages equal to or greater than 1000 V and less than 100 000 V.

3.1.1.2.3 high voltage. A class of nominal system voltages equal to or greater than 100 000 V.

3.1.2 Standard Nominal System Voltages for the United States. These voltages and their associated tolerance limits are listed in ANSI C84.1-1977 [2] for voltages from 120 V through 230 000 V, and in ANSI C92.2-1978 [3]. The nominal system voltages and their associated tolerance limits and notes in the two standards have been combined in Table 15 to provide a single table, listing all the standard nominal system voltages and their associated tolerance limits for the United States. Preferred nominal system voltages and voltage ranges are shown in bold face type while other systems in substantial use which are recognized as standard voltages are shown in medium type. Other voltages may be encountered on older systems but they are not recognized as standard voltages. The transformer connections from which these voltages are derived are shown in Fig 3.

3.1.3 Application of Voltage Classes

3.1.3.1 Low-voltage class voltages are used to supply utilization equipment.

3.1.3.2 Medium-voltage class voltages are used as primary distribution voltages to supply distribution transformers which step the medium voltage down to a low voltage to supply utilization equipment. Medium voltages of 13 800 V and below are also used to supply utilization equipment such as large motors. See Table 22.

3.1.3.3 High-voltage-class voltages are used to transmit large amounts of electric power between transmission substations. Transmission substations located adjacent to generating stations step the generator voltage up to the transmission voltage for transmission to transmission substations in the load area where the transmission voltage is transformed down to a primary distribution voltage to supply distribution transformers which step the primary distribution voltage down to a utilization voltage. Transmission lines also interconnect transmission substations to provide alternate paths for power transmission to improve the reliability of the transmission system.

3.1.4 Voltage Systems Outside of the United States. Voltage systems in other countries generally differ from those in the United States. For example, 416Y/240 V is widely used as a utilization voltage even for residential service. Also the frequency in many countries is 50 Hz instead of 60 Hz, which affects the operation of some equipment such as motors which will run approximately 17% slower. Plugs and receptacles are generally different, which helps to prevent utilization equipment from the United States from being connected to the wrong voltage.

In general, equipment rated for use in the United States cannot be used outside of the United States and equipment rated for use outside of the United States cannot be used in the United States. If electrical equipment made for use in the United States must be used outside the United States, information on the voltage, frequency, and type of plug required should be obtained. If the difference is only in the voltage, trans-

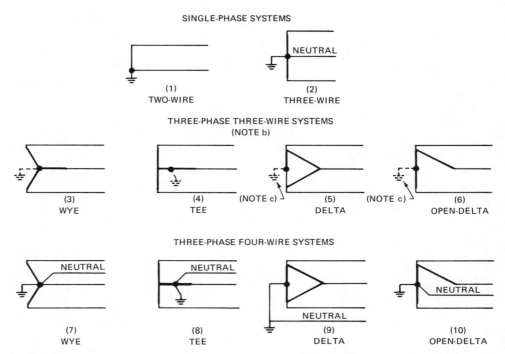

**Fig 3
Principal Transformer Connections to
Supply the System Voltages of Table 15**

NOTES: (a) The above diagrams show connections of transformer secondary windings to supply the nominal system voltages of Table 15. Systems of more than 600 V are normally three-phase and supplied by connections (3), (5) ungrounded, or (7). Systems of 120—600 V may be either single-phase or three-phase and all the connections shown are used in some extent for some systems in this voltage range.

(b) Three-phase, three-wire systems may be solidly grounded, impedance grounded, or ungrounded, but are not intended to supply loads connected phase-to-neutral (as are the four-wire systems).

(c) In connections (5) and (6) the ground may be connected to the midpoint of one winding as shown (if available), to one phase conductor (*corner* grounded) or omitted entirely (ungrounded).

(d) Single-phase services and single-phase loads may be supplied from single-phase systems or from three-phase systems. They are connected phase-to-phase when supplied from three-phase, three-wire systems and either phase-to-phase or phase-to-neutral from three-phase four-wire systems.

Table 15
Standard Nominal System Voltages and Voltages Ranges

Voltage Class	Nominal System Voltage (Note a)			Voltage Range A (Note b)			Voltage Range B (Note b)		
	Two-wire	Three-wire	Four-wire	Minimum Utilization Voltage	Minimum Service Voltage	Maximum Utilization and Service Voltage (Note c)	Minimum Utilization Voltage	Minimum Service Voltage	Maximum Utilization and Service Voltage
Single-Phase Systems									
Low Voltage (Note 1)	120			110	114	126	106	110	127
		120/240		110/220	114/228	126/252	106/212	110/220	127/254
Three-Phase Systems									
			208Y/120 (Note d)	191Y/110	197Y/114	218Y/126	184Y/106 (Note 2)	191Y/110 (Note 2)	220Y/127
			240/120	220/110	228/114	252/126	212/106	220/110	254/127
		240		220	228	252	212	220	254
			480Y/277	440Y/254	456Y/263	504Y/291	428Y/245	440Y/254	508Y/293
		480		440	456	504	424	440	508
		600 (Note e)		550	570	630 (Note e)	530	550	635 (Note e)
Medium Voltage		2400		2160	2340	2520	2080	2280	2540
			4160Y/2400	3740Y/2160	4050Y/2340	4370Y/2520	3600/2080	3950Y/2280	4400Y/2540
		4160		3740	4050	4370	3600	3950	4400
		4800		4320	4680	5040	4160	4560	5080
		6900		6210	6730	7240	5940	6560	7260
			8320Y/4800	(Note f)	8110Y/4680	8730Y/5040	(Note f)	7900Y/4560	8800Y/5080
			12 000Y/6930		11 700Y/6760	12 600Y/7270		11 400Y/6580	12 700Y/7330
			12 470Y/7200		12 160Y/7020	13 090Y/7560		11 850Y/6840	13 200Y/7620
			13 200Y/7620		12 870Y/7430	13 860Y/8000		12 504Y/7240	13 970Y/8070
			13 800Y/7970		13 460Y/7770	14 490Y/8370		13 110Y/7570	14 520Y/8380
		13 800		12 420	13 460	14 490	11 880	13 110	14 520
			20 780Y/12 000	(Note f)	20 260Y/11 700	21 820Y/12 600	(Note f)	19 740Y/11 400	22 000Y/12 700
			22 860Y/13 200		22 290Y/12 870	24 000Y/13 860		21 720Y/12 540	24 200Y/13 970
		23 000			22 430	24 150		21 850	24 340
			24 940Y/14 400		24 320Y/14 040	26 190Y/15 120		23 690Y/13 680	26 400Y/15 240
			34 500Y/19 920		33 640Y/19 420	36 230Y/20 920		32 780Y/18 930	36 510Y/21 080
		34 500			33 640	36 230		32 780	36 510

(Continued on page 77)

Table 15 (Continued)

High Voltage

Nominal System Voltage	Maximum Voltage (Note g)
46 000	48 300
69 000	72 500
115 000	121 000
138 000	145 000
161 000	169 000
230 000	242 000
(Note h)	
345 000	362 000
500 000	550 000
765 000	800 000
1 100 000	1 200 000

NOTES: (1) Minimum utilization voltages for 120–600 V circuits not supplying lighting loads are as follows:

Nominal System Voltage	Range A	Range B
120	108	104
120/240	108/216	104/208
208Y/120	187Y/108	180Y/104
240/120	216/108	208/104
(Note 2) 240	216	208
480Y/277	432Y/249	416Y/240
480	432	416
600	540	520

(2) Many 220 V motors were applied on existing 208 V systems on the assumption that the utilization voltage would not be less than 187 V. Caution should be exercised in applying the Range B minimum voltages of Table 1 and Note (1) to existing 208 V systems supplying such motors.

NOTE: Notes (a) through (h) integrally apply to this table.

(a) Three-phase three-wire systems are systems in which only the three-phase conductors are carried out from the source for connection of loads. The source may be derived from any type of three-phase transformer connection, grounded or ungrounded. Three-phase four-wire systems are systems in which a grounded neutral conductor is also carried out from the source for connection of loads. Four-wire systems in Table 15 are designated by the phase-to-phase voltage, followed by the letter Y (except for the 240/120 V delta system), a slant line, and the phase-to-neutral voltage. Single-phase services and loads may be supplied from either single-phase or three-phase systems. The principal transformer connections that are used to supply single-phase and three-phase systems are illustrated in Fig 3.

(b) The voltage ranges in this table are illustrated in ANSI C84.1-1977 [2], Appendix B.

(c) For 120-600 V nominal systems, voltages in this column are maximum service voltages. Maximum utilization voltages would not be expected to exceed 125 V for the nominal system voltage of 120, nor appropriate multiples thereof for other nominal system voltages through 600 V.

(d) A modification of this three-phase four-wire system is available as a 120/208Y V service for single-phase, three-wire, open-wye applications.

(e) Certain kinds of control and protective equipment presently available have a maximum voltage limit of 600 V; the manufacturer or power supplier or both should be consulted to assure proper application.

(f) Utilization equipment does not generally operate directly at these voltages. For equipment supplied through transformers, refer to limits for nominal systems voltage of transformer output.

(g) For these systems Range A and Range B limits are not shown because, where they are used as service voltages, the operating voltage level on the user's system is normally adjusted by means of voltage regulation to suit their requirements.

(h) Information from ANSI C92.2-1978 [3]. Nominal voltages above 230 000 V are not standardized. The nominal voltages listed are typically used with the associated preferred standard maximum voltages.

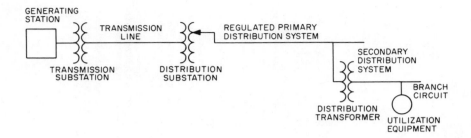

Fig 4
Typical Utility Power Generation, Transmission, and
Distribution System

formers are generally available to convert the supply voltage to the equipment voltage.

3.1.5 Voltage Standard for Canada. The voltage standard for Canada, Preferred Voltage Levels for AC Systems, 0 to 50 000 V — Recommended Reference Standard for Canadian Practice, published by the Canadian Standards Association, differs from the United States standard both in the list of standard nominal voltages and in the tolerance limits.

3.2 Voltage Control in Electric Power Systems

3.2.1 Principles of Power Transmission and Distribution on Utility Systems. To understand the principles of voltage control required to provide satisfactory voltage to utilization equipment, a general understanding is necessary of the principles of power transmission and distribution in utility systems, since most commercial buildings obtain their electric power requirements from the local electric utility company. Figure 4 is a one-line diagram of a typical utility power generation, transmission, and distribution system.

Generating stations are located near convenient sources of fuel and water.

The generated power, except for station requirements, is transformed in a transmission substation at the generating station up to a transmission voltage in the range from 46 000 V to 765 000 V for transmission to major load areas. Transmission lines are classified as unregulated because the voltage is usually controlled only to keep the lines operating within normal voltage limits and to facilitate power flow. ANSI C84.1-1977 [2] specifies only the nominal and maximum values of voltage for systems over 34 500 V.

Transmission lines supply distribution substations equipped with transformers which step the transmission voltage down to a primary distribution voltage, generally from 4160 V to 34 500 V, although 12 470 V and 13 200 V are in the widest use. However, there is an increasing trend toward the use of 34 500 V for primary distribution as the average load density increases.

Voltage control is applied, when necessary, for the purpose of providing satisfactory voltage to the terminals of utilization equipment. The transformers used to step the transmission voltage down to the primary distribution voltage are generally equipped with tap-changing-underload equipment, which changes the ratio

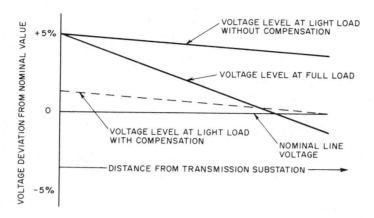

Fig 5
Effect of Regulator Compensation on Primary
Distribution System Voltage

of the transformer under load in order to maintain the primary distribution voltage within a narrow band regardless of fluctuations in the transmission voltage. Separate step or induction regulators may also be used.

Generally the regulator controls are equipped with compensators which raise the voltage as the load increases and lower the voltage as the load decreases to compensate for voltage excursions in the primary distribution system. This prevents the voltage from rising to excessive values during light-load conditions when the voltage drop along the primary distribution system is low. This is illustrated in Fig 5. Note that buildings close to the distribution substation will receive voltages that average higher than those received by buildings at a distance from the distribution substation. Switched or fixed capacitors are also used by utility distribution companies to improve the voltage on the primary feeders.

The primary distribution system supplies distribution transformers which step the primary distribution voltage down to utilization voltages, generally in the range of 120 V through 600 V, to supply a secondary distribution system to which the utilization equipment is connected. Small transformers used to step a higher utilization voltage down to a lower utilization voltage such as 480 V to 280Y/120 V are considered part of the secondary distribution system.

The supply voltages available to a commercial building depend upon whether the building is supplied by a distribution transformer, the primary distribution system, or the transmission system which, in turn, depends on the electric power requirements of the building.

Small buildings up to several hundred kilovoltamperes of load and all buildings supplied from secondary networks are supplied from the distribution transformer. The secondary distribution system consists of the connections from the distribution transformer to the building service and the building wiring.

Medium-size buildings and multibuilding complexes with loads of a few thousand kilovoltamperes are generally connected to the primary distribution system. The building owner provides the section of the primary distribution system within the building, the distribution transformers, and the secondary distribution system.

Large buildings and multibuilding complexes with loads of more than a few thousand kilovoltamperes may be connected to the transmission system. The building owner provides the primary distribution system, the distribution transformers, the secondary distribution system, and may provide the distribution substation.

If power is supplied by local generation at the building, the generators will replace the primary distribution system up to the distribution transformer where generation voltage is over 600 V, and will also replace the distribution transformer where generation is at 600 V or below.

3.2.2 Development of Voltage Tolerance Limits for ANSI C84.1-1977 [2]. The voltage tolerance limits in ANSI C84.1-1977 [2] are based on ANSI/NEMA MG1-1978 [5], which establishes the voltage tolerance limits of the standard low-voltage induction motor at ±10% of nameplate voltage ratings of 230 V and 460 V. Since motors represent the major component of utilization equipment, they were given primary consideration in the establishment of the voltage standard.

The best way to show the voltages in a distribution system is in terms of a 120 V base. This cancels the transformation ratios between systems, so that the actual voltages vary solely on the basis of the voltage drops in the system. Any voltage may be converted to a 120 V base by dividing the actual voltage by the ratio of transformation to the 120 V base. For example, the ratio of transformation for the 480 V system is 480/120 or 4, so 460 V in a 480 V system would be 460/4 or 115 V.

The tolerance limits of the 460 V motor in terms of the 120 V base become 115 V plus 10% or 126.5 V and 115 V minus 10% or 103.5 V. The problem is to decide how this tolerance range of 23 V should be divided between the primary distribution system, the distribution transformer, and the secondary distribution system which make up the regulated distribution system. The solution adopted by American National Standards Committee C84 is shown in Table 16.

The tolerance limits of the standard motor on the 120 V base of 126.5 V maximum and 103.5 V minimum were raised 0.5 V to 127 V maximum and 104 V minimum to eliminate the fractional volt. These values became the tolerance limits for Range B in the standard. An allowance of 13 V was allotted for the voltage drop in the primary distribution system. Deducting this voltage drop from 127 V establishes a minimum of 114 V for utility company services supplied directly from the primary distribution system. An allowance of 4 V was provided for the voltage drop in the distribution transformer and the connections to the building low-voltage wiring. The actual drop will depend on the load, its power factor, and the transformer impedance. Deducting this voltage drop from the minimum distribution voltage of 114 V provides a minimum of 110 V for the utility company supply from 120 V through 600 V. An allowance of 6 V, or 5%, was made for the voltage drop in the building wiring, which is the same as provided in ANSI/NFPA 70-1981 [4], Section 210-19(a), for the maximum voltage drop in build-

Table 16
Standard Voltage Profile for a Regulated Power Distribution System, 120 V Base

	Range A	Range B
Maximum allowable voltage	126(125*)	127
Voltage drop allowance for the primary distribution feeder	9	13
Minimum primary service voltage	117	114
Voltage drop allowance for the distribution transformer	3	4
Minimum low-voltage service voltage	114	110
Voltage drop allowance for the building wiring	6(4†)	6(4†)
Minimum utilization voltage	108(110†)	104(106†)

* For utilization voltages of 120 through 600 V.
† For building wiring circuits supplying lighting equipment.

ing low-voltage wiring. This completes the distribution of the 23 V tolerance zone down to the minimum utilization voltage of 104 V on the 120 V base.

The Range A limits for the standard were established by reducing the maximum tolerance limits from 127 V to 126 V and increasing the minimum tolerance limits from 104 V to 108 V. This spread band of 18 V was then allotted: 9 V for the voltage drop in the primary distribution system to provide a minimum primary service voltage of 117 V; 3 V for the voltage drop in the distribution transformer and secondary connections to provide a minimum low-voltage service voltage of 114 V; and 6 V for the voltage drop in the building wiring to provide a minimum utilization voltage of 108 V.

Four additional modifications were made in this basic plan to establish ANSI C84.1-1977 [2]. The maximum utilization voltage in Range A was reduced from 126 V to 125 V for low-voltage systems in the range from 120 V through 600 V, because there should be sufficient load on the distribution system to provide at least a 1 V drop on the 120 V base under most operating conditions.

This maximum voltage of 125 V is also a practical limit for lighting equipment because the life of the 120 V incandescent lamp is reduced by 42% when operated at 125 V (Table 16). The voltage drop allowance of 6 V on the 120 V base for the drop in the building wiring was reduced to 4 V for circuits supplying lighting equipment, which raised the minimum voltage limit for utilization equipment to 106 V in Range B and 110 V in Range A because the minimum limits for motors of 104 V in Range B and 108 V in Range A were considered too low for satisfactory operation of lighting equipment. The utilization voltages for the 6900 V and 13 800 V systems in Range B were adjusted to coincide with the tolerance limits of ±10% of the nameplate rating of the 6600 V and 13 200 V motors used on these respective systems.

To convert the 120 V base voltage to equivalent voltages in other systems, the voltage on the 120 V base is multiplied by the ratio of the transformer which would be used to connect the other system to a 120 V system. In general, oil-filled distribution transformers for systems below 15 000 V have nameplate ratings which are the same as the standard system nominal voltages, and the ratio of the standard nominal voltages may be used to make the conversion (see ANSI C57.12.20-1974 [1]). However, for primary distribution voltages over 15 000 V, the primary nameplate rating of oil-filled distribution transformers is not the same as the standard system nominal voltages. Also, distribution transformers may be equipped with taps to change the ratio of transformation. Thus if the primary distribution voltage is over 15 000 V, or taps have been used to change the transformer ratio, the actual transformer ratio must be used to convert the base voltage to that of another system. Other types of distribution transformers such as dry-type have the same voltage ratios as the oil-filled distribution transformers used by utility companies.

For example, the maximum tolerance limit of 127 V on the 120 V base for the service voltage in Range B is equivalent, on the 4160 V system, to (4160/120) · 127 = 4400 V to the nearest 10 V. However, if the 4160 V to 120 V transformer is set on the +2.5% tap, the voltage ratio would be 4160 + (4160 · 0.025) = 4160 + 104 = 4264 V to 120 V. The voltage on the primary system equivalent to 127 V on the secondary system would be (4264/120) · 127 = 35.33 · 127 = 4510 V to the nearest 10 V. If the maximum distribution voltage of 4400 V is applied to the 4264 V to 120 V transformer, then the secondary voltage would become 4400/4260 · 120 = 124 V. So the effect of using a +2.5% tap is to lower the secondary voltage range by 2.5%.

3.2.3 System Voltage Tolerance Limits.

The width of the voltage range for all standard nominal system voltages

in the utilization and distribution range of 120 V through 34 500 V is specified in ANSI C84.1-1977 [2] for two critical points on the distribution system: the point of delivery by the supplying utility company and the point of connection to utilization equipment. The voltage tolerance limits at the point of delivery by the supplying utility provide the voltage limits within which the supplying utility must maintain the supply voltage in order to provide satisfactory operation of the user's utilization equipment. The voltage tolerance limits at the point of connection of utilization equipment provide the voltage limits within which the utilization equipment manufacturer must design the utilization equipment to operate satisfactorily. For transmission voltages over 34 500 V, only the maximum voltage is specified, because these voltages are normally unregulated, and only a maximum voltage is required to establish the design insulation level for the line and associated apparatus.

The actual voltage measured at any point on the system will vary depending on the location of the point of measurement and the system load at the time the measurement is made. Fixed voltage changes take place in transformers in accordance with the transformer ratio, while voltage variations occur from the operation of voltage control equipment and the changes in voltage drop between the supply source and the point of measurement due to changes in the current flowing in the circuit.

3.2.4 Voltage Profile Limits for a Regulated Distribution System.

Figure 6 shows the voltage profile of a regulated power distribution system using the limits of Range A in Table 15. Assuming a standard nominal distribution voltage of 13 200 V, Range A in Table 15, shows that this voltage should be main-

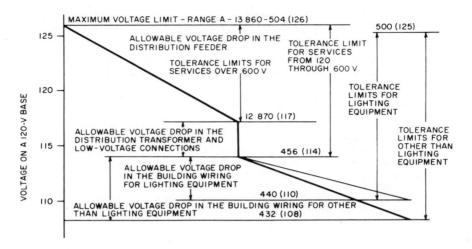

Fig 6
Voltage Profile of the Limits of Range A, ANSI C84.1-1977

tained by the supplying utility between a maximum of 126 V and a minimum of 117 V on a 120 V base. Since the base multiplier for converting from the 120 V system to the 13 200 V system is 13 200/120 or 110, the actual voltage limits for the 13 200 V system are 110 · 126 or 13 860 V maximum and 110 · 117 or 12 870 V minimum.

If a distribution transformer with a ratio of 13 200 V to 480 V is connected to the 13 200 V distribution feeder, Range A of Table 15 requires that the nominal 480 V supply must be maintained by the supplying utility between a maximum of 126 V and a minimum of 114 V on the 120 V base. Since the base multiplier for the 480 V system is 480/120 or 4, the actual values are 4 · 126 or 504 V maximum and 4 · 114 or 456 V minimum.

Range A of Table 15 as modified for utilization equipment from 120 V through 480 V provides for a maximum utilization voltage of 125 V and a minimum of 110 V for lighting equipment

and 108 V for other than lighting equipment on the 120 V base. Using the base multiplier of 4 for the 480 V system, the maximum utilization voltage would be 4 · 125 or 500 V, and the minimum for other than lighting equipment would be 4 · 108 or 432 V. For lighting equipment connected phase-to-neutral, the maximum voltage would be 500 V divided by the square root of 3 or 288 V and the minimum voltage would be 4 · 110 or 440 V divided by the square root of 3 or 254 V.

3.2.5 Nonstandard Nominal System Voltages. Since ANSI C84.1-1977 [2] lists only the standard nominal system voltages in common use in the United States, system voltages will frequently be encountered which differ from the standard list. A few of these may be so widely different as to constitute separate systems in too limited use to be considered standard. However, in most cases the nominal system voltages will differ by only a few percent, as shown in Table 17. A closer examination of the table

Table 17
Nominal System Voltages

Standard Nominal System Voltages	Associated Nominal System Voltages
Low-voltage systems	
120	110, 115, 125
120/240*	110/220, 115/230, 125/250
208Y/120*	216Y/125
240/120*	
240	230, 250
480Y/277*	416Y/240, 460Y/265
480*	440, 460
600	550, 575
Primary distribution voltage systems	
2400	2200, 2300
4160Y/2400	
4160*	4000
4800	4600
6900	6600, 7200
8320Y/4800	
12 000Y/6930	11 000, 11 500
12 470Y/7200*	
13 200Y/7620*	
13 800Y/7970	
13 800*	14 400
20 780Y/12 000	
22 860Y/13 200	
23 000	
24 940Y/14 400*	
34 500Y/19 920*	
34 500	33 000

* Preferred standard nominal system voltages.

shows that these differences are due mainly to the fact that some voltages are multiples of 110 V, others are multiples of 115 V, and a few are multiples of 120 V.

The reasons for these differences go back to the original development of electric power distribution systems. The first utilization voltage was 100 V. However, the supply voltage had to be raised to 110 V in order to compensate for the voltage drop in the distribution system. This led to overvoltage on equipment connected close to the supply, and the utilization equipment rating was raised

to 110 V also. As generator sizes increased and distribution and transmission systems developed, an effort to keep transformer ratios in round numbers led to a series of utilization voltages of 110 V, 220 V, 440 V, and 550 V, and a series of distribution voltages of 2200 V, 4400 V, 6600 V, and 13 200 V.

As a result of the effort to maintain the supply voltage slightly above the utilization voltage, the supply voltages were raised again to multiples of 115 V, which resulted in a new series of utilization voltages of 115 V, 230 V, 460 V, and 575 V and a new series of distribution voltages of 2300 V, 4600 V, 6900 V, and 13 800 V.

As a result of the development of the 208Y/120 V network system, the supply voltages were raised again to multiples of 120 V. This resulted in a new series of utilization voltages of 120 V, 208Y/120, 240 V, 480 V, and 600 V and a new series of primary distribution voltages of 2400 V, 4160Y/2400 V, 4800 V, 12 000 V, and 12 470Y/7200 V. However, most of the existing primary distribution voltages continued in use and no 120 V multiple voltages developed at the transmission level.

In the case of low-voltage systems, the associated nominal system voltages in the right-hand column are obsolete and should not be used. Manufacturers are encouraged to design utilization equipment to provide acceptable performance within the utilization voltage tolerance limits specified in the standard where possible. Some numbers listed in the right-hand column are used in equipment ratings, but these should not be confused with the numbers designating the nominal system voltage on which the equipment is designed to operate. See 3.4.

In the case of primary distribution voltages, the numbers in the right-hand column may designate an older system in which the voltage tolerance limits are maintained at a different level than the standard nominal system voltage, and special consideration must be given to the distribution transformer ratios, taps, and tap settings. See 3.2.7.

3.2.6 System-Voltage Nomenclature. The nominal system voltages in Table 15 are designated in the same way as the designation on the nameplate of the transformer for the winding or windings supplying the system.

(1) *Single-Phase Systems*

120 — Indicates a single-phase two-wire system in which the nominal voltage between the two wires is 120 V.

120/240 — Indicates a single-phase three-wire system in which the nominal voltage between the two-phase conductors is 240 V, and from each phase conductor to the neutral is 120 V.

(2) *Three-Phase Systems*

240/120 — Indicates a three-phase four-wire system supplied from a Δ-connected transformer. The midtap of one winding is connected to a neutral. The three-phase conductors provide a nominal 240 V three-phase system, and the neutral and two adjacent phase conductors provide a nominal 120/240 V single-phase system.

Single number — Indicates a three-phase three-wire system in which the number designates the nominal voltage between phases.

Two numbers separated by Y/ — Indicates a three-phase four-wire system from a Y-connected or tee-connected (T) transformer in which the first number indicates the nominal phase-to-phase voltage and the second the nominal phase-to-neutral voltage

NOTES: (1) All single-phase systems and all three-phase four-wire systems are suitable for the connection of phase-to-neutral load.

(2) See Section 4 for methods of system grounding.

See Fig 3 for transformer connections.

3.2.7 Use of Distribution Transformer Taps to Shift the Utilization Voltage Spread Band.

Except for small sizes, distribution transformers are normally provided with five taps on the primary winding, generally two 2.5% above and below rated voltage and one at rate voltage. These taps permit the transformer ratio to be changed to raise or lower the secondary voltage spread band to provide a closer fit to the tolerance limits of the utilization equipment. There are two general situations which require the use of taps.

(1) Where the primary distribution system voltage spread band is above or below the limits required to provide a satisfactory secondary voltage spread band. This occurs under two conditions:

(a) Where the primary voltage has a slightly different nominal value than the transformer primary nameplate rating. For example, if a 13 200 V — 480 V transformer is connected to a nominal 13 800 V system, the nominal secondary voltage would be 13 800/13 200 · 480 = 502 V. However, if the 13 800 V system was connected to the +5% tap of the 13 200 V — 480 V transformer at 13 860 V, the secondary voltage would be 13 860/13 800 · 480 = 482 V, which is practically the same as would be obtained from a transformer having the proper ratio of 13 200 V — 480 V.

(b) Where the primary voltage spread is in the upper or lower part of the tolerance limits provided in ANSI C84.1-1977 [2]. For example, a 13 200 V — 480 V transformer is connected to a 13 200 V primary distribution system close to the distribution substation so that the primary voltage spread band falls in the upper half of the tolerance

zone for Range A in the standard or 13 200 V to 13 860 V. This would result in a nominal secondary voltage under no-load conditions of 480 V to 504 V. By setting the transformer on the +2.5% tap at 13 530 V, the secondary voltage would be lowered 2.5% to a range of 468 V to 491 V. This would materially reduce the overvoltage on utilization equipment.

(2) Adjusting the utilization voltage spread band to provide a closer fit to the tolerance limits of the utilization equipment. For example, Table 18 shows the shift in the utilization voltage spread band for the +2.5% and +5% taps as compared to the utilization voltage tolerance limits for Range A of ANSI C84.1-1977 [2] for the 480 V system. Table 19 shows the voltage tolerance limits of the old standard 440 V and the new standard 460 V three-phase induction motors. Table 20 shows the tolerance limits for the old standard 265 V and the new standard 277 V fluorescent lamp ballasts. A study of these three tables shows that the normal (100%) tap setting will provide the best fit with the tolerance limits of the 460 V motor and the 277 V ballast, but a setting on the +5% tap will provide the best fit for the 440 V motor and the 265 V ballast. For older buildings having appreciable numbers of both ratings of motors and ballasts, a setting on the +2.5% tap may provide the best compromise.

Note that these examples assume that the tolerance limits of the supply and utilization voltages are within the tolerance limits specified in ANSI C84.1-1977 [2]. This may not be true, so the actual voltages should be measured (preferably with a seven-day recording voltmeter) to obtain readings during the night and over weekends when maxi-

Table 18
Tolerance Limits from Table 15, Range A, in Volts

Nominal System Voltage	Transformer Tap Voltage	Minimum Utilization Voltage	Minimum Service Voltage	Maximum Utilization and Service Voltage
480Y/277	Normal	440Y/254	454Y/262	500Y/288
468Y/270	+2.5%	429Y/248	443Y/255	488Y/281
456Y/263	+5%	418Y/241	431Y/249	475Y/274

Table 19
Tolerance Limits for Standard Three-Phase Induction Motors, in Volts

Motor Rating	10% Minus	10% Plus
460	414	506
440	396	484

Table 20
Tolerance Limits for Standard Fluorescent Lamp Ballasts, in Volts

Ballast Rating	10% Minus	10% Plus
277	249	305
265	238	292

mum voltages occur. These actual voltages can then be used to compare with Tables 18, 19, and 20 to check the proposed transformer ratios and tap settings.

Where a building has not yet been built so that actual voltages can be measured, the supplying utility company should be requested to provide the expected spread band for the supply voltage, preferably supported by a seven-day graphic chart from the nearest available location. If the building owner is to furnish the distribution transformers, recommendations should also be obtained from the supplying utility company on the transformer ratios, taps, and tap settings. With this information a voltage profile can be prepared to check the expected voltage spread at the utilization equipment.

In general, distribution transformers should have the same primary voltage rating as the nominal voltage of the primary distribution system. Taps should be provided at +2.5% and +5% and -2.5% and -5% to provide adjustment in either direction.

Note that the voltage spread at the secondary terminals of the transformer is equal to the voltage spread at the primary terminals plus the voltage drop in the transformer. Taps only serve to move the secondary voltage up or down in the steps of the taps. They cannot correct for excessive spread in the supply voltage or excessive drop in the building distribution system. Therefore, if the voltage spread at the utilization equipment exceeds the tolerance limits of the equipment, action must be taken to improve voltage conditions (see 3.12).

3.3 Voltage Selection

3.3.1 Selection of Utilization Voltage of 600 V and Below. Generally the preferred utilization voltage for large commercial buildings is 480Y/277 V. The three-phase power load is connected directly to the system at 480 V, and fluorescent ceiling lighting is connected phase-to-neutral at 277 V. Small dry-type transformers rated 480 V —

280 Y/120 V are used to provide 120 V for convenience outlets and 208 V, single phase and three phase, for office machinery.

Single-phase transformers with secondary ratings of 120/240 V may also be used to supply lighting and small office equipment. However, single-phase transformers should be connected in sequence on the primary phases to maintain balanced load on the primary system.

Where the supplying utility furnishes the distribution transformers, the choice of voltages will be limited to those the utility will provide. Most utilities provide all of the voltages listed in the standard form 120 V through 480 V, although all may not be available at any one location.

The built-up downtown areas of many large cities are supplied from low-voltage networks (see Section 4). Originally the only voltage available was 208 Y/120 V. Now most utilities will provide spot network installation at 480 Y/277 V for large buildings. For tall buildings, space will be required on upper floors for transformer installations and the primary distribution cables supplying the transformers.

Apartment buildings generally have the option of using either 208 Y/120 V, three-phase, four-wire, or 120/240 V, single-phase since the major load in residential occupancies consists of 120 V lighting fixtures and appliances. The 208 Y/120 V systems should be more economical for large apartment buildings and 120/240 V systems satisfactory for small apartment buildings. However, large single-phase appliances such as electric ranges rated for use on 120/240 V single-phase systems cannot be used efficiently on 208 Y/120 V, three-phase, four-wire systems because the line-to-line voltage is appreciably below the rated voltage of the appliance. Where central air conditioning is provided, a large motor is required to drive the refrigeration compressor. See 3.9.1 on the effects of starting large motors.

3.3.2 Utility Service Supplied from a Primary Distribution System. When a commercial building or commercial complex becomes too large to be supplied at utilization voltage from a single distribution transformer installation, the utility company's primary distribution line must be tapped to supply distribution transformer installations, generally of the dry type, epoxy-filled cast type or non-propagating liquid filled type located inside the building, or oil-filled transformers located outside the building or in transformer vaults.

Whatever primary distribution voltage the supplying utility provides in the area must be accepted. The utility company's primary distribution voltages in the widest use are 12 470 Y/7200 V and 13 200 Y/7620 V. Some utilities may provide a transformation to a lower distribution voltage such as 4160 Y/2400 V which may be more economical for the building wiring system. However, the voltage drop in the additional transformation will increase the voltage spread at the utilization equipment. The voltage spread band should be checked in this case to make sure that it is within satisfactory limits.

Special consideration should always be given when starting larger motors to minimize the voltage dip so as not to affect the operation of other utilization equipment on the system supplying the motor (see 3.9). Larger motors, generally over 150 hp, may be supplied at medium voltage such as 2400 V or 4160 V from a separate transformer to eliminate the voltage dip on the low-voltage system. However, these motors and control may be more expensive and consideration

Table 21
Low-Voltage Motors

Motor Nameplate Voltage	Preferred Horsepower Limits
115	No minimum—15 hp maximum
200	No minimum—200 hp maximum
460 and 575	1 hp minimum—1000 hp maximum
	Medium Voltage Motors
2300	50 hp minimum—6000 hp maximum
4000	100 hp minimum—7500 hp maximum
4500	250 hp minimum—No maximum
6600	400 hp minimum—No maximum

must be given to the fact that the maintenance electricians in commercial buildings may not be qualified to maintain medium-voltage equipment. A contract with a qualified electrical firm may be required for maintenance.

Standard voltages and preferred horsepower limits for polyphase induction motors likely to be used for air conditioning are shown in Table 21.[23]

In recent years many utilities have begun leasing transmission voltages in the range from 15 000 V to 35 000 V for distribution circuits. However, equipment costs in the range of 25 000 V through 35 000 V are quite high, and a transformation down to a lower voltage may prove the most economical. Note that Table 16 provides for only one transformation between the primary and secondary distribution voltages so that a voltage profile taking into account both transformations should be prepared to make sure that the voltage at the utilization equipment will fall within acceptable limits.

3.3.3 **Utility Service Supplied from Transmission Lines.** Normally commercial buildings and building complexes are not supplied directly from utility trans-

mission lines. If details for these supply voltages are required, see IEEE Std 141-1976 [6].

3.3.4 **Utility Policy for Supplying Tenants in Commercial Buildings.** Some utilities have a policy of providing all or part of the building wiring up to the point of connection with individual tenants of commercial buildings in return for the right to provide utilization voltage direct to each tenant. If a commercial building is to be rented to more than one tenant, the local utility should be contacted to determine if they wish to supply electric service direct to the tenants. Such an arrangement will not only save the building owner the cost of the building feeders, but also either the cost of the submetering and billing or the problems of dividing the electric bill among the tenants and the building owner, or including it in the rent.

The cost of electricity to tenants billed individually is higher than the pro rata share of a common bill because of the charges in the lower steps of the rate structure to cover metering, meter reading, and billing. However, individual billing encourages conservation on the part of the tenant.

3.4 **Voltage Ratings for Utilization Equipment.** Utilization equipment is de-

---

[23] Taken from [11], Table 18—5.

fined as electrical equipment which converts electric power into some other form of energy such as light, heat, or mechanical motion. Every item of utilization equipment must have a nameplate listing, among other things, the rated voltage for which the equipment is designed. With one major exception, most utilization equipment carries a nameplate rating which is the same as the voltage system on which it is to be used, that is, equipment to be used on 120 V systems is rated 120 V; for 208 V systems, 208 V; for 240 V systems, 240 V; for 480 V systems, 480 V; and for 600 V systems, 600 V. The major exception is motors and equipment containing motors. See Table 22. Motors are also about the only utilization equipment used on systems over 600 V. Single-phase motors for use on 120 V systems are rated 115 V. Single-phase motors for use on both 208 V and 240 V single-phase systems are rated 230 V. However, a 230 V, single-phase motor supplied at 208 V will have 19% less torque and overload capacity and 23% greater slip. If the driven load is more than 81% of the motor horsepower rating, the next size larger motor may be required if the duty cycle is severe enough to overheat the motor. If 200 V, single-phase motors are available they should be used on 208 V systems.

Prior to the late 1960s, low-voltage three-phase motors were rated 220 V for use on both 208 V and 240 V systems, 440 V for use on 480 V systems, and 550 V for use on 600 V systems. The reason was that most three-phase motors were used in large industrial plants where relatively long circuits resulted in voltages considerably below nominal at the ends of the circuits. Also, utility supply systems had limited capacity and low voltages were common

Table 22
Voltage Ratings of Standard Motors

Nominal System Voltage	Nameplate Voltage
Single-phase motors	
120	115
240	230
Three-phase motors	
208	200
240	230
480	460
600	575
2400	2300
4160	4000
4800	4600
6900	6600
13 800	13 200

during heavy load periods. As a result, the average voltage applied to three-phase motors approximated the 220 V, 440 V, and 550 V nameplate ratings.

In recent years the supplying utilities have made extensive changes to higher distribution voltages. Increased load densities have resulted in shorter primary distribution systems. Distribution transformers have been moved inside buildings to be closer to the load. Lower impedance wiring systems have been used in the secondary distribution system. Capacitors have been used to improve power factors and reduce voltage drop. All of these changes have contributed to reducing the voltage drop in the distribution system and raising the average voltage applied to utilization equipment. By the mid 1960s, surveys indicated that the average voltage supplied to motors on 240 V and 480 V systems was 230 V and 460 V, respectively, and there were increasing numbers of complaints of overvoltage on motors.

At about the same time the Motor and Generator Committee of the National

Electrical Manufacturers Assocation decided that improvements in motor design and insulation systems would allow a reduction of two frame sizes in the standard low-voltage three-phase induction motor. As a part of this rerate program, the nameplate voltage of the low-voltage motors was increased from 220 V, 440 V, and 550 V to 230 V, 460 V, and 575 V, respectively. Subsequently a motor rated 200 V for use on 208 V systems was added to the program. Table 22 shows the present voltage ratings of standard motors as specified in ANSI/NEMA MG1-1978 [5].

The question has been raised why the confusion over the difference between the nameplate rating of equipment and the system nominal voltage cannot be eliminated by making the nameplate rating of utilization equipment the same as the nominal voltage of the system on which the equipment is to be used. The reason is that the performance guarantee for utilization equipment is based on the nameplate rating and not the system nominal voltage. For utilization equipment such as motors where the performance peaks in the middle of the tolerance range of the equipment, better performance can be obtained over the tolerance range specified in ANSI C84.1-1977 [2], by selecting a nameplate rating closer to the middle of this tolerance range.

3.5 Effect of Voltage Variation on Utilization Equipment. Whenever the voltage at the terminals of utilization equipment varies from its nameplate rating, the performance of the equipment and its life expectancy changes. The effect may be minor or serious, depending on the characteristics of the equipment and the amount of voltage deviation from the nameplate rating. NEMA standards provide tolerance limits within which performance will normally be acceptable. In precise operations, however, closer voltage control may be required. In general, a change in the applied voltage causes a proportional change in the current. Since the effect on the load equipment is proportional to the product of the voltage and the current and since the current is proportional to the voltage, the total effect is approximately proportional to the square of the voltage.

However, the change is only approximately proportional and not exact because the change in the current affects the operation of the equipment so the current will continue to change until a new equilibrium position is established. For example, if the load is a resistance heater, the increase in current will increase the temperature of the heater which will increase its resistance which will reduce the current. This effect will continue until a new equilibrium current and temperature is established. In the case of an induction motor, a reduction in the voltage will cause a reduction in the current flowing to the motor causing the motor to start to slow down. This reduces the impedance of the motor causing an increase in the current until a new equilibrium position is established between the current and the motor speed.

3.5.1 Induction Motors. The variations in characteristics as a function of voltage are given in Table 23.

The most significant effects of low voltage are reduction in starting torque and increased full-load temperature rise. The most significant effects of high voltage are increased torque, increased starting current, and decreased power factor. The increased starting torque will increase the accelerating forces on cou-

Table 23
General Effect of Voltage Variations on Induction-Motor Characteristics

Characteristic	Function of Voltage	Voltage Variation	
		90% Voltage	110% Voltage
Starting and maximum running torque	$(Voltage)^2$	Decrease 19%	Increase 21%
Synchronous speed	Constant	No change	No change
Percent slip	$1/(Voltage)^2$	Increase 23%	Decrease 17%
Full-load speed	Synchronous speed slip	Decrease 1.5%	Increase 1%
Efficiency			
Full load	—	Decrease 2%	Increase 0.5—1%
¾ load	—	Practically no change	Practically no change
½ load	—	Increase 1—2%	Decrease 1—2%
Power factor			
Full load	—	Increase 1%	Decrease 3%
¾ load	—	Increase 2—3%	Decrease 4%
½ load	—	Increase 4—5%	Decrease 5—6%
Full-load current	—	Increase 11%	Decrease 7%
Starting current	Voltage	Decrease 10—12%	Increase 10—12%
Temperature rise, full load	—	Increase 6—7 °C	Decrease 1—2 °C
Maximum overload capacity	$(Voltage)^2$	Decrease 19%	Increase 21%
Magnetic noise - no load in particular	—	Decrease slightly	Increase slightly

plings and driven equipment. Increased starting current causes greater voltage drop in the supply circuit and increases the voltage dip on lamps and other equipment. In general, voltages slightly above nameplate rating have less detrimental effect on motor performance than voltages slightly below nameplate rating.

3.5.2 Synchronous Motors. Synchronous motors are affected in the same way as induction motors, except that the speed remains constant, unless the frequency changes, and the maximum or pull-out torque varies directly as the voltage if the field voltage remains constant as in the case where the field is supplied by a generator on the same shaft with the motor. If the field voltage

varies with the line voltage as often occurs with a static rectifier source, then the pull-out torque varies as the square of the voltage.

3.5.3 Incandescent Lamps. The light output and life of incandescent filament lamps are critically affected by the impressed voltage. The variation of life and light output with voltage is given in Table 24. The figures for 125 V and 130 V lamps are also included because these ratings are useful in locations where long life is more important than light output.

3.5.4 Fluorescent Lamps. Fluorescent lamps, unlike incandescent lamps, operate satisfactorily over a range of ±10% of the ballast nameplate voltage rating. Light output varies approximately in

Table 24
Effect of Voltage Variations on Incandescent Lamps

Applied Voltage (volts)	Lamp Rating					
	120 V		125 V		130 V	
	Percent Life	Percent Light	Percent Life	Percent Light	Percent Life	Percent Light
105	575	64	880	55	—	—
110	310	74	525	65	880	57
115	175	87	295	76	500	66
120	100	100	170	88	280	76
125	58	118	100	100	165	88
130	34	132	59	113	100	100

direct proportion to the applied voltage. Thus a 1% increase in applied voltage will increase the light output by 1% and, conversely, a decrease of 1% in the applied voltage will reduce the light output by 1%. The life of fluorescent lamps is affected less by voltage variation than that of incandescent lamps.

The voltage-sensitive component of the fluorescent fixture is the ballast, which is a small reactor or transformer that supplies the starting and operating voltages to the lamp and limits the lamp current to design values. These ballasts may overheat when subjected to above normal voltage and operating temperature, and ballasts with integral thermal protection may be required. See ANSI/NFPA 70-1981 [4], Section 410-73(e).

3.5.5 High-Intensity Discharge Lamps (Mercury, Sodium, and Metal Halide). Mercury lamps using the conventional unregulated ballast will have a 30% decrease in light output for a 10% decrease in terminal voltage. If a constant wattage ballast is used, the decrease in light output for a 10% decrease in terminal voltage will be about 2%.

Mercury lamps require 4 to 8 min to vaporize the mercury in the lamp and reach full brilliance. At about 20% undervoltage, the mercury arc will be extinguished and the lamp cannot be restarted until the mercury condenses which takes from 4 to 8 min unless the lamps have special cooling controls. The lamp life is related inversely to the number of starts so that, if low-voltage conditions require repeated starting, lamp life will be affected adversely. Excessively high voltage raises the arc temperature which could damage the glass enclosure when the temperature approaches the glass softening point.

Sodium and metal halide lamps have similar characteristics to mercury lamps although the starting and operating voltages may be somewhat different. See the manufacturers' catalogs for detailed information.

3.5.6 Infrared Heating Processes. Although the filaments in the lamps used in these installations are of the resistance type, the energy output does not vary with the square of the voltage because the resistance varies at the same time. The energy output does vary roughly as some power of the voltage, slightly less than the square, however. Voltage variations can produce unwanted changes in the process heat available unless thermostatic control or other regulating means are used.

3.5.7 Resistance Heating Devices. The energy input and, therefore, the heat output of resistance heaters varies approximately as the square of the impressed voltage. Thus a 10% drop in voltage will cause a drop of approximately 19% in heat output. This, however, holds true only for an operating range over which the resistance remains approximately constant.

3.5.8 Electron Tubes. The current-carrying ability or emission of all electron tubes is affected seriously by voltage deviation from nameplate rating. The cathode life curve indicates that the life is reduced by half for each 5% increase in cathode voltage. This is due to the reduced life of the heater element and to the higher rate of evaporation of the active material from the surface of the cathode. It is extremely important that the cathode voltage be kept near nameplate rating on electron tubes for satisfactory service. In many cases, this will necessitate a regulated power source. This may be located at or within the equipment, and often consists of a regulating transformer having constant output voltage and limited current.

3.5.9 Capacitors. The reactive power input of capacitors varies with the square of the impressed voltage. A drop of 10% in the supply voltage, therefore, reduces the reactive power by 19%. Where the user has made a sizable investment in capacitors for power factor improvement, he loses a large part of the benefit of this investment.

3.5.10 Solenoid-Operated Devices. The pull of alternating-current solenoids varies approximately as the square of the voltage. In general, solenoids are designed liberally and operate satisfactorily on 10% overvoltage and 15% undervoltage.

3.5.11 Solid-State Equipment. Silicon-controlled rectifiers, transistors, etc, have no thermionic heaters, and thus are not nearly as sensitive to long-time voltage variation as the electron tube components they are replacing. Sensitive equipments are frequently provided with internal voltage regulators, so that they are independent of supply system regulation. These and power solid-state equipment are, however, generally limited regarding peak reverse voltage. They can therefore be adversely affected by abnormal voltages of even microsecond duration in the reverse direction. An individual study of the voltage capabilities of the equipment including surge characteristics, is necessary to determine if abnormal voltage will result in a malfunction.

3.6 Calculation of Voltage Drops. Building wiring designers must have a working knowledge of voltage-drop calculations, not only to meet ANSI/NFPA 70-1981 [4], Sections 210-19 (a) and 215-2 (c) requirements, but also to ensure that the voltage applied to utilization equipment is maintained within proper limits. Due to the vector relationships between voltage and current and resistance and reactance, voltage-drop calculations require a working knowledge of trigonometry, especially for making exact computations. Fortunately most voltage-drop calculations are based on assumed limiting conditions and approximate formulas are adequate.

3.6.1 General Mathematical Formulas. The vector relationships between the voltage at the beginning of a circuit, the

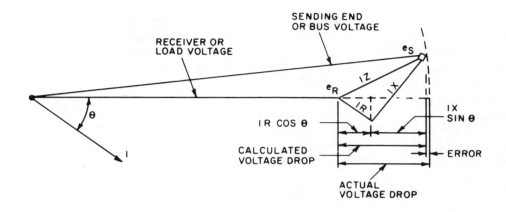

Fig 7
Vector Diagram of Voltage Relations for Voltage-Drop Calculations

voltage drop in the circuit, and the voltage at the end of the circuit are shown in Fig 7. The approximate formula for the voltage drop is

$$V = IR \cos \theta + IX \sin \theta$$

where

V = voltage drop in circuit, line to neutral

I = current flowing in conductor

R = line resistance for one conductor, Ω

X = line reactance for one conductor, Ω

θ = angle whose cosine is the load power factor

$\cos \theta$ = load power factor, in decimals

$\sin \theta$ = load reactive factor, in decimals

The voltage drop V obtained from this formula is the voltage drop in one conductor, one way, commonly called the line-to-neutral voltage drop. The line-to-line voltage drop is computed by

multiplying the line-to-neutral voltage drop by the following constants:

Voltage System	Multiply by
Single phase	2
Three phase	1.732

In using this formula, the line current I is generally the maximum or assumed load current or the current-carrying capacity of the conductor.

The resistance R is the alternating-current resistance of the particular conductor used considering the particular type of raceway in which it is installed. It depends on the size of the conductor (measured in American wire gauge, AWG, for smaller conductors and in thousands of circular mils, kcmil, for larger conductors), the type of conductor (copper or aluminum), the temperature of the conductor (normally 60 °C for average loading and 75 °C, or 90 °C, depending on the conductor rating, for maximum loading), and whether the conductor is installed in magnetic (steel) or nonmagnetic (aluminum or nonmetallic) raceway.

The reactance X also depends on the size and material of the conductor, whether the raceway is magnetic or nonmagnetic, and on the spacing between the conductors of the circuit. The spacing is fixed for multiconductor cable, but may vary with single-conductor cables so that an average value must be used. Reactance occurs because the alternating current flowing in the conductor causes a magnetic field to build up and collapse around each conductor in synchronism with the alternating current. This magnetic field cuts across the conductor itself and the other conductors of the circuit, causing a voltage to be induced into each in the same way that current flowing in the primary of a transformer induces a voltage in the secondary of the transformer. Since the induced voltage is proportional to the rate of change of the magnetic field which is maximum when the current is passing through zero, the induced voltage will be a maximum when the current is passing through zero or, in vector terminology the voltage wave is 90° out of phase with the current wave.

θ is the angle between the load voltage and the load current. Cos θ is the power factor of the load expressed as a decimal and may be used directly in the computation of IR cos θ. Sin θ and cos θ can be obtained from a trigonometric table or calculator.

IR cos θ is the resistive component of the voltage drop and is in phase or in the same direction as the current. IX sin θ is the reactive component of the voltage drop and is 90° out of phase, or displaced from the current. Sin θ is positive when the current lags the voltage (lagging power factor) and negative

when the current leads the voltage (leading power factor).

The approximate calculation of the voltage at the receiving end as shown in Fig 7 is:

$$e_R = e_s - V = e_s - (IR \cos \theta + IX \sin \theta)$$

For exact calculations, the following formula may be used:

$$e_R = \sqrt{e_s^2 - (IX \cos \theta - IR \sin \theta)^2} - (IR \cos \theta + IX \sin \theta)$$

3.6.2 Cable Voltage-Drop Tables. Voltage-drop tables and charts are sufficiently accurate to determine the approximate voltage drop for most problems. Table 25 contains four sections giving the three-phase line-to-line voltage drop per 10 000 A · ft for both copper and aluminum conductors in both magnetic and nonmagnetic conduits. The figures are for single-conductor cables operating at 60 °C. However, the figures are reasonably accurate up to a conductor temperature of 75 °C and for multiple-conductor cable. In most commercial buildings the voltage drops in the high-voltage primary-distribution system will be insignificant in comparison with the low-voltage system voltage drops. However, the table may be used to obtain approximate values. For borderline cases the exact values obtained from the manufacturer for the particular cable should be used. The resistance is the same for the same wire size, regardless of the voltage, but the thickness of the insulation is increased at the higher voltages which increases the conductor spacing resulting in increased reactance causing increasing errors at the lower power factors. For the same reason, the table cannot be used for open-wire or other installations such as trays where

Table 25
Three-Phase Line-to-Line Voltage Drop for 600 V Single-Conductor Cable per 10 000 A · ft 60 °C Conductor Temperature, 60 Hz

Load Power Factor Lagging	Wire Size (AWG or kcmil)																						
	14*	12*	10*	8*	6	4	2	1	1/0	2/0	3/0	4/0	250	300	350	400	500	600	700	750	800	900	1000
Section 1: Copper Conductors in Magnetic Conduit																							
1.00	53	33	21	13	8.4	5.3	3.4	2.6	2.1	1.7	1.4	1.1	0.92	0.78	0.68	0.60	0.50	0.42	0.37	0.35	0.34	0.31	0.28
0.95	50	32	20	13	8.2	5.3	3.5	2.8	2.3	1.9	1.5	1.3	1.1	1.0	0.88	0.81	0.71	0.64	0.59	0.57	0.55	0.52	0.50
0.90	48	30	19	12	8.0	5.2	3.4	2.8	2.3	1.9	1.6	1.3	1.2	1.1	0.95	0.88	0.78	0.71	0.66	0.64	0.62	0.59	0.57
0.80	43	27	17	11	7.3	4.8	3.2	2.6	2.3	1.9	1.6	1.4	1.2	1.1	1.0	0.95	0.85	0.80	0.74	0.73	0.71	0.68	0.66
0.70	38	24	15	9.9	6.6	4.4	3.0	2.5	2.1	1.8	1.5	1.3	1.2	1.1	1.0	0.97	0.88	0.83	0.80	0.78	0.76	0.73	0.71
Section 2: Copper Conductors in Nonmagnetic Conduit																							
1.00	53	33	21	13	8.4	5.3	3.3	2.6	2.1	1.6	1.3	1.0	0.88	0.73	0.62	0.55	0.45	0.38	0.33	0.29	0.28	0.26	0.23
0.95	50	32	20	13	8.2	5.3	3.4	2.7	2.2	1.8	1.5	1.1	1.0	0.92	0.80	0.71	0.62	0.54	0.50	0.47	0.45	0.43	0.40
0.90	48	30	19	12	7.9	5.1	3.3	2.7	2.2	1.8	1.5	1.1	1.1	0.95	0.85	0.76	0.68	0.59	0.55	0.54	0.52	0.48	0.47
0.80	43	27	17	11	7.2	4.7	3.1	2.5	2.1	1.7	1.4	1.1	1.1	0.97	0.88	0.81	0.73	0.66	0.62	0.59	0.57	0.55	0.54
0.70	38	24	15	9.7	6.4	4.3	2.8	2.4	2.0	1.6	1.4	1.1	1.1	0.97	0.88	0.83	0.74	0.69	0.66	0.64	0.62	0.59	0.57
Section 3: Aluminum Conductors in Magnetic Conduit																							
1.00	—	52	33	21	13	8.4	5.2	4.2	3.3	2.6	2.1	1.7	1.4	1.2	1.0	0.91	0.74	0.63	0.55	0.52	0.49	0.45	0.42
0.95	—	50	32	20	13	8.2	5.3	4.2	3.4	2.7	2.3	1.8	1.6	1.4	1.2	1.1	0.94	0.83	0.76	0.73	0.70	0.65	0.62
0.90	—	48	30	19	12	7.9	5.1	4.1	3.4	2.7	2.3	1.9	1.6	1.4	1.3	1.2	0.99	0.88	0.82	0.79	0.76	0.72	0.69
0.80	—	43	27	17	11	7.3	4.7	3.9	3.2	2.6	2.2	1.8	1.6	1.4	1.3	1.2	1.0	0.95	0.88	0.85	0.83	0.80	0.76
0.70	—	37	24	15	10	6.5	4.3	3.6	2.9	2.4	2.1	1.7	1.6	1.4	1.3	1.2	1.1	0.98	0.92	0.89	0.87	0.83	0.80
Section 4: Aluminum Conductors in Nonmagnetic Conduit																							
1.00	—	52	33	21	13	8.4	5.2	4.2	3.3	2.6	2.1	1.7	1.4	1.2	1.0	0.88	0.70	0.59	0.51	0.47	0.44	0.39	0.36
0.95	—	50	32	20	13	8.2	5.2	4.2	3.4	2.7	2.2	1.8	1.5	1.3	1.1	1.0	0.85	0.74	0.67	0.63	0.60	0.56	0.52
0.90	—	48	30	19	12	7.9	5.0	4.1	3.3	2.6	2.2	1.8	1.5	1.3	1.2	1.1	0.89	0.79	0.71	0.68	0.65	0.61	0.57
0.80	—	42	27	17	11	7.2	4.6	3.8	3.1	2.5	2.1	1.7	1.5	1.3	1.2	1.1	0.92	0.83	0.76	0.73	0.71	0.66	0.63
0.70	—	37	24	15	9.9	6.4	4.2	3.4	2.8	2.3	1.7	1.6	1.4	1.3	1.1	1.1	0.92	0.83	0.78	0.75	0.73	0.69	0.66

To convert voltage drop to	Multiply by
Single phase, three wire, line to line	1.18
Single phase, three wire, line to neutral	0.577
Three phase, line to neutral	0.577

*Solid Conductor. Other conductors are stranded.

there is appreciable spacing between the individual phase conductors.

In using the table, the normal procedure is to look up the voltage drop for 10 000 A · ft and multiply this value by the ratio of the actual number of ampere-feet to 10 000. Note that the distance in feet is the distance from the source to the load.

Example 1:

500 kcmil copper conductor in steel (magnetic) conduit

Circuit length 200 ft

Load 300 A at 80% power factor

What is the voltage drop?

Using Table 25, Section 1 the intersection between 500 kcmil and the power factor gives a voltage drop of 0.85 V for 10 000 A · ft:

$$200 \text{ ft} \cdot 300 \text{ A} = 60\ 000 \text{ A} \cdot \text{ft}$$

$$(60\ 000/10\ 000) \cdot 0.85$$

$$= 6 \cdot 0.85$$

$$= 5.1 \text{ V drop}$$

voltage drop, phase to neutral

$$= 0.577 \cdot 5.1$$

$$= 2.9 \text{ V}$$

Example 2:

No 12 aluminum conductor in aluminum (nonmagnetic) conduit

Circuit length 200 ft

Load 10 A at 70% power factor

What is the voltage drop?

Using Table 25, Section 4 the intersection between No 12 aluminum conductor and 0.70 power factor is 37 V for 10 000 A · ft:

$$200 \text{ ft} \cdot 10 \text{ A} = 2000 \text{ A} \cdot \text{ft}$$

$$\text{voltage drop} = (2000/10\ 000) \cdot 37$$

$$= 0.2 \cdot 37 = 7.4 \text{ V}$$

Example 3:

Determine the wire size in Example 2 to limit the voltage drop to 3 V.

Voltage drop in 10 000 A · ft would be $(10\ 000/2000) \cdot 3 = 5 \cdot 3 = 15$ V

Using Table 25, Section 4 move along the 0.70 power factor line to find the voltage drop not greater than 15 V. No 8 aluminum has a drop of 15 V for 10 000 A · ft; so it is the smallest aluminum conductor in aluminum conduit which could be used to carry 10 A for 200 ft of circuit with a voltage drop of not over 3 V, line to line.

3.6.3 Busway Voltage-Drop Charts and Tables. Tables 26 and 27 and Figs 8—10 show voltage drops per 100 ft at rated current (end loading) for the entire range of lagging power factors. The actual voltage drop for a three-phase system at a given load power factor equals

$$\frac{[(\text{rated-load voltage drop}) \cdot (\text{actual load}) \cdot (\text{actual length})]}{(\text{rated load} \cdot 100 \text{ ft})}$$

The voltage drop for a single-phase load connected to a three-phase system busway is 15.5% higher than the values shown in the tables. For two-pole busway serving a single-phase load, the voltage-drop values in Tables 26 and 27 should be multiplied by 1.08.

The foregoing discussion concerning uniformly distributed loading and concentrated load, of course, applies to a busway. Since plug-in types of busways are particularly adapted to serving the distributed blocks of load, care should be exercised to ensure proper handling of such voltage-drop calculations. Thus, with uniformly distributed loading, the values in the tables should be divided by two. When several separate blocks of load are tapped off the run at various points, the voltage drop should be

Table 26
Voltage-Drop Values for Three-Phase Busways with Copper Bus Bars, in Volts per 100 ft, Line to Line, at Rated Current with Entire Load at End
(Divide Values by Two for Distributed Loading)

Rating (amperes)	Power Factor									
	20	30	40	50	60	70	80	90	95	100
Low-voltage-drop ventilated feeder										
800	3.66	3.88	4.04	4.14	4.20	4.20	4.16	3.92	3.60	2.72
1000	1.84	2.06	2.22	2.40	2.54	2.64	2.72	2.70	2.62	2.30
1350	2.24	2.44	2.62	2.74	2.86	2.94	2.96	2.90	2.78	2.30
1600	1.88	2.10	2.30	2.46	2.62	2.74	2.82	2.84	2.76	2.42
2000	2.16	2.34	2.52	2.66	2.78	2.84	2.90	2.80	2.68	2.30
2500	2.04	2.18	2.38	2.48	2.62	2.68	2.72	2.62	2.50	2.14
3000	1.96	2.12	2.28	2.40	2.52	2.58	2.60	2.52	2.40	2.06
4000	2.18	2.36	2.54	2.68	2.80	2.80	2.90	2.80	2.68	2.28
5000	2.00	2.16	2.30	2.40	2.50	2.60	2.68	2.60	2.40	2.10
Low-voltage-drop ventilated plug-in										
800	6.80	6.86	6.92	6.86	6.72	6.52	6.04	5.26	4.64	2.76
1000	2.26	2.56	2.70	2.86	2.96	3.00	3.00	2.92	2.80	2.28
1350	2.98	3.16	3.32	3.38	3.44	3.46	3.40	3.22	3.00	2.32
1600	2.28	2.44	2.62	2.78	2.90	3.00	2.96	2.94	2.88	2.44
2000	2.58	2.78	2.92	3.02	3.10	3.16	3.08	3.00	2.82	2.28
2500	2.32	2.50	2.66	2.76	2.86	2.90	2.86	2.78	2.66	2.18
3000	2.18	2.34	2.48	2.60	2.70	2.74	2.72	2.66	2.58	2.10
4000	2.42	2.56	2.76	2.88	3.00	3.02	3.00	2.96	2.84	2.36
5000	2.22	2.30	2.48	2.60	2.70	2.76	2.74	2.68	2.60	2.16
Plug-in										
225	2.82	2.94	3.04	3.12	3.18	3.18	3.10	2.86	2.70	2.04
400	4.94	5.08	5.16	5.18	5.16	5.02	4.98	4.30	3.94	2.64
600	5.24	5.34	5.40	5.40	5.36	5.00	4.50	2.10	3.62	2.92
800	5.06	5.12	5.16	5.06	5.00	4.74	4.50	3.84	3.32	1.94
1000	5.80	5.88	5.84	5.76	5.56	5.30	4.82	4.12	3.52	1.94
Trolley busway										
100	1.2	1.38	1.58	1.74	1.80	2.06	2.20	2.30	2.30	2.18
Current-limiting ventilated										
1000	12.3	12.5	12.3	12.2	11.8	11.1	10.1	8.65	7.45	3.8
1350	15.5	15.6	15.4	15.3	14.7	13.9	12.6	10.7	9.2	4.7
1600	18.2	18.2	18.0	17.5	16.6	15.6	14.1	11.5	9.5	4.0
2000	20.4	20.3	20.0	19.4	18.4	17.0	13.9	12.1	10.1	3.8
2500	23.8	23.6	23.0	22.2	21.0	19.2	17.2	13.5	10.7	3.8
3000	26.0	26.2	25.8	24.8	23.4	21.5	19.1	15.1	12.0	4.0
4000	29.1	28.8	28.2	27.2	25.6	25.2	21.0	16.6	13.0	4.1

determined for the first section using total load. The voltage drop in the next section is then calculated using the total load minus that which was tapped off at the first section, etc.

Figure 11 shows the voltage-drop curve versus power factor for typical light-duty trolley busway carrying rated load.

Example Using Fig 9: Find the line-to-line voltage drop on a 300 ft run of 800 A plug-in type busway with rated load at 80% power factor.

Solution: Enter the chart at 80% on the horizontal scale. Follow a vertical line to its intersection with the curve for 800 A and extend a line horizontally to its intersection with the vertical scale.

This intersection gives the voltage drop per 100 ft. Multiply this value by 3 to find the voltage drop for 300 ft:

line-to-line voltage drop = 4.5 · 3

= 13.5 V

Table 27
Voltage-Drop Values for Three-Phase Busways with Aluminum Bus Bars, in Volts per 100 ft, Line to Line, at Rated Current with Entire Load at End
(Divide Values by Two for Distributed Loading)

Rating (amperes)	Power Factor									
	20	30	40	50	60	70	80	90	95	100
Low-voltage-drop ventilated feeder										
800	1.68	1.96	2.20	2.46	2.68	2.88	3.04	3.12·	3.14	2.90
1000	1.90	2.16	2.38	2.60	2.80	2.96	3.06	3.14	3.12	2.82
1350	1.88	2.20	2.48	2.74	3.02	3.24	3.44	3.56	3.58	2.38
1600	1.66	1.92	2.18	2.42	2.64	2.84	3.02	3.12	3.16	2.94
2000	1.82	2.06	2.30	2.50	2.70	2.88	3.02	3.10	3.04	2.80
2500	1.86	2.10	2.34	2.56	2.74	2.90	3.04	3.10	3.08	2.78
3000	1.76	2.06	2.26	2.52	2.68	2.86	2.98	3.06	3.04	2.78
4000	1.74	1.98	2.24	2.48	2.70	2.88	3.04	3.08	3.12	2.88
5000	1.72	1.98	2.20	2.42	2.62	2.80	2.92	3.02	3.02	2.80
Low-voltage-drop ventilated plug-in										
800	2.12	2.38	2.58	2.80	3.00	3.16	3.26	3.30	3.24	2.90
1000	2.44	2.66	2.86	3.06	3.22	3.36	3.42	3.38	3.28	2.84
1350	2.22	2.48	2.78	3.00	3.24	3.46	3.60	3.68	3.64	3.30
1600	1.82	2.12	2.38	2.62	2.80	2.96	3.08	3.16	3.14	2.88
2000	2.00	2.30	2.50	2.76	2.92	3.06	3.12	3.18	3.12	2.80
2500	2.00	2.28	2.50	2.70	2.92	3.02	3.12	3.16	3.08	1.78
3000	1.98	2.26	2.44	2.66	2.86	3.00	3.10	3.18	3.14	2.82
4000	1.94	2.20	2.48	2.64	2.86	3.00	3.12	3.18	3.16	2.88
5000	1.90	2.16	2.38	2.58	2.76	2.92	3.06	3.10	3.08	2.52
Plug-in										
100	1.58	2.10	2.62	3.14	3.56	4.00	4.46	4.94	5.10	5.20
225	2.30	2.54	2.76	3.68	3.12	3.26	3.32	3.32	3.26	2.86
400	3.38	3.64	3.90	4.12	4.22	4.34	4.38	4.28	4.12	3.42
600	3.46	3.68	3.84	3.96	4.00	4.04	3.96	3.74	3.52	2.48
800	3.88	4.02	4.08	4.20	4.20	4.14	4.00	3.66	3.40	2.40
1000	3.30	3.48	3.62	3.72	3.78	3.80	3.72	3.50	3.30	2.50
Small plug-in										
50	2.2	2.6	3.0	3.5	3.8	4.1	4.5	4.7	4.8	4.6
Current-limiting ventilated										
1000	12.3	12.3	12.1	11.8	11.2	10.9	9.5	8.0	6.6	3.1
1350	16.3	16.3	16.1	15.6	14.7	13.7	12.1	8.1	8.0	3.1
1600	18.0	17.9	17.7	17.0	16.1	14.9	13.4	10.7	8.6	3.3
2000	22.5	22.4	21.8	21.2	19.9	18.2	16.0	12.7	9.9	3.1
2500	25.0	24.6	23.9	23.1	21.7	19.9	17.5	13.7	10.8	3.0
3000	26.2	25.8	25.1	24.1	22.7	20.8	18.2	14.2	10.9	2.9
4000	31.4	31.0	30.2	28.8	27.4	24.8	21.5	16.5	12.7	2.9

Example Using Fig 10: Find the line-to-line voltage drop on a 200 ft run of 1500 A busway carrying 90% rated current at 70% power factor.

Solution: Enter the chart at 70% on the horizontal scale. Follow a vertical line to its intersection with the curve for 1500 A and extend a line horizontally to its intersection with the vertical scale. This intersection gives the voltage drop for a 100 ft run at rated load. For 200 ft at 90% load,

line-to-line voltage drop = 6.4 · 2 · 0.9

= 11.5 V

Example Using Fig 11: Find the line-to-line voltage drop on a 500 ft run with 50 A load at 80% power factor. The load is concentrated at the end of the run.

100

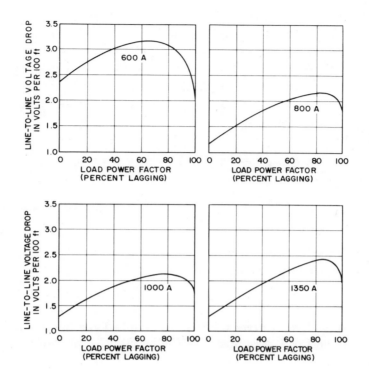

Fig 8
Voltage-Drop Curves for Typical Interleaved Construction
of Copper Busway at Rated Load, Assuming 70 °C
Operating Temperature

Solution: Enter the chart at 80% on the horizontal scale. Follow a vertical line to its intersection with the curve and extend a line horizontally to its intersection with the vertical scale. This intersection gives the voltage drop for 100 ft. For 500 ft,

line-to-line voltage drop = 3.03 · 5

$$= 15.15 \text{ V}$$

3.6.4 Transformer Voltage-Drop Charts. Figure 12 may be used to determine the approximate voltage drop in single-phase and three-phase 60 Hz liquid-filled self-cooled transformers. The voltage drop through a single-phase transformer is found by entering the chart at a kilovoltampere value three times the rating of the single-phase transformer. Figure 12 covers transformers in the following ranges:

Single-phase
  250 kVA—500 kVA, 8.6 kV to 15 kV insulation classes
  833 kVA—1250 kVA, 2.5 kV to 25 kV insulation classes
Three-phase
  225 kVA—750 kVA, 8.6 kV to 15 kV insulation classes
  1000 kVA—10 000 kVA, 2.5 kV to 25 kV insulation classes

101

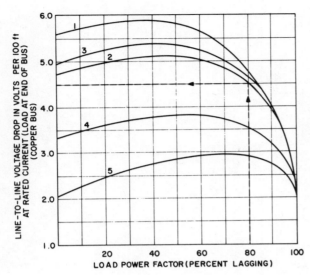

Curve	Bars per Phase	Size of Bar	Rated Current (amperes)
1	1	¼ × 4	1000
2	1	³/₁₆ × 4	800
3	1	¼ × 2	600
4	1	³/₁₆ × 2	400
5	1	³/₁₆ × 1	225

Fig 9
Voltage-Drop Curves for Typical Plug-in Type Busway at
Rated Load, Assuming 70 °C Operating Temperature

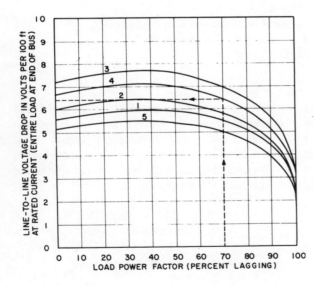

Curve	Bars per Phase	Size of Bar	Rated Current (amperes)
1	1	¼ × 2	600
2	1	¼ × 4	1000
3	2	¼ × 2½	1200
4	2	¼ × 4	1500
5	4	¼ × 3	2300

Fig 10
Voltage-Drop Curves for Typical Feeder Busways Mounted Flat
Horizontally, Assuming 70 °C Operating Temperature

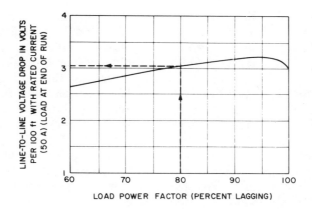

Fig 11
Voltage-Drop Curve Versus Power Factor for Typical Light-Duty Trolley Busway Carrying Rated Load, Assuming 70 °C Operating Temperature

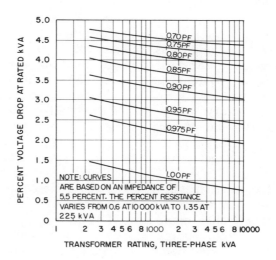

Fig 12
Voltage-Drop Curves for Three-Phase Transformers, 225—10 000 kVA, 5—25 kV

Example Using Fig 12: Find the percent voltage drop in a 2000 kVA three-phase 60 Hz transformer rated 4160 V — 480 V. The load is 1500 kVA at 0.85 power factor.

Solution: Enter the chart at 2000 kVA on the horizontal scale. Follow a vertical line to its intersection with the 0.85 power-factor curve. From this point extend a line horizontally to its intersection with the vertical scale. This intersection gives the percent voltage drop for rated load. Multiply this value by the ratio of actual load to rated load:

percent voltage drop at rated load = 3.67

percent voltage drop at 1500 kVA

$$= 3.67 \cdot 1500/2000 - 2.75$$

actual voltage drop = 2.75% · 480 = 13.2 V

NOTE: Figure 12 applies to 5.5% impedance transformers. For transformers of substantially different impedance, the information for the calculation should be obtained from the manufacturer.

3.7 Improvement of Voltage Conditions.

Poor equipment performance, overheating, nuisance tripping of overcurrent protective devices, and excessive burnouts are signs of unsatisfactory voltage. Low voltage occurs at the end of long low-voltage circuits. High voltage occurs at the beginning of low-voltage circuits close to the source of supply.

In cases of low voltage, the first step is to make a load survey to measure the current taken by the affected equipment, the current in the circuit supplying the equipment, and the current being supplied by the distribution transformer under peak-load conditions to make sure that the low voltage is not due to overloaded equipment. If the low voltage is due to overload, then corrective action must be taken to relieve the over-loaded equipment.

If overload is ruled out or if the utilization voltage is excessively high, a voltage survey should be made, preferably by using recording voltmeters to determine the voltage spread at the utilization equipment under all load conditions and the voltage spread at the utility supply, for comparison with ANSI C84.1-1977 [2], to determine if the unsatisfactory voltage is caused by the plant distribution system or the utility supply. If the utility supply exceeds the tolerance limits specified in ANSI C84.1-1977 [2], the utility should be notified. Most utilities will assist in making this voltage survey by providing the recording voltmeters required to determine the voltage during maximum- and minimum-load conditions.

If low voltage is caused by excessive voltage drop in the low-voltage wiring (over 5%), then additional circuits must be run in parallel with the affected circuits or to supply sections of the affected equipment in order to reduce the voltage drop. If the power factor of the load is low, capacitors may be installed to improve the power factor and reduce the voltage drop. Where low voltage affects a large area, the best solution may be to go to primary distribution if the building is supplied from a single transformer station, or to install an additional transformer in the center of the affected area if the building has primary distribution. Buildings wired at 208Y/120 V or 240 V may be changed over economically to 480Y/277 V if an appreciable section of the wiring system is rated 600 V and motors are dual rated 220 · 440 V or 230 · 460 V.

If the voltage is consistently high or low and the building has primary distribution, the distribution transformer taps may be changed in the direction to improve the voltage (see 3.2.7). If the

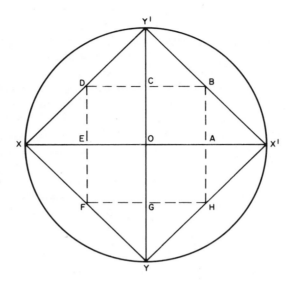

Fig 13
Effect of Low-Voltage Source Location
on Area which Can Be Supplied under
Specific Voltage-Drop Limits

building is supplied from a single distribution transformer furnished by the supplying utility, then a voltage complaint should be made to the utility.

3.8 Voltage-Drop Considerations in Locating the Low-Voltage Power Source. One of the major factors affecting the design of the low-voltage distribution system is the proper location of the low-voltage supply as close as possible to the center of load. This applies in every case, from a service drop from a distribution transformer on the street to a distribution transformer located outside or inside the building. Frequently, building aesthetics or available space require the low-voltage power supply to be installed at a corner of a building, without regard to what this adds to the cost of the building wiring to keep the voltage drop within satisfactory limits.

Reference [7] shows that if a power supply is located in the center of a horizontal floor area at point O, Fig 13, the area that can be supplied from circuits run radially from point O with specified circuit constants and voltage drop would be the area enclosed by the circle of radius O—X. However, conduit systems are run in rectangular coordinates; thus with this restriction, the area which can be supplied is reduced to the square X—Y—X'—Y' when the conduit system is run parallel to the axes X—X'—Y—Y'. But the limits of the square are not parallel to the conduit system, and, to fit the conduit system into a square building with walls parallel to the conduit system, the area must be reduced to F—H—B—D.

If the supply point is moved to the center of one side of the building, which is a frequent situation when the transformer is placed outside the building, the area which can be served with the

Table 28
Areas that Can Be Supplied for Specific Voltage Drops and Voltages

Nominal System Voltage (volts)	Distance (feet)			
	5% Voltage Drop		2.5% Voltage Drop	
	OX	OA	OX	OA
120/240	360	180	180	90
208	312	156	156	78
240	360	180	180	90
480	720	360	360	180

specified voltage drop and specified circuit constants is E—A—B—D. If the supply station is moved to a corner of the building, a frequent location for buildings supplied from the rear or from the street, the area is reduced to O—A—B—C.

Every effort should be made to place the low-voltage supply point as close as possible to the center of the load area. Note that this study is based on a horizontal wiring system and any vertical components must be deducted to establish the limits of the horizontal area which can be supplied.

Using an average value of 30 ft/V for a fully loaded conductor, the distances in Fig 13 for 5% and 2.5% voltage drops are shown in Table 28. For a distributed load, the distances will be approximately twice the values shown.

3.9 Momentary Voltage Variations—Voltage Dip. The previous discussion has covered the relatively slow changes in voltage associated with steady-state-voltage spreads and tolerance limits. However, certain types of utilization equipment impose a heavy load momentarily which causes a dip in the supply voltage and a corresponding dip in the output of lighting equipment, partic-

ulary incandescant lamps. Figure 14 taken from [6], Section 5, Fig 5-3 shows that a voltage dip in the order of 0.5% produces a noticeable change in the light output of an incandescent lamp and when the dip is cyclical, can cause objectionable flicker.

One source of voltage dips in commercial buildings is the starting of large air-conditioning motors on a distribution transformer which also supplies incandescent lights. For example, if the horsepower rating of the motor is 5% of the kilovoltampere rating of the distribution transformer, the starting kilovoltamperes of the motor will be at least 5 times the full-load kilovoltamperes (since motors draw approximately 1 kVA/hp) or more than 25% of the kilovoltampere rating of the transformer). Since the power factor of the motor starting current is quite low, 30 to 40%, the actual impedance of the transformer shown on the transformer nameplate can be used directly to calculate the approximate voltage dip. For a normal 6% impedance transformer, the full-load voltage drop would be 6%, so that at 25% load the voltage drop would be approximately 1.5%. Figure 14 shows that such a voltage dip would be quite noticeable and, if the motor started frequently, could be objectionable. Light load conditions can produce frequent starting for some types of equipment. Other types of intermittently operated equipment such as elevators, X-ray machines, and flashing signs may produce a flicker when connected to lighting circuits.

If the voltage dip combined with the starting frequency approaches the objectionable zone, a more accurate calculation should be made using the actual locked rotor current of the motor as obtained from the manufacturer or indi-

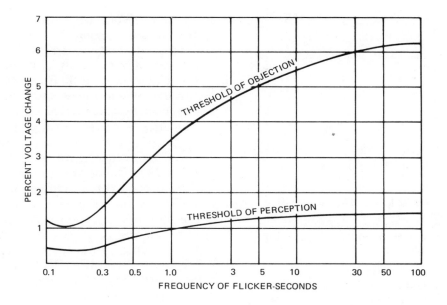

Fig 14
Flicker Chart

cated by the code letter on the motor nameplate. The values for the code letters are listed in ANSI/NEMA MG1-1978 [5] and in ANSI/NFPA 70-1981 [4] Article 430.

If the amount of the voltage dip in combination with the frequency falls within the objectionable range, then consideration should be given to methods of reducing the dip to acceptable values, such as using two or more smaller motors, providing a separate distribution transformer for the motors, or using reduced voltage starting.

If a commercial building is supplied from a single electric utility company service and the building owner or a tenant causes flicker problems for another tenant, the building owner is responsible for correcting the flicker problem although the supplying electric utility may assist in the investigation. If one customer of the electric utility causes flicker on another customer, the affected customer should file a complaint with the local electric utility company. Most electric utility companies have guidelines on what is considered an objectionable flicker but these guidelines vary widely among companies. The problem is that individuals vary widely in their susceptibility to flicker. Tests have indicated that some individuals are irritated by a flicker which is barely noticeable by other individuals. Figure 14 is considered conservative by many utilities but this may be better for the building wiring designer.

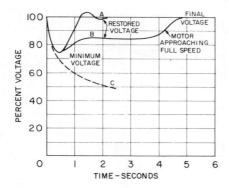

Motor-starting kVA = 100 percent of
generator rating
A — No initial load on generator
B — 50 percent initial load on generator
C — No regulator

**Fig 15
Typical Generator-Voltage Behavior
Due to Full-Voltage Starting of a Motor**

3.10 Calculation of Voltage Dips

3.10.1 Effect of Motor Starting on Standby or Emergency Generators. Figure 15 shows the behavior of the voltage of a generator when an induction motor is started. Starting a synchronous motor has a similar effect up to the time of pull-in torque. The case used for this illustration utilizes a full-voltage starting device, and the full-voltage motor starting kilovoltamperes are about 100% of the generator rating. It is assumed for curves A and B that the generator is provided with an automatic voltage regulator.

The minimum voltage of the generator as shown in Fig 15 is an important quantity because it is a determining factor affecting undervoltage devices and contactors connected to the system and the stalling of motors running on the system. The curves of Fig 16 can be used for estimating the minimum voltage occurring at the terminals of a generator supplying power to a motor being started.

3.10.2 Effect of Motor Starting on Distribution System. It is characteristic of most alternating-current motors that the current which they draw on starting is much higher than their normal running current. Synchronous and squirrel-cage induction motors started on full voltage may draw a current as high as seven or eight times their full-load running current. This sudden increase in the current drawn from the power system may result in excessive drop in voltage unless it is considered in the design of the system. The motor-staring kilovoltamperes, imposed on the power-supply system, and the available motor torque are greatly affected by the method of starting used. Table 29 gives a comparison of several common methods.

3.10.3 Motor-Starting-Voltage Drop— Transformers. Frequently, in the case of purchased power, there are transformers or cables, or both, between the starting motor and the generator. Most of the drop in this case is within the distribution equipment. When all of the voltage drop is in this equipment, the voltage falls immediately (because it is not influenced by a regulator as in the case of the generator) and does not recover until the motor approaches full speed. Since the transformer is usually the largest single impedance in the distribution system, and therefore incurs most of the total drop, Fig 17 has been plotted in terms of motor-starting kilovoltamperes which are drawn if rated transformer secondary voltages were maintained.

3.10.4 Motor-Starting-Voltage Drop — Cables and Busways. The motor-starting-voltage drop due to the impedance of cables and busways may be calculated by means of Figs 8—11 and Tables 26 and 27 using the locked-rotor current and power factor as the load. Note that in

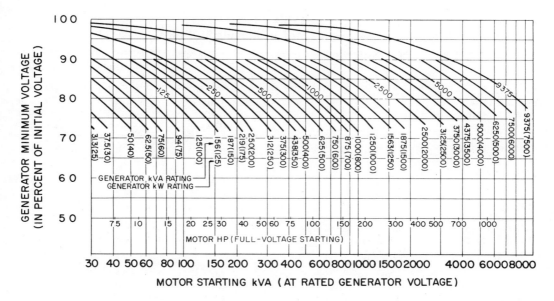

NOTES: (1) The scale of motor horsepower is based on the starting current being equal to approximately 5.5 times normal.
(2) If there is no initial load, the voltage regulator will restore voltage to 100% after dip to values given by curves.
(3) Initial load, if any, is assumed to be of constant-current type.
(4) Generator characteristics are assumed as follows. (a) Generators rated 1000 kVA or less; Performance factor k = 10; transient reactance X_d' = 25%; synchronous reactance X_d' = 120%. (b) Generators rated above 1000 kVA: Characteristics for 3600 r/min turbine generators.

Fig 16
Minimum Generator Voltage Due to Full-Voltage Starting of a Motor

computing the circuit length, only the common section of the circuit between the supplying transformer and the motor and the point at which the voltage drop is being calculated should be used. For very large motors of several hundred horsepower, the voltage drop in the system supplying the transformer may have to be considered.

Table 25 cannot normally be used since the locked-rotor power factor there is below the 70% limit. In this case the voltage drop must be calculated using the approximate formula. However, since the power factor is quite low, the

resistance component is generally negligible and only the reactance component needs to be computed.

3.11 Phase-Voltage Unbalance in Three-Phase Systems

3.11.1 Causes of Phase-Voltage Unbalance. Most utilities use four-wire grounded-wye (Y) distribution systems so that single-phase distribution transformers can be connected phase-to-neutral to supply single-phase load such as residences and street lights. Variations in single-phase loading causes the currents in the three-phase conductors to be

Table 29
Comparison of Motor-Starting Methods

Type of Starter (Settings Given Are the More Common for Each Type)	Motor Terminal Voltage (% Line Voltage)	Starting Torque (% Full-Voltage Starting Torque)	Line-Current (% Full-Voltage Starting Current)
Full-voltage starter	100	100	100
Autotransformer			
80% tap	80	64	68
65% tap	65	42	46
50% tap	50	25	30
Resistor starter, single step (adjusted for motor voltage to be 80% of line voltage)	80	64	80
Reactor			
50% tap	50	25	50
45% tap	45	20	45
37.5% tap	37.5	14	37.5
Part-winding starter (low-speed motors only)			
75% winding	100	75	75
50% winding	100	50	50

NOTE: For a line voltage not equal to the motor rated voltage multiply all values in the first and last columns by the ratio (actual voltage)/(motor rated voltage). Multiply all values in the second column by the ratio $[(\text{actual voltage})/(\text{motor rated voltage})]^2$.

different, producing different voltage drops and causing the phase voltages to become unbalanced. Normally the maximum phase-voltage unbalance will occur at the end of the primary distribution system, but the actual amount will depend on how well the single-phase loads are balanced between the phases on the system. However, a perfect balance can never be maintained because the loads are continually changing, causing the phase-voltage unbalance to be continually changing also. Blown fuses on three-phase capacitor banks will also unbalance the load and cause phase-voltage unbalance. Most distribution transformers used to step the distribution voltage down to a utilization voltage have Δ-connected primaries. Unbalanced primary voltages will introduce a circulating current into the Δ winding tending to rebalance the secondary voltage. Under these conditions, phase-voltage unbalance in the primary distribution system tends to correct itself and should not be a problem.

Commercial buildings make extensive use of four-wire Y utilization voltages to supply lighting loads connected phase-to-neutral. Proper balancing of single-phase loads among the three phases on both branch circuits and feeders is necessary to keep the load unbalance and the corresponding phase-voltage unbalance within reasonable limits.

3.11.2 Measurement of Phase-Voltage Unbalance. The simplest method of expressing the phase-voltage unbalance is to measure the voltages in each of the three phases [8] — [10]. The voltage

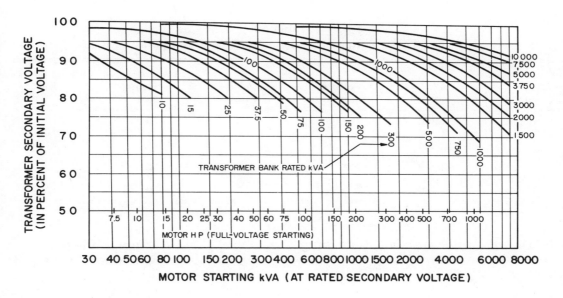

MOTOR STARTING kVA (AT RATED SECONDARY VOLTAGE)

NOTES: (1) The scale of motor horsepower is based on the starting current being equal to approximately 5.5 times normal.

(2) Short-circuit kVA of primary supply is assumed to be as follows:

Transformer Bank kVA	Primary Short-Circuit kVA
0—300	25 000
500—1000	50 000
1500—3000	100 000
3760—10 000	250 000

(3) Transformer impedances are assumed to be as follows:

Transformer Bank kVA	Transformer Bank Impedance (%)
10—50	3
75—150	4
200—500	5
750—2000	5.5
3000—10 000	6

(4) Representative values of primary system voltage drop as a fraction of total drop, for the assumed conditions, are as follows:

Transformer Bank kVA	System Drop / Total Drop
100	0.09
1000	0.25
10 000	0.44

Fig 17
Voltage Drop in a Transformer Due to
Full-Voltage Starting of a Motor

111

Table 30
Effect of Phase-Voltage Unbalance on Motor Temperature Rise

Motor Type	Load	% Voltage Unbalance	% Added Heating	Insulation System Class	Temperature Rise (°C)
U frame 220/440 V	Rated	0	0	A	60
	Rated	2	8	A	65
	Rated	3.5	25	A	75
T frame 230/460 V	Rated	0	0	B	80
	Rated	2	8	B	86.4
	Rated	3.5	25	B	100

unbalance is the maximum deviation from the average of the three-phase voltages:

voltage unbalance

$$= \frac{\left(\begin{array}{c}\text{maximum deviation} \\ \text{from average phase voltage}\end{array}\right)}{\text{(average phase voltage)}}$$

The phase-voltage unbalance may also be expressed in symmetrical components as the ratio of the negative-sequence voltage to the positive-sequence voltage:

voltage unbalance factor

$$= \frac{\text{negative-sequence voltage}}{\text{positive-sequence voltage}}$$

The second formula defines the negative-sequence component of the voltage which is a more accurate indication of the effect of phase-voltage unbalance.

3.11.3 Effect of Phase-Voltage Unbalance. When unbalanced phase voltages are applied to three-phase motors, the phase-voltage unbalance causes additional circulating currents to flow in the motor and generate additional heat loss. Reference [9] includes a table showing an extra heating which may be expected from both U frame and T frame motors

(Table 30). Reference [10] provides a more comprehensive review of the problem.

Sealed compressor motors used in air conditioners seem most susceptible to phase-voltage unbalance. Originally sealed compressor motors were limited to small sizes, but they are now being built in units up to 1000 hp or more. These motors operate with higher current densities in the windings because of the added cooling effect of the refrigerant. Thus the same percent increase in the heat loss due to circulating current, caused by phase-voltage unbalance, will have a greater effect on the sealed compressor motor than it will on a standard aircooled motor.

Since the windings in sealed compressor motors are inaccessible, they are normally protected by thermally operated switches embedded in the windings and set to open and disconnect the motor when the winding temperature exceeds the set value. The motor cannot be restarted until the winding has cooled down to the point at which the thermal switch will reclose.

When a motor trips out, the first step in determining the cause is to check the running current after it has been restarted to make sure that the motor is not overloaded. The next step is to measure the three-phase voltages to determine the amount of phase-voltage unbalance. Table 15 indicates that where the phase-voltage unbalance exceeds 2%, the motor is likely to become overheated if it is operating close to full load.

Some electronic equipment such as computers may also be affected by phase-voltage unbalance of more than 2% or 2.5%. The equipment manufacturer can supply the necessary information.

In general, single-phase loads should not be conncected to three-phase circuits supplying equipment sensitive to phase-voltage unbalance. A separate circuit should be used to supply this equipment.

A large single-phase transformer may be connected in open Δ with a small single-phase transformer to supply a large single-phase load and a small three-phase load. Such installations can produce phase voltage unbalance due to the unequal impedance and loads in the two transformers. If objectionable phase-voltage unbalance occurs in such an installation, a second single-phase transformer should be added to complete the Δ connection or the three-phase load should be connected to a separate three-phase transformer.

3.12 Harmonic Voltages

3.12.1 Nature of Harmonics.
Harmonics are integral multiples of the fundamental frequency. For example, for 60 Hz power systems, the second harmonic would be $2 \cdot 60$ or 120 Hz and the third harmonic would be $3 \cdot 60$ or 180 Hz.

Harmonics are caused by devices which change the shape of the normal sine wave of voltage or current in synchronism with the 60 Hz supply. In general those include three-phase devices in which the three-phase coils are not exactly symmetrical, and single- and three-phase loads in which the load impedance changes during the voltage wave to produce a distorted current wave. This distortion creates harmonics since all harmonics, being integral multiples of the fundamental frequency, must pass through zero at the same points as the fundamental. Therefore, a distorted wave must be made up of a fundamental and harmonics of various frequencies and magnitudes.

Inductive reactance varies directly as the frequency so that the current in an inductive circuit is reduced in proportion to the frequency for a given harmonic voltage. Conversely, capacitive reactance varies inversely as the frequency so that the current in a capacitive circuit is increased in proportion to the frequency for a given harmonic voltage. If the inductive reactance and the capacitive reactance are the same, they will cancel each other and a given harmonic voltage will cause a large current to flow, limited only by the resistance of the circuit. This condition is called resonance and is more apt to occur at the higher harmonic frequencies.

3.12.2 Effect of Harmonics.
The harmonic content and magnitude existing in any power system is largely unpredictable and effects will vary widely in different parts of the same system because of the different effects of different frequencies. Since the distorted wave is in the supply system, harmonic effects may occur at any point on the system where the distorted wave exists and are not limited to the immediate vicinity of the harmonic-producing device. Where power is converted to direct current or some other frequency, harmonics will exist in any distorted alternating component of the converted power.

Harmonics may be transferred from one circuit or system to another by direct connection or by inductive or capacitive coupling. Since 60 Hz harmonics are in the low-frequency audio range, the transfer of these frequencies into communications, signaling, and control circuits employing frequencies in the same range may cause objectionable interference.

3.12.3 Harmonic-Producing Equipment

(1) *Arc Equipment.* Arc furnaces and arc welders supplied from transformers have widely fluctuating loads and produce harmonics. Normally these do not cause very much trouble unless the supply conductors are in close proximity to communication and control circuits or there are large capacitor banks on the system.

(2) *Gaseous-Discharge Lamps.* Fluorescent and mercury lamps produce small arcs and, in combination with the ballast, produce harmonics, particularly the third. Experience shows that the third-harmonic current may be as high as 30% of the fundamental in the phase conductors and up to 90% in the neutral where the third harmonics in each phase add directly since they are displaced one third of a cycle.

Note that ANSI/NFPA 70-1981 [4], Section 210—22 (b) requires that the computed load be based on the total ampere ratings of the units and not on the total watts of the lamp for circuits supplying this type of load.

(3) *Rectifiers.* Half-wave rectifiers which suppress alternate half-cycles of current generate both even- and odd-numbered harmonics. Full-wave rectifiers tend to eliminate the even-numbered harmonics and usually diminish the magnitude of the odd-numbered harmonics.

The major producer of harmonics is the controlled rectifier whose input current waveform is a variation of a square wave which is rich in odd-numbered harmonics. Most rectifiers used in commercial buildings are six-pulse types producing harmonic numbers 5, 7, 11, 13, 17, 19 . . . in steadily decreasing magnitudes with increasing frequency. Controlled rectifiers are often used in adjustable speed drives and regulated power supplies for electronic equipment.

(4) *Rotating Machinery.* Normally the three-phase coils of both motors and generators are sufficiently symmetrical that any harmonic voltages generated are too small to cause any interference.

(5) *Induction Heaters.* Induction heaters use 60 Hz or higher frequency power to induce circulating currents in metals to heat the metal. Harmonics are generated by the interaction of the magnetic fields caused by the current in the induction heating coil and the circulating currents in the metal being heated. Large induction heating furnaces may create objectionable harmonics.

(6) *Capacitors.* Capacitors do not generate harmonics. However, the reduced reactance of the capacitor to the higher frequencies magnifies the harmonic current in the circuit containing the capacitors. In cases of resonance, this magnification may be very large. High harmonic currents may overheat the capacitors. In addition, the high currents may induce interference with communication, signal, and control circuits.

3.12.4 Reduction of Harmonic Interference.

Where harmonic interference exists, the conventional measures such as increasing the separation between the power and communication conductors and the use of shielded communication conductors should be considered. Where capacitor banks magnify the harmonic current, the capacitors should be reduced in size or removed. Where resonant conditions exist, the capacitor bank should be changed in size to shift the resonant point to another frequency or small reactors should be connected in series with the capacitors to de-tune the circuit. Where harmonics pass from a power system to a communication, signal, or control circuit, through a direct connection such as a power supply, filters may be used to suppress or short-circuit the harmonic frequencies.

Objectionable harmonic currents can be isolated with a series-resonant circuit to ground at the harmonic frequency. Locate the resonant circuit near the source of the harmonic (which may be a rectifier unit). This resonant circuit must be sized to carry all the harmonic current the system is generating. See [13].

3.13 Transient Overvoltages. Transient overvoltages (sometimes called spikes) are momentary excursions of voltage outside of the normal 60 Hz voltage wave. Originally, the major sources of transient overvoltages were lightning strokes on or near overhead supply lines and intermittent ground contacts on ungrounded systems. However, in recent years the switching of heavily loaded circuits, especially those involving large amounts of capacitance or inductance, with devices such as vacuum switches, controlled rectifier devices, and current-limiting fuses that chop the alternating-current wave has resulted in a proliferation of transient overvoltages to the extent that they are frequently called electrical noise because of the similarity to the noise in communications circuits which obscures the desired signal.

At the same time, solid-state devices, especially in microminiature sizes, which are introduced into computers, control systems, and other electronic equipment, are very susceptible to transient overvoltages, especially in the reverse direction. Where such electronic equipment is used, every effort must be made to minimize possible sources of transient overvoltages and to protect the equipment against transient overvoltages which may occur with proper surge protective devices.

3.14 References

[1] ANSI C57.12.20-1974, Requirements for Overhead-Type Distribution Transformers 67 000 V and Below, 500 kVA and Smaller and supplement C57.12.20a-1978.

[2] ANSI C84.1-1977, Voltage Ratings for Electric Power Systems and Equipment (60 Hz).

[3] ANSI C92.2-1978, Preferred Voltage Ratings for Alternating-Current Electrical Systems and Equipment Operating at Voltages above 230 kV Nominal.

[4] ANSI/NFPA 70, 1981 National Electrical Code.

[5] ANSI/NEMA MG1-1978, Motors and Generators.

[6] IEEE Std 141-1976, Electric Power Distribution for Industrial Plants.

[7] MICHAEL, D.T. Proposed Design Standard for the Voltage Drop in Building Wiring for Low-Voltage Systems. *IEEE Transactions on Industry and General Applications*, vol IGA-4, Jan/Feb 1968, pp 30–32.

[8] Electrical Utility Engineering Reference Book, vol 3, Distribution Systems. East Pittsburgh, Pa: *Westinghouse Electric Corporation*, 1965.

[9] ARNOLD, R.E. NEMA Suggested Standards for Future Design of AC Integral Horsepower Motors. *IEEE Transactions on Industry and General Applications*, vol IGA-6, Mar/Apr 1970, pp 110—114.

[10] LINDERS, J.R. Effects of Power Supply Variations on AC Motor Characteristics. *Conference Record*, 1971 *IEEE Industry and General Applications Group Annual Meeting*, pp 1055—1068.

[11] FINK, D.G., and BEATY, H.W., *Standard Handbook for Electrical Engineers*, 11th ed. New York: McGraw-Hill, 1978.

[12] Electrical Data Book, Electric Machinery Manufacturing Co, Minneapolis, Minnesota 55413.

[13] STRATFORD, R.P., Rectifier Harmonics in Power System. *Conference Record*, 1978 *IEEE Industry and Applications Society Annual Meeting*.

3.15 Supplementary Standards

[14] IEEE Std 142-1982, Recommended Practice for Grounding of Industrial and Commercial Power Systems.

[15] IEEE Std 242-1975, Recommended Practice for Protection and Coordination of Industrial and Commercial Power Systems.

[16] ANSI/IEEE Std 446-1980, Recommended Practice for Emergency and Standby Power Systems for Industrial and Commercial Applications.

4. Power Sources and Distribution Systems

4.1 General Discussion. In other sections, basic engineering, loads, voltages, apparatus, and circuit protection features for commercial buildings are discussed. This section considers electric power supplies, metering and billing, primary and secondary connections of transformers, system grounding, distribution circuit arrangements, emergency systems and equipment, and power factor correction. It is the responsibility of the engineer to develop an efficient and economical means of receiving electric power and distributing it to each area to be served. This function can be carried out in many ways. His selection of system arrangements, components, and voltages should be engineered to perform the function reliably, safely, and to deliver the power at correct voltages without hazard to personnel, the building, or equipment.

4.2 Electric Power Supply

4.2.1 Selecting a Power Source. In most cases the selection of a power source will be determined by an analysis made by the design engineers or the utility engineers, or both. Economics usually dictate the selection. With the exception of large high-load factor complexes, the costs still favor purchase of electricity for the prime power requirements. On-site total energy and cogeneration systems may, in the future, appear more attractive as the energy situation and environmental restrictions impact the utilities. Standby electric generating equipment may be provided in addition to prime purchased power, to produce emergency power necessary for critical loads upon failure of the prime source or, where permitted, to reduce the peak demand of the utility's source as a means of reducing the monthly utility bill.

The following criteria are of prime importance in the selection of the power source.

(1) *Availability*. Most commercial buildings are located where electric utility service is available or can be made available.

For purchased power the voltage selected and its characteristics, for either primary service or secondary service, is based on the utility's distribution standards and the particular services available in the specific area of the facility or facilities being planned. All utility company charges associated with installation of a new service, or the expansion of an existing service, must be included. Charges should include utility facilities relocation costs to provide working or code clearances, or both, for new construction or alterations to existing facilities. Special or nonstandard service requests can be expensive and failure to notify the owner of additional costs would not give the owner a true picture of the project.

(2) *Reliability*. Generally the reliability, voltage and frequency regulation of electric utility service in many areas of the United States are superior to self-generation. The reliability of utility service is, of course, dependent not only on the generating facilities, but on the exposure of the feeders from the generating plant. See item (4).

(3) *Standby Power*. Standby power, as opposed to prime power and emergency power, is made available in case of failure of the prime source for systems other than emergency systems. The source can either be a second utility company service, on-site generation or back up batteries. Emergency systems provide the minimum required for life safety and must be made available automatically upon failure of the prime power. Standby systems provide service to equipment generally considered essential for facility operation or to prevent loss of critical systems or computer data. Standby power may be either automatically transferred or manually transferred, or may be continuously available as in the case of a UPS system. If standby systems are to furnish loads less than the prime power load of the facility, all equipment must be selected in advance and an additional system of distribution provided, including circuits, panelboards, feeders, transformers, switchboards etc. The standby distribution systems generally

interface with the prime power source at the point of service or at the service entrance and include some means of transfer.

(4) *Purchased Versus Generated Power for Prime Power*. The installation of electric generating equipment should be considered only after a thorough analysis of the total owning and operating cost of each system. Many factors will influence the analysis, such as the type of heating system: (electric, steam, or hot water) and the comparative installation and operating cost of the alternatives. An evaluation should be made of the type of electrical system to be used in the building, which may vary depending on whether it distributes purchased or generated power.

Generated power will have a much larger new investment cost because of larger and higher priced boilers, generating equipment, space requirements, and pollution and noise control devices. Generating equipment will have to be adequate to handle all loads, including starting of all motors, and with ample capacity for necessary maintenance and emergencies, with considerations of additional standby capacity for secondary contingencies. Generated power will have a greater operating cost because of special or additional operating personnel, larger fuel bills, higher maintenance, and possibly the cost of electric utility standby. Generated power may also have a higher cost for taxes, depreciation and insurance. Table 31 compares some of the investment and operating items in the two methods. Federal (EPA and DOE) regulations may impose severe limitations on the building of fossil fuel plants.

(a) *Purchased Power*. Before discussing power costs and availability with the electric utility, the following load data for initial and future requirements should be

Table 31
Cost Comparison of Purchased and Generated Power

Purchased Power	Generated Power
Investment (if applicable)	
Substation or vault	Generating equipment
Metering and service	Additional boiler capacity
Standby equipment	Additional building space
	Heat-recovery equipment
	Additional pollution control
	Water treatment equipment
	* Additional land for fuel storage/handling
Operating costs	
Electric utility billing	Fuels
*Maintenance, labor, and	Maintenance, labor, and supplies
supplies of medium-voltage	Operating labor
equipment	Insurance
	Taxes
	Depreciation
	Standby utility service
	* Ash disposal charges

*If applicable

estimated as accurately as possible.

Demand and connected load

Average usage or load factor

Seasonal variation

Power factor of total load

Required reliability

Motor load including size of largest motor

If the electric utility must install additional facilities to meet the service quality needed, an extra charge by the electric utility may be applied. The annual operating cost can be ascertained from the rate schedule of the utility. If building expansion is likely, check the ability of the purchased-power facilities to take care of the increased load.

(b) *Generated Power.* In addition to the electrical data required when considering purchased power, the hourly, daily, and seasonal steam requirements should be examined in evaluating generated power. In the case where steam or hot water is to be used for heating and cooling, the exhaust steam or heat re-

claimed from gas, oil engines, or gas turbine generating equipment can be utilized at certain times. Only an hour-by-hour study of the coincidence of the electric energy demand and the building requirements for heating and cooling can determine the economics. An extensive study by engineers fully familiar with the comparative economics of purchased electric service versus generated power is essential. A discussion of fuel costs and selection is beyond the scope of this book.

When steam facilities are installed to generate electric energy only, the full investment cost must be charged against the electric system. These costs would include building, boilers, turbine generators and switchgear, fuel handling facilities, water treatment, condensing water and ash handling when required. Under these conditions, generated power cannot generally be justified on an economical basis.

As with purchased power, the engineer should check the feasibility and cost of facilities to grow with incremental expansions in the future. Often large expenditures are required with on-site generation for comparitively small building or load expansions.

4.2.2 Planning for Utility Service. Each utility differs to some degree from every other utility in its service policies and requirements. Therefore, communication should be established with the local supplying utility through the customer service department or electric marketing department as soon as possible so that local requirements can be incorporated in the building plans. The utility engineers will need the following information.

(1) Plot plan of the area showing the building or buildings, both present and future, roadways, and other structures

(2) Preferred point of delivery for electric service

(3) Estimated connected load, maximum demand and power factor, and any requirements for future increases

(4) Preferred voltage

(5) Any special equipment such as large motors which may disturb the supply system when started, or equipment which may be affected by supply system disturbances

(6) Any requirements for emergency or standby power

(7) A one-line diagram of the service equipment and, for primary service, the primary distribution system including size and ratings of switches and protective devices

(8) A load tabulation which indicates the portions of the total load designated for each of the following: lighting, receptacles, heating, motors, miscellaneous power, and clean power.

The utility will provide the following information when requested.

(1) Rate or rates available

(2) Voltage or voltages available and normal tolerance limits

(3) Point of delivery of electric service if preferred point is not acceptable

(4) Line route from the property line to the point of delivery for any portion of the line installed by the utility

(5) Any charges for service, including cost of any underground portion of the line

(6) Requirements for connections at the point of delivery

(7) Requirements for metering

(8) Available short-circuit capacity of the supply system and time-current coordination information.

(9) Space requirements for a transformer station if required and furnished by the utility

(10) Any utility requirements for service entrance equipment

(11) Any special local requirements and local exceptions to ANSI/NFPA 70-1981 [2] [24] applying to utility-associated equipment

(12) Any limitations on the starting of large motors

(13) Recommended ratios and taps for transformers provided by the builder

(14) Availability and cost of an alternate or standby electric supply

(15) Available historical data as to reliability of transmission and distribution feeders from which the new service will be derived.

4.2.3 Electric Utility Rates. Each electric utility has a series of rate schedules for supplying power to customers under various conditions. To arrive at the most economical condition for obtaining power, a comparison of these rates should be made. The following are factors that usually form the basis of establishing

---

[24] Numbers in brackets correspond to those in the references at the end of this section.

rates and for evaluating them.

(1) Maximum demand in kilowatts or kilovoltamperes

(2) Energy consumption in kilowatt-hours

(3) Adjustment for low power factor

(4) Voltages available

(5) Transformer or substation ownership

(6) Fuel cost adjustment clause

(7) Demand interval

(8) Minimum bill stipulations

(9) Multiple-metering provisions

(10) Auxiliary or standby service charges

(11) Seasonal or time of day service rates or charges, or both

(12) Prompt payment savings

(13) Provision for off-peak loads and interruptible loads

(14) Limitations on resale of power to tenants by building owner

(15) Incentives for utility service direct to tenants

(16) Total electric construction

(17) Elimination of multiple-metering service points

A few of the above major factors are discussed below.

(1) *Demand or Fixed Charges.* Demand charges cover all generally predictable utility costs such as depreciation, interest, and insurance. Capital investments for land, buildings, generating equipment and switchgear, transmission lines and structures, transformation and distribution equipment are depreciated over the estimated or specified life of the equipment. Demand charges reflect the investment required by the electric utility to serve the customer's maximum rate of consumption (demand). The demand is usually determined by a demand meter.

(2) *Energy or Variable Charges.* Energy charges include such items of cost as fuel, operating labor, maintenance,

raw materials such as chemicals, and all taxes. Rates must cover all costs involved in operating the electric utility plus a reasonable profit.

(3) *Power Factor Adjustment.* Some electric utilities penalize the customer if the power factor of the load drops below a stipulated value, and some utilities provide a credit for high power factors.

(4) *Voltage.* Most buildings are supplied by the electric utilities at utilization voltages such as 208Y/120 V or 480Y/277 V, which are directly usable by the load equipment. When the building area or load becomes too large to be supplied at utilization voltage due to excessive cost or excessive voltage drop, or both, the building must be supplied at a distribution voltage, which might be typically 4160 V or 13 200 V. In some instances voltages as high as 34 500 V have been supplied to transformers in building vaults. With a customer-owned distribution system, the associated equipment including transformers must be installed by the builder. The only exceptions are where the utility provides all or part of the primary system, including transformers, in return for the right to sell power directly to tenants, or where the utility and the local inspecting authority will permit more than one service in the building. In the latter case, the utility may treat each service as a separate customer and the bills will be higher than a single bill (see 4.4).

(5) *Fuel Cost.* Since the major component of the cost of electric energy is the cost of fuel over which the utility has little or no control, most utility rates have a fuel adjustment clause which provides an adjustment to the energy charges for the average current cost of fuel based on the actual heat content.

(6) *Other Factors.* The remainder of the rate factors listed as having a bearing

on the customer's bill are usually well defined on the utility's rate schedule or will be provided on request by the utility's electric sales department.

Some utilities offer preferential rates or other incentives in order to promote a larger and more diversified load base.

Since the oil embargo of 1973, and as a direct consequence of the National Energy Act of 1978, a change in conventional electric utility rate practices is unfolding in a new concept called *rate reform*. Essentially *rate reform* usually incorporates various

(1) Inverted rates

(2) Flat rates

(3) Lifeline rates

(4) Marginal cost pricing (MCP)

(5) Long range incremental costing (LRIC)

(6) Construction work in progress (CWIP)

(7) Interruptible rates

(8) Time of day pricing

(9) Reduction in the number of declining blocks

(10) Modification of fuel rate adjustments to the fuel rates

(11) Deletion of electric heat discounts

When comparing new *reform* rates versus old rates for customers it is usually difficult to determine the individual impact of rate reform and rate increase. Each customer will be affected differently based on the type and schedule of operation and the extent of load management equipment existing or planned.

4.3 Interrelated Utility and Project Factors that Influence Design. Factors which influence design of the electric systems are many and are covered in other sections. However, this brief check list contains some of the factors which may be helpful in planning system design.

(1) Type, size, shape, and occupancy purposes of the building or buildings

(2) Voltages and voltage tolerance limits of the electric utility system available at the building site

(3) Electrical rate plans available from the electric utility company

(4) Availability of aerial or underground service and of radial, loop, or network sources from the electric utility company

(5) Type and rating of building utilization equipment

(6) Economics of utilization voltage distribution as compared to medium-voltage distribution

(7) Necessity of including a change to a higher voltage, such as changing from a 208Y/120 V system to 480Y/277 V or going to medium-voltage distribution in a modernization project

(8) Complete or partial replacement of old or obsolete equipment in a modernization project

(9) Application of modern lighting and space conditioning principles to modernization projects

(10) Reliability of the source or sources of supply. Consistency in maintaining needed reliability throughout all the system is essential to the overall solution. The engineer should carefully evaluate for reliability each part of his design as well as that of the electric utility feeders and their sources. For instance, some incoming feeders may be tapped for other customers; some may be exposed to hazards or have a history of outages so that they need to be reinforced or backed up with an alternate set of feeders. The engineer should consider all possibilities of planned and inadvertent outages to determine the justification for such reinforcements or alternate feeders. Fortunately most electric utility feeders have a high degree of reliability. The electric utility's future plans

for all feeders involved in the building service should be considered.

(11) Economics of distribution system as a consequence of available fault levels of utility services and customer furnished limiters such as transformers, reactors, and current limiting protective devices.

A mutual and congenial understanding and appreciation of each other's problems is highly desirable at all stages of negotiation between the electric utility and the building owner and engineer. This is true for both new buildings and expansions. Most commercial building projects, of course, pose no conflicts of interest between the parties. But there are projects where it would be a great advantage to one party if existing plans, designs, procedures, or rules could be modified. Modifications required might include the customer's need for consideration of a more attractive rate structure, a more economical incoming voltage, or location of incoming service equipment at a more desirable point. The electric utility may want the customer to limit motor-starting currents, or to provide more convenient access routes for service, or to install more adequate fault-protection equipment, which can be of mutual advantage. Neither the electric utility nor the customer's engineer can appreciate the importance of each other's problem without an open and cooperative exchange of information at all stages of the project.

4.4 Electric Utility Metering and Billing. An understanding of utility metering and billing practices is important for evaluating service arrangements. Practices vary depending upon local utility and regulatory body requirements. The design, usage, and load characteristic for a given application should be carefully weighed before selecting service voltage and

metering characteristics. If large momentary high-demand loads, high-seasonal loads, or low power factor loads are involved, billing penalties may be involved. On the other hand, high load-factor or high power factor loads may merit a billing allowance or credit.

It is considered good practice to consult the electric utility supplying service early in the design stages. Late utility negotiations may result in increased costs or delays in service, or both. A complete discussion of service, metering, and billing requirements is always in order, no matter how preliminary. This should provide time for the consideration of various proposals and the selection of the one best suited to a given application.

4.4.1 Metering by Type of Premises. Availability of a particular kind of metering and billing generally depends upon the nature and characteristics of the premises, type of load involved, and local utility and regulatory body requirements. Because of the important influences of the metering scheme on the economics and design of the distribution system, especially in multiple-occupancy buildings, an early decision on the system to be employed is essential.

(1) A single-occupancy building, such as a hospital, a school, or an office building occupied by a single tenant, will be metered by the utility at the service entrance with a watthour-demand meter. With multiple services and where permitted by the electric utility, watthour-meter readings may be added together to take advantage of lower rates, and the demands on two or more services may be totalized so that the customer may benefit from their diversity.

(2) Multiple-occupancy buildings, such as apartment houses, shopping centers, condominiums, and large office buildings,

are generally equipped with an individual meter for the owner and for each tenant, except in cases where light and power are included in the tenant's rent, when a master metering may be utilized. Such buildings may, in some localities, be *submetered* with the owner buying power at wholesale rates on electric utility master metering and reselling it to his tenants at legally prescribed rates using his own meters.

(3) Where tenants are individually metered, either by the electric utility or by the owner, it is important to provide sufficient flexibility in the metering arrangement to facilitate metering changes as tenant changes occur.

4.4.2 Metering by Service-Voltage Characteristics. Metering of the incoming service may be located either on the high-voltage or on the low-voltage side of the transformer, depending on the terms of the contract with the electric utility. When the metering is on the high-voltage side of the transformer, the losses of the transformer will be metered and charged to the customer. In some cases the customer is given a discount in his billing to offset this loss. Requirements vary according to individual utility regulations.

4.4.3 Meter Location. Subject to agreement with the utility, meters may be installed indoors at the customer's secondary distribution point, in a suitable meter room, or in a separate control house which may also contain control and associated primary service switchgear. Outdoor installation including pole mounting, exterior wall attachment, or pad mounting may also be used subject to utility approval. It is good practice to review meter locations with the utility early in the design. In general, utilities require accessibility for meter reading and maintenance purposes, and suitable meter protection.

4.4.4 Meter Mounting, Control, and Appurtenant Equipment. Utilities publish regulations and requirements covering meter mounting, control, and associated equipment. Utility billing metering may be broadly grouped into three categories: (1) self-contained metering, (2) instrument transformer metering, and (3) special metering. Description and general requirements for each category follow.

(1) *Self-Contained Metering.* Meters are connected directly to the system wiring being metered. The customer is generally required to furnish and install a suitable meter mounting box and associated wiring, conduit, devices, fittings, and bonding. This metering is normally used up to a maximum load of 400 A for low-voltage systems.

(2) *Instrument Transformer Metering.* The utility will require the use of instrument transformers between the system wiring and the meter wiring where the service rating exceeds currents or voltages in the order of 200 A or 480 V. The customer is generally required to furnish and install the instrument transformer cabinet or mounting assembly, meter box, conduit and fittings, and bonding. The utility generally furnishes instrument transformers. The instrument transformers are installed and meter-wiring connections may be made by the utility or the customer, according to the utility requirements.

(3) *Special Metering.* This includes totalizer metering, impulse metering, telemetering, reactive component or power-factor metering, etc. For complete requirements in all such cases, the utility should be consulted.

4.4.5 Types of Metering. Summarized below are various types of metering, including application. Utility requirements cover the types of metering available for a given application.

(1) *Master Metering.* This is a single metered electric service to multiple-occupancy premises. Tenant service costs are either included in the rent as a flat charge or determined by submeters, depending upon local utility regulations.

(2) *Multiple Metering.* A separate meter is established for each tenant's requirements in a multiple-occupancy building. Each tenant is separately metered and billed by the utility.

(3) *Primary Metering.* This is medium-voltage metering up to 72 000 V. The customer generally owns and maintains service transformer(s) and meter mounting equipment. Metering is generally owned and maintained by the utility.

(4) *Secondary Metering.* Under 600 V the utility generally owns and maintains service transformers, metering transformers, and metering. The customer generally owns meter wiring and meter mounting equipment.

(5) *Totalized Metering.* Coincident demand or multiple-services are metered by either impulse or integrating demand metering to provide diversified demand registration equivalent to that for a single meter. It is generally required by the utility when the service to a single stab is impractical.

(6) *Impulse Metering.* It is used to determine coincident demand. Meter registration is effected by use of electric impulses. Each impulse is a function of load and time. Impulses are received from several sources (that is, metering points) and counted by a totalizing meter. The totalizing meter integrates the received impulses over a given period of time (characteristically the demand interval) to provide a readout of the total demand. Printed tape demand meters and totalizing demand meters utilize impulse metering. Magnetic tapes are now being used.

(7) *Compensated Metering.* This is applicable to primary metered service to a single transformer bank. Rather than primary metering, secondary metering together with a transformer-loss compensator is used. The transformer-loss compensator is calibrated to compensate for the service transformer losses; hence, meter registration includes service transformer losses equivalent to that of a primary meter, saving the cost of high-voltage instrument transformers.

(8) *Submetering.* Additional metering is installed on a building distribution system for the purpose of determining demand or energy consumption, or both, for certain building load subdivisions, and where the same metering is preceded by a master billing meter. Where submetering is required for billing tenants of a commercial building, the metering may be at the medium- or low-voltage feeder distribution point if all loads on the feeder are for one customer. Where a feeder supplies more than one customer or where power costs are to be accurately apportioned among various departments, the metering must be installed at each unit substation or low-voltage feeder.

(9) *Subtractive Metering.* This is an application of submetering. Readings of submeters are subtracted from associated master meter readings for billing purposes.

(10) *Coincident Demand.* See (5)

(11) *Telemetering.* Metering impulses are transmitted from one location to another for the purpose of meter reading at a remote location. It is generally used to totalize two or more distant locations.

(12) *Power Factor Metering.* Either reactive kilovoltamperes, or kilovoltamperes and kilowatts are metered to determine the power factor for utility

billing purposes. Coincident or cumulative metering is used, depending upon utility rate schedule. Low power factor loads are often subject to a billing penalty, whereas high power factor loads may merit a billing discount. The exact schedules can be obtained from the supplying utility.

4.4.6 Utility Billing. It is customary for utilities to meter and bill each customer individually. Utility rates usually consider fixed and variable cost requirements to provide service. Hence *rate schedules* generally take the form of a *block* rate, wherein incremental service costs usually vary as a function of customer usage. Electric-service costs generally comprise two components, the demand charge and the energy charge. The demand charge is based upon the maximum rate of electric-service usage as determined by the demand meter. The energy charge is based upon the total energy consumption as determined by the kilowatthour meter. Many utility rate structures do not include a demand charge for smaller size loads (generally less than 50 kW demand) because of the added cost for the demand meter. Many utilities have been granted permission to add provision for variable cost factors to their rates. Examples include purchased fuel differential and real-estate-tax differential costs. Under these provisions, the utility may pass along to its customers increased costs; however, the utility must pass along decreased costs as well.

(1) *Master Metering, Rent Inclusion*. This service offers a saving to the owner in first cost for metering equipment. Savings in operating costs depend upon type of multiple-occupancy building and applicable utility rates. The owner has to determine tenant electric service costs, usually as a flat sum for purposes of incorporation in tenant lease or rental agreements. The flat rate encourages wastage of power by the tenants.

(2) *Multiple Metering and Billing*. This service generally requires a higher first cost to the owner for multiple-occupancy buildings over the cost of master metering. On the other hand, the utility is responsible for collecting all tenant electric-service costs.

(3) *Conjunctional Billing*. Large commercial or institutional customers having several buildings within the territory served by a given utility should explore the availability of conjunctional billing. This consists of adding together the readings of two or more individual billing meters for purposes of a single billing. Because of the usual practice of decreasing rates for larger demands an energy usage conjunctional billing can result in lower billing than individual billing. Conjunctional billing will generally result in a higher bill than a master or totalizing meter because the maximum demand readings on the individual meters rarely occur simultaneously so that the arithmetic sum will be greater than the simultaneous sum, unless a provision is made for coincident demand measurement.

(4) *Power Factor Billing*. If the type of load to be installed in the commercial building will result in a poor power factor, then an evaluation should be made to determine if power factor improvement can be justified to avoid penalty payments or other related costs.

(5) *Flat Billing*. Certain applications involve service to loads of a fixed characteristic. For such loads, the supplying utility may offer no-meter or flat connected service. Billing is based upon time and load characteristics. Examples include street lighting, traffic signals, and area lighting.

(6) *Off-Peak Billing*. This is reduced billing for service utilized during utility

off-peak periods such as water-heating loads. The utility monitors and may control off-peak usage through control equipment or special metering.

(7) *Standby Service Billing*. Also known as breakdown or auxiliary service, this service is applicable to utility customers whose electric requirements are not supplied entirely by the utility. In such cases, billing demand is determined either as a fixed percentage of the connected load or by meter, whichever is higher. This applies to loads which are electrically connected to some other source of supply and for which breakdown or auxiliary service is requested.

(8) *Backup Service Billing*. This service is provided through more than one utility circuit, solely for a utility customer's convenience. The utility customer customarily bears the cost of establishing the additional circuit and associated supply facilities. Each backup service is generally separately metered and billed by the utility.

(9) *Demand Billing*. Usually this represents a significant part of electric service billing and a good understanding of kilowatt demand metering and billing is important. An electric demand meter measures the average rate of use of electric energy over a given period of time, usually 15 min, 30 min, or 1 h time intervals. A demand register records the maximum demand since the last reading. The demand register is reset when read for billing purposes.

(10) *Minimum Billing Demand*. A utility customer may be subject to minimum demand billing, generally consisting of (a) a fixed amount or (b) a fixed percentage of the maximum demand established over a prior billing period. This type of charge usually applies to customers with high instantaneous demand loads, such as users of welding or X-ray equipment, customers whose operations are seasonal, or those who have contracted for a given service capacity. Equipment requirements and service usage schedules should be carefully reviewed to reduce or avoid minimum billing demand charges.

(11) *Load-Factor Billing*. The ratio of average kilowatt demand to peak kilowatt demand during a given time period is referred to as the load factor. Many utilities offer a billing allowance or credit for high load-factor usage, a qualification usually determined by evaluating how many hours during the billing period the metered demand was used. As an example of such a credit, the utility may provide a reduced rate for the number of kilowatthours that are in excess of the maximum (metered) demand multiplied by a given number of hours (after 360 h for a 720 h month or a 50% load factor).

(12) *Interruptible or Curtailable Service*. Another form of peak load shaving used by the utilities is interruptible or curtailable service. Primarily available for large facilities with well defined loads that can be readily disconnected, the utility offers the customer a billing credit for being able to request a reduction of demand to a specified contract level during a curtailment period. The monthly credit for each billing month is determined by applying a demand charge credit to the excess of the maximum measured demand used for billing purposes over the contract demand. Should the customer fail to reduce his measured demand, during any curtailment period at least to the contract demand, severe financial penalties are incurred.

4.5 Transformer Connections. Commercial building utilization of low-voltage three-phase systems of recent vintage in the United States fall into either of two nominal voltage levels: 208Y/120 V or 480Y/277 V. Either of these systems can supply three-phase or single-phase loads; both frequently exist in the typical commercial building. The transformer connection used to derive these voltages is almost exclusively delta-wye (Δ—Y) or a specially constructed Y—Y transformer commonly used in pad-mounted transformers. The Δ primary cancels out virtually all third-harmonic components which may be introduced in electrical transformation equipment or in lighting ballasts. The secondary Y connection provides a tap for the neutral and a convenient grounding point as described in 4.7.1.

Where power loads are fed from a separate transformer, the Δ—Δ connection is excellent from the harmonic and unbalanced-load standpoints, but a convenient balanced grounding point is not provided (and in some instances may not be desired). There is little need to consider any other connections under normal circumstances in new commercial-building electrical systems.

Where systems are to be expanded, existing conditions may dictate the use of other connections than Δ—Y or Δ—Δ. It is important to understand that certain transformer connections are less desirable than others for given applications; and that some connections such as three single-phase transformers supplying a three-phase four-wire unbalanced load from a three-wire supply can actually be destructive (in terms of a floating neutral). Occasionally, service requirements of the utility may dictate the use of a system with a four-wire Y primary. The following paragraphs state a few of the limitations of other connections for the special circumstances where the preferred connections listed above cannot be used.

Where it is desired to use a Y primary and a Y secondary, consideration should be given to either building the three-phase bank, using a shell type core construction which will carry zero-sequence flux, or providing an additional tertiary winding. This connection minimizes ferroresonance and is therefore often preferred for transformers supplied from long runs of cable.

The primary or secondary windings of a three-phase transformer can be connected either Δ or Y. It is recommended that at least one of the windings be Δ connected to provide a path for third-harmonic currents to circulate.

The Y four-wire primary with the Y three- or four-wire secondary and the Y four-wire primary with the Δ three-wire secondary are not to be recommended for use without proper engineering consideration. If one leg of the primary line is lost, the presence of the neutral will provide three-phase flux conditions in the core. The phase that has lost its primary will then become a very high-reactance winding, resulting in fringing flux conditions. The flux will leave the core and enter the surrounding magnetic materials such as the clamping angles, tie rods, enclosure, etc. This produces an effective induction heater and results in a high secondary voltage appearing across the load of the faulty phase. This induction effect in a matter of seconds can destroy the transformer. It is also possible that, should the fault occur by means of one of the primary lines grounding, the primary winding at fault could then act as a secondary and feed back to the ground, thereby causing high current to flow in this part of the circuit. These conditions are inherent with this type of connection. Whether the transformer is of the dry or liquid type makes no difference.

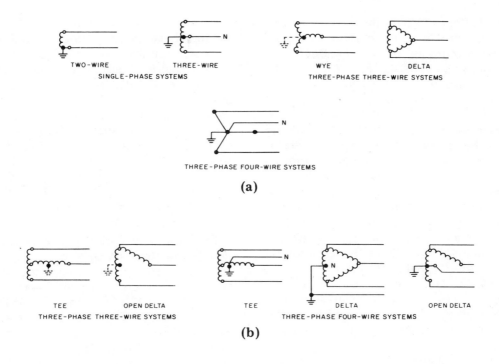

Fig 18
(a) Most Commonly Used Transformer Secondary Connections
(b) Less Commonly Used Transformer Secondary Connections

4.6 Principal Transformer Secondary Connections. Systems of more than 600 V are normally three-phase Y or Δ ungrounded or Y grounded. Systems of 120 V to 600 V may be either single phase or three phase. Three-phase three-wire systems are usually solidly grounded or ungrounded, but may be impedance grounded. They are not intended to supply loads connected phase to ground. Three-phase four-wire solidly grounded Y systems are used in most modern commercial buildings. Single-phase services and single-phase loads may be supplied from single-phase systems or from three-phase systems. They are connected phase-to-phase when supplied from three-phase three-wire systems and either phase-to-phase or phase-to-neutral from three-phase four-wire systems (see Fig 18).

Transformers may be operated in parallel and switched as a unit, provided that the overcurrent protection for each transformer meets the requirements of ANSI/ NFPA 70-1981 [2], Section 450. To obtain a balanced division of load current, the transformers should have the same rated percent impedance and be operated on the same voltage-ratio tap.

4.7 System Grounding. The recommended practice for most systems involves grounding of one conductor of the supply and ANSI/NFPA 70-1981 [2] requires grounding of certain systems, as described below. The conductor connected to ground is called the *grounded* conductor and should be distinguished from the *grounding* conductor, which is the conductor used to connect noncurrent-carrying conductive parts of electrical equipment to ground to prevent these parts from acquiring a potential above ground as a result of an insulation failure and causing injury to a person who might contact them. System grounding has the following advantages.

(1) It limits the voltages due to lightning, line surges, or unintentional contact with higher voltage lines and stabilizes the voltage to ground during normal operation.

(2) It limits or prevents the generation of transient overvoltages by changes in the electrostatic potential to ground caused by an intermittent ground on one of the conductors of an ungrounded system.

(3) In combination with equipment grounding, it can be designed to provide a safe method of protecting electric distribution systems by causing the overcurrent or ground-fault protective equipment to operate to disconnect the circuit in case of a ground fault.

4.7.1 Grounding of Low-Voltage Systems (600 V and Below). ANSI/NFPA 70-1981 [2] requires that the following low-voltage systems be grounded.

(1) Systems which can be grounded so that the voltage to ground of any ungrounded conductor does not exceed 150 V. This makes grounding mandatory for the 208Y/120 V three-phase, four-wire system and the 120/240 V single-phase three-wire system.

(2) Any system where load is connected between any ungrounded conductor and the grounded conductor. This extends mandatory grounding to the 480Y/277 V, three-phase, four-wire system and the 240/120 V, three-phase, four-wire system. The 240/120 V, three-phase, four-wire system is not recommended for commercial buildings, but it is supplied by some utilities.

(3) ANSI/NFPA 70-1981 [2] has special requirements for grounding direct-current systems and alternating-current systems under 50 V.

The grounded conductor is called the neutral on three-phase Y-connected systems and single-phase three-wire systems since it is common to all ungrounded conductors. ANSI/NFPA 70-1981 [2] requires the grounded conductor to be identified to prevent confusion with the ungrounded conductors.

A few utilities provide 240 V and 480 V three-phase three-wire systems with one phase grounded (corner grounded). This type of grounding is not recommended for commercial buildings and should be accepted only if a suitable alternative system will not be provided.

ANSI/NFPA 70-1981 [2] requires that separately derived systems be grounded in accordance with its rules. An example of a separately derived system is one in which a transformer is used to derive another voltage. The best examples are the transformation from a 480 V system to 208Y/120 V or 240/120 V to supply 120 V load.

An exception to the grounding requirements is permitted for health-care facilities (ANSI/NFPA 70-1981 [2], Article 517) where the use of a grounded system might subject a patient to electrocution or a spark might ignite an explosive atmosphere in case of an insulation failure (see Section 16).

Three-phase three-wire systems of 240 V, 480 V, and 600 V are not required to be grounded, but these systems are not recommended for commercial buildings. If they are used, consideration should be given to resistance grounding to obtain the advantages of grounding and limit the damage which can be caused by a ground fault. Note that with resistance grounding, load cannot be connected from phase to neutral.

4.7.2 Grounding of Medium-Voltage Systems (Over 600 V). Medium-voltage systems are encountered in commercial buildings when the building becomes too large to be supplied from a single transformer station and the utility primary distribution voltage must be taken through or around the building or buildings to supply the various transformers. Many utility distribution systems are solidly grounded to permit single-phase transformers to be connected phase-to-neutral to supply residences and other small loads, although ungrounded or impedance grounded systems may occasionally be encountered. The designer must accept whatever grounding system the supplying utility provides.

About the only time that the designer has a choice in the grounding of medium-voltage systems is where the supplying utility provides a voltage over 15 000 V and the designer elects to step this voltage down to a lower voltage to distribute through the building, or where large motors (several hundred horsepower) are required such as in large air-conditioning installations and it is more economical to use an intermediate voltage such as 4160 V.

Under these conditions one practice is to use a Y connected system and ground the neutral through a resistance which is low enough to stabilize the system voltages but high enough to limit the ground-

fault current to a value which will not cause extensive equipment damage before the protective devices can operate. See IEEE Std 142-1982 [1] for details. Since the ground-fault current is limited, ground-fault protection must be installed in addition to the phase over-current protection to disconnect the circuit in case of a ground fault.

4.7.3 Ground Fault Circuit Interrupters (Personnel Protectors). The so-called *people* or *personnel* protector (GFCI) is a very sensitive device responding to ground leakage current of 2 mA to 10 mA with 5 mA as a typical design standard. The two circuit wires are both passed through the core of a window current transformer; any difference between these two currents results in a current in the current transformer secondary. Any difference between the two line currents represents a leakage current to ground; the small current transformer output current, through amplification, trips the integral circuit breaker de-energizing the circuit. While the GFCI is sensitive to very small ground leakage currents, such as that flowing when a person touches an energized circuit, it does no more than a similar circuit breaker lacking the ground fault detection circuit in preventing high current from flowing through a ground fault. It simply interrupts the very small ground faults as rapidly as it does very large ones. This system, which is designed primarily for protection of people, should not be confused with the ground-fault protection whose primary function is to detect ground faults at larger magnitude and which is primarily designed to limit the destructive effects of a ground fault, as contrasted with shock hazard. ANSI/NFPA 70-1981 [2] requires GFCI protection for a number of receptacle locations where a significant shock hazard could exist.

4.8 Distribution Circuit Arrangements. Many factors should be considered in the design of the electric power distribution system for a modern commercial building. Some of the most important factors that will influence system design and circuit arrangement are the characteristics of the electric service available at the building site, the characteristics of the load, the quality of service required, the size and configuration of the building, and costs.

Electric service for commercial buildings is available from secondary-network systems in the downtown areas of many large cities in the United States. This service is usually provided from the general distributed street network at a nominal voltage of 208Y/120 V. In cases where the kilovoltampere damand of the building load is sufficiently high to justify the establishment of a spot-network system, service may be available at 480Y/277 V instead of 208Y/120 V. If the building is very large, the electric utility may establish spot-network substations on intermediate floors in the building as well as at the basement level.

If a commercial building is small enough to be supplied from a single transformer station, the recommended practice is to allow the utility to install the transformer and purchase power at utilization voltage. Commercial-building personnel are often not qualified to operate and maintain medium-voltage equipment, and any option to provide the transformer in return for a reduction in the rate must justify the expense and risk involved in owning the transformer. Where the building is too large to be supplied from a single transformer station located at a point suitable to the supplying utility,

power may be purchased at the utility distribution voltage and taken through or around the building to supply the transformers stepping down to utilization voltage. ANSI/NFPA 70-1981 [2] and utility policy, with some exceptions, provide for only one service to a building, and utilities, as a general rule, will not provide transformers suitable for installation indoors unless in utility approved vaults. In cases where commercial buildings have more than one tenant, some utilities will furnish the medium-voltage system and transformers in return for the right to sell power direct to the tenants, and for buildings supplied from a utility network.

Five basic circuit arrangements are used for the distribution of electric power for commercial facilities. They are the radial, primary-selective, secondary-selective, secondary-network and loop circuit arrangements. The following discussion of these circuit arrangements covers both medium-voltage and low-voltage circuits. The reader should recognize that the medium-voltage circuits and substations may be owned by either the utility company or the building owner, depending upon the electric rates, local practice, and requirements of the particular electric utility serving the specific building site.

In the remainder of this section, where circuit breakers are shown in the diagrams, fused equipment may be a design choice. In either case proper design considerations including fault protection, interlocking, automatic or manual control, training, experience, availability, and capabilities of operating and maintenance personnel must be fully evaluated in developing a safe and reliable system. See Section 9 for a discussion of electrical protection.

4.8.1 Radial Feeders. If power is brought into a commercial building at utilization voltage, the simplest and the lowest cost means of distributing the power is to use a radial circuit arrangement. Since the majority of commercial buildings are served at utilization voltage, the radial circuit arrangement is used in the great majority of commercial buildings. The low-voltage service-entrance circuit comes into the building through service entrance equipment and terminates at a main switchgear assembly, switchboard, or panelboard. Feeder circuits are provided to the loads or to other subswitchboards, distribution cabinets, or panelboards.

If power is purchased at a medium voltage, one or more transformers may be located to serve low-voltage radial circuits. Circuit breakers or fused switches are required on both the medium- and low-voltage circuits in this arrangement except where ANSI/NFPA 70-1981 [2] permits the medium-voltage device to serve for the secondary protection.

Figure 19 shows the two forms of radial circuit arrangements most frequently used in commercial buildings. Under normal operating conditions, the entire load is served through the single

incoming supply circuit, and in the case of medium-voltage service, through the transformer. A fault in the supply circuit, the transformer, or the main bus will cause an interruption of service to all loads. A fault on one of the feeder or branch circuits should be isolated from the rest of the system by utilizing *selectively coordinated* main and branch circuit protective devices. Under this condition, continuity of service is maintained for all loads except those served from the faulted branch circuit.

Continuity of service to the loads in commercial buildings is very important from a safety viewpoint as well as with regard to the normal activities of the occupants of the building. The safety aspect becomes more critical as the height of the building and number of people in the building increase. This requirement for continuity of service often requires multiple paths of power supply as opposed to the single path of power supply in the radial circuit arrangement. However, modern distribution equipment has demonstrated sufficient reliability to justify the use of the radial circuit arrangement in many commercial buildings. Where the risk and consequence of service loss is small, branch circuits and feeders to panels or switchboards that serve branch feeders are almost invariably radial feeders. As the demand or the size of the building or both increase, several smaller secondary substations rather than one large secondary substation may be required to maintain adequate voltage at the utilization equipment. Each of the smaller substations may be located close to the center of the load area that it is to serve. This arrangement, shown in Figs 20 and 21, will provide better voltage conditions, lower system losses, and after a less expensive installation than the arrangement using rel-

Fig 19
Radial Circuit Arrangements

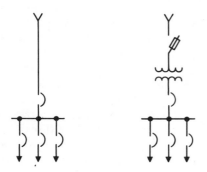

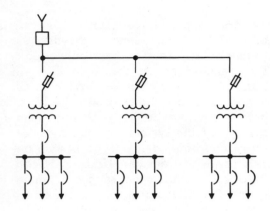

Fig 20
Radial Circuit Arrangement

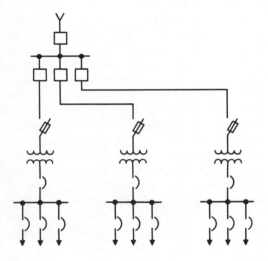

Fig 21
Radial Circuit Arrangement

atively long high-capacity low-voltage feeder circuits.

The relative economics of the radial circuit arrangements using low-voltage or medium-voltage feeders will vary with the building size, demand, cost of floor space, and billing rates. The medium-

voltage systems require an investment in transformers, medium-voltage protective devices, medium-voltage cable, and possibly some rentable floor space for substation locations. On the other hand, the investment in feeder and riser circuits of a low-voltage system of the same capacity may become excessive if voltage drop limitations are to be met.

A fault in a primary feeder, in the arrangement shown in Fig 20 will cause the main protective device to operate and interrupt service to all loads. If the fault were in a transformer, service could be restored to all loads except those served from that transformer. If the fault were in a primary feeder, service could not be restored to any loads until the source of trouble had been eliminated. Since it is to be expected that more faults will occur on the feeders than in the transformers, it becomes logical to consider providing individual circuit protection on the primary feeders as shown in Fig 21. This arrangement has the advantage of limiting outages due to a feeder or transformer fault, to the loads associated with the faulted equipment. The cost of the arrangement of Fig 21 will usually exceed the cost of the arrangement of Fig 20.

4.8.2 Primary-Selective Feeders. The circuit arrangements of Fig 22 provide means of reducing both the extent and duration of an outage caused by a primary-feeder fault. This operating feature is provided through the use of duplicate primary-feeder circuits and load-break switches that permit connection of each secondary substation transformer to either of the two primary-feeder circuits. Each primary-feeder circuit must have sufficient capacity to carry the total load in the building. Suitable interlocks for each pair of fused switches or circuit breakers are usually required.

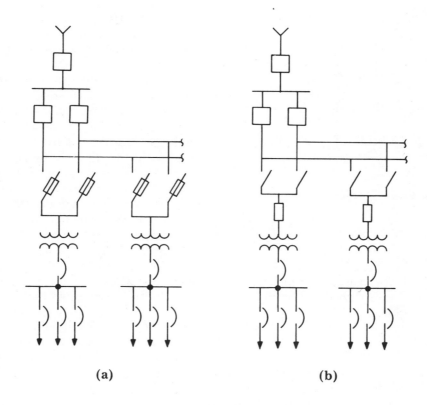

(a)   (b)

Fig 22
Primary Selective Circuit
Arrangements

Under normal operating conditions, the appropriate switches are closed in an attempt to divide the load equally between the two primary-feeder circuits. Should a primary-feeder fault occur, there will be an interruption of service to only half of the load. Service can be restored to all loads by switching the de-energized transformers to the other primary-feeder circuit. The primary-selective switches are usually manually operated and outage time for half the load is determined by the time it takes to accomplish the necessary switching. An automatic throwover switching arrange-

ment can be used to reduce the duration of interruption of service to half of the load. However, the additional cost of the automatic feature may not be justified in many applications. If a fault occurs in a secondary substation transformer, service can be restored to all loads except those served from the faulted transformer.

The higher degree of service continuity afforded by the primary-selective arrangement is realized at a cost that is usually 10% to 20% above the cost of the circuit arrangement of Fig 20 because an additional primary circuit and the primary

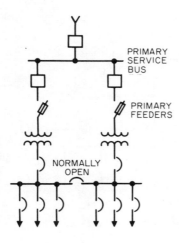

Fig 23
Secondary-Selective Circuit Arrangement
(Double-Ended Substation with
Single Tie)

switching equipment at each secondary substation is needed. The cost of the primary selective arrangement, using manual switching, will sometimes be less than the radial arrangement.

A variation of the circuit arrangements shown in Fig 22 utilizes three primary-selective feeders comprised of two normal feeders and one standby feeder. Each feeder is sized between half and two thirds the total load and supplies one third of the total load under normal conditions. Under emergency conditions, with a primary cable fault, the load on the faulted cable can be transferred to the standby feeder. Depending on the capacity of the standby feeder, load can be transferred to the remaining normal feeder or left on the standby feeder until the cause for failure is corrected.

4.8.3 Secondary-Selective Feeders. Under normal conditions, the secondary-selective circuit arrangement of Fig 23 is operated as two separate radial systems.

The secondary bus-tie circuit breaker or switch in the double-ended substation is normally open.

The load served from a secondary substation should be divided equally between the two bus sections. If a fault occurs on a primary feeder or in a transformer, service is interrupted to all loads associated with the faulted feeder or transformer. Service may be restored to all secondary buses by first opening the main secondary switch or circuit breaker associated with the faulted transformer and primary feeder, and then closing the bus-tie device. The two transformer secondary switches or circuit breakers in each substation should be interlocked with the secondary bus-tie device in such a manner that all three cannot be in the closed position simultaneously. This prevents parallel operation of the two transformers and thereby minimizes the interrupting duty imposed on the secondary protective devices. It also eliminates the possibility of interrupting service to all loads on the bus when a fault occurs in either a primary feeder or a transformer. To prevent closing the tie on a faulted switchgear bus a main-tie-main interlock scheme may be provided to lock out the tie whenever a secondary main has interrupted a downstream fault.

The cost of the secondary-selective circuit arrangement will depend upon the spare capacity in the transformers and primary feeders. The minimum transformer and primary-feeder capacity will be determined by essential loads that must be served under standby operating conditions. If service is to be provided for all loads under standby conditions, then each primary feeder should have sufficient capacity to carry the total load, and each transformer should be capable of carrying the total load on both substation buses.

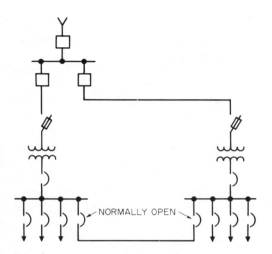

Fig 24
Secondary-Selective Circuit Arrangement
(Individual Substations with
Interconnecting Ties)

This type of circuit arrangement will be more expensive than either the radial or primary-selective circuit arrangement, but it makes restoration of service to all essential loads possible in the event of either a primary-feeder or transformer fault. The higher cost results from the duplication of transformer capacity in each secondary substation. This cost may be reduced by load-shedding non-essential feeders.

A modification of the secondary-selective circuit arrangement is shown in Fig 24. In this arrangement there is only one transformer in each secondary substation, but adjacent substations are interconnected in pairs by a normally open low-voltage tie circuit. When the primary feeder or transformer supplying one secondary substation bus is out of service, the essential loads on that substation bus can be supplied over the tie circuit. The operating aspects of this system are somewhat complicated if the two substations are separated by distance.

4.8.4 Secondary Network. Many buildings with radial distribution systems are served at utilization voltage from utility secondary-network systems. The network supply system assures a relatively high degree of service reliability. The utility network may take the form of a distributed network or a spot network. If the building demand is in the order of 750 kVA or higher, a spot network will often be established to serve the building. In buildings where a high degree of service reliability is required, and where spot-network supply may not be available, the distributed secondary-network circuit arrangement is often used. This is particularly true of institutional buildings such as hospitals. The network may take the form of several secondary substations interconnected by low-voltage circuits. However, the most common practice is to use some form of the spot-network circuit arrangement.

A simple spot-network, such as shown in Fig 25, consists of two or more identical transformers supplied by separate primary-feeder circuits. The transformers

Fig 25
Simple Spot-Network
Circuit Arrangement

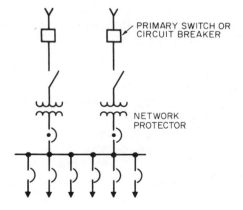

are connected to a common low-voltage bus through network protectors and are operated in parallel. A network protector is a special heavy-duty electrically operated power air circuit breaker controlled by network relays in such a way that the circuit breaker automatically opens when power flows from the low-voltage bus toward the transformer. When voltages in the system are such that power would flow toward the low-voltage bus from the transformer, it will close automatically.

Network protectors are normally equipped with protective relays which operate for faults in the network transformer or primary voltage feeder only. The network is often operated on the assumption that network faults will *burn* open.

Network protectors without supplementary protection do not meet the requirements of ANSI/NFPA 70-1981 [2] for overcurrent, ground-fault, or short-circuit protection. Provided the network protector is customer furnished and considered as the *service disconnect*, protection of the network or collector bus may be added by providing sensing devices, including ground fault detection, to trip the network protectors.

A conventional electrically operated circuit breaker with long-time and short-time, overcurrent and ground-fault protection plus network relays can meet the ANSI/NFPA 70-1981 [2] requirements. However, the full reliability of the network may be compromised since selectivity between overcurrent and ground trip units and the network relays is difficult to obtain.

Under normal operating conditions, the total load connected to the bus is shared equally by the transformers. Should a fault occur in a transformer or on a primary feeder, the network protector associated with the faulted transformer or feeder will open on reverse-power flow to isolate the fault from the low-voltage bus. The remaining transformer or transformers in the substation will continue to carry the load and there will be no interruption of service to the loads, except for a voltage dip during the time that it takes for the protective equipment to operate.

If only two transformers are used in a spot-network substation, each transformer must be capable of carrying the total load served from the low-voltage bus. The amount of spare transformer capacity in the substation can be reduced by using a primary-selective switching arrangement with each transformer, or by using three or more transformers. If the primary-selective switching arrangement is used, the total load can be about 160% of the nameplate rating of one of the transformers. In case of a lost primary this produces an overload on one transformer until the remaining transformer can be switched to the other feeder. Although this procedure is not recommended, if it is used, secondary breakers and rating, and setting of the overcurrent protective devices must be selected to permit this extraordinary method of operation.

The interrupting duty imposed on the low-voltage protective devices in a spot-network substation is higher than in radial, primary-selective, or secondary-selective substations having the same load capability because of the spare transformer capacity required in the spot-network substation and because the transformers are operated in parallel. The spare transformer capacity, the network protectors, and the higher interrupting duty will make the spot-network arrangement more expensive than the other arrangements.

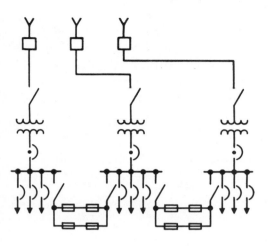

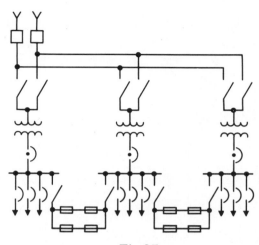

Fig 26
Secondary-Network Circuit Arrangement

Fig 27
Primary-Selective Secondary-Network
Circuit Arrangement

The true secondary network as op-posed to the spot network is shown in Fig 26. In this arrangement there is only one transformer in each secondary sub-station, and the substations are inter-connected by normally closed low-voltage tie circuits. The tie circuits permit inter-change of power between substations to accommodate unequal loading on the substations and to provide multiple paths of power flow to the various load buses. In normal operation, the substa-tions are about equally loaded and the current flowing in the tie circuits is rela-tively small. However, if a network pro-tector opens to isolate a transformer on primary-feeder fault, the load on the as-sociated bus is then carried by the adja-cent network units and is supplied over the tie circuits. This arrangement provides for continuous power supply to all low-voltage load buses, even though a pri-mary-feeder circuit or a transformer is taken out of service.

In the network arrangement of Fig 27, if there were three incoming primary-feeder circuits and three transformers, the combined capacity of two of the transformers may be sufficient to carry the entire load on the three substations on the basis that only one feeder is out of service at one time. Generally, these transformers would all have the same rat-ings. With this arrangement, as with the spot-network arrangement, a reduction in spare transformer capacity can be achieved if a primary-selective switching arrangement is used at each substation transformer. However, if three or more primary-feeder circuits are available, the reduction in transformer capacity achieved through the use of a primary-selective arrangement may be small.

Cable ties or busway ties, as shown in Figs 26 and 27, will require careful con-sideration of load distribution during contingencies and of the safety aspects with regard to backfeeds. Key or other mechanical-interlocking switches or cir-cuit breakers may be essential. The ap-plication of cable limiters shown in Figs 26 and 27 is discussed further in Section 5.

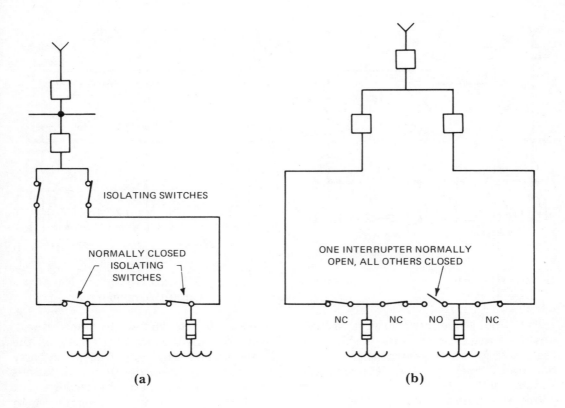

ISOLATING SWITCHES

NORMALLY CLOSED
ISOLATING
SWITCHES

ONE INTERRUPTER NORMALLY
OPEN, ALL OTHERS CLOSED

NC   NC   NO   NC

(a)   (b)

Fig 28
Looped Primary Circuit Arrangement
(a) Closed Loop (Obsolete) (b) Open Loop

When the electric service is the secondary of a network system and the system characteristics fall within the ANSI/NFPA 70-1981 [2] requirements for ground-fault protection, a selectivity problem exists that requires a detailed analysis tempered with sound judgment.

4.8.5 Looped Primary System. The looped primary system, Fig 28, is basically a two-circuit radial system with the ends connected together to form a continuous loop. Early versions of the closed-loop system as shown in Fig 28(a) were designed to be operated with all loop isolating switches closed. Although

it is relatively inexpensive, this system has fallen into disfavor because its apparent reliability advantages are offset by the interruption of all service by a fault anywhere in the loop, by the difficulty of locating primary faults, and by safety problems associated with the non-load-break isolating switches.

Newer open-loop versions shown in Fig 28(b), designed for modern underground commercial and residential distribution systems, utilize fully rated air, oil and vacuum interrupters. Equipment is available in voltages up to 34.5 kV with interrupting ratings for both continuous load and fault currents to meet most system requirements. Certain equipment can close in and latch on fault currents, equal to the equipment interrupting values, and still be operational without maintenance.

With the elimination of the major disadvantages of the older loop systems and the current demand for decentralized systems with low-profile pad mounted equipment and greater reliability than the simple radial system, the open-loop primary system has become a viable distribution solution.

The major advantages of the open-loop primary system over the simple radial system is the isolation of cable or transformer faults, or both, while maintaining continuity of service for the remaining loads. With coordinated transformer fusing provided in the loop tap position, transformer faults can be isolated without any interruption of primary service. Primary cable faults will temporarily drop service to half of the connected loads until the fault is located, then, by selective switching, the unfaulted sections can be restored to service leaving only the faulted section to be repaired.

Disadvantages of the loop systems are the increased costs to fully size cables, protective devices and interrupters to total capacity of the load (entire load on one feeder), and the time delay necessary to locate the fault, isolate the section, and restore service.

The development of load-break cable terminators and the regulatory requirements for total underground utility distribution have led to the use of loop-loop primary distribution circuit arrangements.

For systems having a large quantity of small capacity transformers the loop-loop design is the lowest cost of the loop circuit arrangements. It has the same dis-

advantage as the old designs since it is possible to connect a cable terminator into a cable fault.

Figure 29 shows a loop-loop system where pad mounted loop sectionalizing switches are provided in the main loop and load break cable terminators are provided in the secondary loops. The main loop is designed to carry the maximum system load whereas secondary loops are fused to handle load concentrations smaller than the total system capacity. For additional discussion on looped systems see Section 7.

4.9 Emergency and Standby Power Systems. Emergency electric services are required for protection of life, property, or business where loss might be the result of interruption of the electric service. The extent of emergency services required depends on the type of occupancy, the consequences of a power interruption, and the frequency and duration of expected power interruptions.

Municipal, state, and federal codes define minimum requirements for emergency systems for some types of public buildings and institutions. These shall be adhered to; but economics or other advantages may result in making provisions beyond these minimums. See ANSI/ NFPA 70-1981 [2], Section 700, 701, and 702.

The following presents some of the basic information on emergency and standby power systems. For additional information and design details see ANSI/ IEEE Std 446-1980 [3].

4.9.1 Lighting. Exit lights and emergency lights sufficient to permit safe exit from the building should be supplied from an emergency source of power in buildings where the public may

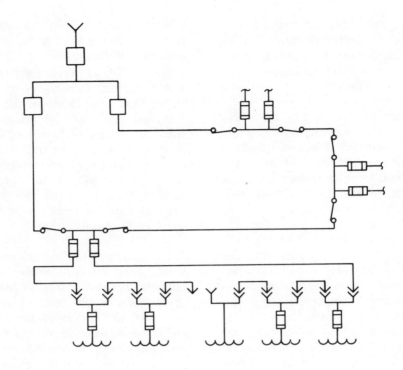

Fig 29
Looped Primary Circuit Arrangement

congregate. This includes auditoriums, theaters, hotels, large stores, sports arenas, etc. Local regulations should always be referred to for more specific requirements. If the emergency lighting units are not used under normal conditions, power must be available to them immediately upon loss of the normal supply. If the emergency lights are normally in service and served from the normal supply, provisions must be made to transfer them automatically to the emergency source when the normal supply fails.

Sufficient lights should be provided in stairs, exits, corridors, and halls so that failure of any one unit will not leave any area dark or endanger persons leaving the building. Adequate lighting and rapid automatic transfer to prevent a period of darkness is important in public areas. Human safety is improved and the chance of pilfering or damage to property is minimized.

Some codes require that emergency sources of power for lighting be capable of carrying their connected loads for at least one and one half hours. There are cases where provisions should be made for providing emergency service for much longer periods of time, such as health-care facilities, communication, police, and fire-fighting services. One- to three-hour capacity is more practical and in many installations 5 h to 6 h capacity is provided. During a severe storm or catastrophe the demands on hospitals, communications, police, and fire-fighting facilities will be increased. A third source of power to achieve the desired reliability may be required.

When installation of a separate emergency power supply is not warranted but some added degree of continuity of service for exit lights is desired, they may be served from circuits connected ahead of the main service-entrance switch for some occupancies. This assures that switching of loads and tripping due to faults in the building electric system will not cause loss of the exit lights. However, this arrangement does not protect against failures in the electric utility system.

4.9.2. Power Loads. An emergency source for supplying power loads is required where loss of such load could cause extreme inconvenience or hazard to personnel, loss of product or material, or contamination of property. The type and size of the emergency system must be determined through consideration of the health and convenience factors involved and whether the utilization affects hospitals, communication systems, alarm systems, police, or fire-fighting facilities.

Laboratories where continuous processes are involved or where chemical or nuclear experimentation is carried on are very demanding insofar as power and ventilating system requirements are concerned. Loss of adequate power for ventilation could permit the spread of poisonous gases or radioactive contamination throughout the building and even cause loss of life. A building contaminated from radioactive waste could be a total loss or require expensive clean-up measures. Many processes or experiments cannot tolerate a power loss which would interrupt cooling, agitation, or other facilities.

Emergency power for fire pumps should be provided where water requirements cannot be met from other sources.

Emergency power for elevators should also be considered when elevators are necessary to evacuate buildings or the cost seems warranted to avoid inconvenience to the public. This does not mean that full capacity is essential for all elevators.

4.9.3 Power Sources. Sources for emergency power may include batteries, local generation, or a separate source over separate lines from the electric utility. The quality of service required, the amount of load to be served, and the characteristics of the load will determine which type of emergency supply is required.

4.9.3.1 Batteries. Batteries offer an extremely reliable source of energy but require regular attention and care. Also, their capacity in ampere-hours is limited. Inspection and tests of individual cells must be made at regular intervals to assure that electrolyte level and correct charge are maintained. If lead-acid batteries are used, they must be of the sealed glass jar type. Ample space should be provided together with adequate ventilation of the battery room. Batteries of the nickel-lead-alkaline type or the nickel-cadmium type may be used provided that characteristics of each battery type and the load are considered; these batteries are more suitable for standby service and require less attention than the lead-acid units.

Battery-charging equipment will be determined by the battery characteristics and the type of load being served. The capacity of the equipment will depend on the size of load and the time such load must be supplied. For low charging rates, trickle type chargers may be satisfactory. For higher charging rates, electronically controlled chargers are generally used. Heavier loads may require motor-generator sets or heavy-duty silicon rectifiers, the latter presently being the more advantageous choice from the standpoint of efficiency and convenience.

For a simple and effective lighting installation to permit building evacuation, small package units containing light, battery, charger, and relay are suggested. See ANSI/NFPA 70-1981 [2], Article 700 for installation requirements. Normal supply voltage also actuates a relay to de-energize the light on the emergency unit. When the normal supply fails, the relay drops out and the light is energized from the self-contained battery. Provision is made for testing the unit.

4.9.3.2 Local Generation. Local generation is advisable where service is absolutely essential for lighting or power loads, or both, and where these loads are relatively large and distributed over large areas. Several choices are available in the type of prime mover, voltage of the generator, and method of conection to the system. Various alternates should be considered. The prime mover can be driven by steam, gas, gasoline, diesel, or commercial bottled gases.

For generators over 500 kW, gas-turbine driven units may be a favorable choice. This type of unit has acceptable efficiency at full load, but it is much less efficient than other types of drives at partial load. Gas-turbine-driven units do not start as rapidly as other drives but are reliable and require a minimum of attention.

Fuel storage requirements must be determined after considering the frequency and duration of power outages, the types of emergency loads to be served, and the ease of replenishing fuel supply. Some installations may require that a supply sufficient for three months be maintained while a one-day supply may be adequate for others.

Some codes require an on-site fuel supply capable of operating the prime mover at full demand load for at least 2 h.

Generator selection can only be made after a careful study of the system to which it is connected and the loads to be carried by it. The voltage, frequency, and phase relationships of the generator should be the same as in the normal system. The size of the generator will be determined by the load to be carried with consideration given to the size of individual motors to be started. The speed and voltage regulation required will determine the accuracy and sensitivity of regulating devices. When a generator is required to carry emergency loads only during power outages and must not operate in parallel with the normal system, the simplest type of regulating equipment is usually adequate. For parallel operation good quality regulators and governors are needed to assure proper and reactive active power loading of the generator. If the generator is small in relation to the system, it is usually preferable to have a large drooping characteristic in the governor and considerable compensation in the voltage regulator so that the local generator will follow the larger system rather than try to regulate it.

4.9.3.3. Dual-Service Connection. Where the local utility can provide two or more service connections over separate lines and from separate generation points so that system disturbances or storms are not apt to affect both supplies simultaneously, local generation or batteries may not be justified. A second line for emergency power should not be relied upon, however, unless total loss of power can be tolerated on rare occasions. The alternate feeder can either serve as standby with primary switching

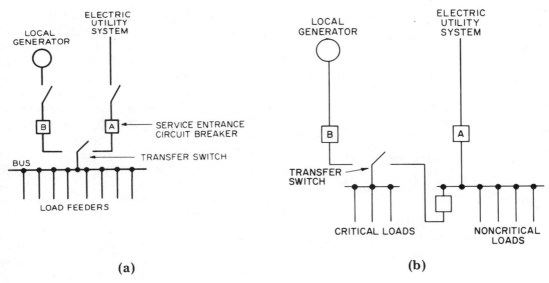

(a)

(b)

Fig 30
Typical Transfer Switching Methods

or have its own transformer with secondary switching.

4.9.4 Transfer Methods. Figure 30(a) shows a typical switching arrangement when a local emergency generator is used to supply the entire load upon loss of normal power supply. All emergency loads are normally supplied through breaker A. Breaker B is open and the generator is at rest. When the normal supply fails, the transfer switch undervoltage relay is de-energized and after a predetermined time delay closes its contacts. The time delay is introduced so that the generator will not be started unnecessarily for transient voltage dips. Where the alternate source is a generator, sufficient time before transfer must be allowed to permit the generator to reach full speed before application of load.

It should be noted that the arrangement shown in Fig 30(a) does not provide complete protection against power disruption within the building.

Figure 30(b) shows a typical switching arrangement in which only the critical loads are transferred to the emergency source, in this case a generator. For maximum protection the transfer switch is located close to the critical loads. For further discussion on application of automatic transfer switches, see Section 5.

Figure 31 shows two separate sources of power external to the buildings that have the necessary reliability to satisfy the needs for emergency power. Relaying is provided to transfer the load automatically to either source if the other one fails. The control is so arranged that a transfer will not take place unless one source (alternate or normal) is energized. If the alternate supply is not able to carry the entire load, provision should be made to drop noncritical loads when the transfer takes place.

Figure 32 shows a secondary selective substation where each incoming service feeds its respective bus and the loads

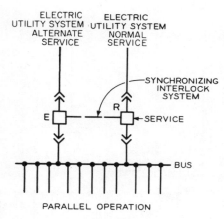

Fig 31
Service Transfer Using
Circuit Breakers

connected to the bus. Should one service fail, the de-energized bus can be connected to the other bus via the tie breaker.

Normally the two R circuit breakers are closed and the tie circuit breaker is open. The three circuit breakers are interlocked to permit any two to be closed, but prevent the three circuit breakers being closed simultaneously.

Many other arrangements of switching and relaying are possible, but the four cases shown illustrate the principles. A careful study of each system should be made to determine the exact needs and critical features and to select an arrangement that meets the requirements consistent with sound economy and applicable codes.

4.9.5 Special Precautions. Transfer of resistance and most power and lighting loads can be made as rapidly as desired, depending on the type of service required and the available switching equipment. Note that there will be a momentary outage in the order of a fraction of a second when the transfer switch operates to transfer the load from the main to the emergency source. Therefore, all equipment connected to the emergency source must be reviewed to determine the effect of the momentary outage and what precautions should be taken. The short outage is hardly noticeable with incandescent and fluorescent lights.

However, mercury, ordinary sodium, and metal halide lamps will drop out so

Fig 32
Secondary Transfer Using
Circuit Breakers

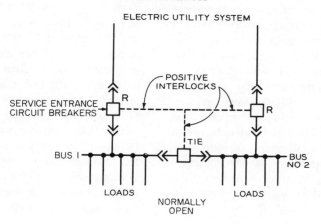

they should not be used for emergency lighting. Motors with undervoltage releases not equipped with built-in time delay and relay controls will drop out. Computers may shut down and some data may be lost.

Application of motor loads on limited power sources such as standby and emergency generators presents certain problems to system designers usually not existent on large utility furnished services. The most critical problem is the sizing of the generator system to be capable of running all normally connected lighting, power, and motor loads plus starting the largest motor load. Information required in making the determination of generator size includes the kVA and kW of each load plus the starting kVA and kW of the largest motor. When the largest motor is the predominate load the generator may be twice the size of the motor. Consultation with the manufacturer of generator systems is advised when large motor loads are contemplated. Other problems include:

(1) Regenerative loads developed by operation of elevators

(2) Nuisance breaker tripping and possible motor damage incurred when switching between two out-of-phase energized power sources

(3) Means and methods of temporarily shedding motor loads prior to transfer to avoid overloading the limited power source

(4) Proper grounding of the generator neutral to avoid extraneous ground-fault tripping of the normal source breaker

(5) SCR type loads which may interfere with operation of unprotected exciter-regulators

4.9.6 Transfer of Power. When local generation is operated in parallel with the utility system, there will generally be a momentary voltage dip as the local generator starts to supply the utility line until the reverse current or reverse power relays operate to disconnect the utility line. The voltage should not go all the way to zero as is the case with a transfer switch, but it can go low enough to cause equipment to drop out; so its effect must be considered. Faults on the utility system and in the building can also cause voltage dips. The relay system to protect both the utility supply and the building will require approval of the utility.

Normally the transfer from a normal to an alternate or emergency supply is accomplished automatically. The return to the normal supply can be automatic or manual. Manual return should be considered when the unexpected restoration of the normal supply would cause equipment to restart and endanger either the equipment or persons. For example, elevators should not be permitted to restart automatically because rescue operations may be in progress to free trapped passengers.

4.9.7 Uninterruptible Supplies. Where uninterruptible power is required, one of the following types may be used.

(1) A storage battery, rectifier, and inverter may be employed. The inverter (or electric motor-generator) is fed normally from the rectifier which is, in turn, fed from the alternating-current source. The inverter or motor-generator set reconverts the direct current supplied by the rectifier to alternating current for load utilization. The storage battery floats across the direct-current line until such time as the main alternating-current power fails, when it will feed into the line supplying the motor-generator set or the inverter. Such systems may operate for minutes to permit orderly shutdown of equipment, or for hours. This system

also has the advantage of supplying inter-ference-free power to the load.

(2) If uninterruptible power is required for extended periods, the most economical procedure is to provide for a relatively short period of battery operation followed by the application of an emergency generator. If even a momentary interruption of a few cycles cannot be tolerated, the generator can be used to operate the charger for the battery, provided the charger is of sufficient capacity.

(3) Where some drop in voltage and frequency are permissible, a flywheel-type alternator runs as a motor across the line. Upon power failure, a clutch engages the drive of a gasoline or diesel engine, bringing it up to speed to drive the alternator. The inertia of the flywheel and alternator will keep up, within limits, system voltage and frequency until the prime mover takes over.

Because of their relatively high cost, uninterruptible power supplies are usually limited to loads which cannot stand even a momentary interruption, such as certain computers or hospital life-supporting systems. Consultation with the suppliers of data-processing equipment or computers is essential; for example, some computers have their own motor-generators which will minimize transient problems while other equipment will be relatively unaffected by the transfer time of a transfer switch.

4.10 Power Factor Correction and Voltage Regulation. Voltage regulation and power factor correction are closely related. Desired voltage regulation in many cases is costly to obtain. Larger conductors to reduce voltage drop under load are in many cases the proper solution. However, power factor correction may also be justified for four reasons:

(1) To improve voltage regulation

(2) To lower the cost of electric energy where the electric utility rates vary with the power factor at the metering point

(3) To reduce the energy losses in conductors

(4) To utilize the full capacity of transformers, switches, circuit breakers, buses, and conductors for active power only, thereby lowering the capital investment and annual costs.

4.10.1 Voltage Regulation. The goal of good voltage regulation is controlling the voltage of the system so that it will stay within a practical and safe range of voltage tolerances under all design loads. Voltage at any utilization equipment should be within the guaranteed operative range of the equipment. Type and size of wires or cables, types of raceways, reactances of transformers and cables, selection of motor-starting means, circuit design, power factor correction, means and degree of loading, all affect voltage regulation.

Voltage regulation in any circuit, expressed in percentage, is

$$\frac{(\text{no-load voltage} - \text{full-load voltage}) \cdot 100}{\text{no-load voltage}}$$

Where it is not economical to control voltage drop through conductor sizing, circuit design, or other means, voltage regulators may be needed. Several types of voltage regulators, either automatic or manual, are available for all types and sizes of loads from individual electronic devices to the equipment for an entire laboratory or department store. Voltage regulators are frequently used by the electric utility companies in their distribution-system feeders and are seldom needed within commercial buildings, except for electronic equipment. Normally the power and light distribution system within large commercial buildings can be designed economically and adequately without the use of large voltage regulators.

4.10.2 Power Factor Correction. If the type of load to be installed in the commercial building will result in a poor power factor, then an evaluation should be made to determine if installing capacitors can be justified either to stay within the power factor range specified by the electric utility to avoid penalty payments or to obtain a reduction in the power bill.

Where large machines like blowers or refrigeration or air compressors are to be installed, a study should be made to determine whether it would be economical to install a synchronous motor and utilize it for power factor correction. The cost of the synchronous motor with its controller should be compared with the cost of a squirrel-cage motor with its simpler controls plus separate static capacitors.

4.11 System Reliability Analysis. One of the questions often raised during the design of the power distribution system is how to make a quantitative comparison of the failure rate and the forced downtime in hours per year for different circuit arrangements including radial, primary-selective, simple spot-network, and secondary-network circuits. This quantitative comparison could be used in tradeoff decisions involving the initial cost versus the failure rate and the forced downtime per year. The estimated cost of power outages at the various distribution points could be considered in deciding which type of circuit

arrangement to use. The decisions could thus be based upon total owning cost over the useful life of the equipment rather than the first cost.

4.11.1 Reliability Data for Electrical Equipment. In order to calculate the failure rate and the forced downtime per year, it is necessary to have reliability data on the electric utility supply and each piece of electrical equipment used in the power distribution system. One of the best sources for this type of data is the extensive survey of equipment reliability included in ANSI/IEEE Std 493-1980 [4] which presents average data for all equipment manufacturers and a variety of applications.

4.11.2 Reliability Analysis and Total Owning Cost. Statistical analysis methods involving probability of failure may be used to make calculations of the failure rate and the forced downtime for the power distribution system. The methods and formulas used in these calculations are given in ANSI/IEEE Std 493-1980 [4].

4.12 References

[1] IEEE Std 142-1982, Recommended Practice for Grounding of Industrial and Commercial Power Systems.

[2] ANSI/NFPA 70-1981 National Electrical Code.

[3] ANSI/IEEE Std 446-1980, Recommended Practice for Emergency and Standby Power Systems for Industrial and Commercial Applications.

[4] ANSI/IEEE Std 493-1980, Recommended Practice for the Design of Reliable Industrial and Commercial Power Systems.

5. Power Distribution Apparatus

5.1 General Discussion. Electrical systems for commercial installations encompass a wide variety of electrical apparatus. There are numerous choices to be made between similar equipment either with overlapping functions or which are direct substitutes but with varying advantages or degrees of acceptability to a particular application. The engineer making these basic decisions should consider all facets of the actual project including, but not limited to, protection, coordination, initial cost including installation, operational personnel and cost, maintenance facilities and cost, availability and cost of space, and the procurement time to meet objectives. Equipment connecting directly to the serving electrical utility should be compatible with the utility's requirements.

General descriptions of apparatus frequently used in these electric systems follow in this order:

(1) Transformers
(2) Medium- and high-voltage fuses
(3) Metal-enclosed 5 kV to 34.5 kV power-interrupter switchgear
(4) Metal-clad 5 kV to 34.5 kV power circuit-breaker switchgear
(5) Metal-enclosed 600 V power switchgear and circuit breakers
(6) Metal-enclosed distribution switchboards
(7) Primary unit substations
(8) Secondary unit substations
(9) Panelboards
(10) Molded-case air circuit breakers
(11) Low-voltage fuses
(12) Service protectors
(13) Enclosed switches
(14) Bolted-contact switches
(15) Network protectors
(16) Outdoor structural substations
(17) Automatic transfer switches
(18) Load transfer devices and interlocks
(19) Remote control contactors
(20) Motor control apparatus
(21) Conductors (insulated-wire busway)
(22) Raceways
(23) Subsurface duct and manhole systems
(24) Aerial distribution systems

ANSI and NEMA standards are listed in 5.21 and indicated in the description of each item where applicable.

5.2 Transformers. Transformers in commercial installations are normally used to change a voltage level from a utility distribution voltage to a voltage that is usable within the building and are also used to reduce building distribution voltage to a level that can be utilized by specific equipment. Applicable standards are the ANSI C57 series and NEMA TR series.

5.2.1 Transformer Types. The following types of transformers are normally used in commercial buildings:

(1) Substation

(2) Primary unit substation

(3) Secondary unit substation (power center)

(4) Network

(5) Pad mounted

(6) Indoor distribution

Many other types of transformers are manufactured for special applications such as welding, constant voltage supply, and high-impedance requirements. Discussion of the special transformers and their use is beyond the scope of this standard.

(1) *Substation Transformers.* Used with outdoor switchyards, they are rated 750 kVA—5000 kVA for single-phase units and 750 kVA—25 000 kVA for three-phase units. The primary voltage range is 2400 V and up. Taps are usually manually operated while de-energized, but automatic load tap changing may be obtained. The secondary voltage range is 480 V—34 500 V. Primaries are usually Δ-connected and secondaries are usually Y-connected because of the ease of grounding their secondary neutral. The insulation and cooling medium is usually oil. High-voltage connections are on cover-mounted bushings. Low-voltage connections may be cover bushings or an air terminal chamber.

(2) *Primary Unit Substation Transformers.* Used with their secondaries connected to medium-voltage switchgear or motor control they are rated 1000 kVA—10 000 kVA and are three-phase units. The primary voltage range is 6900 V—138 000 V. The secondary voltage range is 2400 V—14 400 V. Taps are usually manually operated while de-energized, but automatic load tap changing may be obtained. Primaries are usually Δ-connected. The insulation/cooling medium may be oil, nonfire-propagating liquids, air, cast coil, or gas. The high-voltage connections may be cover bushings, an air terminal chamber, throat, or flange. The low-voltage connection is a throat or flange.

(3) *Secondary Unit Substation Transformers.* Used with their secondaries connected to low-voltage switchgear or switchboards, they are rated 112.5 kVA—2500 kVA and are three-phase units. The primary voltage range is 2400 V—34 500 V. The taps are manually operated while de-energized. The secondary voltage range is 208 V—480 V. The primaries are usually Δ-connected and secondaries are usually Y-connected. The insulation and cooling medium may be oil, nonfire-propagating liquids, air, cast coil, or gas. The high-voltage connections may be cover bushings, an air terminal chamber, a throat, or a flange. The low-voltage connection is a throat or flange.

(4) *Network Transformers.* Used with secondary network systems, they are rated 300 kVA—2500 kVA. The primary voltage range is 4160 V—34 500 V. The Taps are manually operated while de-energized. The secondary voltages are 216Y/125 V and 480Y/277 V. The insulation and cooling medium may be oil, nonfire-propagating liquid, air, cast coil, or gas. The primary is Δ-connected, the secondary is Y-connected. The high-voltage connection is generally a network switch (on-off-ground). The secondary connection is generally an appropriate

network protector, or a low-voltage power air circuit breaker designed to provide the functional equivalent of a network protector.

ANSI C57.12.40-1975 [10][25] applies to liquid-immersed subway- and vault-type network units. A subway-type unit is suitable for frequent or continuous operation while submerged in water, a vault-type unit is suitable for occasional submerged operation.

(5) *Pad-Mounted Transformers.* They are used outside buildings where conventional unit substations might not be appropriate and are either single-phase or three-phase units. As they are of tamper resistant construction, they do not require fencing. Primary and secondary connections are made in compartments which are located adjacent to each other but separated by barriers from the transformer and each other. Access is through pad-locked hinged doors designed so that unauthorized personnel cannot enter either compartment without destroying the enclosure. Where ventilating openings are provided, tamper resistant grills are used. Gauges and accessories are in the low-voltage compartment. These units are rated 75 kVA—2500 kVA. The primary voltage range is 2400 V—34 500 V. Taps are manually operated while de-energized. The secondary voltage range is 208 V—480 V. Primaries may be Δ- or special construction Y-connected and secondaries are usually Y-connected. The insulation and cooling medium may be oil, nonfire-propagating liquid, air, cast coil, or gas. The high-voltage connection is in an air terminal chamber which may contain just pressure- or disconnecting-type connectors or may have a disconnecting device, either fused or unfused. The connections

---

[25] Numbers in brackets correspond to those in the references at the end of this Section.

may be for either single or looping feed. The low-voltage connection is usually by cable at the bottom, but it may also be by bus duct.

The dry-type pad-mounted transformer does not have the inherent fire hazards of the oil-filled pad-mounted transformer and frequently the dry-type pad-mounted transformer is mounted on the roofs of buildings to be as near to the load center as possible.

ANSI C57.12.22-1980 [9] applies to oil-immersed units with primary voltages of 16 340 V and below.

(6) *Indoor Distribution Transformers.* Used with panelboards and separately mounted, they are rated 1 kVA—333 kVA for single-phase units and 3 kVA—500 kVA for three-phase units. Both primaries and secondaries are 600 V and below (the most common ratio is 480—208Y/120 V). The cooling medium is air (ventilated or nonventilated). Smaller units have been furnished in encapsulated form. High- and low-voltage connections are pressure-type connections for cables. Impedances of distribution transformers are usually lower than of substation or secondary unit substation transformers.

Indoor distribution transformers are also available at higher primary voltages. NEMA TR27-1965 (R1976) [49] is applicable for primary volages above 600 V up to 15 000 V. Cast-coil transformers are available at primary voltages up to 34 500 V.

Many special types of indoor distribution transformers are available for specific applications. Manufacturers' catalogs should be consulted in all cases. The one most often encountered in commercial building projects is the isolating transformer with separate windings as opposed to the auto transformer which has a single winding. Its applications include motor control, computers, X-ray machines, laboratories, lighting, and hospital anesthetizing locations. The term *isolating transformer* refers to the type of application in which the primary circuit must be insulated from the secondary circuit. It is synonymous with *insulating transformer*, since any insulating transformer will isolate the primary circuit from the secondary circuit. However, many isolating applications require an electrostatic shield. This is a grounded conductor between the primary and secondary windings which reduces the interference or power disturbances which could otherwise be transmitted from one circuit to the other by transformer action.

5.2.2 Transformer Specifications. The following factors should be considered in specifying transformers:

(1) Kilovoltampere rating

(2) Voltage ratings, ratio, and method of connection, that is, Δ or Y

(3) Voltage taps and whether manual or automatic load tap changing is required

(4) Impedance value

(5) Type of cooling and temperature rise, including insulation material class

(6) Voltage insulation class and basic impulse level

(7) Grounding requirements

(8) Permissible sound level

(9) Matching characteristics if they are to be paralleled with other transformers.

(1) *Kilovoltampere Rating.* Table 32 gives the preferred kilovoltampere ratings of both single- and three-phase transformers according to ANSI/IEEE C57. 12.00–1980 [27].

(2) *Voltage Ratings and Ratio.* All the preferred kilovoltampere ratings in Table 32 are obviously not available as standard at all voltage ratings and ratios. In general, the smaller sizes apply to

Table 32
Preferred Kilovoltampere Ratings

Single-Phase	Three-Phase
3	9
5	15
10	30
15	45
25	75
37.5	112.5
50	150
75	225
100	300
167	500
250	750
333	1000
500	1500
833	2000
1250	2500
1667	3750
2500	5000
3333	7500
5000	10 000

lower voltages and the larger sizes to higher voltages. Voltage ratings and ratios should be selected in accordance with available standard equipment indicated in manufacturers' catalogs. This is recommended, if at all possible, both from the viewpoint of cost and time for initial procurement, and for ready replacement in the event of necessity.

Generally, a three-phase transformer secondary voltage should be selected at 480 V. This has become standard and is compatible with three-phase motors which are now rated 460 V standard. Under normal circumstances, a 460 V rating for the transformer secondary should not be selected unless the load is predominantly older motors rated 440 V, and located close to the transformer.

(3) *Voltage Taps.* Taps are used to change the ratio between the high- and low-voltage windings. Manual de-energized tap changing is usually used to compensate for differences between the transformer ratio and the system nomi-

nal voltages. The tap selected in the transformer should be based upon maximum no-load voltage conditions. For example, a standard transformer rated 13 200 V—480 V may have four 2.5% taps in the 13 200 V winding (two above and two below 13 200 V). If this transformer is connected to a system whose maximum voltage is 13 530 V, then the 13 530 V to 480 V tap could be used to provide a maximum of 480 V at no load.

Tap changers are classified as follows:

(a) Under load. Taps can be changed when the transformer is energized and loaded. These are used to compensate for excessive variations in the supply voltage. They are infrequently associated with commercial building transformers except as part of outdoor substations over 5000 kVA. Load tap changers can be controlled automatically or manually.

(b) No Load. Taps can be changed only when the transformer is de-energized. Tap leads are brought to an externally operated tap changer with a handle capable of being locked in any tap position. This is a standard accessory on most liquid- and sealed-type transformers. On very small liquid-filled transformers and most ventilated dry-type transformers, the taps are changed by moving internal links made accessible by a removable panel on the enclosure.

Adequate taps should be specified to maintain proper voltage level at the customer's bus if the electric utility's voltage should fluctuate widely. If the voltage changes expected are frequent, then consideration should be given to specifying that taps should be capable of being automatically changed under load to maintain a constant secondary bus voltage. In commercial buildings where a small part of the load is voltage sensitive a locally installed device for that load might be a more practical and economical solution. Automatic voltage regulators or constant voltage transformers should be considered as alternatives to load tap changers.

(4) *Typical Impedance Values for Power Transformers.* Typical impedance values for power transformers are given in Table 33. These values are at the self-cooled transformer kilovoltampere ratings and are subject to a tolerance of ±7.5%, as set forth in ANSI/IEEE C57.12.00-1980 [27]. Nonstandard impedances may be specified with a nominally higher cost: higher impedances to reduce available fault currents or lower impedances to reduce voltage drop under heavy-current low-power factor surge conditions. Consult manufacturers' bulletins for impedances of small transformers as they can vary considerably.

(5) *Insulation Temperature Ratings.* Transformers are manufactured with various insulation material systems as shown in Table 34. Performance data with reference to conductor loss and impedances should be referred to a temperature of 20 °C over the rated average conductor temperature rise as measured by resistance. While Table 34 represents the limiting standard requirements, transformers with lower conductor losses and corresponding lower temperature rises are available, where longer life expectancy and reduced operating costs are desired. Insulation material classes A, B, F, and H are used in dry-type transformers. Insulation material classes A and A (modified) are used in oil- and nonfire-propagating liquid-filled transformers.

(6) *Insulation Classes.* Voltage insulation classes are listed in Table 35.

(7) *Sound Levels.* Permissible sound levels are listed in Tables 36 and 37.

5.2.3 Transformer Construction. Trans-

Table 33
Transformer Approximate Impedance Values

High-Voltage Rating (volts)	Design Impedance (percent)	
	Low Voltage, Rated 480 V	Low Voltage, Rated 2400 V or Higher
Power Transformers		
2400 to 22 900	5.75	5.5
26 400, 34 400	6.0	6.0
43 800	6.5	6.5
67 000		7.0

Rated kVA	Design Impedance (percent)
Secondary Unit Substation Transformers	
112½ through 225	Not less than 2
300 through 500	Not less than 4.5
Above 500	5.75
Network Transformers	
1000 and smaller	5.0
Above 1000	7.0

Table 34
Insulation Temperature Ratings

Average Conductor Temperature Rise* ($^\circ$C)	Maximum Ambient Temperature ($^\circ$C)	Hot-Spot Temperature Differential* ($^\circ$C)	Total Permissible Ultimate Temperature* ($^\circ$C)
55	40	10	105
65	40	15	120
80	40	30	150
115	40	30	185
150	40	30	220

*Maximum at continuous rated load.
**Dry-type transformers using a 220 $^\circ$C insulation system can be designed for lower temperature rises (115 $^\circ$C or 80 $^\circ$C) to conserve energy, increase life expectancy, and provide some continuous overload capability.

Table 35
Voltage Insulation Classes and Dielectric Tests

Nominal System Voltage (kV)	Insulation Class	Dry Transformers		Oil-Immersed Distribution Transformers		Oil-Immersed Power Transformers	
		Basic Impulse Level (kV)	Low-Frequency Test (kV)	Basic Impulse Level (kV)	Low-Frequency Test (kV)	Basic Impulse Level (kV)	Low-Frequency Test (kV)
1.2	1.2	10	4	30	10	45	10
2.4	2.5	20	10	45	15	60	15
4.8	5.0	30	12	60	19	75	19
8.32	8.7	45	19	75	26	95	26
14.4	15.0	60	31	95	34	110	34
23.0	25.0	110	37	125	40	150	50
34.5	34.5	150	50	150	50	200	70

NOTE: Ventilated dry-type transformers and cast coil transformers can be built to match the basic impulse level of the oil-immersed distribution transformers.

formers are constructed in several different types which are discussed below. This section is generally applicable to transformers of the liquid-filled, ventilated-dry, or gas-filled dry types. Liquid-insulated and gas-filled transformers have their windings brought out to bushings or to junction boxes on the ends or top of the transformers. Ventilated dry-type transformers usually have their windings terminated within the enclosure of the transformer to either stand-off insulators or bus bar terminals.

(1) *Oil-Filled Transformers.* Oil-filled transformers are constructed with the windings encased in an oil-tight tank filled with mineral insulating oil. Oil-filled transformers should be avoided inside commercial buildings unless proper precautions are taken by building a truly fireproof and explosion-proof vault for the transformers. See the requirements of ANSI/NFPA 70-1981 [33]. The oil provides insulation between the various sections of the winding and between the

windings and the tank, and serves as a cooling medium, absorbing heat from the windings and transferring it to the outside of the tank. To increase the transfer of heat to the air, tanks are provided with cooling fins (to increase the area of the radiating surface) or with external cooling tubes or radiators. The hot oil circulates through the radiators transferring the heat picked up in the transformer windings to the radiator and then to the surrounding air.

Fans are sometimes installed to force air over the radiators in order to increase the full-load rating by approximately 15% on transformers rated 750 kVA—2000 kVA and 25% on transformers rated 2500 kVA—10 000 kVA.

It is essential that the oil in the transformer be maintained clean and free from moisture. Moisture can enter the transformer through leaks in the tank covers or when moisture-laden air is drawn into the transformer. Transformers can draw air into the tanks through *breathing* action resulting from changes

Table 36
Sound Levels for Dry-Type Transformers, in Decibels

Equivalent Two-Winding kVA	Self-Cooled Ventilated 1	Self-Cooled Sealed 2	Forced-Air-Cooled Ventilated* 3
0—9	45	45	
10—50	50	50	
51—150	55	55	
151—300	58	57	
301—500	60	59	
501—700	62	61	
701—1000	64	63	
1001—1500	65	64	
1501—2000	66	65	
2001—3000	68	66	
3001—4000	70	68	
4001—5000	71	69	
5001—6000	72	70	
6001—7500	73	71	
0—1167			67
1168—1667			68
1668—2000			69
2001—3333			71
3334—5000			73
5001—6667			74
6668—8333			75
8334—10 000			76

NOTES: (1) Columns 1 and 2 — class AA rating, column 3 — class FA and AFA rating.
(2) As given in NEMA ST 20-1972 (R1978) [32], Part IV, Table 4-4, page 26, sound levels for dry-type units rated 1.2 kV and less differ from those given here.

*Does not apply to sealed-type transformers.

in volume of oil and air in the tank with changes in temperature. Most modern transformers are tightly sealed and do not breathe if they are maintained free from leaks.

Insulating oil, through the normal aging process, develops a small amount of acid which, if allowed to increase above well established limits, can cause damage to insulation in the transformer. Yearly testing to determine the dielectric breakdown voltage of the oil (a low dielectric test indicates presence of water or other foreign material) and neutralization number (a high neutralization number indicates presence of acid in the oil) by a competent testing laboratory will greatly prolong the life of the transformer. Oil samples should be withdrawn under carefully controlled conditions as directed by the group making the oil test. In some areas, this service is available from the electric utility.

(2) *Nonfire-Propagating Liquids.* With discontinuance of the use of askarel (PCB) for the cooling fluid of liquid-

Table 37
Sound Levels for Single- and Three-Phase
Oil-Cooled Transformers, in Decibels

Equivalent Two-Winding kVA	Without Fans	With Fans
0—300	56	
301—500	58	
501—700	60	70
701—1000	62	70
1001—1500	63	70
1501—2000	64	70
2001—3000	65	71
3001—4000	66	71
4001—5000	67	72
5001—6000	68	73
6001—7500	69	73
7501—10 000	70	74

filled transformers for indoor installations, other liquids are now being employed such as polyalpha olefins and silicones. Per ANSI/NFPA 70-1981 [33], for indoor installations the fire point of transformer fluids shall be not less than 300 °C and if the fluid is flammable special provisions to prevent the spread of fire are required.

(3) *Ventilated Dry-Type Transformers.* Ventilated dry-type transformers are constructed in much the same manner as oil-filled transformers, except that the insulating oil is replaced with air and larger clearances or different insulating materials are used to compensate for the lower dielectric strength of air. The ventilated dry-type transformer is provided with a sheet-metal enclosure surrounding the winding for mechanical protection of the windings and the safety of personnel. Ventilating louvers are installed in the enclosure to permit thermal circulation of air directly over the winding for cooling. Fans are sometimes installed to force air directly over the windings in order to increase the full-load rating by approximately 33%. These types of transformers are normally installed indoors and require periodic cleaning of the complete core and coil assembly and an adequate supply of clean ventilating air. These transformers are gaining acceptance in the 15 kV and 34.4 kV class, and can be built to match the basic impulse levels of oil-immersed transformers and with special modification for use outdoors. Precautions before energizing are recommended after lengthy shutdown or periods when the insulation has been subjected to moisture.

(4) *Sealed Dry-Type Transformers.* Sealed dry-type transformers are constructed essentially the same as the ventilated dry-type transformers. The enclosing tank is sealed and operated under positive pressures. It may be filled with nitrogen or other dielectric gas. Heat is transferred from the winding to the gas within the transformer housing and from there to the tank and to the surrounding air.

The sealed (gas-filled) dry-type transformer can be installed both outdoors and indoors and in areas where corrosive or dirty atmosphere would make it impossible to use a ventilated dry-type transformer. These units are not available with fans.

(5) *Cast-Coil Type Transformers.* Cast-coil dry-type transformers are constructed with primary and secondary windings encapsulated (cast) in reinforced epoxy resin. Because of the cast-coil construction, they are ideal in applications where moisture or air-borne contaminants, or both, are a major concern. This type of construction is available with primary voltage ratings through 34.5 kV class and BIL ratings through 200 kV BIL. They are ideal alternates for liquid-filled or gas-filled units in indoor or roof-top applications. They are nonflammable and pose no

environmental hazard and may be used indoors without the precautions normally associated with liquid-filled transformers. They may be forced-air-cooled to increase their self-cooled ratings by 50%.

(6) *Totally Enclosed, Nonventilated Dry-Type Transformers.* Totally enclosed, nonventilated dry-type transformers are constructed essentially the same as the ventilated dry-type transformer. The enclosure, while not sealed, contains air, so the units have the same basic impulse insulation level capabilities as ventilated dry-type transformers. The totally enclosed, nonventilated dry-type transformer can be installed both indoors and outdoors and in areas where corrosive or dirty atmosphere would make it impossible to use a ventilated dry-type transformer. These units are available with fan cooling for a minimum of 25% increase in capacity.

5.3 Medium- and High-Voltage Fuses. Medium- and high-voltage fuses are part of many commercial power distribution systems. Applicable standards are ANSI C37.46-1980 [8] and NEMA SG-2-1976 [43].

Modern fuses suitable for the range of voltages encountered, fall into two general categories: distribution fuse cutouts and power fuses.

(1) *Distribution Fuse Cutouts.* According to the ANSI definition, the distribution fuse cutout has the following characteristics:

(a) Dielectric withstand [basic impulse insulation level (BIL)] strengths at distribution levels

(b) Application primarily on distribution feeders and circuits

(c) Mechanical construction basically adapted to pole or crossarm mounting except for the distribution oil cutout

(d) Operating voltage limits correspond to distribution system voltage

Characteristically, a distribution fuse cutout consists of a special insulating support and a fuse holder. The fuse holder, normally a disconnecting type, engages contacts supported on the insulating support and is fitted with a simple, inexpensive fuse link. The fuse holder is lined with an organic material, usually horn fiber. Interruption of an overcurrent takes place within the fuse holder by the action of de-ionizing gases, liberated when the liner is exposed to the heat of the arc established when the fuse link melts in response to the overcurrent.

(2) *Power Fuses.* According to the ANSI definition, the power fuse is identified by the following characteristics:

(a) Dielectric withstand (BIL) strengths at power class levels

(b) Application primarily in stations and substations

(c) Mechanical construction basically adapted to station and substation mountings.

Power fuses have other characteristics that differentiate them from distribution fuse cutouts in that they are available in higher voltage, current, and interrupting-current ratings, and in forms suitable for indoor and enclosure application as well as all types of outdoor applications.

A power fuse consists of a fuse support plus a fuse unit, or alternately a fuse holder which accepts a refill unit or fuse link.

Power fuses are classified as either expulsion type or current-limiting type, depending on the method used to interrupt overcurrents.

Expulsion-type power fuses, like distribution fuse cutouts, interrupt overcurrents through the de-ionizing action of the gases liberated from the lining of the interrupting chamber of the fuse by the heat of the arc established when the fusible element melts. Current-limiting type power fuses interrupt overcurrents by subjecting the arc established by melting of the fusible element to the mechanical restriction and cooling action of a powder or sand filler surrounding the fusible element.

The earliest forms of expulsion-type power fuses, being outgrowths of distribution fuse cutouts, were fiber lined. Such fuses had limited interrupting capacity and could not be used within buildings or in enclosures, and thus led in the 1930s to the development of solid-material expulsion-type power fuses. These fuses utilize densely molded solid boric-acid powder as lining for the interrupting chamber. This solid-material lining liberates noncombustible, highly deionized steam when subjected to the arc established by melting of the fusible element. Solid-material expulsion-type power fuses have higher interrupting capacities than fiber-lined expulsion-type power fuses of identical physical dimensions, produce less noise, need lesser clearance in the path of the exhaust gases, and, importantly, can be applied with normal electrical clearances indoors, or in enclosures when equipped with exhaust control devices. (An exhaust control device is available which provides silent operation by containing all arc-interruption products. An alternate, smaller, more economical device vents a greatly reduced volume of effectively cooled gases with no appreciable or objectionable noise.) These advantages, plus their availability in a wide range of current and interrupting ratings, have led to the wide use of solid-material expulsion-type power fuses in utility, industrial, and commercial power-distribution systems.

Current-limiting type power fuses were introduced in the United States almost simultaneously with solid-material expulsion-type power fuses. One form of current-limiting fuse is used with and coordinated with high-voltage motor starters for high-capacity 2400 V and 4160 V distribution circuits in large industrial complexes. Another form of current-limiting fuse is used with voltage transformers, distribution transformers, and small power transformers connected to high-capacity 2400 V to 34 500 V distribution circuits.

Current-limiting power fuses operate without expulsion of gases, because all the arc energy of operation is absorbed by the powder or sand filler surrounding the fusible element. They provide current limitation if the overcurrent value greatly exceeds the fuse ampere rating, thereby reducing the stresses and possible damage in the circuit up to the fault. These fuses can be applied indoors or in enclosures, and require only normal electrical clearances.

5.3.1 Fuse Ratings

(1) *High-Voltage Fiber-Lined Expulsion-Type Power Fuses.* This category has its principal usage in outdoor applications at the subtransmission voltage level. This fuse is available in current ratings and three-phase symmetrical short-circuit interrupting ratings as given in Table 38.

(2) *High-Voltage Solid-Material Boric-Acid Fuses.* High-voltage solid-material boric-acid fuses are available in two styles.

(a) The fuse-unit style in which the fusible element, interrupting element, and operating element are all combined in an insulating tube structure, with the entire unit being replaceable

(b) The fuse-holder and refill-unit

Table 38
Maximum Continuous Current and Interrupting Ratings for
Horn Fiber-Lined Expulsion-Type Fuses

Rated Maximum Voltage (kV)	Continuous Current Ratings (Amperes) (Maximum)				Maximum Interrupting Rating* (kA) rms Symmetrical
8.3	100	200	300	400	12.5
15.5	100	200	300	400	16.0
25.8	100	200	300	400	20.0
38.0	100	200	300	400	20.0
48.3	100	200	300	400	25.0
72.5	100	200	300	400	20.0
121	100	200			16.0
145	100	200			12.5
169	100	200			12.5

*Applies to all continuous current ratings.

Table 39
Maximum Continuous Current and Interrupting Ratings for
Solid-Material Boric-Acid Fuses (Fuse Units)

Rated Maximum Voltage (kV)	Continuous Current Ratings (Amperes) (Maximum)			Corresponding Maximum Interrupting Ratings (kA) rms Symmetrical		
17		200		14.0		
27		200		12.5		
38	100	200	300	6.7	17.5	33.5
48.3	100	200	300	5.0	13.1	31.5
72.5	100	200	300	3.35	10.0	25.0
121	100	250		5.0	10.5	
145	100	250		4.2	8.75	

style in which only the refill unit is replaced after operation

Solid-material boric-acid fuses in the fuse-unit style are principally for use outdoors at subtransmission and distribution voltages. The fuse units are specifically designed for use in outdoor poletop or station-style mountings, as well as indoor mountings installed in metal-enclosed interrupter switchgear, indoor vaults, and pad-mounted gear. Indoor mountings incorporate an exhaust-control device that contains most of the arc-interruption products and virtually eliminates noise accompanying a fuse operation. These exhaust-control devices do not require a reduction of the interrupting rating of the fuse.

Indoor mountings for use with fuse units rated up to 27 kV maximum can be furnished with an integral hookstick-operated load-current interrupting device, thus providing for single-pole live switching in addition to the fault inter-

Table 40
Maximum Continuous Current and Interrupting Ratings for
Solid-Material Boric-Acid Fuses (Refill Units)

Rated Maximum Voltage (kV)	Continuous Current Ratings (Amperes) (Maximum)			Corresponding Maximum Interrupting Ratings (kA) rms Symmetrical		
2.75	200	400	720*	7.2	37.5	37.5
4.8	200	400	720*	17.2	37.5	37.5
8.25	200	400	720*	15.6	37.5	29.4
15.5	200	400	720*	14.0	34.0	25.0
25.8	200	300		12.5	21.0	
38	200	300		6.25	17.5	

*Parallel fuses

Table 41
Maximum Continuous Current and Interrupting Ratings for
Current-Limiting Fuses

Rated Maximum Voltage (kV)	Continuous Current Ratings (Amperes) (Maximum)				Corresponding Maximum Interrupting Ratings (kA) rms Symmetrical			
2.75	225	450*	750*	1350*	50.0	50.0	40.0	40.0
2.75/4.76		450*				50.0		
5.5	225	400	750*	1350*	50.0	62.5	40.0	40.0
8.25		125	200*			50.0	50.0	
15.5	65	100	125*	200*	85.0	50.0	85.0	50.0
25.8		50	100*			35.0	35.0	
38		50	100*			35.0	35.0	

*Parallel fuses

rupting function provided by the fuse.

These fuses are available with current and interrupting rating as given in Table 39.

The solid-material boric-acid fuse in the fuse-holder and refill-unit style can be used either indoors or outdoors at medium- and high-distribution voltages. These fuses are also available with integral load-current interrupting devices for single-pole live switching. The fuses are available in current and interrupting ratings as given in Table 40.

(3) *Current-Limiting Power Fuses.* Current-limiting power fuses suitable for the protection of auxiliary power transformers, small power transformers, and capacitor banks are available with current and interrupting ratings as given in Table 41.

Fuses for the protection of medium-voltage transformers are available with interrupting ratings to 80 kA (symmetrical) at 5.5 kV, 120 kA at 15.5 kV, and 44 kA at 25.8 kV and 38 kV.

Current-limiting fuses suitable only for

Table 42
Maximum Continuous Current and Interrupting Ratings for
Current-Limiting Fuses (Motor Starters)

Rated Maximum Voltage (kV)	R Designation	Continuous Current Ratings (Amperes) (Maximum)	Maximum Interrupting Rating (kA) rms Symmetrical
2.54	50 R	700	50.0
2.75/5.5	—	750	50.0
5.0	50 R	700	50.0
7.2	18 R	390	50.0
8.3	6 R	170	50.0

use with high-voltage motor starters are available with current and interrupting ratings as given in Table 42.

5.3.2 Fuse Applications

(1) *Power Supply*. Where a commercial project is served by a utility at a subtransmission voltage of 23 kV— 161 kV and a transformer substation provides in-plant service at utilization voltage or primary distribution voltage, power fuses can be used as an economical primary-side overcurrent protective device for transformer banks with 1500 kVA maximum rating.

With their high short-circuit interrupting capability and high-speed operation, power fuses will protect the subtransmission circuit by clearing faults at the transformer. In addition, power fuses can provide back-up protection in the event of malfunction of the transformer secondary overcurrent protective device.

In addition to providing overcurrent protection to the main power transformers, power fuses are used to provide protection for instrument transformers and for capacitor banks.

(2) *Power Distribution*. The principal functions of overcurrent protective devices at these primary voltages are:

(a) To interrupt high values of overcurrent

(b) To act as back-up protection in

the event of a malfunction of the next downstream protective device

(c) To open circuits under overcurrent conditions

Modern medium-voltage power fuses can be used to provide this protection for virtually all types and sizes of distribution systems. Such fuses used with properly coordinated and designed load-interrupter switches may be applied outdoors, in vaults, or in metal-enclosed interrupter switchgear.

5.4 Metal-Enclosed 5 kV to 34.5 kV Interrupter Switchgear. Metal-enclosed interrupter switchgear can be used to provide switching capability and overcurrent protection through the use of interrupter switches and power fuses. It can also be used for ground-fault protection of resistance-grounded systems if properly applied. Rated maximum voltages are 4.76 kV, 8.25 kV, 15.0 kV, 15.5 kV, 25.8 kV and 38.0 kV, with main bus ratings of 600 A, 1200 A, or 2000 A. Interrupting ratings are determined by the power fuses, for which maximum ratings are given in Tables 40 and 41. Power fuses are available in a wide range of current ratings and are offered in a selection of time-current characteristics to provide proper coordination with other protective devices and with the thermal characteristics of the power transformer. The interrupter switches, which may be manually or automatically operated, are rated 200 A, 600 A, or 1200 A, continuous and interrupting. An applicable standard for metal-enclosed interrupter switchgear is ANSI/IEEE C37.20-1969 [22], 6.4 and NEMA SG5-1975 [47], Part 11.07.

Metal-enclosed interrupter switchgear does not incorporate a reclosing feature, since reclosing is rarely desirable in power systems for commercial buildings, where the conductors are commonly arranged in cable trays or enclosed in raceways or busways. The rare faults that occur in such installations require significant repair before re-energization.

Metal-enclosed interrupter switchgear can be used in high-continuity distribution circuits, such as the conventional (two-switch) and the split-bus (three-switch) primary-selective systems.

Interrupter switchgear is usually less expensive than metal-clad power switchgear (see 5.5). This would permit the engineer to improve service continuity by providing more radial feeders per dollar of equipment cost using interrupter switchgear.

5.4.1 Automatic Control Devices.
Automatic control devices can be incorporated in metal-enclosed interrupter switchgear in conjunction with motor powered switch operators to provide high service continuity through primary-selective systems, by initiating the automatic transfer of sources providing service to the main bus (or buses) in the event of a fault or outage on one of the sources. Optional features include provisions for manual or automatic back transfer (with open or closed transition), time delay on transfer, and lockout on faults.

Interrupter-switch manufacturers can also provide an open-phase relay system which initiates circuit interruption to protect loads from single-phasing which may occur as a result of broken conductors or fuse operations in the source-side circuit. This relay can also be applied to protect against single-phasing due to load-circuit fuse operations.

5.4.2 Auxiliary Equipment and Features.
Metal-enclosed interrupter switchgear may include (in addition to interrupter switches and power fuses) instrument transformers, meters, and other auxiliary devices, including motor-type power operators for remote operation of the interrupter switches (or operation of the switches in an automatic transfer scheme, when used in conjunction with an automatic control device). The power fuses may be equipped with blown-fuse indicators, for positive visual checking of fuses while in their mountings.

5.4.3 Capability Required.
Metal-enclosed interrupter switchgear should comply with ANSI/NFPA 70-1981 [33], Section 710-21(e) which requires that interrupter switches, when used in combination with fuses or circuit breakers, safely withstand the effects of closing, carrying, or interrupting all possible currents up to the assigned maximum short-circuit rating. (See also ANSI/IEEE C37.20-1969 [22], 6.4.8.) Fault-interrupting ratings are not required for interrupter switches inasmuch as the associated fuses should be selected to interrupt any faults that may occur.

5.5 Metal-Clad 5 kV to 34.5 kV Power Circuit-Breaker Switchgear.
Metal-clad switchgear is available with voltage ratings of 4.16 kV—34.5 kV and with circuit breakers having interrupting ratings from 8.8 kA at 4.16 kV to 40 kA at

34.5 kV as standard. Continuous-current ratings are 1200 A, 2000 A, 3000 A and 3750 A. Applicable standards include ANSI/IEEE C37.20-1969 [22], Section 6.2, and NEMA SG 5-1975 [47], Part 9.03, for power switchgear and ANSI/IEEE C37.04-1979 [12] through ANSI C37.12-1969 (R1974) [6], NEMA SG 4-1968 [46] and ANSI/IEEE C37.100-1981 [26] for power circuit breakers.

Metal-clad switchgear has a power circuit breaker as the main circuit interrupting and protective device. Major parts of the primary circuit such as circuit switching or interrupting devices, buses, potential transformers, and control power transformers are completely enclosed by grounded metal barriers. Circuit instruments protective relays, and control switches are mounted on a hinged control panel or occasionally on a separate switchboard remote from the switchgear. The power circuit breaker is readily removable and has self-coupling disconnecting primary and secondary contacts. Potential transformers and control power transformer fuses may be provided in drawout assemblies to permit the safe changing of fuses.

Automatic shutters to shield the stationary primary contacts when the circuit breaker is removed are provided as well as other necessary interlocking features to ensure proper sequence of operation. The drawout feature facilitates inspection and maintenance of the circuit breaker. In addition, it permits the quick replacement of any circuit breaker with a spare, and therefore, provisions for bypassing it during circuit-breaker maintenance periods are generally not required. The circuit-breaker compartments have separable main and secondary disconnect contacts to achieve connected, test, and disconnect positions. The test position provides a feature whereby the circuit

breaker may be electrically exercised while disconnected from the main power circuit. The disconnect position allows the circuit breaker to be disconnected from the main power and control supply, locked, and stored in its cubicle.

Metal-clad switchgear can provide the switching, isolation, protection, and instrumentation of all the incoming, bus tie, and feeder circuits. All parts are housed within grounded metal enclosures, thereby providing a high degree of safety for both personnel and equipment. All line conductors are opened simultaneously in the event of circuit-breaker tripping. A wide variety of parameters can be programmed into the tripping function.

The insulation used in the vital points of the metal-clad switchgear is of the potential tracking resistant type and may be flame retardant. Thus the equipment presents a very minimum fire hazard and is suitable for indoor installations without being placed in a vault. For outdoor equipment, a weatherproof enclosure is provided over the same switchgear components as are used for the indoor switchgear assemblies. Protected aisle construction to permit maintenance in inclement weather can also be provided.

5.5.1 Power Circuit Breakers. The power circuit breaker most generally applied on 2.4 kV, 4.16 kV, 7.2 kV, and 13.8 kV systems is of the air-magnetic type.

Recently, vacuum-type power circuit breakers have been made available. Due to the possibility of vacuum circuit breakers creating a higher-than-normal voltage during interruption precaution is required. Vacuum circuit breakers offer the advantage of a *contained arc*. Air-blast minimum oil, and SF_6 circuit breakers are available at all medium voltages.

Vacuum and SF_6 breakers offer the advantage of faster clearing time than air magnetic breakers. SF_6 does this without the potential voltage transient effects of vacuum breakers. For a tabulation of standard ratings of circuit breakers for metal-clad switchgear, see ANSI C37.06-1979 [1] and ANSI C37.6-1971 (R1976) [2].

5.5.2 Instrument Transformers and Protective Relaying. All of these circuit breakers utilize relays which are operated by current and voltage transformers. This combination provides a wide range of protection which is field adjustable. With protective relaying, full tripping selectivity can usually be obtained between all of the circuit breakers in the equipment in case of faults.

5.5.3 Control. Power circuit breakers are electrically operated devices and must be provided with a source of control power. Control power can be obtained from a battery or from a control power transformer located within the switchgear.

5.5.4 Main-Bus Continuous-Current Selection. Main-bus continuous-current ratings are available to match the continuous-current ratings of the associated power circuit breakers. By the proper physical arrangement of the source and load circuit breakers or bus taps, it is possible to engineer the lowest bus current requirements consistent with system capacity. For example, it may be necessary to have a 2000 A source circuit breaker (or breakers), yet only require a 1200 A main bus. Regardless of the lower bus capacity at different points, the bus is designed and rated for the present and future current capacity at the maximum point. It would not be tapered for reducing current capacity.

5.6 Metal-Enclosed 600 V Power Switchgear and Circuit Breakers

5.6.1 Drawout Switchgear. Metal-enclosed drawout switchgear using power air circuit breakers is available for protection and control of low-voltage circuits. Rigid ANSI standards dictate the design, construction, and testing to assure reliability to the user. Industry standards are ANSI/IEEE C37.20-1969 [22], ANSI/IEEE C37.13-1981 [21], NEMA SG 5-1975 [47], Part 9.02, and NEMA SG 3-1975 [44].

Unlike distribution switchboards where a broad variety of protecting devices or panelboards can be incorporated, the main, tie, and feeder positions in power switchgear are limited to drawout power circuit breakers. Drawout power switchgear is more adaptable and procurable with complex control circuitry such as sequential interlocking, automatic transfer, or complex metering. Power automatic transfer functions are often custom designed using electrically operated power circuit breakers rather than pre-engineered automatic transfer devices or panels. In some instances this can be more economical. This is especially true when one or more of the power circuit breakers will exist for other reasons. The test position in drawout construction provides the opportunity to operate the automatic transfer equipment without connection to the power buses.

This class of switchgear is available in both indoor and outdoor construction. The latter usually is constructed to provide a sheltered aisle with an overhead circuit-breaker removal device. An integral roof-mounted circuit-breaker removal device is also available for indoor construction.

The individual power air circuit breakers are in compartments isolated from each other and from the bus area. Compartments accommodate circuit breakers in ANSI sizes of 225 A, 600 A, 1600 A, 2000 A, 3000 A, and 4000 A arranged in multiple high construction. Some manufacturers offer 800 A, 2500 A, and 3200 A instead of 600 A, 2000 A, and 3000 A ratings. The power air circuit breakers can be electrically or manually operated and equipped with added devices such as shunt trip, undervoltage, auxiliary switches, etc. They are available either with conventional electromagnetic overcurrent direct-acting tripping devices or static tripping devices.

The drawout circuit breakers and compartments have separable main and secondary disconnect contacts to achieve connected, test, disconnect, and fully withdrawn positions. The test position provides a feature whereby the circuit breaker may be exercised while disconnected from the main power circuit. The disconnect position allows the circuit breaker to be disconnected from the main power and control supply, locked, and stored in its compartment. In the fully withdrawn position, the circuit breaker is exposed for inspection and adjustments and may be removed from the switchboard for replacement or inspection.

Separate compartments are provided for required meters, relays, instruments, etc. Potential and control power transformers are usually mounted in these compartments to be front accessible. Current transformers may be mounted around the stationary power primary leads within the circuit breaker compartment (front accessible) or in the rear bus area.

The rear section of the switchboard is isolated from the front circuit breaker section and accommodates the main bus, feeder terminations, small wiring, and terminal blocks. Bus work is usually aluminum, designed for an allowable temperature rise of 65 °C above an average 40 °C ambient. A copper bus is available at an added cost. Circuit-breaker terminals are accessible from the rear of the switchboard. Cable lugs or bus way risers are provided for top or bottom exits from the switchgear. Control wiring from the separable control contacts of the circuit breaker is extended to terminal blocks mounted in the rear section. These blocks accommodate remote control and intercompartment and frame wiring by the manufacturer.

5.6.2 Low-Voltage Power Air Circuit Breakers. Low-voltage power air circuit breakers are long-life, quick-make (via stored energy manual or electrical closing mechanism), quick-break switching devices with integral inverse-time overload or instantaneous trip units. These circuit breakers also have a short-time (30 Hz) rating permitting the substitution of short-time tripping devices in place of the instantaneous tripping feature. Interrupting ratings for each circuit breaker depend on the voltage of the system to which it is applied (that is, 240 V, 480 V, 600 V, alternating current, 60 Hz) and on whether it is equipped with an instantaneous or short-time tripping feature as part of the circuit-breaker assembly or equivalent panel-mounted protective relays. It is this short-time rating of the circuit breakers that permits the designer to develop selective systems. These circuit breakers are open-construction assemblies on metal frames, with all parts designed for accessible maintenance, repair, and ease of replacement. They are intended for service in switchgear compartments or other enclosures of dead front construction at 100% of their rating in a 40 °C ambient without compensation or derating. Tripping units are field adjustable over a wide range and are completely interchangeable within their frame sizes.

Static-type tripping units are available from most manufacturers. Static trip units may provide an additional degree or number of steps in selectivity when only a small margin of spread exists between optimum protective settings for connected loads downstream and utility or other existing protective-device settings upstream. Static devices readily permit the inclusion of ground-fault protection as part of the circuit-breaker assembly.

A low-voltage power circuit breaker can be used by itself or with integral current-limiting fuses in drawout construction or separately mounted fuses to meet interrupting current requirements up to 200 000 A symmetrical rms. When part of the circuit breaker, the fuses are combined with an integral mounted blown fuse indicator and breaker trip device to open all three phases.

Power air circuit breakers may be used for the control and protection of large low-voltage motors. They can be equipped to provide disconnect, running overload, and short-circuit protection. They are generally not suitable where operation is highly repetitive.

5.6.3 Selection of Circuit-Breaker Tripping Characteristics. The degree of service continuity available from a low-voltage distribution system depends on the degree of coordination between circuit-breaker tripping characteristics. Two methods of tripping coordination are in general use, each representing a different degree of service continuity and of initial cost. These methods, or systems, combine circuit-breaker ratings and tripping characteristics as follows:

(1) *Non-Selective System.* A system in which all circuit breakers have adequate interrupting capacity for the fault current available at the point of application.

All circuit breakers are equipped with long time-delay and instantaneous overcurrent trips. Ordinarily only the circuit breaker nearest the fault will open. The main circuit breaker will trip, however, when the fault current exceeds its instantaneous trip setting, and service continuity may be lost.

(2) *Selective System.* In a selective system the main circuit breaker is equipped with overcurrent trip devices having long-time-delay and short-time-delay functions. The feeder circuit breakers are equipped with overcurrent trip devices having long-time-delay and instantaneous functions, unless they are required to be selective with other protective devices nearer the load. In this case, the feeders are equipped with trip devices having both long and short time delay.

In a selective system, only the circuit breaker nearest the fault trips. Service continuity is thus maintained through all other circuit breakers. The selective system offers a maximum of service continuity, with a slightly higher initial cost for the short-time functions instead of the standard instantaneous function.

5.7 Metal-Enclosed Distribution Switchboards. Metal-enclosed distribution switchboards are frequently used in commercial buildings at 600 V and below for service entrance, power, or lighting distribution, and as the secondary sections of unit substations. A wide range of protective devices and single- or multisection assemblies are available for large services from 400 A to 4000 A. While 4000 A equipment is available, the use of smaller services is recommended. NEMA PB 2-1978 [42] is applicable.

Equipment ground-fault protection is recommended where the switchboard is applied on grounded-Y systems. It is required on electrical services of more than 150 V to ground for any service disconnecting means rated 1000 A or more. See ANSI/NFPA 70-1981 [33], Section 230.95 for minimum requirements.

Automatic transfer between main and emergency sources is generally provided by pre-engineered transfer switches or pre-engineered transfer devices, rather than custom engineered circuitry and relay operation of main circuit breakers.

5.7.1 Components. The following components are available:

(1) Service protectors

(2) Molded-case circuit breakers, group or individual mounted

(3) Fusible switches

(4) Motor starters

(5) Low-voltage ac power circuit breaker (generally limited to main or tie position)

(6) Bolted contact pressure switches

(7) Transfer devices or switches

(8) Instrumentation, metering, and relaying

Instrumentation and metering include the utility company metering equipment, voltmeters, ammeters, wattmeters, and similar instruments, and the potential and current transformers as required.

5.7.2 Construction Features

(1) *Front Accessible — Front Connected*

Designed to be installed against a wall

All mechanical and electrical connections are made from the front

Multisection switchboards have backs lined up

Switchboards are enclosed on all sides except the bottom

Maximum rating 2000 A

Drawout low-voltage ac power circuit breakers are not available as branch devices

Load side risers are not available

(2) *Rear Accessible — Front Connected*

Designed to be free standing

Designed for rear accessibility

All main connections are made from the rear

All normal maintenance to the main bus is performed from the rear

All line and load connections for branch devices are made from the front

Cross bus is located behind the branch devices and is accessible only from the rear

Multisection switchboards have fronts lined up

Capable of accepting all components

(3) *Rear Accessible — Rear Connected*

Designed to be free standing

Designed for rear accessibility

All main connections are made from the rear

All normal maintenance to the main bus is performed from the rear

All line and load connections for branch devices are made from the rear

All cross bus and line and load connections for branch devices are accessible only from the rear

Multisection switchboards have fronts lined up

Capable of accepting all components

5.8 Primary Unit Substations. Primary unit substations are best described by their function, that is, to transform power from high or medium voltages down to a voltage above 1000 V, and to provide protection and control for the lower voltage feeder circuits. Primary unit substations are most often used today in commercial buildings to convert a 13.2 kV or 13.8 kV service to 4160 V or 2400 V for large motors. They may be used, however, to provide the 13.2 kV—13.8 kV (or 4160 V) service when power is being purchased at a higher voltage.

These unit substations are physically and electrically coordinated, indoor or outdoor, combinations of primary unit substation type transformers and power-interrupter switchgear or power circuit-breaker switchgear. NEMA 201-1970 (R1976) [50] is applicable.

The incoming and secondary sections of primary unit substations are available in arrangements to suit the many variations of power distribution circuits, as described in 5.9.

For detailed information on the transformer section refer to 5.2. Similarly, for detailed information on the switchgear section refer to 5.3, 5.4, and 5.5.

5.9 Secondary Unit Substations. Secondary unit substations are best described by their function, that is, to transform power from the 2300 V to 15 000 V range down to 600 V or less and to provide protection and control for low-voltage feeder circuits. Secondary unit substations consist of coordinated incoming-line, transformer, and low-voltage sections. Each of these major sections is available in several forms for both indoor and outdoor application and to suit the many variations of power distribution circuit arrangements. NEMA 210-1970 (R1976) [51] is applicable.

5.9.1 Basic Circuits. Four basic circuits are most widely used in the following order:

(1) Simple radial
(2) Secondary selective system
(3) Primary selective system
(4) Network system

The above are described in Section 4.

5.9.2 Incoming-Line Sections. For use with the simple radial, secondary-selective, or spot-network system, this section will generally consist of one fused two-position (open—close) 5 kV or 15 kV air

interrupter switch. This is the same device as discussed in 5.4. Liquid-filled interrupter switches, fusible oil cutouts, or simply an air-filled terminal chamber may satisfy the application.

For use on primary-selective systems, either two key-interlocked metal-enclosed air interrupter switches with a common set of fuses, or one metal-enclosed air interrupter switch and one set of fuses in series with a key-interlocked (line 1— line 2) selector switch may be used. The use of two air interrupter switches requires the additional floor space of the additional cubicle, but isolates the normal and alternate primary supplies. The combination of the air interrupter switch and the selector switch is accomplished in a single cubicle.

5.9.3 Transformer Section. This unit transforms the incoming power from the higher primary to lower secondary voltage. Ratings, voltages, and connections are as covered in 5.2.1. The transformer is mechanically and electrically coordinated to the incoming (primary) line section and to the secondary section.

5.9.4 Secondary Section. This section provides the protection and control for the low-voltage feeder circuits. It may consist of a drawout power-circuit-breaker type switchgear assembly, a metal-enclosed distribution switchboard, a panelboard mounted in or on the transformer section, or a single secondary protective device. Aluminum bus work has replaced copper. The latter is available, but at additional cost.

For detailed information on the secondary switchgear section refer to 5.6 and 5.7.

5.10 Panelboards. Electrical systems in commercial buildings usually include panelboards utilizing fusible or circuit-breaker devices, or both. They are generally classified into two categories:

(1) Lighting and appliance panels

(2) Power distribution panels

Panelboard mounting of motor starter units may also be involved. NEMA PB1-1977 [41] and UL 67-1979 [52] are applicable.

5.10.1 Lighting and Appliance Panelboards. These panels have more than 10% of the overcurrent devices rated 30 A or less, for which neutral connections are provided. The number of overcurrent devices (branch circuit poles) is limited to a maximum of 42 in any one box. When the 42 poles are exceeded, two or more separate boxes are required. A common front for multiple boxes is usually available. Narrow-width box constructions are used to fit into a 10 in or 8 in structural wide-flange beam where mounting of a panelboard on a building column is appropriate. Column extensions and pull boxes are available for this application.

Ratings of these panels are single-phase, two-wire 120 V or three-wire 120/240 V; 120/208 V, three-phase, three-wire 208 V, 240 V, or 480 V; and three-phase, four-wire 208Y/120 V or 480Y/277 V.

5.10.2 Power Distribution Panelboards. This type includes all other panelboards not defined as lighting and appliance panelboards. The 42 overcurrent protective device limitation does not apply. However, care should be exercised not to exceed practical physical limitations such as standard box heights and widths available. Common fronts for two or more boxes are often impractical from a weight and installation standpoint due to the size of this type of panelboard.

Ratings are single-phase, two- or three-wire; three-phase, three- or four-wire; 120/240 V through 600 V alternating current, 250 V direct current; 50 A—1600 A, 1200 A maximum branch.

5.10.3 Motor-Starter Panelboards. Rather than individual mounting, a small number of motor starters can be grouped into a panelboard. Motor starter panelboards consist of combination units utilizing either molded-case or motor circuit protector fusible disconnects. The combination starters are factory wired and assembled. Class A provides no wiring external to the combination starter; class B provides control wiring to terminal blocks furnished near the side of each unit. Where a large number of motors are to be controlled from one location or additional wiring between starters and to master terminal blocks is required, conventional motor control centers are most commonly used.

5.10.4 Multisection Panelboards. Both lighting and appliance panelboards or power distribution panelboards requiring more than one box are called multisection panelboards. Unless a main overcurrent device is provided in each section, each section must be furnished with main bus and terminals of the same rating for connecting to the one feeder. The three methods commonly used for interconnecting multisection panelboards are as follows.

(1) *Gutter Tapping.* Increased gutter width may be required. Tap devices are not furnished with the panelboard.

(2) *Subfeeding.* A second set of main lugs (subfeed) are provided directly beside the main lugs of each panelboard section, except the last in the lineup.

(3) *Through Feeding.* A second set of main lugs (through feed) are provided on the main bus at the opposite end from the main lugs of each section, except the last in the lineup. This method has the undesirable feature of allowing the current of the second panelboard section to flow through the main bus of the first section.

5.10.5 Panelboard Data. To assist the engineer planning an installation, manufacturers' catalogs provide a wide choice of panelboards for specific applications. Some very important rules governing the application of panelboards are described in ANSI/NFPA 70-1981 [33].

(1) *Six-Circuit Rule.* ANSI/NFPA 70-1981 [33], Section 230-71 provides that a device may be suitable for service entrance equipment when not more than six main disconnecting means are provided. In addition, a disconnecting means shall be provided for the ground conductor.

(2) *30-Conductor Rule.* ANSI/NFPA 70-1981 [33], Section 362-5 states that wireways shall not contain more than 30 conductors at any cross section, unless the conductors are for signaling or motor control. It further states that the total cross-sectional areas of all the conductors shall not exceed 20% of the internal cross section of the wireway. Column panels or panels fed by a single wireway are limited to three main conductors and 27 branch and neutral conductors (12 circuit panelboard, single phase, three wire). When the neutral bar is mounted in a column-panel pullbox, this will be changed to two main conductors and 28 branch circuits (28 circuit panelboard).

(3) *Gutter-Tap Rule.* ANSI/NFPA 70-1981 [33], Section 240-21 states that overcurrent devices shall be located at the point where the conductor to be protected receives its supply. Exception No 5 to this paragraph permits omission of the main overcurrent device if (a) the smaller conductor has a current-carrying capacity of not less than the sum of the allowable current-carrying capacities of the one or more circuits or loads supplied and (b) the tap is not over 10 ft long and does not extend beyond the panelboard it supplies. Gutter taps are permitted under this ruling.

5.11 Molded-Case Circuit Breakers. Standard designs of molded-case circuit breakers (MCCB) are quick-make and quick-break switching devices with both inverse-time and instantaneous trip action. They are encased within rigid nonmetallic housings and vary greatly in size and ratings. Standard frames are available with 30 A — 4000 A current and 120 V — 600 V ac and 125 V — 250 V dc ratings.

The smaller breakers are built in one-, two-, or three-pole construction and are sealed units without adjustable instantaneous trips. The larger ratings are usually available in three- or four-pole frames only and have interchangeable and adjustable instantaneous trip units. With modifications and new developments the manufacturers' catalogs should be consulted to obtain the MCCB best suited for user requirements.

The current domestic standards are: NEMA AB1-1975 [37] and UL 489-1980 [53].

(1) *Requirements.* With few exceptions the manufacturing, ratings, and performance requirements are the same for both standards. Typically, MCCBs are submitted for UL witness testing which is repeated periodically for certification. The required switching tests are conducted sequentially with a set of MCCBs according to a listed schedule. The test samples undergo all tests which include overload, endurance, and short circuit.

(2) *Accessories.* Molded-case circuit breakers (MCCB) are usually operated manually but solenoids are available for remote tripping and electrical motor operators are available for remote operation with the larger frames. Other attachments are auxiliary contacts for signal-

ling and undervoltage devices to trip the MCCB on reduced system potential. All MCCB designs employ a trip-free mechanism which prevents injury to an operator who closes a breaker into a fault. The larger frames have ground fault designs utilizing external current transformers and relays to energize a shunt trip within the MCCB.

(3) *Application.* Ambient temperature and system frequency must be considered for all MCCBs. Unlike the power air circuit breakers, MCCBs usually require a 20% current derating when installed in enclosures. Several manufacturers offer 100% rated MCCBs with 600 A frames and larger. With few exceptions, conventional MCCB designs cannot be coordinated for selectivity. These breakers employ rapid mechanisms having little inertia, and interrupting times at maximum fault levels are usually one cycle or less.

MCCBs employing electronic trip units and current transformers can be applied for selective coordination and their short-time ratings vary with designs. Many of these modern designs have internal ground fault detection which improves system protection.

5.11.1 Types of Molded-Case Circuit Breakers. These devices are available in the following general types.

(1) *Thermal-Magnetic.* This type employs temperature sensitive bimetals which provide inverse or time-delayed tripping on overloads, and coils or magnet and armature designs for instantaneous tripping.

(2) *Dashpot.* This type utilizes a pneumatic or hydraulic scheme for overload protection and may have a coil for instantaneous operation.

(3) *Magnetic Only.* These breakers employ only instantaneous tripping and are used in welding or motor-circuit applica-

tion. ANSI/NFPA 70-1981 [33] recognizes adjustable magnetic types only for motor-circuit applications.

(4) *Integrally Fused.* Especially designed current limiting fuses are housed within the molded case for extended short-circuit application in systems with 100 kA or 200 kA available, and interlocks are provided to ensure that the MCCB trips when any fuse operates.

(5) *Current Limiting.* This type employs electromagnetic principles to effectively reduce the let-through magnitudes of current and energy ($I^2 t$). Their ratings and number of effective operations are available in manufacturers' literature, and some designs are UL approved.

(6) *High Interrupting Capacity.* Many manufacturers offer this type for application in systems having high fault currents. They employ stronger, high temperature molded material, but retain the standard breaker dimensions.

5.11.2 Use of Molded-Case Circuit Breakers. Molded-case circuit breakers are suitable in various equipments and installations.

(1) Individual enclosures

(a) Wall mounted dust resistant, NEMA Types 1A and 12 (See ANSI/NEMA ICS 6-1978 [30].)

(b) Outdoor raintight, NEMA Type 3 (See ANSI/NEMA ICS 6-1978 [30].)

(c) Hazardous, NEMA Types 4, 5, 7 and 9 (See ANSI/NEMA ICS 6-1978 [30].)

(2) In panelboards and distribution switchboards

(3) In switchgear having rear connected, bolt-on, plug-in, or draw-out features

(4) In combination starters and motor control centers

(5) In automatic transfer switches

In case (5), molded-case circuit breakers may be used as part of automatic transfer switches to serve as service or feeder disconnects and to provide overcurrent protection. They may also be used as part of the automatic transfer switch when found suitable for this particular task and when operated by appropriate mechanisms in response to initiating signals, such as loss of voltage, etc. If MCCBs are used to combine both functions, an external manual operator should be provided for independent disconnections of both the normal and alternate supplies. Particularly in the larger sizes (current ratings), consideration should be given to the anticipated number of operations to which the equipment will be subjected because MCCBs are not designed for highly repetitive duty.

5.12 Low-Voltage Fuses. A fuse may be defined as an overcurrent protective device with a circuit-opening fusible part that is heated and severed by the passage of overcurrent through it. The fusible element of a fuse opens in a time that varies inversely with the magnitude of current that flows through the fuse. The time-current characteristic depends upon the rating and type of fuse.

Non-time-delay fuses are fuses that have no intentional built-in time delay. They are generally employed in other than motor circuits or in combination with circuit breakers where the circuit breaker provides protection in the overload current range and the fuse provides protection in the short-circuit current range.

Time-delay fuses have intentional built-in time delay in the overload range. This time-delay characteristic often permits the selection of fuse ratings closer to full-load currents.

Time-delay fuses are widely used as they have adequate time delay to permit their use as motor overcurrent running protection. Dual-element time-delay fuses provide protection for both motors and circuits and make it possible to use a fuse whose current rating is not far above the full-load current of the circuit. The fuse will permit starting-inrush current of a motor, but stands ready to open the circuit on long continued overcurrent.

5.12.1 Fuse Ratings. Low-voltage fuses have current, voltage, and interrupting ratings which should not be exceeded in practical application. In addition, some fuses are also rated according to their current-limiting capability as established by Underwriters Laboratories, Inc (UL) standards and are so designated by a class marking on the fuse label (classes L, K1, K5, RK1, RK5, J, etc). Current-limiting capabilities are established by UL according to the maximum peak current let-through and the maximum I^2t let-through of the fuse upon clearing a fault.

(1) *Current Rating.* Current rating of a fuse is the maximum direct current or alternating current in amperes at rated frequency which it will carry without exceeding specified limits of temperature rise. Current ratings which are available range from milliamperes up to 6000 A.

(2) *Voltage Rating.* Voltage rating is the alternating- or direct-current voltage at which the fuse is designated to operate. Low-voltage fuses are usually given a voltage rating of 600 V, 300 V, 250 V, or 125 V alternating current or direct current, or both.

(3) *Interrupting Rating.* Interrupting rating is the assigned maximum short-circuit current (usually alternating current) at rated voltage which the fuse will safely interrupt. Low-voltage fuses may have interrupting ratings of 10 000 A,

50 000 A, 100 000 A, or 200 000 A symmetrical rms.

5.12.2 Current Limitation. A current-limiting fuse is one which allows less than available current to flow into a fault for a relatively low ratio of available to rated current of the fuse. It is designed so that in the current-limiting range a high enough arc voltage is developed as a fusible element melts to prevent the current from reaching the magnitude it otherwise would reach. The action is so fast that the current does not reach peak value in the first half-cycle (see Fig 33).

The current-limiting action limits the total energy flowing into a fault and thus minimizes mechanical and thermal stresses in the elements of the faulted circuit.

5.12.3 NEC Categories of Fuses. The National Electrical Code (NEC), ANSI/NFPA 70-1981 [33] recognizes two principal categories of fuses, plug fuses and cartridge fuses. In addition, the NEC mentions the following fuses: time-delay fuses, current-limiting fuses, noncurrent-limiting fuses, fuses over 600 V, primary fuses.

(1) *Plug Fuses.* Plug fuses are rated 125 V and are available with current ratings up to 30 A. Their use is limited to circuits rated 125 V or less, and they are usually employed in circuits supplied from a system having a grounded neutral and no conductor in such circuits operating at more than 150 V to ground. The NEC, ANSI/NFPA 70-1981 [33], requires type S plug fuses in all new installations of plug fuses because they are tamper resistant. A nonremovable adapter which screws into a standard Edison screw base limits the size of the type S plug fuse which can be inserted.

(2) *Cartridge Fuses.* Table 43 shows cartridge fuses and fuse-holder case sizes according to current and voltage. All fuses recognized by the NEC, ANSI/NFPA 70-1981 [33] which have interrupting ratings exceeding 10 000 A must be marked on the fuse label with the designated interrupting rating. Fuses rated 10 000 A may be so designated.

5.12.4 UL Listing Requirements. The UL standard covering fuses requires the following for listing.

(1) Fuses must carry 110% of their rating continuously when installed in the test circuit specified in the standard.

(2) Fuses of 0—60 A rating must open

Fig 33
Current-Limiting Action of Fuses

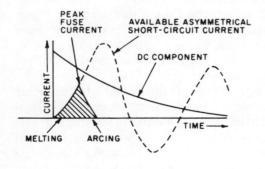
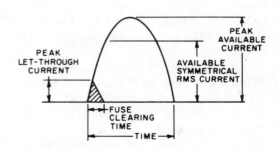

Table 43
Fuse Classification
(Cartridge fuses and fuseholders
shall be classified as listed here.)

Not Over 290 Volts	Not Over 300 Volts	Not Over 600 Volts
0—30	0—30	0—30
31—60	31—60	31—60
61—100	61—100	61—100
101—200	101—200	101—200
201—400	201—400	201—400
401—600	401—600	401—600
601—800	601—800	601—800
801—1200	801—1200	801—1200
1201—1600	1201—1600	1201—1600
1601—2000	1601—2000	1601—2000
2001—2500	2001—2500	2001—2500
2501—3000	2501—3000	2501—3000
3001—4000	3001—4000	3001—4000
4001—5000	4001—5000	4001—5000
5001—6000	5001—6000	5001—6000

NOTE: Fuses shall be permitted to be used for voltages at or below their voltage ratings.

within 1 h and fuses of 61—600 A rating within 2 h when carrying 135% of rating in the specified test circuit. Fuses rated above 600 A must open within 4 h when carrying 150% of rated current in the test circuit.

(3) Different current and voltage ratings of fuses must have specified physical dimensions which prevent interchangeability.

(4) Fuses listed as having an interrupting rating in excess of 10 000 A must have their interrupting rating shown on the fuse.

5.12.5 Fuses Carrying Class Letter. The UL (in conjunction with NEMA) has established standards for the classification of fuses by letter rather than by type. The class letter may designate interrupting rating, physical dimensions, degree of current limitation (maximum peak let-through current), and maximum clearing energy (ampere-squared seconds)

under specific test conditions, or combinations of these characteristics. The descriptions of these classes are as follows.

(1) *Class G Fuses, 0—60 A.* Class G fuses are miniature fuses rated 300 V, primarily developed for use on 480Y/277 V systems for connection phase to ground. These fuses are available in ratings up to 60 A and carry an interrupting rating of 100 000 A symmetrical rms. Case sizes for 15 A, 20 A, 30 A and 60 A are each of a different length. Fuse holders designed for a specific case size will reject a larger fuse.

Class G fuses are considered to be time-delay fuses according to UL if they have a minimum time delay of 12 s at 200% of their current rating.

(2) *Class H Fuses, 0—600 A.* Class H fuses have dimensions previously listed in the NEC. These fuses are often referred to as code fuses. Although these fuses are not marked with an interrupting rating, they are tested by UL on circuits which deliver 10 000 A alternating current and may be marked 10 000 A IC. They are rated 600 V or 250 V. The two fuses which are recognized as class H fuses are (a) one-time fuses (nonrenewable) and (b) renewable fuses.

The ordinary one-time cartridge fuse is the oldest type of cartridge fuse in common use today. It utilizes a zinc or copper link and has limited interrupting capabilities. The use of the one-time fuse is decreasing due to its limited interrupting rating and lack of intentional time-delay.

Renewable fuses are similar to one-time fuses, except that they can be taken apart after interrupting a circuit and the fusible element replaced. Renewal links are usually made of zinc. Their ends are clamped or bolted to the fuse terminals.

(3) *Class J Fuses, 0—600 A.* Class J

fuses have specific physical dimensions which are smaller than the 600 V class H fuses. Class H fuses cannot be installed in fuseholders designed for class J fuses. Class J fuses are current limiting and carry an interrupting rating of 200 000 A symmetrical rms. UL has also established maximum allowable limits for peak let-through current and let-through energy $I^2 t$ which are slightly less than those for class K1 fuses of the same current rating.

Time-delay standards have not been established for class J fuses; therefore none are listed by UL as time-delay fuses. Fuses having class J dimensions are available with varying degrees of time delay in the overload range.

(4) *Class K Fuses, 0—600 A.* Class K designates a specific degree of peak let-through current and maximum clearing $I^2 t$. Present class K fuses have the same dimensions of class H fuses, but have interrupting ratings higher than 10 000 A, that is, 50 000 A, 100 000 A, or 200 000 A symmetrical rms. To date UL has established three levels designated K1, K5, and K9, with class K1 having the greatest current-limiting ability and K9 the least.

Manufacturers of electrical equipment furnish equipment having a withstand rating based on a rejection class fuse. To be listed also as time-delay fuses, class K fuses are required by UL to have a minimum time delay of 10 s at 500% of rated current.

(5) *Class R Fuses, 0—600 A.* Class RF designates a class K fuse with a rejection feature on one end. All class RK fuses have a 200 000 A IC. Class K5 fuses become class RK5 and class K1 fuses become class RK1 fuses when rejection features are added.

(6) *Class L Fuses, 601—6000 A.* Class L fuses have specific physical dimensions

and bolt-type terminals. They are rated 600 V and carry an interrupting rating of 200 000 A symmetrical rms. Class L fuses are current limiting and UL has specified maximum values of peak let-through current and $I^2 t$ for each rating. Standards for time-delay characteristics, in the overload range, have not been established for class L fuses. Most available class L fuses have a minimum time delay of approximately 4 seconds at 500% of rated current. Class L fuses are not listed by UL as time-delay fuses.

(7) *Supplementary Fuses.* There are other fuses with special characteristics and dimensions designed for supplementary overcurrent protection, some of which conform to UL standards.

5.12.6 Cable Limiters (Protectors). Cable limiters are available on the market today for use in multiple-cable circuits to provide short-circuit protection for cables. Cable limiters are rated up to 600 V with interrupting ratings as high as 200 000 A symmetrical rms. They are rated according to cable size, that is, 4/0, 500 kcmil, etc, and have numerous types of terminations.

These limiters are designed to provide short-circuit protection for cables. They are used primarily in low-voltage networks or in service-entrance circuits where more than two cables per phase are brought into a switchboard. A typical one-line diagram representing a cable limiter installation is shown in Fig 34. (Note that for isolation of a faulted cable the limiters must be located at each end of each cable.) The limiter does not provide overload protection as described in ANSI/NFPA 70-1981 [33], Section 240. It does not have the characteristics associated with fuses, but will limit the extent of the fault while preserving service to the balance of the system.

178

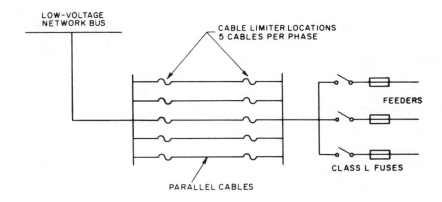

**Fig 34
Typical Circuit for Cable-Limiter Application**

5.13 Service Protectors. A service protector is a nonautomatic circuit-breaker-type switching and protective device with integral current-limiting fuse. Stored energy operation provides for manual or electrical closing. Switching under normal or abnormal current conditions up to at least twelve times continuous current ratings of the service protector is permissible. It is capable of closing and latching against fault currents up to 200 000 A symmetrical rms. During fault interruption, the service protector will withstand the stresses created by the let-through current of the fuses. Therefore, for all operating conditions, including normal load, overload, and fault switching up to the maximum interrupting capacity, this dual-protective device will adequately open the circuit.

Downstream equipment is subject only to the let-through current of the fuses. Protection against single phasing is included in the design of service protectors.

Service protectors are generally available at continuous current ratings of 800 A, 1200 A, 1600 A, 2000 A, 3000 A, 4000 A, 5000 A and 6000 A for use on 240 V and 480 V alternating current, in both two-pole and three-pole construction. They are used in both wall-mounted and free-standing compartments as well as in switchboards. Service protectors are often used with ground-fault protective equipment since their circuit-breaker type of construction gives total fault clearing time of under 3 Hz after shunt tripping by the ground-fault detector.

Manufacturers' catalogs should be consulted for complete ranges of equipment features and specific applications. ANSI/IEEE C37.29-1981 [23] and NEMA SG3-1975 [44] are applicable.

5.14 Enclosed Switches. Enclosed switches are switches with or without fuse holders, completely enclosed in metal, operable without opening the enclosure, and with provisions for padlocking in the OFF position. See NEMA KS1-1975 [38] and ANSI/UL 98-1979 [34].

Table 44
Switches Rated 100 HP and Less

Overload Current Requirements for AC Switches

Switch Rating (Horsepower)	No of Operations/FLA*
50 and less	10 at 10X
	40 at 6X
51 — 100	50 at 10X

*Full-load amperes of motor

5.14.1 NEMA Requirements. The NEMA requirements are as follows:

(1) *General-Duty (Type GD).* General-duty switches are available in 30 A—600 A ratings and are intended for light service where usual load conditions prevail in systems not exceeding 240 V alternating current and are for use with Class H fuses.

They are capable of interrupting 600% of full-load current at rated voltage a total of 50 times.

(2) **Heavy-Duty (Type HD).** Heavy-duty switches are available from 30 A—3600 A ratings and are intended for systems not exceeding 600 V alternating current. They may also be suitable for 600 V direct current. Various designs may accommodate Class H, J, R or L fuses and approved kits are available to convert these switches for use with different fuse types.

The ac interrupting ratings are based on their equivalent horsepower ratings and test requirements are as given in Tables 44 and 45.

All HD switches approved for dc motors must successfully interrupt 400% of the full-load current 50 times at rated voltage.

(1) Underwriters Laboratories (UL) has three switch categories:

(a) General use without a horsepower rating

(b) General use with a horsepower rating

(c) Fused motor-circuit type

(2) Underwriters Laboratories (UL) interrupting requirements are as follows:

(a) General use switches must operate 50 times at 150% of nominal current

(b) Horsepower rated switches have requirements similar to those of NEMA for ratings above 100 hp. Ratings less than 100 hp must interrupt load currents of approximately 160% nominal current 50 times

Table 45
Switches Rated Above 100 HP

Overload Current Requirements for AC Switches—10 Operations

Switch Rating (Horsepower)	240 V	480 V	600 V
125	1870	936	750
150	2160	1080	864
200	2880	1440	1150
250	3610	1810	1450
300	—	2170	1730
350	—	2480	2020
400	—	2860	2290
500	—	3540	2830

5.14.2 Application

(1) *Current.* Switches must have a current rating of at least 125% of the expected continuous load current

(2) *Frequency.* Unless otherwise noted all ac rated switches are approved for 60 Hz systems only

(3) *Temperature.* Both NEMA and UL stipulate a maximum temperature limit of 30 °C rise throughout the conductor path when operated without fuses and carrying rated current, except for switches for use with 400 A and 600 A Class J fuses and all switches used with Class L fuses which are permitted a maximum temperature rise of 60 °C when carrying 80% of their nameplate rating

(4) *Fused Switches.* Switches approved with fuses are short-circuit tested at various magnitudes of fault current to determine the switch's capability to either withstand let-through currents of the fuses or interrupt those current values which do not cause instantaneous fuse melting

Ground Fault. Switches approved as disconnects when used with ground fault detectors employing a solenoid to open the switch must be carefully coordinated with fuses to ensure that the switch operates only within its interrupting capacity.

Underwriters Laboratories (UL) approved switches with ratings from 30 A—1200 A for this application and are tested according to UL 1053-1976 [55] and are classified as follows:

(1) Class 1 service requires switches to be capable of interrupting at least 12 times their nameplate ratings

(2) Class 2 service requires provisions to prevent opening of the switch on fault currents exceeding the normal ratings mentioned previously in this section.

5.15 Bolted-Pressure Switches

5.15.1 Manual Operations. A bolted-contact switch consists of movable blades and stationary contacts with arcing contacts and a simple toggle mechanism for applying pressure to both the hinge and jaw contacts in a manner similar to a bolted bus joint when the switch is closed. The operating mechanism consists of a spring which is compressed by the operating handle and released at the end of the operating stroke to provide quick make and break switching action.

5.15.2 Electrical Trip. The electrical-trip bolted-contact switch is basically the same as the manually operated switch, except that a stored energy latch mechanism and a solenoid trip release are added to provide simple and economical automatic electrical opening. These switches are designed specifically for use with ground-fault protection equipment and have a contact interrupting rating of 12 times continuous rating. They are capable of tripping at 55% of normal voltage and the opening time is approximately 6 Hz.

Both manually-operated and electrical-trip switches are designed for use with class L current-limiting fuses. They are available in ratings of 800 A, 1200 A, 1600 A, 2000 A, 2500 A, 3000 A, and 4000 A, 480 V, alternating current, will carry 100% of rating, and are suitable for use on circuits having available fault currents of 200 000 A symmetrical rms. Both manually-operated and electrical-trip switches are available for switchboard mounting or in individual wall-mounted and free-standing enclosures.

Bolted contact switches are covered by UL 977-1980 [54], CSA Std C22.2 No 2-1980 [35] and NEMA KS 2-1976 [39]. Manufacturers' catalogs should be consulted for complete range of equipment features and application information.

5.16 Network Protectors. The network protector is a heavy-duty power air circuit breaker with special relaying designed to permit paralleling the outputs of a number of transformers, fed from different primary feeders, to a collector bus. Protectors are used in spot-network substations or secondary networks (see 4.8.4). The network protector serves to prevent backfeeding from the collector bus through the protector and through the transformer into the primary feeder. Such a backfeed could result from a fault in the high-voltage feeder, from another load on the primary line at a time when the line is disconnected from the utility power station, or even from the exciting current of the transformer when the utility feeder circuit breaker opens. When proper voltage is restored to a feeder, the network protector will close, permitting the re-energized feeder to accept its share of the load.

The network protector has no forward overcurrent protection other than fuses which are designed to open slowly under extremely heavy short-circuit currents. Originally the concept was to permit faults in network cables to burn themselves clear and to allow all overcurrent devices downstream plenty of time to operate. The modern approach is to install cable limiters, as described in 5.12.5, at each end of each cable to isolate cable faults. The network protector fuses are intended to operate only to remove a protector and transformer from the secondary bus in the event of a relay or protector trip mechanism malfunction. This is to prevent backfeeding a faulted primary feeder or a network transformer.

The network protector has two plug-in relays, the master and phasing relays, which trip the protector circuit breaker if the power flow is from the collector to a transformer, and reclose the circuit breaker when the transformer secondary voltage is slightly above and leading the collector bus voltage. The settings of these relays involve a 360-degree vector diagram (power in or out, lead or lag) as well as magnitude settings for differences of collector and transformer voltages. Although it may not always be desirable, these relays can be set to open the circuit breaker on reverse magnetizing current of the transformer. Adjustable desensitizing time relays may be used to avoid nuisance tripping.

The network protector is withdrawable and in certain ratings is available in a drawout design which is very like a drawout circuit breaker. Other ratings require internal disconnection by maintenance personnel to withdraw the circuit-breaker element.

An external handle can be used to lock the protector open, which is essential in preventing backfeeds during maintenance of the high-tension feeders. Protectors are available as wall, switchboard, or transformer mounted units bused directly from the transformer. Dustproof, dust-tight, drip-proof and submersible enclosures permit location in any available part of the building.

Network protector applications where nondedicated feeders opposed to network feeders must be used require special consideration which is beyond the scope of this standard.

Network protectors can be fitted with external control to trip and lock out in response to overcurrent, ground, or heat sensing relaying. Reclosing relaying of the protector must always be under the ultimate control of the protector master relay.

Occasionally, in a commercial building elevators or other loads capable of regenerating into the system could cause the protectors to open because of reverse power. A relay can be added which de-sensitizes the system to reverse currents of this type.

NEMA SG 3.1-1962 (1976) [45] covers the application and rating standards for the network protectors. Network protectors have essentially no overload capacity while the transformer associated with the protector has heavy overload capabilities. Therefore the protector will usually be rated on a current basis higher than the transformer full-load rating. Protectors are rated at 125 V, 240 V, 480 V, or 575 V with a maximum current of 5000 A.

5.17 Outdoor Structural Substations.

Outdoor structural-type substations are required for incoming voltages above 34.5 kV. They are also used at 23 kV and 34.5 kV, but because of the size of the equipment and spacing required between current-carrying parts, they may not be appropriate. The substation may be owned by either the electric utility or the customer, depending upon the type of contract executed for purchased power.

This type of substation may consist of

(1) Enclosure
(2) Structure
(3) Switching equipment
   (a) Circuit breakers
   (b) Outdoor distribution fuse cut-outs
   (c) Outdoor fuse combinations and group-operated switch
(4) Metering equipment
(5) Relaying and control
(6) Grounding equipment
(7) Lightning protection
(8) Power transformers

5.17.1 Enclosures. All outdoor substations should be enclosed with a suitable wall or fence. When a wire fence is used, it usually consists of woven mesh, 7 ft high, with several strands of barbed wire at the top. In addition to a small gate, which is used for normal traffic, a single or double large gate should be provided for handling of large equipment. All metal fences and gates should be adequately grounded. Enclosures and substation locations should be checked to see that normal use, adjacent stored material, or parked vehicles would not defeat the safety features of the fence, and, all gates must be provided with adequate locks which are easy to inspect.

5.17.2 Structure. Outdoor structures are usually of steel, aluminum, or wood. A comparison of the advantages or disadvantages of each type is given in Table 46.

5.17.3 Switching Equipment. The selection and application of circuit breakers and relays, fuses, and disconnect switches should always be discussed with electric utility engineers so that their specifications and requirements can be taken into account during the early design stages of a substation. The application of fuses and protective relaying requires careful study and understanding of the supply and distribution system. The services of equipment manufacturers or utility company representatives may be of assistance. The building electrical design engineer should take the lead and responsibility for the overall coordination of the protective system.

Before selecting switching equipment, the basic impulse withstand level (BIL) of the substation should be established. Equipment, insulators, etc, should be selected with a BIL sufficiently high so that the lightning arresters will provide protection. The BIL of equipment can be determined from manufacturers' literature. High-voltage switching equipment is needed to transfer loads, to iso-

Table 46
Comparison of High-Voltage Substation Structures

	Steel	Aluminum	Wood
Appearance	Good	Good	Poor
Rigidity	Good	Good	Poor
Life	Long	Long	Short
Needs foundations	Yes	Yes	No*
Adds to basic insulation level	No	No	Yes
Resistance to poor atmosphere	May be poor	May be poor	Usually good
Cost of maintenance	Good	Good	Poor
Cost	High	High	Low

*Cribbing at pole base(s) may be needed where (a) soil conditions require and (b) guys must be minimized.

late circuits for maintenance, and to remove faulty equipment from the power source. This equipment may be circuit breakers, load break switches, or fuses.

(1) *Circuit Breakers.* Circuit breakers for high-voltage service are normally the oil type. They have certain advantages as circuit-interrupting devices. They are convenient to operate and can be operated from a remote point. Service can be restored faster than by replacing a blown fuse. Their operation under fault conditions can be controlled more accurately and they are available in higher interrupting capacities.

In specifying circuit breakers, the following must be considered:

(a) Rated voltage

(b) Maximum operating voltage

(c) Minimum voltage for rated interrupting kilovoltamperes

(d) Impulse withstand voltage

(e) Frequency

(f) Continuous-current rating

(g) Momentary-current rating

(h) Four-second-current rating

(i) Three-phase interrupting symmetrical capacity

(j) Interrupting current rating at rated voltage

(k) Interrupting time in cycles

(l) Method of operating closing and tripping

(m) Number of poles

(n) Bushing-type current transformers required

For a tabulation of circuit-breaker ratings, see ANSI C37.06-1979 [1] and C37.6-1971 (R1976) [2].

(2) *Outdoor Distribution Fuse Cutouts.* These units are available in ratings up to 38 000 V and are used outdoors. Standard ratings are shown in ANSI C37.42-1969 (R1974) [7]. Fuse cutouts are normally used for disconnection and protection of small pole-mounted transformers. Most fuse cutouts drop open (drop-out feature) when the fuse blows. The fuse cutout is normally operated with an insulated hook stick which permits the operator to stand clear of the exhaust path when closing the fuse. This type of equipment will successfully interrupt magnetizing (no-load) current on small transformers. It should never be used to interrupt transformer load current except with a stick-operated portable load-break tool. A number of commercially available fuse cutouts are equipped with attachment hooks for such a tool.

(3) *Outdoor Group-Operated Switch and Fuse Combination.* This combination of a three-pole group-operated switch and fuses is widely used for disconnection and protection of medium-sized transformers for medium and high voltages. The conventional disconnect switch has very little current-interrupting capacity and when used to interrupt transformer magnetizing current or line-charging current, it should be equipped with arcing horns and should be mounted on top of the structure with no energized conductors placed above the switch. The arcing horns assist in interrupting small currents, and the arc should have room to blow freely upward. Another means of interrupting transformer magnetizing and line-charging current is the arc restrictor (buggy whip). This device whips out the arc when the switch is opened.

Some high-voltage outdoor switches are equipped with a nozzle which blows a blast of high-pressure air between the movable and stationary contact of the switch while the switch is being opened. This device is called an air blast attachment and aids in the interruption of currents with an air-break switch.

To provide greatly improved switching capability, interrupter switches are used instead of air-break disconnects. These devices interrupt transformer magnetizing current, load current, and relatively low levels of fault current without external arcing, and hence may be used in any position and on compact structures.

Vacuum load-break attachments are available which can be added to the air-break switch. With the vacuum load-break device, the switch can be depended upon to break load currents up to 1200 A.

The majority of these switches are

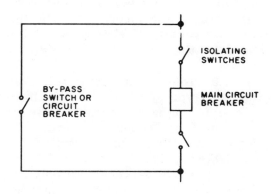

**Fig 35
Bypass of Circuit Breaker for
Maintenance Purposes**

manually operated with an operating mechanism accessible from ground level with the switches mounted on top of the outdoor structure. Motor-operated mechanisms are available which provide for operating interrupter switches or air-break disconnects from a battery or from a low-voltage alternating-current power supply. The motor-driven operators also permit automatic control of interlocked switches even from a remote location.

ANSI/IEEE C37.30-1971 [24] and NEMA SG 6-1974 (R1979) [48] are applicable.

Figure 35 shows a typical method of installing isolating switches and a by-pass switch or circuit breaker in conjunction with the main circuit breaker so that the main circuit breaker may be isolated for maintenance and testing purposes.

5.17.4 Metering Equipment. The general subject of service-entrance equipment is covered elsewhere. However, the metering current and potential transformers are usually supplied as a part of the substation and are located on either the high-voltage or the low-voltage buses, depending on whether power is to be purchased at the incoming or the distribution voltage. This decision is influenced by the terms of the utility contract.

5.17.5 Protective Relaying and Control. The type of relaying for the high-voltage incoming lines will usually be determined in conjunction with the electric utility and will depend on the method of supply.

When power is delivered over only one circuit, it is possible that relaying for the line will not be installed at the customer's end, protection being provided by the utility at the source end only. If the supply is from two or more lines, the protection depends on whether the lines are operated in parallel.

When the incoming lines are operated in parallel, current directional relays are usually installed to isolate a faulted feeder. This may also require the automatic closing of a tie circuit breaker between the line circuits in order to maintain service on both transformer banks.

If the incoming lines are not operated in parallel, then usually one circuit is considered the normal circuit and the second one the emergency or standby circuit. In case of a fault on the normal feeder, both ends must be opened before the load is transferred to the emergency feeder. This can be accomplished by the tripping of the circuit breakers both on the supply end and at the load end. When two circuits are operated in this manner, relaying is usually arranged to close the circuit breaker automatically on the emergency feeder with a loss of voltage on the normal feeder.

On most transformers only overcurrent protection is supplied, usually by means of circuit breakers with relays or with fuses. If the transformer has a capacity of perhaps 7500 kVA—10 000 kVA or

above, differential relay protection may be justified. This protects against internal faults of all magnitudes.

Transformers may also be protected against internal faults by means of pressure relays that operate with a rapid increase in gas pressure caused by an internal fault in the transformer.

High-voltage circuit breakers are normally controlled electrically, so the control devices may either be located locally at the circuit breaker or at a remote control point within the building.

5.17.6 Grounding

(1) *System Grounding.* The necessity or desirability for grounding of the supply circuits will be determined by the electric utility. Where grounding is indicated, the type of grounding is also the utility's responsibility. The decision concerning grounding of the medium- or low-voltage systems is the customer's responsibility, and the subject is discussed in Section 4.

(2) *Equipment and Structure Grounding.* Because of the possibility that a breakdown in insulation, fallen conductors, etc, may accidentally energize normally nonconducting portions of the substation, all structures, equipment enclosures, conduits, operating handles, fences, etc, should be effectively grounded. To keep the potential between the noncurrent-carrying parts and the earth to a minimum, the grounding conductors should have a current-carrying capacity to carry the maximum fault current that may flow; and the connection between the grounding electrode and the earth should have a resistance as low as possible. Refer to IEEE Std 142-1982 [29] and IEEE Std 80-1976 [36].

5.17.7 Lightning Protection. The insulation level of overhead lines is necessarily considerably higher than the insulation level of terminal apparatus such as transformers, switchgear, pot-heads, etc, which comprise the service entrance to buildings. Such overhead lines are vulnerable to overvoltage, principally from direct or induced lightning voltages and switching surges. These overvoltages can have values varying from several times the impulse and low-frequency withstand strength of the terminal apparatus down to very low values.

It is a fundamental characteristic of traveling voltage waves that they tend to increase in voltage when they arrive at equipment having a surge impedance higher than that of the incoming line. The magnitude of such incoming waves will approximately double at the terminals of a transformer or an open power circuit breaker. Because of this characteristic, equipment connected by cable to overhead circuits generally requires arrester protection at each end of the cable to guard against the possibility of transient overvoltages.

Protection against direct strokes is usually provided at outdoor substation installations in the form of grounded masts or overhead ground wires stretched above the installation to intercept lightning strokes which might otherwise terminate on the lines or apparatus.

In addition to direct-stroke protection it is essential that the entrance equipment such as transformers, circuit breakers, etc, be protected from traveling waves by the installation of lightning arresters having protective characteristics below the impulse insulation strength of the terminal apparatus. It is recommended that lightning arresters be installed as close as possible to the high-voltage terminals of the power transformer, and that other equipment requiring surge protection be grouped as close as possible to the arresters.

The station-type arrester provides by far the best protective level and the highest surge discharge ability and is usually recommended for important installations. The intermediate class is used for less important installations and for line protection where the cost of the station-type arrester would be unaffordable. NEMA LA 1-1976 [40], ANSI/IEEE C62.1-1981 [28], and ANSI C62.2-1969 [11].

5.18 Automatic Transfer Switches. Automatic transfer switches of double-throw construction are primarily used for emergency and standby power generation systems rated 600 V and less. These transfer switches do not normally incorporate overcurrent protection and are designed and applied in accordance with ANSI/NFPA 70-1981 [33], particularly Articles 230, 517, 700, 701 and 702. They are available in ratings from 30 A to 4000 A. For reliability, most automatic transfer switches rated above 100 A are mechanically held and are electrically operated from the power source to which the load is to be transferred.

An automatic transfer switch is usually located at the main or secondary distribution bus which feeds the branch circuits. Because of its location in the system, the abilities which must be designed into the transfer switch are unique and extensive as compared with the design requirements for other branch circuit devices. For example, special consideration should be given to the following characteristics of an automatic transfer device.

(1) Its ability to close against high inrush currents

(2) Its ability to carry full rated current continuously from the normal and emergency sources

(3) Its ability to withstand fault currents

(4) Its ability to interrupt at least six times the full load currents

(5) Additional electrical spacing and insulation as needed for two unsynchronized power sources

The arrangements shown in Fig 36 provide protection against failure of the utility service. In addition to utility failures, continuity of power to critical loads can also be disrupted by:

(1) An open circuit within the building area on the load side of the incoming service

(2) Overload or fault condition

(3) Electrical or mechanical failure of the electric power distribution system within the building

It therefore is desirable to locate transfer switches close to the load and have the operation of the transfer switches independent of overcurrent protection. Many engineers advocate the use of multiple transfer switches of lower current located near the load rating rather than one large transfer switch at the point of incoming service. A typical transfer scheme using multiple transfer switches is shown in Fig 36.

Automatic transfer switches operate rapidly with the total operating time usually less than 0.5 s, depending upon the rating of the transfer switch. Therefore, transferring motor loads may require special consideration in that the residual voltage of the motor may be out of phase with that of the power source to which the motor is being transferred. On transfer, this phase difference may cause serious damage to the motor, and excessive current drawn by the motor may trip the overcurrent protective device. Motor loads above 50 hp with relatively low load inertia in relation to torque requirements, such as pumps and compressors, may require special controls. Automatic transfer switches can be provided with accessory controls that disconnect motors prior to transfer and reconnect them after transfer when the residual voltage has been substantially reduced. Automatic transfer switches can also be provided with in-phase monitors that prevent retransfer to the normal source until both sources are synchronized.

Another approach is to use a three-position transfer switch with accessory controls that allow the switch to pause in the neutral position, while the residual voltage decays substantially, before completing the transfer.

Other accessories include time delays of 1 s to 3 s to ignore harmless momentary power dips and adjustable time delays of 2 min to 30 min on retransfer to allow the normal source voltage to stabilize before assuming the load. Furthermore, consideration should be given to the minimum voltage at which the load will operate satisfactorily to determine if the automatic transfer switch should be provided with close differential voltage protection. Additional accessories may include time delay on transfer to emergency, test switches, auxiliary contacts, lockout relays and switching neutral contacts as may be needed for ground-fault sensing.

5.19 Load Transfer Devices and Interlocks. This section is confined to commercial-building circuitry of low and medium voltages. These devices are for transfer of critical loads from normal (such as purchased power) to emergency (such as another incoming feeder or feeders, or a standby mechanically driven alternator or alternators) in the event of failure of the normal source. To ensure continuity of power at the points of utilization, one or more such devices

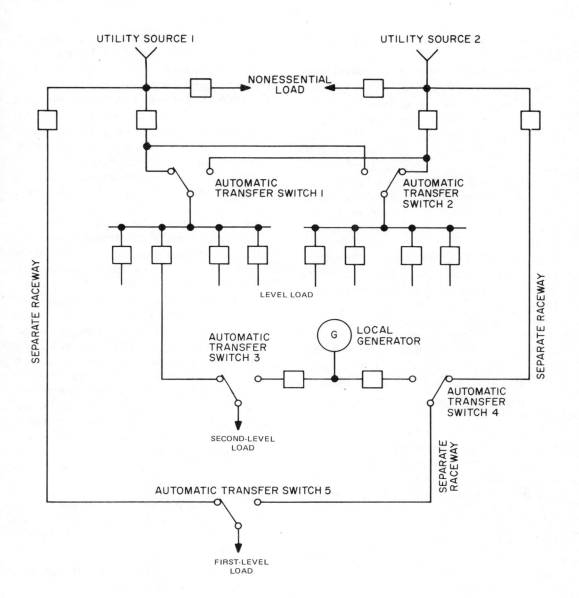

**Fig 36
Multiple Automatic Transfer Switches
Providing Varying Degrees of
Emergency Power**

should be considered according to the various reliability classifications of utilization equipment and their logical economical circuit groupings.

The loads to be transferred should be selected according to the importance of the reliability classification of the utilization equipment or groups of equipment involved, considering the duration and frequency of interruptions allowable. Certain configurations for various levels are shown in Fig 36. The degree of reliability requirements may be classified as follows:

(1) *Level 1.* Critical loads involving safety to life and property and where emergency power is legally required

(2) *Level 2.* Loads of less critical nature but where emergency or stand-by power is legally required

(3) *Level 3.* Loads where power outage may cause discomfort or damage to a product, process, or building facility

Examples of the above level classifications are as follows:

(1) Health-care facilities, many military systems, aeronautical safety, and certain communication systems

(2) Some transportation systems, critical controls for heating, ventilating, cooling pump water, and emergency lighting

(3) A second block of emergency lighting, computer systems, manufacturing process, residential and commercial farming facilities

The selection of the load transfer devices should be based upon the required reliability classification of the utilization equipment. It should be based upon an intimate knowledge of the various characteristics and limitations of the devices for the particular application, considering the characteristics of each utilization component to be served. The electrical engineer should give particular consideration to motor loads, to be sure that both the transfer device and the stand-by power source have enough capacity for the large, low-power-factor currents they may impose following the transfer. Critical loads should utilize automatic load transfer devices with adequate monitoring and control relays.

The required characteristics of the transfer devices should include the capability to successfully and repeatedly make and break the load currents at their various make and break power factors, carry rated current continuously when enclosed (with due regard to possible deterioration of contacts under arcing conditions) and withstand successfully through-fault currents while other circuit protective devices are clearing the faults. Also, emphasis should be placed on accessibility and ease of thorough inspection of contact elements after repeated subjections to such operations, load, and fault currents. The ability of the device to withstand repeated operations required for the application is a basic requirement.

Load transfer devices are available in the following forms:

(1) Automatic transfer switches available in ratings from 30 A to 4000 A to 600 V (Fig 17) and to 1200 A in medium-voltage class

(2) Automatic power circuit breakers consisting of two or more power circuit breakers which are mechanically or electrically interlocked, or both, rated 600 A to 3000 A, in both low- and medium-voltage classes.

(3) Manual transfer switches (600 V) available in current ratings from 30 A to 200 A

(4) Nonautomatic transfer switches available in ratings from 30 A to 4000 A,

to 600 V, manually controlled and electrically operated

(5) Manual or electrically operated bolted pressure switches (600 V) fusible or nonfusible available from 800 A to 6000 A

Circuit breakers may or may not require the energy from an electric storage battery for operation. While batteries make the transfer equipment independent of ac power, they require periodic maintenance. Extreme care should be exercised to assure that transfer control and operation do not in any way detract from either overcurrent protection or readily accessible disconnect means.

Magnetically operated transfer switches operate very rapidly because of their double-throw feature. Solenoid-operated circuit breakers also operate very quickly. Motor-operated circuit breakers are slower.

If power circuit breakers or molded-case circuit breakers serve both functions of load transfer and service entrance devices, they must include overcurrent protection and readily accessible disconnect means. Frequently remote manual and automatic trip means are justified for fire or other dangerous situations.

Medium-voltage load transfer devices generally are circuit breakers arranged for manual, manual-electric, or automatic-electric operation with mechanical or electrical interlocks, or both.

Automatic transfer of power sources may be bidirectional or unidirectional with manual reset from emergency to normal positions. The fully automatic bidirectional system should include a long time delay for the transfer back from emergency to normal sources to ensure recovery of the normal source to a stable situation.

A transfer device may include override time delay, or other means, to avoid transfer during a short-circuit condition in the transferable branch. In many systems a short-circuit on the utilization circuit to be transferred can be made to automatically lock out the transfer from normal to emergency source and from emergency to normal until reset by hand.

Loss-of-potential from the normal power supply should start the emergency alternator unit or units. Emergency manual or automatic fire shutdown circuits of all electric power in a building should lock out (with a manual reset feature) the automatic startup of the emergency unit and lock out the automatic transfer device system, except in special situations, such as fire pumps, where several degrees of reliability classifications may be applied.

Loss-of-potential alarm may be provided on an emergency utility supply to initiate action to restore the emergency supply. An alarm should be provided to indicate transfer operation so that action may be initiated to restore the normal supply.

A transfer switch operation causes a momentary outage on the equipment transferred, consequently the circuitry should be reviewed with regard to permitting automatic restart. In some instances, manual restart may be preferred. If the normal and emergency sources can be paralleled momentarily, the transfer switch may be equipped with a closed-transition mode so it can be operated without such an outage.

In order to properly ground the neutral of the service source and alternate sources, it may be necessary to switch the neutral along with the phase conductors. Switching of the neutral conductor may also simplify ground-fault sensing.

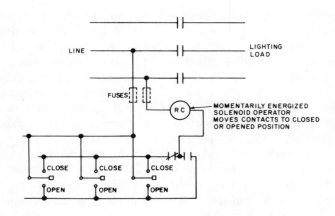

Fig 37
Mechanically Held, Electrically Operated Lighting
Contactor Controlled by Multiple Momentary
Toggle-Type Control Stations

5.20 Remote-Control Contactors

5.20.1 Remote-Control Lighting Contactors. Lighting contactors are used for controlling large blocks of lighting. They are generally used in sizes from 25 A to 225 A and are mounted in panelboards or separate enclosures. Standard control voltages are 120 V, 240 V, 277 V and 480 V. They make it possible to turn blocks of light fixtures on or off from convenient locations or from one central location. In addition to convenience of control, installation savings can be realized by reducing the length of power cable runs.

Lighting contactors are actuated electromagnetically and are either magnetically or mechanically held. Magnetically held lighting contactors are usually controlled by an on-off single-pole single-throw toggle switch and will drop open upon loss of control voltage.

Mechanically held lighting contactors will not change contact position upon drop or loss of control voltage. The operating coil is energized only during the opening or closing operation, thereby eliminating coil hum and coil power drain. In addition to time switches and photoelectric cell relays, a mechanically held lighting contactor can be controlled from any number of control stations, as shown in Fig 37.

Auxiliary relays may be used with lighting contactors to accommodate long runs between the lighting contactor and the control switch, for two-wire control, and for control by sensitive contact devices.

5.20.2 Remote-Control Switches for Power Loads. Remote-control switches provide convenient and accessible control of power circuits from any number of control stations. They are mechanically held and therefore will not change contact position upon loss of control voltage.

Remote-control switches are available in sizes from 30 A to 1200 A, suitable for 600 V alternating-current service and are designed primarily for inductive loads. They may be used for lighting or noninductive loads that exceed the capacities of smaller mechanically held lighting contactors. Standard control voltages are 120 V, 240 V, 277 V, and 480 V, alternating current.

The simplicity and reliability of these switches are mainly due to the unique operating mechanism. Without the use of hooks, latches, or semipermanent magnets, the contacts are positively locked in position. The solenoid coil in the operating mechanism is energized only during the instant of operation. Auxiliary contacts in the switch automatically disconnect the coil when the switch has operated, thus eliminating continuous energization of the operating coil. The same operating power is used to open or close the switches, and controlling stations do not break any load current.

Remote-control switches suitable for all classes of load are capable of carrying rated current continuously without contact deterioration or overheating. They are capable of closing against high inrush currents without contact welding or excessive contact erosion. They can interrupt locked-rotor motor currents or 600% overload at 0.40—0.50 power factor.

Remote-control switches are often installed in panelboards required to withstand fault currents in excess of 10 000 A. As a component of the panelboard, the remote-control switch must be capable of withstanding the magnetic stresses and thermal effects of the maximum available fault current.

Remote-control switches are used where disconnection of circuits is a matter of safety to life or property. Wherever electrical power is being distributed over a wide area, remote-control switches also provide economy and convenience. With their use, electrical layout can be designed without regard to the accessibility of the disconnect switches, thus simplifying the distribution system and making it more flexible for future expansion.

Distribution panels can be located to provide direct feeders and short branch circuits resulting in minimum line-voltage drops. Small conductors can be used for the control stations, and an unlimited number of stations can be used for each remote-control switch, providing additional convenience and economy. In addition to push-button control stations, remote-control switches can be operated by time switches, photo-electric cells, central control stations, break glass stations, and auxiliary relays.

5.21 References

[1] ANSI C37.06-1979, Preferred Ratings and Related Required Capabilities for AC High-Voltage Circuit Breakers Rated on a Symmetrical Current Basis (Consolidated edition[26]).

[2] ANSI C37.1971 (R 1976), Schedules of Preferred Ratings for AC High-Voltage Circuit Breakers Rated on a Total Current Basis.

[3] ANSI C37.7-1960 (R 1976) (reaffirmed with change in title), Interrupting Rating Factors for Reclosing Service for AC High-Voltage Circuit Breakers Rated on a Total Current Basis.

[4] ANSI C37.9-1953 (R 1976) (reaffirmed with change in title), Test Code for AC High-Voltage Circuit Breakers Rated on a Total Current Basis [includes supplement C37.9a-1970 (R 1971)].

[26] ANSI documents are available from The American National Standards Institute, 1430 Broadway, New York, NY 10018.

[5] ANSI C37.11-1975, Requirements for Electrical Control for AC High Voltage Circuit Breakers Rated on a Symmetrical Current Basis and a Total Current Basis.

[6] ANSI C37.12-1969 (R 1974), Guide Specifications for AC High-Voltage Circuit Breakers Rated on a Symmetrical Current Basis and a Total Current Basis.

[7] ANSI C37.42-1969 (R 1974), Specifications for Distribution Enclosed, Open, and Open-Link Cutouts.

[8] ANSI C37.46-1980, Specifications for Power Fuses and Fuse Disconnecting Switches.

[9] ANSI C57.12.22-1980, Requirements for Pad Mounted Compartmental-Type, Self-Cooled, Three-Phase Distribution Transformers with High-Voltage Bushings High-Voltage 34 500 Grd Y/19 920 V and Below; 2500 kVA and Smaller.

[10] ANSI C57.12.40-1975, Requirements for Secondary Network Transformers, Subway and Vault Types (Liquid Immersed).

[11] ANSI C62.2-1969, Guide for Application of Valve-Type Lightning Arresters for AC Systems.

[12] ANSI/IEEE C37.04-1979, Rating Structure for AC High-Voltage Circuit Breakers Rated on a Symmetrical Current Basis (Consolidated edition, Including supplements C37.04a, C37.04b, C37.04c and C37.04d).

[13] ANSI/IEEE C37.09-1979, Test Procedure for AC High-Voltage Circuit Breakers Rated on a Symmetrical Current Basis (Consolidated edition).

[14] ANSI/IEEE C37.010-1979, Application Guide for AC High-Voltage Circuit Breakers Rated on a Symmetrical Current Basis (Consolidated edition).

[15] ANSI/IEEE C37.011-1979, Application Guide for Transient Recovery Voltage for AC High-Voltage Circuit Breakers Rated on a Symmetrical Current Basis (Consolidated edition) (Revision of C37.0721-1971).

[16] ANSI/IEEE C37.012-1979, Application Guide for Capacitance Current Switching of AC High-Voltage Circuit Breakers Rated on a Symmetrical Current Basis (Consolidated edition) (Revision of C37.0731-1973).

[17] ANSI/IEEE C37.1-1979, Standard Definition, Specification and Analysis of Manual, Automatic, and Supervisory Station Control and Data Acquisition.

[18] ANSI/IEEE C37.2-1979, Standard Electrical Power System Device Function Numbers.

[19] ANSI/IEEE C37.4-1953 (R 1976), Definitions and Rating Structure for AC High-Voltage Circuit Breakers Rated on a Total Current Basis [includes supplements C37.4a-1958 (R 1966) and C37.4b-1970 (R 1971) (reaffirmed with change in title) and C37.4c-1980].

[20] ANSI/IEEE C37.5-1979, Guide for Calculation of Fault Currents for Application of AC High-Voltage Circuit Breakers Rated on a Total Current Basis.

[21] ANSI/IEEE C37.13-1981, Low-Voltage AC Power Circuit Breakers Used in Enclosures.

[22] ANSI/IEEE C37.20-1969 (R 1981), Switchgear Assemblies Including Metal-Enclosed Bus (Includes Supplements a, b, c, d). (Consolidated Edition 1974).

[23] ANSI/IEEE C37.29-1981, Standard for Low-Voltage AC Power Circuit Protectors Used in Enclosures.

[24] ANSI/IEEE C37.30-1971 (R 1977), Definitions and Requirements for High-Voltage Air Switches, Insulators, and Bus Supports.

[25] ANSI/IEEE C37.30a-1975, Supplement to C37.30-1971.

[26] ANSI/IEEE C37.100-1981, Definitions for Power Switchgear (Includes definitions from C37.03-1964).

[27] ANSI/IEEE C57.12.00-1980, General Requirements for Liquid-Immersed Distribution, Power, and Regulating Transformers.

[28] ANSI/IEEE C62.1-1981, Surge Arresters for Alternating-Current Power Circuits.

[29] IEEE Std 142-1982, Grounding of Industrial and Commercial Power Systems.

[30] ANSI/NEMA ICS 6-1978, Enclosures for Industrial Controls and Systems.

[31] ANSI/NEMA MG2-1977, Safety Standard for Construction and Guide for Selection, Installation and Use of Electric Motors and Generators.

[32] ANSI/NEMA ST 20-1972 (R 1978), Dry-Type Transformers for General Applications.

[33] ANSI/NFPA 70-1981, National Electrical Code.[27]

[34] ANSI/UL 98-1979, Safety Standard for Enclosed Switches.

[35] CSA Std C22.2, No 2-1980, Canadian Electrical Code, Part 2, *Safety Standards for Electrical Equipment*, Electric Signs.[28]

[36] IEEE Std 80-1976, Guide for Safety in AC Substation Grounding.

[37] NEMA AB1-1975, Molded-Case Circuit Breakers.[29]

[38] NEMA KS1-1975, Enclosed Switches.

[39] NEMA KS2-1976, Bolted-Pressure Contact Switches.

[40] NEMA LA1-1976, Surge Arresters.

[41] NEMA PB1-1977, Panelboards.

[42] NEMA PB2-1978, Deadfront Distribution Switchboards.

[43] NEMA SG2-1976, High-Voltage Fuses.

[44] NEMA SG3-1975, Low-Voltage Power Circuit Breakers.

[45] NEMA SG3.1-1962 (R 1976), Network Protectors.

[46] NEMA SG4-1975 (R 1980), Alternating-Current High-Voltage Circuit Breakers.

[47] NEMA SG5-1975, Power Switchgear Assemblies.

[48] NEMA SG6-1974 (R 1979), Power Switching Equipment.

[49] NEMA TR27-1965 (R1976), Commercial, Institutional and Industrial Dry-Type Transformers.

[27] The National Electrical Code is published by the National Fire Protection Association, Batterymarch Park, Quincy, MA 02269. Copies are available from the Sales department of American National Standards Institute, 1430 Broadway, New York, NY 10018.

[28] In the US, CSA Standards are available from the Sales department of American National Standards Institute, 1430 Broadway, New York, NY 10018. In Canada they are available at the Canadian Standards Association (Standards Sales), 178 Rexdale Blvd, Rexdale, Ontario, Canada M9W 1R3.

[29] NEMA publications are available from the National Electrical Manufacturers Association (NEMA), 2101 L. Street, NW, Washington, DC 20037.

[50] NEMA 201-1970 (R 1976), Primary Unit Substations.

[51] NEMA 210-1970 (R 1976), Secondary Unit Substations.

[52] UL67-1979, Electric Panelboards.[30]

[53] UL489-1980, Molded-Case Circuit Breakers and Circuit Breaker Enclosures.

[54] UL977-1980, Fused-Power Circuit Devices.

[55] UL1053-1976, Ground-Fault Sensing and Relaying Equipment.

------

[30] UL publications are available from Underwriters Laboratories, Inc , Publication Stock, 333 Pfingsten Road, Northbrook, Illinois 60062.

6. Controllers

6.1 General Discussion. Controls play an important and growing role in commercial buildings. They are used in heating, lighting, ventilation, air conditioning, elevators, etc. Most commercial buildings require some form of automatic or programmed control. Controls cover so many fields that it is nearly impossible to separate them from the discussions of the systems that they control. However, this section covers those controls primarily associated with motors.

Heating, ventilating, air conditioning, refrigeration, pumping, elevators, and conveyors require the use of motors. They can be operated manually or automatically to respond and perform the function for which they are intended. Furthermore, protection must be afforded the motor and the electric supply system. A motor controller causes the motor to respond to a signal from a pilot device and provides the required protection.

Most integral horsepower motors used in commercial buildings are of the squirrel-cage design and are powered from three-phase alternating-current low-voltage distribution systems. Controllers to be applied on distribution systems up to 600 V are given horsepower and current ratings by the National Electrical Manufacturers' Association (NEMA). These ratings range from 2 hp for Size 00 to 1600 hp for Size 9, based on 480 V system. See ANSI/NEMA ICS6-1978 [5][31]. Unless special provisions are made to interrupt higher current, standard controllers are tested for interruption of current equal to ten times the full-load current of their maximum horsepower rating.

Medium-voltage starters are standardized from 2500 V to 7200 V. The interrupting rating is standardized for unfused Class E1 controllers from 25 MVA to 75 MVA and for fused Class E2 controllers for 160 MVA to 570 MVA. For special applications voltage between 600 V and 2500 V are utilized, for example, 830 V for pump panels, 1050 V for mining equipment, and 1500 V for special pumps. Standards for controllers between 600 V and 1000 V are in preparation.

Controllers exist for special purposes expecially in the air-conditioning and heating industry. Other special control-

---

[31] The numbers in brackets correspond to those in the references at the end of this Section.

lers include lighting contactors, transfer switches, etc.

6.2 Starting. The primary function of a motor controller is starting, stopping, and protecting the motor to which it is connected.

A magnetically operated contactor connects the motor to the power source to perform this function. This contactor is designed for a large number of repetitive operations in contrast with the typical circuit-breaker application. Energizing its operating coil with a small amount of control power causes it to close its contacts, connecting each line of the motor to the power supply. If the controller is to be the reversing type, two contactors are used to connect the motor with the necessary phase relation for the desired shaft rotation.

Full-voltage starting of the motor requires only that the contactor connect the motor terminals directly to the distribution system. Starting a squirrel-cage motor from standstill by connecting it directly across the line may allow inrush currents of approximately 500 to 600% of rated current at a lagging power factor of 35 to 50%. The inrush current of motors rated 5 hp and below usually exceed 600% of the rated current. Small motors, for example, 0.5 hp may have inrush currents of 10 times full load motor current. Newer energy efficient motors may even draw higher currents. For applications such as ventilating fans or small pumps, this type of starting is not objectionable. As a result, most of these controllers are full-voltage types. However, some applications such as large compressors for air-conditioning and pumping installations may require motors as large as several thousand horsepower. For many of the larger motors, the starting inrush current may be great enough to cause voltage dips which may affect the building's lighting system.

Electric utilities also have restrictions on starting currents, so that voltage fluctuations can be held to prescribed limits. Before applying large motors, starting limitations should be checked with the utility. Some type of starting which limits the current may be necessary. Some couplings or driven equipment have limitations on torque which may be safely applied. Such maximum torque limits may require reduced voltage starting.

Many kinds of reduced voltage starters are in common use. Figures 38, 39 and 40 show the principles of the most common reduced voltage starters for squirrel-cage motors. In addition, the contactor sequence and control diagrams show the speed versus torque and voltage versus current characteristics.

6.2.1 Part-Winding Starters. Part-winding starting of motors reduces inrush current drawn from the line to about 60% of locked rotor current and reduces torque to about 45% of full-voltage starting torque. This type of starting requires connecting part of the winding to the supply lines for the first step and connecting the balance in an additional step to complete the acceleration. Although special motors can be designed with any division of winding that is practicable, the usual motor used with part-winding starting has two equal windings.

The total starting time should be set for about 2 to 4 s. Due to the severe torque dip during the transfer, the transition time should be short and at approximately half speed. The branch circuit protection is usually set at 200% of each winding current. Part-winding starters are comparatively low cost but are only used for light starting loads such as high-speed fans or compressors with relief or unloading valves.

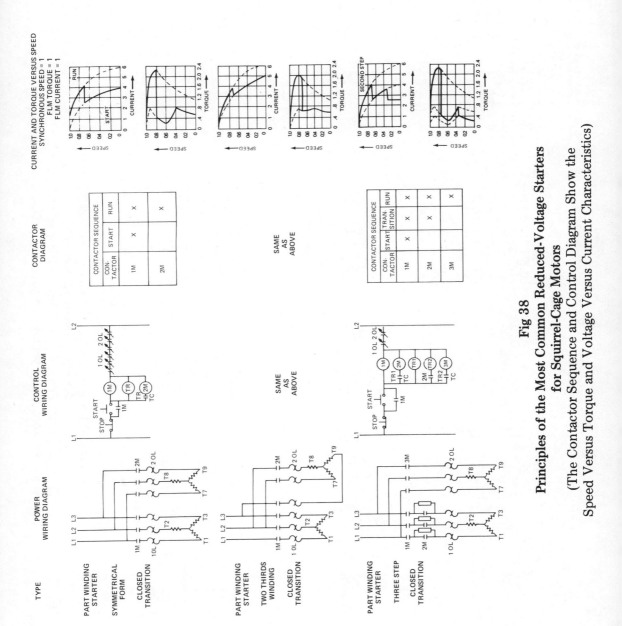

**Fig 38
Principles of the Most Common Reduced-Voltage Starters
for Squirrel-Cage Motors
(The Contactor Sequence and Control Diagram Show the
Speed Versus Torque and Voltage Versus Current Characteristics)**

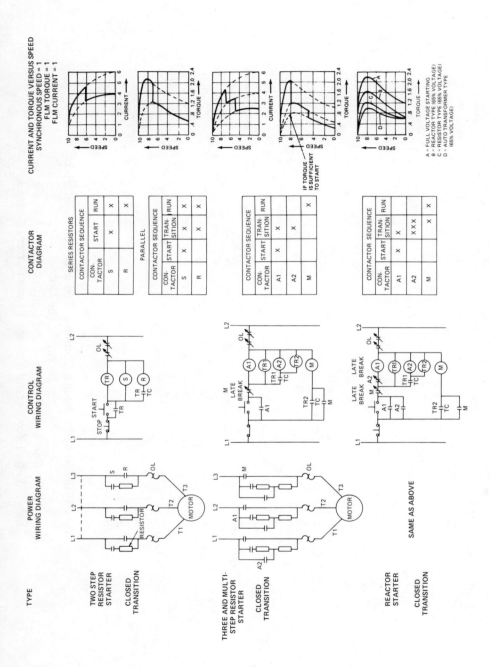

Fig 39
Principles of the Most Common Reduced-Voltage Starters
for Squirrel-Cage Motors
(The Contactor Sequence and Control Diagram Show the
Speed Versus Torque and Voltage Versus Current Characteristics)

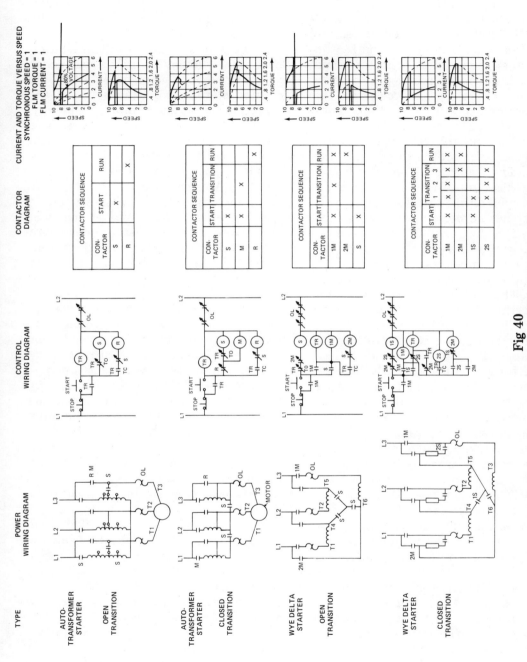

Fig 40

**Principles of the Most Common Reduced-Voltage Starters
for Squirrel-Cage Motors**

(The Contactor Sequence and Control Diagram Show the
Speed Versus Torque and Voltage Versus Current Characteristics)

6.2.2 Resistor or Reactor Starters. The simplest reduced voltage starting is obtained through a primary reactor or resistor. The voltage impressed across the motor terminals is then reduced by the drop across the reactor, or resistor, and the inrush current is reduced in proportion. When the motor has accelerated for a predetermined interval, a timer initiates the closing of a second contactor to short the primary resistor, or reactor, and connect the motor to the full line voltage. The transition from starting to running is smooth since the motor is not disconnected during this transition.

Also, the impressed voltage on the motor is a function of the speed at which it is running. Since the current decreases as the motor accelerates, this decreases the drop across the resistor or reactor. The starting torque of the motor is a function of the square of the applied voltage. Therefore, if the initial voltage is reduced to 50%, the starting torque of the motor will be 25% of its full-voltage starting torque. If the drive has high inertia, such as centrifugal air-conditioning compressors, a compromise must be made between the starting torque necessary to start the compressor in a reasonable time and the inrush current which may be drawn from the system.

Resistor and reactor type reduced-voltage starters provide closed transition and can be used with standard motors. The resistors are usually selected for 5 s on and 75 s off. Other conditions require specially selected resistors. A three-step resistor type starter usually does not start rotating the motor until the end of the first step. At the second step, the starting torque is 45 to 50% of normal starting torque. The time setting is also usually between 3 and 4 s. The branch-circuit protection is the same as for full-voltage starters. This is also true where reactor type reduced-voltage starters are used. They are difficult to adjust however, and generally are only used for larger high-voltage motors.

Reactor type reduced-voltage starters have somewhat better torque speed characteristics than resistor type starters, but resistor starters are less expensive and therefore used much more frequently. Resistor type starters have the disadvantage that the wattage dissipation during startup can be costly for large motors which are started frequently.

6.2.3 Autotransformer Starter. An autotransformer starter has characteristics which are similar, but at the same time more efficient than the resistor-reactor starter. Since an autotransformer controller reduces the voltage by transformation, the starting torque of the motor will vary directly as the line current, even though the motor current is reduced directly with the voltage impressed on the motor. The formula generally used for the starting current drawn from the line with an autotransformer is: the product of the motor locked-rotor current in amperes at full voltage times the square of the fraction of the autotransformer tap, plus one-fourth of the full-load current of the motor. Based on this formula, a motor with 100 A full-load current and 600 A locked-rotor current, when started on the 50% tap of an autotransformer, would only draw 175 A inrush current from the line. This is in contrast to the 300 A drawn from the line in a reactor or primary resistor type starter. If the voltage is reduced to 25% on starting, the torques will be identical on the reactor, primary resistor, and autotransformer starters.

However, on the autotransformer starter, the torque of the motor does not increase with acceleration but remains essentially constant until the transfer is made from starting to running voltage. Also, with an autotransformer type starter using a five-pole start contactor and a three-pole run contactor, the motor is momentarily disconnected from the line on transfer from the start to the run connection. This open transition may result in some voltage disturbance.

To overcome the objection of the open-circuit transition, a circuit know as the Korndorfer connection is in common use. This type of controller requires a two-pole and a three-pole start contactor instead of the five-pole. The two-pole contactor opens first on the transition from start to run, opening the connections to the neutral of the autotransformer. The windings of the transformer are then momentarily used as series reactors during the transfer. This allows a closed circuit transition without losing the advantages of the autotransformer type of starter. Although it is somewhat more complicated, this type of starter is frequently used on high-inertia centrifugal compressors to obtain the advantages of low line current surges and closed-circuit transition. Standard motors can be used with autotransformer starters. The time setting should be 3 to 4 s and 4 to 5 s respectively for open and closed transition autotransformer starters.

6.2.4 Wye-Delta (Y—Δ) Starters. Contactors 1M and 2M as shown in the above circuits for wye-delta starters carry 58% of the motor load; whereas contactors 1S and 2S carry $33\frac{1}{3}$% of the motor load. The NEMA rating of a wye-delta starter is higher than that of a full-voltage starter having the same contactor. In closed transition, contactor 2S is usually one size smaller than 1S. An overload relay is included in each phase, and set at 58% of the full-load motor current. The time setting should be somewhat longer than for part-winding starters; that is,

3 to 4 s on open transition and 3 to 5 s on closed transition autotransformer and Y—Δ starters.

The branch circuit protection has to be selected very carefully for open transition starters. The magnetic trip unit should not trip below 15 times full-load motor current or even higher to avoid tripping on the severe current peak at the transition. The current peak is especially high on autotransformer starters, but it could also be 13 to 14 times full-load motor current on open transition Y—Δ starters. On closed transition Y—Δ and autotransformer starters, the standard branch circuit protective device is selected in the same manner as for full-voltage starters. Autotransformer starters are mostly used in the US for ventilators, conveyors, machine tools, pumps, and compressors without relief valves. Wye-delta starters are extensively utilized in Europe, and in the US, particularly for large air-conditioning units.

6.2.5 Series Parallel Starters. Series parallel controllers are available which initially connect the two windings of each phase in series in a conventional wye arrangement. Since this is maximum impedance, the inrush current is about 25% of full-voltage locked-rotor current and the torque is 25% of maximum starting torque. The second step removes one winding from each phase and allows the motor to run on the other winding, the same as the first step of a part-winding controller. The third and final step connects the balance of the winding to the supply lines to effect the parallel connections for normal operation.

6.2.6 Solid-State Starters. Solid-state reduced-voltage starters provide a smooth, stepless method of acceleration for standard squirrel-cage motors. Three methods of acceleration are available:

(1) Constant current acceleration

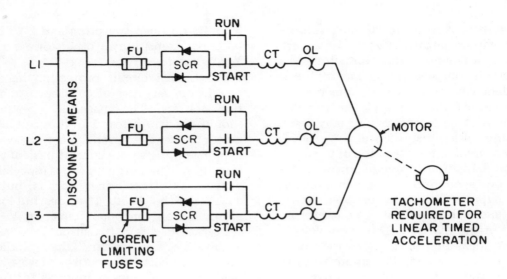

Fig 41
Simplified Wiring Diagram of a
Solid-State Controller

where the motor is accelerated to full speed at a field selectable preset current level

(2) Current ramp acceleration where the voltage is gradually increased to provide smooth stepless acceleration under varying loads

(3) Linear timed acceleration where the motor is accelerated at a linear rate which is field adjustable.

A tachometer feedback circuit is required for the latter type of acceleration. A solid-state control circuit provides control for the silicon-controlled rectifiers which are used to provide the variable voltage to the motor. A schematic diagram of the power circuit is shown in Fig 41. Contactors are often used in the power circuit to provide isolation between the motor and the load.

A typical enclosed solid-state reduced-voltage starter with fusible disconnect is shown in Fig 42. Solid-state starters are particularly suitable for applications which require extremely fast or a large number of operations, or both, (several million under load). In addition to starting motors, solid-state controllers are also used for speed control of ac motors, further discussed in 6.14.

6.2.7 Cost Comparison. Figure 43 shows a relative cost comparison of some of the more commonly used reduced-voltage starters. Though not shown, cost of solid-state controllers varies considerably depending on rating and features. They are generally more expensive than electromagnetic controllers.

Wound-rotor or slip-ring motors may be used on large installations which justify their cost. The primary control for these motors is the same as for full-voltage squirrel-cage motor starters and the secondary controller inserts resistance in the three phases of the rotor for starting. At standstill, a motor of this type is essentially a transformer and therefore, resistance in the secondary is the equivalent of a high-impedance load on a transformer. Current drawn from the line is limited by the amount of resistance inserted in the secondary.

6.3 Protection. Protection of motor branch circuits is divided between protection on running overload and short-circuit protection. Running overloads are overloads up to locked-rotor current which is usually six times, sometimes eight to ten times full-load motor current.

Since 1972, NEMA standards have recommended three overload relays (one per phase) for running overload protection. Most overload relays have thermal elements which are heated by interchangeable heaters in series with the motor or fed through current transformers. The ratings of these heaters are determined by the full-load motor current. Most European and some US relays do not have interchangeable heaters but have bimetals which are heated either by the current itself or by heat conduction. The movement of the bimetal can be adjusted to utilize the relay for a certain current range (usually 1:1.5). In the US, these relays are mainly used for protection of special motors (that is, hermetically-sealed motors) which require an extremely fast-acting relay.

NEMA standards divide overload relays into classes; Class 30, 20, and 10. The class is defined by the maximum time in seconds in which the relay must function on six times its ultimate trip current ($\leqq 1.25$ full-load motor current for motors having a service factor of 1.15, and $\leqq 1.15$ full-load motor current for motors having a service factor of 1.0).

All thermal responsive elements have an inverse time characteristic. This

**Fig 42
Typical Solid-State Controller with
Fusible Disconnect**

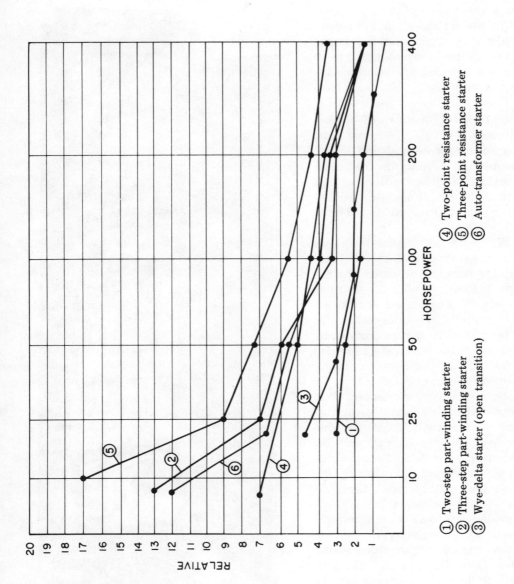

Fig 43
Reduced-Voltage Starters

means that for small overcurrents, considerable time elapses before tripping occurs. However, at high overcurrent (locked-rotor current) tripping occurs in a shorter time. Class 20 overload relays are used for protection of T frame motors and Class 30 for the older U frame motors. No minimum trip time is standard. Overload relays must have sufficient therma capacity to allow the motor to start.

Ambient compensated and noncompensated overload relays are available. Ambient compensated relays should be selected if the motor is in a controlled environment, and the starter is in a non-controlled environment (for example, submersible pumps). In other applications it is usually not advisable to compensate for ambient temperature at the starter because the protected device (motor, cable) is not ambient compensated.

Short-circuit protection for the total motor branch circuit (cable, starter, motor, disconnect device) is provided by the short-circuit protective device (SCPD). This device can be a fuse or a circuit breaker. Table 47 lists the most important short-circuit protective devices, their selection and their protection range to a motor branch circuit.

In many cases, the crossover point between the SCPD and the overload relay characteristic is beyond its limit of self protection of the relay, also often beyond the interrupting capacity of the contactor. However, the SCPD is scheduled to clear the fault while limiting damage to the other components and preventing a safety hazard.

The wire is protected by the overload relay in motor branch circuits on small overloads since an oversized fuse or thermomagnetic circuit breaker does not protect the wire under these conditions. On fault currents, the SCPD with the lowest crossover point between the overload relay and SPCD characteristic is the best fault protection. On high-current faults, any device approved by UL for the particular available short-circuit current is permissible.

In many cases, motors have inherent protectors which are placed in the winding of the motor. These protectors are sensitive to the temperature itself, to the rate of rise of the temperature, or to a combination of rate of rise and current. Usually more than one sensor should be placed in the motor. They are especially advantageous if a motor is used for intermittent duty. Since these devices are built into the motor, they must be furnished with the motor. Such protectors do not protect the motor branch circuit in case of a fault on the line side of the motor.

If severe undervoltage occurs on the distribution system, the motor controllers will normally disconnect the motor from the line. If the motor is under the control of a single contact pilot device such as a thermostat or pressure switch, it may be allowed to restart when the normal voltage is restored. This type of control is called under-voltage release. If there is a large number of motors on the feeder, the simultaneous restarting of all of these motors on return of the voltage could draw an unacceptably large inrush current from the line. Motors under the manual control of an operator are generally not allowed to restart until the operator pushes the button to energize the controller for each individual motor. Control which remains de-energized until actively restarted is referred to as under-voltage protection. On some occasions it may be desirable to measure the duration of the voltage dip and if the under-voltage lasts less than some predetermined time, the motor is not discon-

Table 47
The Most Important Short-Circuit Protective Devices;
Their Selection and Their Protection Range to a Motor Branch Circuit

Kind of Short-Circuit Protective Device	Rating or Setting (NEC)	Protection at Low-Current Faults 6 to 13 Times FLMC	Protection at Medium-Current Faults 12 to 30 Times FLMC	Protection at High-Current Faults 30 Times FLMC
Single element Class H fuses	200 to 400%[5] FLMC	Overload relay[1]	Protection questionable[2]	Up to 5 kA Sizes 1–2 Up to 10 kA Sizes 3–4
Single element Class J fuses	200 to 400%[5] FLMC	Overload relay[1]	Protection questionable[2]	Available up to 100 kA (480 V)[3,6]
Time delay RK5 fuses	150% to 225%[5] FLMC	Overload relay[1]	Protection questionable	Available up to 100 kA (480 V)[3]
Instantaneous trip type circuit-breaker	700 to 1300% FLMC	Overload relay[1]	SCPC Protects	Available up to 22 kA (480 V)
Instantaneous trip type circuit-breaker current limiting or with limiter	700 to 1300% FLMC	Overload relay[1]	SCPD Protects	Available up to 100 kA (480 V)
Motor short-circuit protector	700 to 1300% FLMC Crosses OLR Characteristic	Overload relay[1]	SCPD Protects	Available up to 100 kA (600 V)
Thermal-magnetic circuit-breaker standard type	200 to 250%[5] FLMC	Overload relay[1]	Protection questionable[4]	Available up to 10 kA May be available at higher IC
Thermal-magnetic circuit-breaker, current limiting or with limiter or current limiting fuse	200 to 250%[5] FLMC	Overload relay[1]	Protection questionable[4]	May be available for high IC

FLMC: Full-load motor current
OLR: Overload relay
CB: Circuit breaker
IC: Inrush current

SCPC: Short-circuit protective device
MSCP: Motor short-circuit protector
CL: Current limiting

NOTES: (1) NEMA Standards require that overload relays must be self-protected up to 10 times FLMC but many starters are self-protected up to at least 13 times FLMC (13 times FLMC is the maximum permissible trip setting of instantaneous trip-type circuit breakers and the maximum current at which an MSCP becomes faster than the overload relay).

(2) In comparing the time current characteristics according to ANSI/NFPA 70-1981 [2] of properly selected fuses or thermomagnetic circuit breakers with the characteristics of an overload relay, it can be seen that these curves do not intersect in a sharp angle. This presents a problem in coordination. For example, if a starter has to interrupt more than 10 times its rating and an overload relay trips before the SCPD acts at more than 13 times its rating, neither the overload relay may withstand (until it trips) nor the contractor may be able to interrupt the short-circuit current. (See Fig 44.)

(Footnotes for Table 47 continued on page 209)

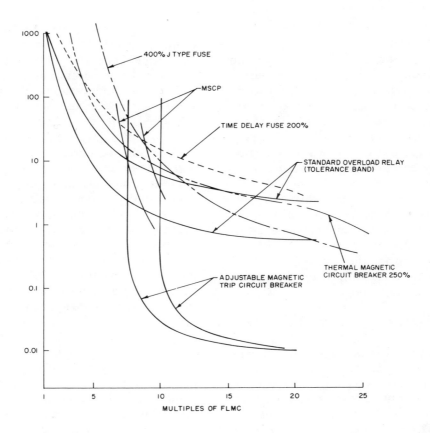

Fig 44
Coordination of the Overload Relay with Dual Element
Fuses, MSCPs, Thermomagnetic and Adjustable Magnetic
Breakers in the Low Overcurrent Range.

Footnotes for Table 47 *(Continued)*

(3) Some manufacturers have UL listed combination starters up to 100 kA with RK5 fuses at 225% FLMC rating or J type fuses with 400% FLMC rating.

(4) Some manufacturers have successfully tested a combination starter having a circuit breaker whose normal interrupting rating is 14 kA, at 480 V to safely interrupt a 22 kA fault at the load side of the starter. Such a combination, if placed in one enclosure, is recognized for 480 V, 22 kA by NEMA and UL.

(5) ANSI/NFPA 70-1981 [2] allows use of fuses and thermomagnetic breakers up to the indicated ratings to avoid nuisance tripping due to motor inrush current.

(6) The usage of RK5 non-time-delay single element fuses up to 400% is permitted but is rarely used for more than 10 kA available fault current.

nected. This feature is called time delay undervoltage protection.

When motors are transferred from one source to another, that is, from an emergency generator to the utility, both the motor controller and motor are often momentarily de-energized. If the utility voltage is out of phase with motor residual voltage, the motor will often draw excessive current and generate high transient torques which may cause nuisance tripping of breakers and possibly cause damage to the motor or its load. To overcome this problem automatic transfer switches often include inphase monitors that prevent transfer until the residual voltage of the motor and the utility voltage are nearly synchronized, or delay transfer until the residual voltage has decayed to a safe level.

Many other special protection means for motors exist. Small voltage imbalances can cause very high currents in the rotor circuit. The condition is even more severe if the motor runs on one phase. After a three-phase motor is stopped, it usually cannot restart if one phase is interrupted. This could be especially dangerous for elevators. Phase-failure relays and single-phase protection can be provided. In addition, ground-fault relays are sometimes used for disconnecting large motors in case of a ground fault.

In recent years, solid-state overload relays have been developed in which motor damage curves are more closely matched. They lend themselves to special motor branch circuit protection applications and will be discussed later.

For applications with extremely long starting time, the overload relay may be bypassed during the starting period.

6.4 Special Features. Many features and additional components can be added to motor controllers to make the motors and drives perform particular functions or to provide additional protection to the motors and systems. Unloading devices on compressors are frequently interlocked with the motor control to assure that the motor will not attempt to start if the compressor is loaded, and will unload the compressor if the load exceeds the available motor torque. Vanes and dampers in air systems, and valves in liquid systems, may be similarly interlocked with the motor controller. If the building requires conveying systems to handle materials, elaborate interlocking between motor controllers can assure proper sequencing of the starting and stopping of the conveyor drives.

Mechanically held contactors may be added to perform additional functions, such as motor feeder disconnect, and controlling other associated loads. The contacts are power driven, that is, for example by a single-solenoid mechanism, into either the open or closed position and positively locked so loss of control voltage cannot cause them to change position.

Combinations of switches, relays, and contactors may be used to automatically or manually alternate operation of multiple pump motors or operate them simultaneously to equalize the running time on each and still provide sufficient capacity for maximum load conditions.

6.5 Control Systems. There are different arrangements available for control systems in commercial buildings. The physical configuration varies depending upon the complexity of the system as indicated below.

6.5.1 Panelboard Type Constructions. The fusible disconnect or circuit breaker for each branch circuit is placed in one enclosure. The handle is either attached to the door or to the disconnect device and the door cannot be opened if the disconnect device is closed. The handle can be padlocked in the OFF position and the disconnect device cannot be closed if the door is open. The starter and, if required, the control transformer, pushbuttons, and pilot devices are often placed in another enclosure on the panel. The two enclosures are interlocked so that the starter is only accessible if the disconnect device is in the OFF position. The disconnect device is usually connected to the bus-bar system of a switch type panelboard and the two enclosures are connected by conduit or cable. Instead of separate enclosures, a unitized combination starter in one enclosure is also used and connected directly to the bus-bar system.

6.5.2 Separate Enclosures. The disconnect devices may be put in separate enclosures and the starters, including pilot devices etc, in the other enclosure. These devices are connected by conduit or cable. The starter and the specific type of disconnect device should be tested and approved as a combination. Each device must be approved by itself except if the disconnect device is a fusible switch and the available short-circuit current is 5000 A or less for Size 1 and 2 starters or 10 000 A for Size 3 or 4 starters.

6.5.3 Combination Starters. The disconnect device, pilot device, starter, and control transformer are all placed in one enclosure and mounted on the wall or machine. Usually the handle is attached to the disconnect device. The handle interlock is the same as in 6.5.1. Each combination starter must be approved for the available short-circuit current. More recently unitized combination starters (combination of disconnect device, starter, handle, pilot devices in one unit) have

been introduced. All protective devices and combination reversing, reduced voltage, and multispeed starters are available. Combination starters capable of withstanding up to 100 kA of short-circuit current capacity at 480 V are available.

6.5.4 Simplified-Control Centers. Some manufacturers furnish standard combination starters in a modular construction, thus, width and length are multiples of a basic dimension. These enclosures are placed in front of a steel construction and wired to a bus system that is usually on the bottom or top of the steel structure. The height of the structure including the bus system is 90 in.

6.5.5 Complex Panels. Complex panels are often placed in special large enclosures. Sometimes these panels have group protection which means that one disconnect device is used for several motors of one machine. Such panels are especially suitable for special equipment such as programmable controllers, relays, electronic devices, timers, resistors, etc. This construction is frequently used for control of large machines or production lines rather than in commercial buildings.

6.5.6 Motor-Control Centers. Motor-control centers are preferred for applications involving central control of multiple motors. Applicable standards include UL845-1980 [6] and ANSI/NEMA ICS 2-322-1978 [4]. The motor-control center consists of a number of basic vertical structures. Each vertical structure has a vertical bus system connected to a horizontal bus system. The horizontal bus system is either behind, on the top, or on the bottom of the vertical bus system. The total height of the motor control center is 90 in. Each vertical section has a number of basic units consisting of combination or unitized combination starters with or without control transformers, combination reversing or multi-speed reduced-voltage starters, etc. These units are prefabricated and can be plugged into the vertical bus structure. This structure is braced for withstanding high-fault currents (42 kA and in exceptional cases 100 kA). Combination starters larger than Size 4 are usually bolted to the vertical bus structure. All compartments are multiples of a basic dimension (usually 3 in or 6 in). The total width of each vertical section is not standardized though most manufacturers use a basic width of 20 in with wider sections available.

Motor-control centers generally consist of a factory fabricated structural metal frame that houses the buses and the various (controllers) and their auxiliaries. Larger motor-control centers are shipped to the job site in *shipping sections*, thus requiring the structural sections to be field bolted and the buses to be connected. In order to ensure proper alignment of the structural frame and the buses, it may be necessary to install levelling channels at the front and the rear of the motor-control centers. These steel channels are placed so the webs are essentially flush with the surface of the floor and the legs embedded in the concrete. Thus, the channels serve as a level base for sections of the motor-control centers that would not be available from the usual concrete floor.

Incoming and outgoing conduits require special consideration when specifying or selecting motor-control centers. For example:

(1) When space and aesthetics permit, conduits may be run horizontally on a ceiling (or other elevated structure) above the motor-control centers. The conduits are then elbowed or bent downward to their point(s) of entry into the top of the motor-control centers. Re-movable top plates are usually provided for convenient conduit entrances.

(2) In some instances conduits may be run horizontally and directly into an upper section of the motor-control centers. In that case it may be necessary to include a box-like metal structure above, and attached to, the standard motor-control centers. Here again, plates are required to permit convenient conduit entry.

(3) In some cases conduits are embedded in floors. Bending the conduit from the embedded points into the motor-control centers may be impractical and a waterproofed trench below the motor-control centers may be necessary. Conduits enter the trench and the conductors are formed (or bent) to proper position.

Metering is often installed in motor-control centers. Because no two motor-control motors are identical, there are no standards for the number or types of meters. However, consideration should be given to voltmeters and respective voltmeter switches, ammeters for each motor starter or one ammeter, in the main bus, or both, kilowatt meters, power-factor meters, and running-time meters.

In addition to the above, when preparing specifications for motor-control centers consideration should be given to the following:

(1) Provide identification labels for each motor starter indicating the motor or device controlled; devices, such as meters, cirucit breakers and fuses; and each compartment, enclosing devices or operational parts.

(2) Provide additional space for anticipated growth or maintenance, or both.

(3) Provide spare parts, such as entire starters, circuit breakers, or parts there-

of that are subject to periodic replace-ments.

(4) Provide spare fuses when fuses are specified as protective devices. At least one complete set of properly identified fuses should be included. Fur-ther, in the case of cartridge type fuses, a suitable fuse puller may be desirable.

(5) Provide lockout tags/padlocks to ensure safe disconnection and service of motors and other equipment con-trolled by the motor starters. Highly visible tags and sometimes padlocks are necessary to prevent closure of the dis-connect device while personnel are servic-ing the motors or equipment.

(6) Enclosure types should be speci-fied depending upon location and serv-ice. The National Electrical Manufac-turers Association (NEMA) has des-ignated various types of structural enclosures, for example: NEMA 1 for indoor (essentially dust free and non-hazardous areas), NEMA 12 for areas where dust collects. See ANSI/NEMA ICS6-1978 [5].

(7) Buses should be specified as either copper or aluminum, depend-ing upon the most advantageous bus for the particular purpose. The proper conductor terminating or connecting devices, or both, should also be specified.

The NEMA designations of classes and types of motor-control centers are listed in Table 48.

A control center can be located on each floor of a building to accommodate the motors on that floor, or all motor controllers may be grouped in a central location. The control center is self-sup-porting and may have units mounted on the front and back. Others may have units mounted only on the front. The motor-control center may be mounted on the wall. Control centers which are approximately 20 in or less in depth are usually more stable if a building is ex-posed to earthquakes.

The short-circuit capability of a motor-control center is determined by both the structure and the individual controllers.

Table 48
NEMA Control Center Classification

Type	Class I	Class II
A	No terminal boards Connection diagrams for unit starters	
B	Terminal boards mounted on unit including power on size 3 and smaller and all control Connection diagrams for unit starters	Same as class 1, type B, with interwiring between starters and sections Connection diagram of complete control assembly
C	Master Section terminal boards including power for size 3 and smaller and all control No wiring between starters sec-tions, or master terminal boards Connection diagrams for unit starters and master terminals	Same as class 1, type C, with interwiring between starters, sections, or master terminal boards Connection diagrams of complex control assembly

The lowest component capability is the short-circuit rating of the entire center. Under a bolted fault condition, UL and NEMA allow specified damage to the unit to which the fault was applied as long as:

(1) The fault current has been interrupted

(2) A dielectric test on the line side of the unit is passed

(3) The operating handle can open the unit door

(4) The line connections are undamaged

(5) The door is not blown open

The short-circuit rating must be equal to or greater than the available fault current on the line terminals of the motor-control center including the motor contributions. If current limiting means (reactors, current-limiting circuit breakers or fuses) are used, only the components ahead of these current-limiting means must be capable of withstanding the available short-circuit current. For further information see Section 9.

6.6 Low-Voltage Controllers and Starters. The most common controller is the across-the-line magnetic type starter. An electromagnet, energized by either the line voltage or a lower voltage from a control transformer, closes the contacts of the contactor. The control voltage is usually 120 V though other voltages between 24 V and 600 V are used. Low voltage has the disadvantage of needing larger control wires due to the increase of the control circuit current. Also, at voltages of 24 V and below, continuity may be a problem. Therefore, at low-control voltage, it is often advisable to use two parallel contacts at the auxiliary device or sliding type contacts to secure continuity. On the other hand, high-control voltage requires greater insulation integrity and more precaution for the safety of personnel. Therefore, the most common control voltage is 120 V.

An overload relay is placed on the load side of the contactor having overload protection in each phase. Overload relays may be nonambient or ambient compensated. Some overload relays have separate indicating contacts to which a light or other alarm can be wired to indicate tripping. Overload relays are trip free; most of them must be reset manually, but some are available for automatic reset. The latter feature cannot be used if the starter is energized by a switch which closes the control circuit permanently since reset would, in effect, constitute a restart.

The starters and contactors in Table 49 are standardized by NEMA.

In addition, there are three classes of overload relays standardized by NEMA.

Table 49
NEMA Standardized Starters and Contactors

Size	00	0	1	2	3	4	5	6	7	8	9
Rated Current Closed A	9	18	27	45	90	135	270	540	810	1215	2250
Rated Current Open A	10	20	30	50	100	150	300	600	900	1350	2500
hp at 480 V		2	5	10	25	50	100	200	400	600	900	1600
hp at 240 V		1.5	3	7.5	15	30	50	100	200	300	450	800

Overload relays must trip at 1.25 times full-load motor current and in 10 (Class 10), 20 (Class 20), or 30 (Class 30) seconds or faster at 6 times ultimate trip current. Motors having a service factor 1 should be protected with an overload which trips ultimately at 115% full-load motor current.

In addition to these NEMA rated starters, special purpose controllers exist for definite purposes such as heating equipment, air conditioning, lighting, capacitors, switching, etc. They are mostly current rather than horsepower rated and must fulfill special test requirements. For example, a contactor for air conditions must be able to interrupt for 6000 times the following: 6 times rated current at 240 V; 5 times at 480 V; 4 times at 600 V. There are also manually operated controllers used mostly up to Size 1 maximum. They are similar to circuit breakers but usually cannot interrupt the short-circuit current. They must have a much longer life than circuit breakers because they must switch rated motor currents more frequently.

6.7 Multispeed Controllers.
Magnetic motor starters can be connected to multispeed motors to obtain different motor velocities.

(1) The windings of each phase can be connected in two different ways so that the stator winding has half the number of poles (twice the speed) in one position compared to the other. The poles can be further connected in different ways to obtain:

(a) Equal power at the two speeds (double parallel star connection at low speed; series delta at high speed).

(b) Equal torque (delta at low speed; double parallel star at high speed).

(c) Variable torque (double torque at double speed; series wye at low speed; double parallel wye at high speed). Attention must be paid so that a high torque and high overcurrent do not occur if the motor winding is switched suddenly from high speed to low speed. Therefore, the high-speed winding must first be disconnected.

(2) If the required speed ratio is different than 2:1, two speed motors must have two separate windings.

(3) By combining (1) and (2), motors having 3 and 4 speeds can be developed. Controllers are available for these motors.

(4) Pole Amplitude Modification (PAM) motors are available for arbitrarily selected speed ratios (for example, 3.2). Conventional two-speed motor controllers are used with these motors.

6.8 Fire-Pump Control.
Fire-pump controllers are frequently installed in commercial buildings to control the operation of fire pumps to maintain water pressure in standpipes and sprinkler systems, under the heavy water demand of a fire.

Fire pump controllers differ from the usual combination motor controller in that there is only limited overcurrent protection built into the controller, generally providing only locked-rotor overcurrent protection for the fire-pump motor. Further, short-circuit protection is usually provided with circuit breakers or integrally-fused circuit breakers, but where fuses are acceptable, they must be capable of carrying locked-rotor current continuously. The standard to which fire-pump controllers are generally required to conform is ANSI/NFPA 20-1978 [1], Chapter 7.

These controllers may provide either manual operation (nonautomatic) or combined manual and automatic operation. The manual controller starts and stops the fire pump by means of a start-stop station on the controller. Provisions are available for remote start stations but remote stop stations are not permitted. In addition, an emergency mechanical operator, externally operable, is provided to close and hold in the contactor mechanically. The combined manual and automatic controller, in addition to the means herein described, starts and stops the fire pump from a pressure switch within the controller, starting when the water pressure at the fire pump discharge header drops to the predetermined setting and stopping when the pressure rises to the higher predetermined setting, but only after a minimum running time. Various starting means such as across-the-line, reduced voltage, part winding, and Y—Δ are available.

These controllers are generally wired to the power source near the service entrance or in some cases are provided with a separate service. ANSI/NFPA 20-1978 [1] mandates that fire-pump controllers have a withstand rating at least equal to the maximum short-circuit current which can flow to the controller.

6.9 Medium-Voltage Starters and Controls. It may be practical to operate larger motors (100 hp and above) in commercial buildings at medium motor voltages of 2300 V, 4000 V, 6600 V and in exceptional cases to 13 800 V. The resulting lower currents result in less line disturbance upon starting, and since the motors are not on the same distribution network as the lighting and other low-voltage devices, any reflected voltage drop due to their starting will be minimal.

Controllers for medium-voltage motors are divided into two NEMA classifications. Class E1 controllers employ their contacts for both starting and stopping the motor and interrupting short circuits

or faults exceeding operating overloads. Class E2 controllers employ their contacts for starting and stopping the motor and employ fuses for interrupting short circuits or faults exceeding operating overloads. Class E1 controllers must interrupt up to 50 MVA fault currents and Class E2 use medium voltage Class R fuses for protection.

These controllers function similarly to low-voltage controllers though they are quite different in mechanical design. They are available as full voltage or reduced voltage, induction or synchronous-motor starters, and as mutiple-speed starters. The contactors may be air-break, oil-immersed type or vacuum type. For motors operating at 13.8 kV, power air or vacuum circuit breakers are usually the switching devices.

A disconnecting switch or multiple disconnecting type fuses can be used to provide motor-disconnecting means. Control circuits are usually 120 V and supplied through control transformers. The extent of relaying can vary greatly, depending upon the degree and type of protection required. Compact fused disconnect interlocked drawout starters with built-in protective relaying are available for motors rated up to 5000 hp. Medium-voltage controllers must withstand high basic impulse levels (BIL) for example, 60 kV for 5 kV controllers, since they are often installed close to the service entrance point. However, standards allow the controllers to be designed to an appreciable lower BIL level if surge arresters are installed.

6.10 Synchronous-Motor Starters. To start a synchronous motor it must be brought up to synchronous speed, or nearly so, with the direct-current field de-energized, and at or near synchronism the field must be energized to pull the motor into step. A small induction motor may be mounted on the shaft of the synchronous motor for bringing it up to speed. The induction motor must have fewer poles than the synchronous motor, so that it may reach the required speed. If the exciter which supplies the field is mounted on the motor shaft, it may be used as a direct-current motor for starting, provided that a separate direct-current supply is available to energize it. However, since most synchronous motors are polyphase and are provided with a damper winding, the common practice is to start them as squirrel-cage induction motors, the torque being supplied by the induced current in the damper winding. Like squirrel-cage motors, they may be connected directly to the line or started on reduced voltage. When they are started on reduced voltage from an autotransformer, the usual practice is to close the starting contactor first, connecting the stator to the reduced voltage, then, at a speed near synchronism, to open the starting contactor and close the running contactor, connecting the stator to full line voltage. A short time later the field contactor is closed, connecting the field to its supply lines. The field may be energized before the running contactor has closed, which will result in a little less line disturbance, but the pull-in torque will be lower.

Instead of an autotransformer to supply the reduced voltage, any of the methods for starting squirrel-cage motors may be used. These include starting resistance in the stator circuit, starting reactance in the stator circuit, and combinations of reactance and autotransformer. The Korndorfer system of autotransformer connection is also possible.

6.11 Direct-Current Motor Controls. Direct-current motors are started by either full-voltage or reduced-voltage starting methods. Generally full-voltage starting is limited to motors of 2 hp or less because of very high starting currents. Reduced-voltage starting is accomplished by inserting a resistance in series with the armature winding. As counter EMF builds up in the armature, the external starting resistance can be gradually reduced and then removed as the motor comes up to speed, either by a current relay or by timers in steps. All resistance should be removed from the circuit as soon as the motor reaches full speed. Motor characteristics and the resistors are different for series and for shunt motors. Speed control of direct-current motors can be accomplished by varying resistance in the shunt or series fields or in the armature circuit. Reversing is accomplished by reversing the flow of current through either the armature or the field.

6.12 Pilot Devices. There are manual and automatic pilot devices which initiate control of motors.

6.12.1 Manually Operated Devices. These are pushbuttons (ON—OFF) with normally open and normally closed contacts; selector switches having two, three, or four positions; or master switches (mostly cam switches). Selector switches open and close the coil circuit of the contactor and often connect to an automatic switch (see 6.12.2) in a third position. Selector switches having spring return keep a certain operational mode only as long as the switch is held in a given position. Most pushbuttons can change the state of the circuit only as long as they are *pushed*, though there are other pushbuttons (infrequently used) that have maintained contacts. A contactor coil can hold itself in after the ON pushbutton closes a *normally open* contact by providing an auxiliary contact *(normally open)* for the starter which closes a circuit parallel to the *normally open* contact of the pushbutton.

Various pushbuttons and indicating lights are included in pushbutton stations and control panels.

There are three standardized pushbutton lines: standard duty (only for limited varieties) heavy duty; and the most universal line, *oil tight.*

There are two lines of standardized oil-tight pushbuttons. In each line, the diameter of the hole in the cover is identical for all operators. The oil-tight pushbutton has all possible variations available.

The most popular line of oil-tight pushbuttons has a 30 mm diameter mounting hole, whereas the smaller line has a 22.5 mm diameter hole. There are a great number of operators available such as mushroom-head buttons, key operated switches, various handles, push-to-test pilot lights, pilot lights for full voltage, or with transformers on low voltage, etc. In addition to switches, many variations are available for foot operation.

Some typical pushbutton elementary circuit connections for alternating-current full-voltage starters are shown in Fig 45.

6.12.2 Automatically Operated Devices. A great number of automatic pilot devices are available to switch a coil circuit depending on the actuating medium.

(1) Conventional limit switches convert a mechanical motion into an electrical control signal. The moving object comes into direct contact with the limit switch actuator. These limit switches have various actuators depending on what kind of movement controls the state of the limit switch contacts. Very often limit switches must have an extremely long life and must be oil tight. They have normally open and normally closed contacts.

(2) Proximity limit switches are becoming more popular. They operate when an object approaches a sensor. The advantage is that physical contact with the object is not necessary; therefore, more accuracy and longer life are possible. The system can be triggered by changing the magnetic field, especially if the approaching object has an iron contact or section. Another method is to change the inductance, capacitance or resistance, when the object approaches the sensor. Photoelectric means are also used.

(3) Float switches are a type of limit switch actuated by the level of a liquid.

(4) Pressure switches respond to pressure changes of a gas or liquid. If the maximum pressure is 500 lb/in^2 and the medium is not harmful to a diaphragm, the switches are diaphragm operated. If the required pressure approaches 2500 lbs/in^2, the pressure switches are usually bellows operated. If the maximum pressure is still higher, piston-type pressure switches are used.

(5) Sail switches are air-pressure actuated devices inserted in air ducts. For example, consider a motor driven fan providing essential flow of air within a duct. Should the motor continue to run, though the fan is mechanically disconnected from the motor by a broken fan belt, the sail switch would sense the change in air flow and provide a signal to actuate an alarm and stop the motor.

(6) Temperature switches use a sensor or bulb to sense the temperature of the surrounding medium. The change in pressure in the sensing element actuates the contacts.

(7) If the control circuit is more complex and requires logic, many possibilities exist to develop a logic diagram.

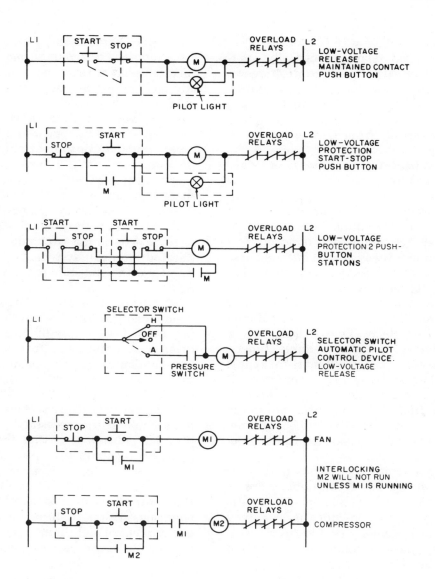

**Fig 45
Typical Push-Button Control Circuits**

(a) The simplest way is by using electro-mechanical relays. Electrically operated, electrically held control relays (small contactors) have 4, 8, and up to 12 normally open or normally closed contacts. These contacts are very often field convertible from normally open to normally closed.

(b) Mechanically held relays or relays which need only be momentarily energized to change position. They have up to 12 normally open or normally closed contacts, and maintain their position even after control power is removed.

(c) Timing Relays or Timers. These are relays having delayed contacts. Timers are available with adjustable delay time from a few cycles to three minutes and more. They are sometimes field-convertible from delay ON to delay OFF.

In addition to these simple timers, there exist, for example, timers for repeat cycling, which repeat a certain ON-OFF pattern (interval timer). These timers begin a timing period after an initial switch is actuated. After a selected time, the output switch is restored to its initial position.

Timers operate on various principles, such as motor driven, dashpot, thermal-solid-state, or pneumatic. Pneumatic timers generally do not have a very high repeat accuracy ($\pm$ 15% to $\pm$ 20%), but they have a moderately long life (approximately 1 million operations), and are low in cost. The actuating time element can be varied by a needle valve or the length of a groove (linear timer). Solid-state timers have higher accuracy and longer life.

(7) Complex systems having a large number of logic elements frequently use solid-state elements.

(a) Solid-state relays are usually similar to solid-state contactors but have only pilot duty rating.

(b) Hard-wired solid-state logic systems require a different wiring technique than is used for relay logic. A relay usually has one input (coil) and several outputs, (normally open, normally closed, with or without time delay contacts) whereas, solid-state logic can have several inputs and one output or combinations thereof.

The basic elements obtained by transistor logic are AND, OR, NOR, NAND, Memory, Time Delay, Retention Memory, etc. The advantage is that solid-state devices do not wear out. Very complicated control systems will become simpler with solid-state devices than with relays.

6.12.3 Programmable Controllers. Programmable controllers are microcomputer-based solid-state devices which are often programmed in a format similar to the familiar relay-logic *ladder diagram*. These controllers utilize digital logic consisting of input and output interface, the central processor, the memory, the program, and the necessary power supply. The advantage is that they can be easily reprogrammed in case of a change of the sequence. Further, many functions which are difficult to obtain with standard relays can be added, such as counting, arithmetic functions, and various sequential and timing functions. Programmable controllers are generally less expensive than relays if the logic is complex. To energize valves, contactors and starters, etc, small solid-state output switches are generally used for interfacing.

The central processing unit checks by scanning perforated tapes or switch positions or input/output (I/O) cards and the control plan which is stored in memory.

There are Read Only Memories (ROM), Read and Write Memories, and Random Access Memories (RAM). Certain types of ROMs may be changed by ultraviolet light or other electronic means. They are programmable and called Programmable Read Only Memories (PROM). Read and Write Memories are usually RAMs which can be programmed and changed. However, unless backed up by a battery supply, they will lose their state if power is lost. The complexity of the programs determines the size of memory required.

The most commonly used central processing unit today is the microprocessor. With sufficient memory, the new programmable controllers are actually small computers which play an important role in energy management of commercial buildings.

Besides relay logic *ladder diagrams*, (still the most common program language), Boolean algebra, and other logic formats (AND, OR, etc) and various high-level computer languages can be used.

The program can be studied and checked on computer terminals utilizing cathode ray tubes. Trouble shooting is made easier with a diagnostic device to check the status of the controller elements. In addition, most PCs have self diagnostic routines to detect errors.

Electromechanical control devices or tranducers can also be used as input sources for programmable controllers. Tranducers measure properties and convert these properties (pressure, temperature, speed, power, etc) to an electrical output which is applied to the input of the controller.

The input and output devices usually work on 120 V alternating current, but supply a low-level output (approximately 5 V dc to the controller. Tranducers are usually interchangeable between manufacturers in that they have standard outputs. Printed circuit (PC) cards have indicated lights to show if they are in an ON or OFF status. Each output circuit can have a separate fuse.

6.13 Speed Control of DC Motors. The shunt motor is excited by a field with a constant or adjustable voltage. In a series motor both armature and field are in series, and therefore have identical currents. The compound motor contains both a shunt and series field. The speed of the motor is proportional to the counter EMF (CEMF) and inversely proportional to field strength. The torque is proportional to the field strength and the armature current. Thus, the speed of the motor can be regulated by either changing the armature current or field current. Both currents can be controlled by voltage adjustment which can be accomplished for the dc motor by resistance or solid-state voltage controls. Since $CEMF = V - I_a R_a$,

where

$CEMF$ = counter generated voltage
V = applied armature voltage
I_a = armature current
R_a = armature resistance

it can be seen that the motor speed will always vary to provide a CEMF to match this equation.

The speed of the shunt motor can be increased by changing the field strength, which means adding resistance to the field circuit. Braking of shunt motors can be obtained by switching a resistor into the isolated armature circuit causing the motor to act as a generator, or by applying current in the opposite direction through various control means. The speed of series motors can be regulated by adding resistors in the armature and field circuit or as is most often done, by adding resistors in parallel to the armature to direct current.

Numerous circuits have been utilized for dc motor control. Motor-generator (Ward-Leonard) control found widespread usage until a few years ago. The output voltage of a shunt-connected generator which applies power to the drive motor is adjusted by field control, and thereby changes the motor speed. Today, speed of dc motors is mainly adjusted by phase control of silicon controlled rectifier converters (SCR) that operate directly off the ac input lines. The SCR is a controlled-rectifier type PNPN semi-conductor with a gate on the second positive layer (P). After being triggered, the SCR stays in the ON state until the main circuit is either interrupted or reverse voltage is applied.

In order to keep the ripple to a minimum, full-wave bridge type rectification with 2 SCRs, and 2 diodes is used for unidirectional armature control for single phase ac. For three-phase ac, 6 SCRs or 3 SCRs and 3 diodes are used. In 3-phase controls, triggering occurs by utilizing three independent trigger circuits which are powered from the secondary windings of one three-phase transformer in Y, thus producing the same delay on all three phases. The dc voltage is dependent on $\cos x$ where x is the delay angle to trigger the SCR. If $x = 180°$, the voltage is zero; if $x = 0$ the output voltage is maximum. The motor can be reversed in different ways:

(1) By reversing the polarity of the field windings, using either electromagnetic contactors or solid-state devices.

(2) By reversing the polarity of the armature voltage with contactors.

(3) By reversing the polarity of the armature voltage using two SCR converters, one connected to produce reversed voltage polarity.

(4) By reversing the generator field voltage in a Ward-Leonard system.

6.14 Speed Control of AC Motors. The simplest speed control is possible with wound-rotor motors. Since the speed of a wound-rotor motor at a given output torque depends on the secondary resistance, an adjustable resistance in the rotor circuit provides adjustable shaft speed. Rotor resistance is usually controlled by contactors or by a water resistor. While this is a simple way to control ac motor speed, its disadvantage is the waste of power in the secondary resistor at reduced speed.

Solid-state power-electronic controllers are coming into increasing use for speed (and directional) control of ac motors. They are used with synchronous and squirrel-cage induction motors, and are rapidly replacing wound-rotor type, drives because they have much better low-speed efficiency. In one type, the ac-dc-ac or *dc link* control, an SCR converter turns 1-phase or 3-phase ac into adjustable-voltage dc, which is then reconverted into adjustable-frequency and adjustable-voltage ac by an SCR inverter. The frequency controls the speed of the motor, and the voltage is controlled to maintain a constant ratio of voltage to frequency (as required to prevent overexcitation of the motor). Both braking and reversal are accomplished by reversing phase rotation of the output ac. Controls ac-dc-ac are used with both induction and synchronous motors, and can provide very wide speed ranges including speeds above the synchronous speed of the motor at 60 Hz.

The other popular type of adjustable-frequency controller is the cycloconverter, in which an adjustable-frequency and voltage ac output for the motor is synthesized directly from the input ac waveforms, with no intervening dc step. Cycloconverter drives are more economical than dc-link drives when very low speeds (approximately 10% of 60 Hz synchronous speed) are needed.

6.15 Power System Harmonics from Adjustable-Speed Motor Controls. Both dc and ac adjustable-speed drives using solid-state techniques (SCR converters, inverters, or cycloconverters) have non-sinusoidal, square-edged ac input current waveforms. These currents may be considered to contain harmonic-frequency components (that is, current components at multiples of power frequency) which propagate through the power system feeding the drive. The effects are usually harmless, but can be troublesome. For instance, a harmonic component can excite a resonant condition between a power-factor-correcting capacitor bank and the inductance of the power system, causing damaging overvoltages to appear at or near the capacitors. In the rare cases when such harmonics problems occur, they can readily be eliminated by such simple methods as changing a capacitor bank rating. Electrical consultants and electrical equipment suppliers can provide valuable advice on the prevention or cure of harmonics problems.

6.16 References

[1] ANSI/NFPA 20-1980, Centrifugal Fire Pumps.

[2] ANSI/NFPA 70-1981, National Electrical Code.

[3] ANSI/NEMA ICS 1-1978, Industrial Controls and Systems.

[4] ANSI/NEMA ICS 2-1978, Devices, Controllers, and Assemblies for Industrial Control.

[5] ANSI/NEMA ICS 6-1978, Enclosures for Industrial Control and Systems.

[6] UL 845-1980, Electrical Motor Control Centers.

7. Services, Vaults, and Electrical Equipment Rooms

7.1 Incoming Lines and Service Laterals. When a commercial building is remote from the electric utility's distribution circuits, the electric supply feeder may be extended across private property to the building service entrance or to a transformer stepdown substation located either inside or outside the building.

From the standpoint of safety, these feeders should generally be installed, owned, and maintained by the electric utility, even if the owner is required to make a financial contribution. This is especially important where high and medium voltages are involved, because qualified personnel to install and maintain these circuits are generally not employed in commercial-building operations. The policy of the electric utility may however, require ownership and maintenance by the customer.

In planning it is important to route the incoming circuit to avoid clearance conflicts with existing or future underground or overhead structures. Poles located in areas subject to vehicular traffic may require curbs or barriers for protection. Where open-wire lines pass near buildings, adequate clearances must be provided to avoid accidental contact by occupants, maintenance or inspection personnel, and firemen.

Because of the potential hazard of electric conductors, the utility company is not only expected to build lines that are free from mechanical failure, but in many states is required by law to meet minimum construction standards. Many of these regulations are well beyond the minimum acceptable rules for clearances and strengths of electric lines specified in ANSI C2-1981 [1].[32]

Overhead lines are normally built in accordance with the local electric utility standards and the requirements of ANSI C2-1981 [1].

7.1.1 Overhead Service. For small buildings supplied at utilization voltages, the overhead service lateral is generally terminated at a bracket on the building at sufficient height to provide the re-

---

[32] The numbers in brackets correspond to those in the references at the end of this Section.

quired ground clearance given in provisions of ANSI C2-1981 [1]. Larger buildings may be served by open-wire lines along the rear, or at one side of the property, terminating at a transformer stepdown station outside the building, or at a cable terminal pole where the stepdown station is inside the building.

An alternative method is to use aerial cable (insulated cables — shielded where applicable — supported by a grounded messenger) attached to poles.

Open-wire lines may consist of copper or aluminum conductors attached to glass, fiberglass, or procelain insulators which in turn are supported on pins mounted on wood crossarms or on pole brackets attached to wood poles set in the ground. The conductors may also be suspended from the crossarms on suspension-type insulators. The latest edition of the ANSI C2-1981 [1] spells out clearances of all types.

7.1.2 Lines Over and On Buildings. The installation of open-wire lines over buildings should be avoided as poor practice, as they interfere with the activities of firemen and maintenance, or security personnel, and present a safety hazard. Where it is impossible to avoid passing close to and over a building, the circuit should be enclosed in a grounded metallic conduit encased in concrete on the roof. The concrete may be omitted for voltages below 300 V.

Grounded metallic-sheathed aerial cable may be used where the line must pass over buildings. If open wire is installed over buildings, minimum clearances for personnel must be maintained over all areas that are accessible to personnel. Clearances must meet provisions of ANSI C2-1981 [1] or the rules of the local code enforcement authority.

For voltages up to and including 5 kV, weatherproof covered wire is generally used. However, the insulating value of weatherproof covering, especially that used prior to 1950, may be practically nil. Above 5 kV, bare wire is sometimes used for open-wire lines. Where safety hazards exist, aerial fully-insulated cables are recommended and quite frequently used by electric utilities. The design of an open-wire line depends upon factors such as the following:

(1) *Safety*

(a) Safety to the public, providing the necessary clearances from line to buildings, railroad tracks, driveways, etc.

(b) Safety to personnel who may operate and maintain the line, involving adequate climbing and working space on the pole, spacing between conductors on the crossarm, and interphase spacing between items of equipment on the pole.

(c) Mechanical strength, involving consideration of wind and ice loads, diameter of pole, size and strength of wire, etc.

(2) *Insulation*

(a) Protection against lightning surges. This is handled by shielding the line from direct strokes and induced surges through the use of lightning arresters, one or more shield wires installed above the power conductors, and by greater insulation.

(b) Insulation against voltage surges caused by power switching. The above mentioned preventive measures also apply for switching surges.

For the details of designing a line, refer to various electrical handbooks, to the local electric utility company, and to experts proficient in this professional work. Flashover characteristics of insulators can be obtained from manufacturers' catalogs; however, this does not necessarily provide the coordination of the insulation level required.

Rights-of-way grants may be required from the owner for lines on private property and permits are required from the responsible governmental authority for lines on public property.

7.1.3 Weather and Environment Considerations.

In designing any outdoor structure, weather forces must be considered. A building is designed to withstand wind on its walls and a snow load on its roof. Similarly, an overhead electric line should be designed to withstand a wind load on the poles and conductor, as well as an ice load on the conductor. The severity of the weather factor varies with location throughout the United States, and reference may be made to the general loading map in ANSI C2-1981 [1].

In damp, foggy, or dirty atmospheres, contamination of insulator surfaces becomes a problem, and special insulators having an unusually long leakage distance should be used to prevent leakage currents across the surface of the insulators.

7.1.4. Underground Service.

Certain conditions may require underground construction. Examples include: conflicts with overhead structures that cannot be by-passed with aerial construction, load density, local ordinances, or regulatory requirements governing construction in new residential subdivisions. Aesthetics may also be a factor.

An underground system is relatively free from many of the problems associated with an overhead system. However, in case of failure the time and expense of locating and repairing cable trouble in an underground system is considerably greater.

Underground systems almost always cost substantially more than equivalent overhead systems. This is especially true of conduit and manhole underground systems. Direct-burial type underground systems such as underground residential distribution (URD) and commercial and industrial park underground distribution (CIPUD) enable a considerable saving for new developments over the cost of an equivalent conduit and manhole system.

(1) URD is a direct-burial single-phase distribution system used by utilities for new residential subdivisions. Organic insulated and jacketed cables are used together with premolded or encapsulated splices and termination devices. Metal-clad transformers mounted on surface pads may be utilized, or transformers and switching devices may be installed in underground *"cans"* or directly buried rather than in cast manholes.

(2) CIPUD is a direct-burial three-phase system used by utilities for commercial-industrial park distribution. The cable system is looped through metal-clad transformers or switchgear mounted either on surface pads or installed in suitable below-grade boxes or vaults. CIPUD (and URD) cable systems are generally run behind curbing in grass areas to minimize paving costs as well as to provide accessibility. Sleeves are generally installed under paved crossings to eliminate the need for breaking and restoring paving when installing or removing cable or making repairs.

(3) Fault indicators assist in enabling rapid determination and repair of faults on CIPUD (and URD) cable systems. Installed at cable termination locations, the fault indicator displays a signal whenever fault current has passed through its sensor. The device either resets automatically after the system is reenergized or is reset manually.

(4) Maintenance and operation of CIPUD (and URD) systems should be trusted only to the utility or to specially

trained personnel for both safety and operational considerations. These systems are not suitable for the high-density loads of urban centers because of the direct-burial aspect and the limited load and fault handling capability of the equipment.

7.1.5 Service-Entrance Conductors Within Building. Regardless of voltage, when service-entrance conductors must pass through the building to the service equipment, a safety hazard presents itself because that part of the circuit in the building may not be protected against short circuits or arcing faults. Where the distance is 10 ft or less, the hazard is considered to be minimal. Circuits over 10 ft long from the point of entrance through the building wall to the service equipment should be installed in encased raceway in at least two inches of concrete. Greater additional protection is provided by the use of metallic conduit suitably encased in concrete.

This protects the building by confining any fire or arcing (because of a short circuit) within the concrete envelope. Safety to property and life, too, is always enhanced by encasing the incoming line in concrete inside the building. Concrete-encased raceway may be installed along ceilings, under the basement floor, or on the roof. Although bus duct is sometimes used for service entrance conductors it is very difficult to provide protection for arcing faults. Neutral grounds normally employed at both the main switchboard and the service transformer(s) make detection complicated. From services fed by spot networks primary sensing is generally ineffective. Overheat detectors spotted frequently over the busway have proven valuable but cooperation from the serving utility to interrupt the utility protection devices at their equipment is necessary.

Incoming electric lines within buildings should be installed to meet or exceed the minimum requirements of ANSI C2-1981 [1] as modified by state laws and local ordinances. Any questions concerning the application of these rules should be taken up with the local code-enforcing authority. It should be recognized that most of these codes cover minimum requirements and are not intended to be recommended design criteria.

Cable systems should be routed to avoid high ambient temperatures caused by steam lines, boiler rooms, etc. Where cable is lead-sheathed, duct runs through cinder beds should be avoided unless the duct system is encased in a sufficiently thick envelope of concrete to make it impervious to the acid condition prevalent in cinder beds. The lead sheath should also be jacketed to protect against corrosion. Precaution should be taken with polyethylene, cross-linked polyethylene, and other organic jacketed cables to prevent chemical degradation of the jacketing where hydrocarbons may be present such as in fueling areas, marsh lands, landfill areas, and similar locations.

Manholes and pull boxes are required in long duct runs to facilitate pulling and splicing the cables. Spare ducts should be installed to provide for the contingency where a faulted cable becomes frozen in a duct and cannot be removed for replacement. This also simplifies installation of future cables required for load growth. A duct system should not be laid in the same trench with gas or sewer service.

7.2 Service-Entrance Installations. Service-entrance equipment is the portion of the system between the customer-owned service equipment and the utility's service drop or lateral. Service-entrance equipment includes the main service control or cutoff for the electric supply and consists of one or more circuit breakers, or switches and fuses, and accessories as well as the metering equipment.

Service-entrance switching equipment is generally paid for by the customer. Some of its design features are frequently influenced or controlled by the electric utility company. The billing metering instruments are nearly always owned and maintained by the electric utility company. The current and potential transformers used exclusively for billing metering purposes may be furnished by the electric utility company at either the customers or the electric "utility" company's expense. Refer to Section 4 for details on utility metering.

The relationship of service-entrance-equipment design and characteristics to the incoming lines or feeders and to the distribution switchgear or switchboard are of vital importance to the customer and to the electric utility company. Therefore it is important that the engineer serve both the customer and the electric utility company by developing a design which satisfies customer requirements without interfering with the quality of the electric service to other customers of the electric utility.

7.2.1 Number of Services. The number of services supplied to a building or a group of buildings will depend upon several factors.

(1) The degree of reliability required for the installation as related to the reliability of the power source. Where service reliability is important, multiple service or standby service, with load transfer arrangements between various parts of the building distribution system, may be indicated. In some cases, economic considerations may dictate a partial decrease in service availability and the sacrifice of nonessential loads during the emergency. If more than one service is required by the customer, an excess service charge may be assessed.

(2) The magnitude of the total load. Since the capacity of an individual service take off is limited by the utility to a maximum current value, additional take offs must be provided as required to meet the building demands.

(3) The availability of more than one system voltage from the utility. If more than one voltage is available, the utility may, for example, supply 208Y/120 V for lighting and receptacles at one or more service-entrance points and 480 V for power.

(4) The physical size of the building or the distances separating buildings comprising a single facility. Tall buildings, buildings occupying a large ground area, and widely separated smaller buildings will often be supplied from multiple services.

(5) ANSI/NFPA 70-1981 [3] and local code requirements including items such as fire walls.

(6) Additional capacity to serve future loads. This practice is costly and careful study of the economics should be made to justify it, rather than redesign in the future to meet more exactly the needs.

7.2.2 Physical Arrangement. The physical arrangement of the service entrance will vary considerably depending upon the type of distribution system employed by the utility and the type of building being served. In some cases, the utility will supply service from one or more transformer vaults located directly outside the buildings with bus stabs through the basement wall. This is the usual arrangement for buildings of moderate height in heavily loaded areas of large cities. Utility-company transformer vaults are sometimes located within the building itself, in the basement, and on the upper floors of tall buildings. Underground service by means of cable from a manhole or a pole in the street is sometimes provided while in other cases overhead services may be available.

Service-entrance-equipment rooms in all cases should be easily accessible, dry and well-lighted, and should comply in all respects with the requirements of the electric utility and local code authorities having jurisdiction.

7.2.3 Low-Voltage Circuits. The service-entrance equipment is one of the most important parts of the electric supply system for buildings because it is through this equipment that the entire load of the building is served. The service-entrance equipment installed initially should either be adequate for all future loads or be designed that it can be supplemented or replaced without interfering with normal operation of the building which it serves.

Because the service entrance is part of the building and involves equipment belonging to the utility company, the choice of entrance equipment and voltage should be a cooperative decision between the building electrical design engineer and the local electric utility. This should be accomplished early to allow the building designer to adapt his design to the present and future supply plans of the utility and to enable the utility to supply power to the building in a manner which considers initial and future requirements. The electrical design engineer should furnish load and other data to the electric utility company to assist it in determining the effect of the building load on its system and, where necessary, plan for the expansion of its facilities. The utility will, at this time, inform the engineer as to the type of services available, their voltages, and whether overhead or underground electric service is to be used. The building engineer should assist the electrical designer in determining the point of service entrance and termination. Other data pertinent to the system design, such as short-circuit current or kilovoltamperes available at the service entrance, service reliability, costs, space requirements for poles, substations, transformer vaults, metering equipment, inrush-current limitations, design standards, and similar information should be obtained from the utility.

A check list of items to be considered in connection with the electric service follows:

(1) Complete characteristics of loads to be served

(a) Kilovoltampere demand, both initial and future, at various utilization voltages

(b) Service continuity requirements

(c) Voltage requirements and limitations of voltage variations

(d) Special loads, such as X-ray machines and computers

(e) Superimposing of carrier current on to the electric system for signals, clocks, or communications

(f) Largest motor inrush current

(g) Significant low-power factor loads

(h) Loads (such as solid-state adjustable speed motor drives) which generate harmonics.

(2) Complete characteristics of all types of service available from the utility under their electrical rate structure

(a) Voltages available and voltage spread

(b) Billing demand clauses of the rates

(c) Rates and special clauses, such as exclusive service, standby service, power factor penalty, and fuel cost adjustments

(d) Possible need for equipment to transform, regulate, or otherwise modify the characteristics of the available service to meet the requirements of the building

(3) Physical and mechanical requirements of the service entrance

(a) Number of locations at which service may be supplied

(b) Type of service entrance: overhead or underground cable, or bus

(c) Points of service termination, including information as to which parts of the service installation will be owned, installed, and maintained by the utility

(d) Location and type of metering equipment, including provisions for totalizing demand and for submetering, where permitted, and provisions for mounting and wiring the electric utility's meters and metering transformers

(e) Space and other requirements for utility vaults, poles, and similar equipment, and access provisions for its installation, maintenance, testing, and meter reading

(f) Avoidance of structural interferences (particularly critical when using bus duct)

(4) Electrical requirements for service entrance

(a) System capacity and fault capability, both present and future (future may be defined as 15 or 20 years hence)

(b) Requirements for the coordination of overcurrent protective devices

(c) Utility-approved types of service and metering equipment, utility grounding methods and requirement for the coordination of ground-fault protection for grounded service systems

(5) Schedule data

(a) The date service and preliminary construction schedule will be required

(b) The dates when full estimated initial load and full load will be required

(c) Temporary construction service requirements

7.2.4 Load-Current and Short-Circuit Capacity. The design of service equipment depends not only upon the continuous-current requirements of the circuit and, hence, the current capacity of the entrance equipment, but also on the short-circuit current available at the service bus. Low-voltage equipment will, therefore, be divided into three categories: low-capacity circuits, medium-capacity circuits, and high-capacity circuits.

(1) *Low-Capacity Circuits.* For service entrances with current ratings less than 600 A, which are fed by individual transformers, problems of entrance equipment may be minor. However, the engineers should check the electric utility for short-circuit duty because some of the newer transformers have very low impedance, resulting in high short-circuit currents. Where several services are supplied from one transformer bank, short-circuit duty may be 15 000 A— 60 000 A (depending upon the back-up system and transformer impedance). Where the available fault current is less than 10 000 A a wide choice of equipment exists. Where initial investment is an important factor, the simplest fused disconnect switch or molded-case circuit breaker may be used. Ground-fault protection is at the option of the designer in lower current installations, and coordination problems are usually not serious.

(2) *Medium-Capacity Circuits.* While there is no standard definition, typical medium-capacity services may have short-circuit duties ranging from 10 000 A to 50 000 A. The available fault current can

be determined only by the utility. The impedance of power transformers may run in the order of five percent while that of distribution transformers, so common in this type installation, may be two percent or less. At the latter impedances the short-circuit current may exceed fifty times the normal load rating.

At these levels, a number of fused-type devices are available as well as certain high-interrupting capacity unfused and fused circuit breakers. While high-interrupting capacity breakers, breakers with current limiters, and switches with current-limiting fuses may protect the basic device, it is imperative that the engineer examine closely the device limiting characteristics to assure that the limiting effect will protect downstream interrupting devices. Unless curves or manufacturer's data, or both, assure such protection it is incumbent on the engineer to include additional current-limiting devices downstream in the system.

For grounded-Y electrical services of more than 150 V to ground, but not exceeding 600 V phase to phase, ground-fault protection is required by ANSI C2-1981 [1] for any disconnecting means rated 1000 A or more. The electrical design engineer may elect to use it at lower levels.

(3) *High-Capacity Circuits.* All service entrances which have an available short-circuit capability in excess of 65 000 A can be considered high-capacity service entrances. Buildings which are fed from alternating-current secondary networks or very large buildings which are fed from a number of parallel transformers are in this class. It is imperative in these installations that breakers (with or without current limiters) or switches with current-limiting fuses suitable for the available fault current be utilized when possible. Multiple sources may prove effective in lowering available fault current. Reactors, current-limiting bus, and current-limiting cable systems have been used to reduce fault current available at the protecting devices but are becoming less common as device interrupting capacities have been increased.

7.2.5 Limiting Fault Current

(1) *Multiple Sources.* The simplest method to eliminate excessively large available short-circuit currents is to divide the electric circuits of the building into several independent parts, where practicable, and where permitted by the utility, with each part fed by one three-phase transformer or by a group of transformers. If the entire load of a large building takes six parallel transformers to supply it, it may be possible, by integrating the design of the building's electric system and the utility company's supply system, to divide this into two groups of three transformers each or into six single transformer loads.

This method is less effective for smaller buildings fed directly from the utility alternating-current secondary network, since serving the building at more than one entrance may not reduce the available short-circuit current on any entrance where network ties exist.

In order to reduce the available short-circuit current at the load side of a large service switch or circuit breaker, the service can be divided into six or fewer smaller service-disconnecting switches or circuit breakers equipped with current-limiting fuses. While each of these devices is capable of handling the available short-circuit current at the service point, the let-through current is greatly reduced by using the smaller current-limiting fuses than would be the case for a single large service with current-limiting fuses. This arrangement permits the use of equipment having lower interrupting ratings further downstream, and may be considered where permitted by local codes.

A variation of the *"divide and conquer"* technique of fault-current reduction consists in designing additional cable reactance into the cable circuits connecting each network transformer to a common bus or in designing additional reactance into the cable circuits from each individual transformer bus to separate service-entrance circuit breakers within the building. Reactance of these cable circuits is controlled by spacing between the phases which are individually put into nonmetallic conduit buried in concrete. Normally, open tie circuit breakers between the various service-entrance buses are interlocked with the main service-entrance circuit breakers. This arrangement permits serving a very heavy load with only moderate amounts of available fault current and good flexibility in case of outages of transformers or primary feeders. However, the local code and the utility company should first be checked to determine if this arrangement will be acceptable.

Where high-capacity low-voltage service is the only choice provided by the utility or is existing, consideration should be given to use of step-up indoor transformers to limit short-circuit current while facilitating distribution.

(2) *Reactors* (Also see 9.6.2). Reactors can be put in the main service connection to reduce the fault current to the capacity of the service-entrance main and feeder circuit breakers. These reactors are often of high continuous-current rating which makes them quite large. Reactors in the main supply can reduce fault currents to about 60 000 A. This is within the rating of the largest circuit breakers, so the use of 200 000 A circuit breakers equipped with fuses is not required. It is unnecessary and usually not economical to reduce the fault current further with reactors.

Smaller reactors can also be used to feed a smaller group of circuit breakers which may be part of the service-entrance equipment to reduce the short-circuit current. Small enclosed reactors are available in current ratings up to 800 A, which may be economically applied to feed groups of smaller circuit breakers. The use of these reactors permits complete selectivity in the coordination of circuit breakers for tripping under fault conditions and permits the use of circuit breakers of reduced interrupting capacity.

Enclosures or guards must be considered for reactors accessible to personnel or when exposed to physical damage.

Where reactors, or reactance of any type, are used, consideration should be given to the voltage drop introduced into the circuit.

(3) *Current-Limiting Bus.* Current-limiting bus is designed with all the bars of one phase installed side by side instead of interlacing bars of different phases as with the usual low-reactance bus. Current-limiting bus is effective in reducing very large short-circuit currents such as 200 000 A down to levels of approximately 100 000 A, but is usually less effective in reducing the short-circuit currents to values much lower than this. Where the local code regulations permit the installation of the service-entrance equipment at a point some distance from the entrance into the building, the use of current-limiting bus presents a solution to the problem of reducing the available short-circuit current, provided that transition units for phase transposition are used frequently enough to balance the reactances and resultant voltages.

However, most codes require that the service disconnect switch or circuit breaker be applied closest to the point where the utility company enters the building. In this case the service-entrance switch or circuit breaker must be adequate for the available short-circuit duty. Current-limiting bus is not feasible unless the runs are long enough to provide the desired reactance. It can sometimes be used between the fully rated main service switch or circuit breaker and downstream switchboards if the run is of sufficient length.

(4) *Current Limiting by Cables.* Current-limiting effects can also be achieved by means of cable in which the spacing between phases is controlled by design. If cables are used ahead of the service disconnect or circuit breaker, local electrical codes may require that they be installed in conduit which is buried in concrete. If permitted by codes and the utility, separation of the cables for each phase in three equally-spaced non-metallic conduits provides a relatively high current-limiting reactance, while also reducing the chances of a phase-to-phase cable fault to nearly zero. Equal (delta) spacing is necessary to keep the reactances balanced, while non-metallic conduit is required to prevent destructive heating by currents induced in the conduit.

Local codes may require the use of a separate vault for service-entrance current-limiting bus where no separate protective device exists ahead of the current-limiting bus.

(5) *Current-Limiting Fuses.* Current-limiting fuses start to limit current at approximately twenty to thirty times their current rating and to clear the circuit before the current has reached its peak value on the first half-cycle. Their current-limiting action results from interrupting the fault current before it can increase to its maximum asymmetrical value during the first half-cycle. The fault current which flows through the fuse while it is melting and interrupting the circuit is called let-through current. Its peak magnitude depends directly on both the continuous rating of the fuse and the fault current available in the system if the fuse were not in the circuit. The higher the continuous rating of the fuse, the more current it will let through.

The curves of Section 9 illustrate typical peak let-through based upon available symmetrical fault current for various current-limiting fuse sizes. To convert these peak let-through values to a symmetrical down-stream interrupting rating that will match a given circuit breaker, it is necessary to use a dividing factor of 1.8 to 2.3, depending on the X/R ratio of the distribution system. If in doubt as to the exact symmetrical let-through values that can be tolerated, one should check the circuit-breaker manufacturer for recommendations based upon actual test values obtained in high-power laboratories.

The use of large current-limiting fuses with the service-entrance switch or circuit breaker may protect the entrance equipment, but does not necessarily protect the lower rated equipment beyond the service entrance such as motor-control centers, distribution switchboards, or panelboards. These circuits can be protected by smaller downstream current-limiting fuses to reduce the let-through, or by the use of current-limiting bus, cable runs, or reactors as previously discussed.

The current-limiting fuse has one of the disadvantages of the instantaneous-trip circuit breakers. Complete selectivity between mains and feeders is impossible over the entire range of fault current, unless the ratio of the fuse ratings is more than two or three to one. Unbalanced faults may blow only one fuse in a fused switch, leaving the entire load operating single phase. With the large amount of motor load in modern buildings, this action can have serious consequences. To prevent this, the fuse may be mounted in a built-in circuit breaker or service protector and fuse combination with inherent anti-single-phase protection.

7.2.6 Ground-Fault Protection. ANSI/NFPA 70-1981 [3], Section 230-95 covers ground-fault protection requirements for equipment. Ground faults, if not interrupted quickly, can cause disastrous burndown of equipment, particularly on high-capacity 480Y/277 V circuits. Fully coordinated ground-fault protection schemes are recommended on such systems. Achievement of proper ground-fault coordination may require a combination of fixed time delay, inverse time delay, etc, between branch feeders and the main service. The designer should check the effectiveness of coordination in all such cases.

7.3 Vaults and Pads for Service Equipment

7.3.1 Vaults. Service transformers and associated switching and protective equipment are often located in vaults. Special precautions should be taken to remove the heat given off by the transformers. Equipment vaults should be located so that they can be ventilated to the outside atmosphere without the use of flues and ducts where practical. Natural ventilation is considered to be the most reliable means of ventilation. The total net area of the ventilator should be not less than two square feet per one hundred kilovoltampere of the installed transformer capacity. Additional ventilation may be required by local codes or electric utilities. Where the load peaks in the summer and where the average outdoor temperature of 24 h periods in the summer exceeds 30 °C the ventilator area should be increased, or an auxiliary means of removing the heat from the vaults, such as fans, should be used.

Where long vertical ventilating shafts from the vault to the top of the building must be used, it is necessary to have a larger vent area to compensate for the added resistance to the flow of air. For such shafts to overcome air friction a fan shall be installed with its discharge directed toward the shaft opening to increase the velocity of the air through the ventilation shaft. The fan should have a cord and plug to facilitate its replacement and it may be either single phase or three phase. A light at the entrance door should indicate fan failure. Controls for the fan should not be permitted in the vault and should be accessible only to authorized personnel.

Suitable screening should be used to prevent birds or rodents from entering the shaft. The ceiling, walls, and floor must be of fire-resistant construction. Reinforced concrete is preferred. Where oil-insulated transformers are used and persons occupy the area adjacent to the vault wall, or where an explosion may otherwise damage a building wall, the vault walls should be sufficiently strong to withstand an explosion. The hazard may also be reduced by limiting the ratio of vault volume in cubic feet to net ventilated area in square feet. Tests indicate that where this ratio does not exceed 50/1, an 8 inch reinforced concrete wall will suffice.

Any opening from a vault into a building must be provided with a tight fitting

Underwriters Laboratories approved three-hour fire door. The vault should be free of all foreign pipes or duct systems. A sump, with protective cover or grate, should be provided in each vault floor, to catch and hold any oil or liquid spillage. The floor should be pitched to the sump. Door sills must be of sufficient height to retain all of the oil from the largest transformer. Fire dampers may be required at air-duct openings.

Grade-level gratings are suitable for underground vaults and will also suffice for a combination access hatchway and ventilation well when the vault is in the basement of a building adjacent to an outside wall. Gratings for sidewalk service vaults should be made strong enough to support the wheels of trucks and must satisfy local code requirements. Gratings of net free air area equal to 63 to 70% of gross grate area are available commercially to meet various loading requirements. Gratings for roadway service should comply with class H 20 highway loadings.

Where multiple banks of transformers supplied from different sources are used, they should be installed in separate compartments to prevent fire in one compartment from affecting adjacent transformers. Switchgear associated with the transformer should also be separately enclosed so that a transformer in trouble can be isolated without entering the transformer compartment and to prevent transformer trouble from involving the switchgear. Consult local codes, ANSI/NFPA 70-1981 [3], and the local utility for specific vault-construction requirements.

7.3.2 Outdoor Pads. Metal-clad pad-type three-phase transformers and switching equipment have gained recognition over the past few years. Designed for installation on surface pads, metal-clad components are an economical and safe means for providing service. Cables enter and leave via the bottom of the component, hence presenting no energized parts to create a hazard. While no supplementary enclosure is required, an enclosure may be provided for unusual or aesthetic purposes. Traffic protecting posts should be provided in near areas available to vehicles. If placed in a vault, provision for the insertion and withdrawal of the metal-clad unit by crane should be allowed. Landscaping or architectural fencing may be used for concealment. Additional space may be required for operation and maintenance by the utility.

7.3.3 Safety. Except for outdoor metal-clad pad-type service transformers, outdoor substations should be enclosed with walls or fences. Adequate aisles should be provided for safe operation and maintenance. Proper clearances, both vertically and horizontally, should be maintained.

All equipment, operating handles, fences, etc should be adequately grounded. See IEEE Std 142-1982 [2] for a complete discussion of grounding requirements and methods. High-voltage warning signs should be prominently displayed. Enclosures, equipment, operating handles, etc, should be locked.

Substations should not be located near windows or roofs where live parts may be reached, or where a fire in the substation could be communicated to the building. Local utilities, authorities, and insurance carriers may require oil-filled transformers to be located not less than certain minimum distances from building openings unless suitably baffled.

Indoor substations should have the same general safety considerations as an outdoor substation, even though they are usually metal-clad. They should have a separate enclosure or should be placed in separate locked rooms and be accessible to authorized personnel only.

Multiple escape means to the outdoors or to other parts of the building should be provided from vaults, and located in front of and to the rear of switchgear and rows of transformers. These emergency escape means should be hinged doors with panic bars on the inside of the doors for quick direct escape in the event of trouble.

7.4 Network Vaults for High-Rise Buildings. The electric demands generated by large modern metropolitan office or commercial high-rise buildings almost invariably require the installation of a multiple number of transformers in close proximity to the structure. Each transformer is connected to a common low-voltage bus through a network protector. Many new buildings have power supplied at two or more locations, one beneath the sidewalk and others in the building or perhaps on the roof. Typically, these installations could provide up to 6000 kVA at 208 V or up to 15 000 kVA at 480 V at one point of service (see 7.2.5).

Design of major network installations divides naturally into two parts. First it is necessary to establish a utilization voltage, number of transformers, and number of service points. It is then necessary to match utility standards with customers' building designs. The design must satisfy customer and utility requirements and also meet municipal regulations, all within a framework of economics. The ability to install, maintain, or replace a component of the supply system without interruption of service is the backbone of network design.

7.4.1 Network Principles. To more fully appreciate the subject of specifying and designing network installations, it is first necessary to understand the principles of a network system. The network is designed to meet power demands on a contingency basis. This is to say that with a predetermined number of components (for example, transformers) out of service, full-load capability is maintained. This is accomplished by operating the remaining equipment above its nameplate rating and allowing slightly reduced service-voltage levels.

Networks are generally designed as first- or second-contingency systems. First-contingency networks generally utilize two or three primary feeders. Second-contingency networks may utilize three, four, six, or more primary feeders. If full-load capability can be maintained with two sets of components (that is, primary feeders) out of service, the system is defined as a second-contingency network. If full-load capability can be maintained with only one set of components out of service, the system is defined as a first-contingency network.

True contingency design also requires that the primary feeder supply system as well as the substations and switching stations ahead of them be built and operated with the distribution system in mind.

The last implication in contingency design is that all network equipment, including the associated high- or low-voltage cable ties, is sufficiently isolated. Then, failure or destruction of a single component in the system will cause only first-contingency operation until repairs or replacements can be effected.

7.4.2 Preliminary Vault Design. The initial step is to prepare a simple sketch of the proposed installation by means of a standard vault-equipment arrangement, showing any adaptations required in the building structure or any interferences with existing obstructions located beneath the sidewalk. The standard designs are similar for sub-sidewalk or in-building locations. Designs should include the following considerations.

(1) Sufficient space should be available within reasonable proximity to customer load centers.

(2) Subsurface conditions should be favorable. It is desirable to avoid the added expense of pilings or footings.

(3) Equipment should be of such design that environmental factors (for example, water) present no serious problems. As an example, underground transformers may be oil cooled by natural convection with an all-welded construction and corrosion-resistant finish. On the other hand, interior transformers may be 220 °C insulation class, aircooled open dry-type units, which provide the safety of nonexplosive nonflammable equipment.

(4) Conformance to municipal regulations should include

(a) General structural design with sidewalk loads which may be in the order of 600 lb/ft^2 or highway loads

(b) Location and size of ventilation and access panels

These factors provide only for the basic adequacy of an installation at a particular location.

7.4.3 Detailed Vault Design. Many other specific considerations are involved in the safe and reliable design of network installations. Major considerations for properly designed vaults follow.

(1) *Ventilation.* Ventilation shall be directly to the atmosphere and sized at two square feet of open area per one hundred kilovoltampere of transformer capacity as a minimum. The electric utility and local codes should be consulted for more stringent requirements. Forced ventilation, if required, should

be a minimum of three cubic feet per minute per kilovoltampere of transformer capacity, unless a higher rating is required by local codes or the utility.

(2) *Construction.* Below-grade vaults shall be reinforced concrete for strength and explosion confinement. Vaults are constructed to be as watertight as possible, but drainage (when permitted and practical) is also provided to eliminate stagnant or casual water accumulation.

(3) *Ventilation Ratio.* The ventilation ratio relates construction and ventilation from previous paragraphs. To avoid excessive pressures in the event of a secondary explosion in a vault containing oil-filled equipment, the ratio of vault volume to net ventilation area should be as small as practical. Such a ratio should be less than fifty cubic feet per square foot of open ventilation area, typically thirty cubic feet per square foot.

(4) *Access.* Direct rapid access is required at any time for maintenance or emergency operating crews.

(5) *Isolation and Protection.* The effects of equipment failure can be reduced by either isolation or protective relaying, or both. In the extreme, the following effects may result from such failure:

(a) Oil-filled equipment:

Explosion, tank rupture

Fire, smoke

Danger of secondary explosion (reexplosion of volatile vapors generated by destruction of class A and B materials)

(b) High-flame point liquid (when permitted by applicable codes):

Violent tank rupture, a form of explosion

(c) Dry type — open:

Smoke with very limited fire possibilities

Very limited possibility of secondary explosion

(d) Dry type — sealed:

Normally not considered hazardous

While such failures are rare, the possibilities cannot be neglected. Of course the location of the equipment in relation to people will determine the overall degree of hazard. For example, an askarel transformer failing in a sidewalk vault may be relatively innocuous, compared to a similar failure in an electric room adjacent to a public area.

Some utilities depend almost entirely upon isolation for safety. At the higher voltages, utility practice may preclude the use of protective relaying which would detect low-level faults. In these instances strong masonry vaults which are vented to the outside are depended upon to contain the effects of a fault. In larger installations, transformers are placed in individual vaults, and the network protectors and collector buses may be similarly isolated. Where the vaults or network area are part of the building distribution system rather than utility owned, protective relaying which will cover all zones is required. Such protection will, as a minimum, include overcurrent and ground-fault protection. It may in addition include differential protection, transformer hot-spot and sudden pressure trip, or alarm. Heat or arc sensing or smoke detection in the room or vault, or even inside larger pieces of equipment, may be provided. None of these, however, can relieve the need for physical isolation which may be required to protect the building occupants and the public from the effects of such failure. The use of machine-room floors or other heavy equipment areas for location of the electric rooms or vaults further enhances the protection afforded. Great care must be exercised to ensure that no smoke or fumes can under any condition enter the normal building ventilation system.

(6) *Apparatus Arrangement.* It is important to provide what might superficially be called wasted space for operation and maintenance of the following:

(a) Drawout of switchgear

(b) Replacement of fuses or cables

(c) Access to equipment accessories

(d) Cleaning

(e) Air circulation

(f) Access to equipment for replacement purposes without disturbing other equipment

(g) Cable pulling and installation

(7) *Miscellaneous*

(a) Heavy-duty roof structures for vehicular traffic (class H20) loading

(b) Interference by curb cuts or driveways

(c) Future street widenings or grade changes

(d) Improved drainage

(e) Spare conduits within buildings

(f) Duct arrangement for separation of primary and secondary feeders

(g) Effects of unnecessarily long cable ties on voltage regulation

(h) Balanced equipment loading

(i) Access for heavy test sets to the vault switchgear

(j) Normal illumination for routine inspection and maintenance with power supply receptacles for additional lighting and test equipment use. Consideration should be given to having part or all of this supply on the building emergency source during outage conditions.

7.5 Service Rooms and Electric Closets. Service and distribution equipment is generally located in electric rooms, while sub-distribution equipment is generally located in electric closets. These areas should be as close to the areas they serve as is practicable.

The rooms should be sized so that there is sufficient access and working space about all electrical equipment to permit ready and safe operation and maintenance. The doors should be of sufficient size to permit easy installation or removal of the electrical equipment contained therein.

7.5.1 Space Requirements. To provide flexibility for future expansion and growth, the electric rooms and closets should be sized somewhat larger than the minimum criteria dictated by ANSI/NFPA 70-1981 [3].

The minimum clear working space in front of electrical equipment is spelled out in ANSI/NFPA 70-1981 [3]. Additional utility working space may be required.

7.5.2 Illumination. Adequate illumination should be provided for all such areas in accordance with NEC provisions.

7.5.3 Ventilation. Ventilation should be provided to limit the ambient temperature of the room to 40 °C (104 °F). When a transformer other than a signal-type transformer is installed in an electric closet or room, some local codes require that a system of mechanical ventilation be provided. Refer to 7.3.1 and 7.4.3 for mechanical ventilation requirements.

7.5.4 Foreign Facilities. Electric closets or rooms are not to be used for storage purposes. Certain local codes prohibit a raceway, wiring panel, or device of a telephone system to be installed in the area. The same codes would even be more stringent on the running of water, gas, or other nonelectrical pipes or ventilating ducts through the rooms or closets. Where the local codes are not specific on this item, the dictates of good judgment or practice should apply. Cold-water pipes can drip due to condensation.

Sleeves and slots or other openings should be provided for cable and busway entrances. Those which are not in use should be sealed with pipe caps, plugs, or barriers. All openings with cables should be sealed with approved putty-like or other materials. Fire stops must be provided in accordance with code requirements where busways or wiring troughs pass between floors or fire rated walls. Sills or elevated sleeve openings may be used to prevent seepage of liquids around cables or busways.

7.6 References

[1] ANSI C2-1981, National Electrical Safety Code.

[2] IEEE Std 142-1982, Grounding of Industrial and Commercial Power Systems.

[3] ANSI/NFPA 70-1981, National Electrical Code.

8. Wiring Systems

8.1 Introduction. Wiring systems in commercial buildings use cable and busway systems. A typical building will have both, as each has advantages for particular applications. The first part of this section, 8.1 through 8.14, discusses cable; the latter part, 8.15 through 8.33 discusses busway.

8.2 Cable Systems. The primary function of cables is to carry energy reliably between source and utilization equipment. In carrying this energy there are heat losses generated in the cable that must be dissipated. The ability to dissipate these losses depends on how the cables are installed, and this affects their ratings.

Cables may be installed in raceway, underground in duct or direct buried, assembled on a messenger, in cable bus, or as open runs of cable.

Selection of conductor size requires consideration of load current to be carried and loading cycle, emergency overloading requirements and duration, fault clearing time, capacity of the short-circuit source voltage drop, ambient temperatures that may exist for three hours or more, and frequency of the system, for example, 400 Hz, for the particular installation conditions.

Insulations can be classified in broad categories as solid insulations, taped insulations, and special-purpose insulations. Cables incorporating these insulations cover a range of maximum and normal operating temperatures and exhibit varying degrees of flexibility, fire resistance, and mechanical and environmental protection.

Installation of cables requires care to avoid excessive pulling tensions that could stretch the conductor or insulation shield, or rupture the cable finish when pulled around bends. The minimum bending radius of the cable or conductors should not be exceeded during pulling around bends, at splices, and particularly at terminations to avoid damage to the conductors. On complex cable runs the engineer should calculate the sidewall pressure exerted on the insulation to ensure that damage will not occur. The engineer may also have to check the conductor jamming ratios in runs of unusual conductor combinations and complexity.

Provisions must be made for the proper terminating, splicing, and grounding of cables. There are minimum clearances between phases and between phase and ground for the various voltage levels. The terminating compartments must be heated to prevent condensation from forming. Condensation or contamination on high-voltage terminations could result in tracking over the terminal surface with possible flashover.

Many users test cables after installation and periodically test important circuits. Test voltages are usually direct current of a level recommended by the cable manufacturer for his particular cable. Usually this test level is well below the direct-current strength of the cable, but it is possible for accidental flashovers to weaken or rupture the cable insulation due to the higher transient overvoltages that can occur from reflections of the voltage wave. ANSI/IEEE Std 400-1980 [9][35] provides a detailed discussion on cable testing.

The application and sizing of cables rated 300 V through 35 kV is governed by ANSI/NFPA 70-1981 [4]. Cable use may also be covered in state and local regulations recognized by the local electrical inspection authority having jurisdiction in a particular area.

The various tables in this section are intended to assist the electrical engineer in laying out and understanding, in general terms, his cable requirements.

8.3 Cable Construction

8.3.1 Conductors. The two conductor materials in common use are copper and aluminum. Copper has historically been used for conductors of insulated cables due primarily to its desirable electrical and mechanical properties. The use of aluminum is based mainly on its favorable conductivity-to-weight ratio, (the

---

[35] The numbers in brackets correspond to those in the references at the end of this Section.

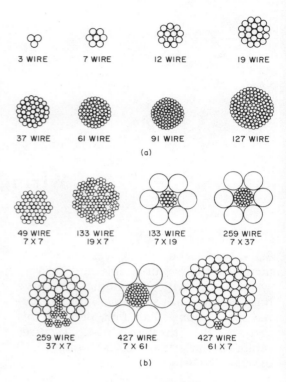

Fig 46
Conductor Stranding
(a) Concentric Lay Strands
(b) Concentric Rope-Lay Strands

highest of the electrical conductor materials) its ready availability, and the lower cost of the primary metal.

The need for mechanical flexibility usually determines whether a solid or a stranded conductor is used, and the degree of flexibility is a function of the total number of strands. ANSI/NFPA 70-1981 [4] requires conductors of size No 8 and larger to be stranded. A cable is defined as a stranded conductor, either bare or insulated, or an assembly of insulated conductors.

Stranded conductors are available in various configurations such as standard

Table 50
Properties of Copper and Aluminum

	Copper Electrolytic	Aluminum EC
Conductivity, % IACS† at 20 °C	100.0	61.0
Resistivity, $\Omega \cdot$ cmil/ft at 20 °C	10.371	17.002
Specific gravity at 20 °C	8.89	2.703
Melting point, °C	1083	660
Thermal conductivity at 20 °C, $(cal \cdot cm)/(cm^2 \cdot °C \cdot s)$*	0.941	0.58
Specific heat, $cal/(g \cdot °C)$*		
for equal weights	0.092	0.23
for equal direct-current resistance	0.184	0.23
Thermal expansion, in; equal to constant $\times 10^{-6} \times$ length in inches $\times$ °F	9.4	12.8
steel = 6.1		
18-8 stainless = 10.2		
brass = 10.5		
bronze = 15		
Relative weight for equal direct-current resistance and length	1.0	0.50
Modulus of elasticity, $(lb/in^2) \times 10^6$	16	10

*cal here denotes the gram calorie.
†International annealed copper standard.

concentric, compressed, compact, rope, and bunched, with the latter two generally specified for flexing service. Bunched stranded conductors, not illustrated, consist of a number of individual strand members of the same size which are twisted together to make the required area in circular mils for the intended service. Unlike the individual strands in the concentric-lay strands illustrated in Fig 46, no attempt is made during manufacture of bunch strand to control the position of each strand with respect to another. This type of conductor is usually found in portable cords.

8.3.2 Comparison Between Copper and Aluminum. Aluminum requires larger conductor sizes to carry the same current as copper. For equivalent ampacity aluminum cable is lighter in weight and larger in diameter than copper cable. The properties of these two metals are given in Table 50.

The 36% difference in thermal coefficients of expansion and the different electrical nature of the oxide films of copper and aluminum require consideration in connector designs. An aluminum oxide film forms immediately on exposure of fresh aluminum surface to air.

Under normal conditions it slowly builds up to a thickness in the range of 3—6 nanometers and stabilizes at this thickness. The oxide film is essentially an insulating film or dielectric material and provides aluminum with its corrosion resistance. Copper produces its oxide rather slowly under normal conditions, and the film is relatively conducting, presenting no real problem at connections.

Approved connector designs for aluminum conductors essentially provide increased contact areas and lower unit stresses than are used for copper cable connectors. These terminals possess adequate strength to ensure that the compression of the aluminum strands exceeds their yield strength and that a brushing action takes place which destroys the oxide film to form an intimate aluminum contact area yielding a low-resistance connection. Recently developed aluminum alloys provide improved terminating and handling as compared to electrical conductor (EC) grades.

Water must be kept from entering the strand space in aluminum conductors at all times. Any moisture within a conductor, either copper or aluminum, is likely to cause corrosion of the conductor metal or impair insulation effectiveness.

8.3.3 Insulation. Basic insulating materials are either organic or inorganic, and there are a wide variety of insulations classed as organic. Mineral-insulated cable employs the one inorganic insulation (MgO) that is generally available.

Insulations in common use are:

(1) Thermosetting compounds, solid dielectric

(2) Thermoplastic compounds, solid dielectric

(3) Paper-laminated tapes

(4) Varnished cloth, laminated tapes

(5) Mineral insulation, solid dielectric granular

Most of the basic materials listed in Table 51 must be modified by compounding or mixing with other materials to produce desirable and necessary properties for manufacturing, handling, and end use. The thermosetting or rubberlike materials are mixed with curing agents, accelerators, fillers, and antioxidants in varying proportions. Crosslinked polyethylene is included in this class. Generally, smaller amounts of materials are added to the thermoplastics in the form of fillers, anti-oxidants, stabilizers, plasticizers, and pigments.

(1) *Insulation Comparison.* The aging factors of heat, moisture, and ozone are among the most destructive of organic based insulations, so the following comparisons are a gauge of the resistance and classification of these insulations.

(a) *Relative Heat Resistance.* The comparison in Fig 47 illustrates the effect of a relatively short period of exposure at various temperatures on the hardness characteristic of the material at that temperature. Basic differences between thermoplastic- and thermosetting-type insulation, excluding aging effect, are evident.

Fig 47
Typical Values for Hardness Versus Temperature

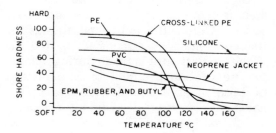

Table 51
Commonly Used Insulating Materials

Common Name	Chemical Composition	Properties of Insulation	
		Electrical	Physical
Thermosetting			
Crosslinked polyethylene	Polyethylene	Excellent	Excellent
EPR	Ethylene propylene rubber (copolymer and terpolymer)	Excellent	Excellent
Butyl	Isobutylene isoprene	Excellent	Good
SBR	Styrene butadiene rubber	Excellent	Good
Oil base	Complex rubberlike compound	Excellent	Good
Silicone	Methyl chlorosilane	Good	Good
TFE*	Tetrafluoroethylene	Excellent	Good
ETFE**	Ethylene Tetrafluoroethylene	Excellent	Good
Neoprene	Chloroprene	Fair	Good
Class CP rubber†	Chlorosulfonated polyethylene	Good	Good
Thermoplastic			
Polyethylene	Polyethylene	Excellent	Good
Polyvinyl chloride	Polyvinyl chloride	Good	Good
Nylon	Polyamide	Fair	Excellent

*For example, teflon or halon.
**For Example, tefzel.
†For example, hypalon.

(b) *Heat Aging.* The effect on elongation of an insulation (or jacket) when subjected to aging in a circulating air oven is an acceptable measure of heat resistance. The air oven test at 121 °C called for in some specifications is severe, but provides a relatively quick method of grading materials for possible use at high conductor temperatures or in hot-spot areas. The 150 °C oven aging is many times more severe and is used to compare materials with superior heat resistance. Temperature ratings of insulations in general use are shown in Table 52.

(c) *Ozone and Corona Resistance.* Exposure to accelerated conditions, such as higher concentrations of ozone (as standardized by ICEA, EM-60 [12] for butyl, 0.03% ozone for 3 h at room temperature), or air oven tests followed by exposure to ozone, or exposure to ozone at higher temperatures, aid in measuring the ultimate ozone resistance of the material. Insulations exhibiting superior ozone resistance under accelerated conditions are silicone, rubber, polyethylene, crosslinked polyethylene, ethylene/propylene rubber (EPR), and polyvinyl chloride. In fact, these materials are, for all practical purposes, inert in the presence of ozone. However, this is not the case with corona discharge.

The phenomenon of corona discharge produces concentrated and destructive thermal effects along with formation of ozone and other ionized gases. Although corona resistance is a property associated with cables over 600 V, in a properly designed and manufactured cable, damaging corona is not expected to be present at operating voltage. Materials exhibiting less susceptibility than polyethylene and crosslinked polyethylene to such discharge activity are the ethylene/propylene rubbers.

Table 52
Conductor Temperatures

Insulation Type	Maximum Voltage Class (kV)	Maximum Operating Temperature (°C)	Maximum Overload* Temperature (°C)	Maximum Short-Circuit Temperature (°C)
Paper (solid-type) multiconductor and single conductor, shielded	9	95	115	200
	29	90	110	200
	49	80	100	200
	69	65	80	200
Varnished cambric	5	85	100	200
	15	77	85	200
	28	70	72	200
Polyethylene (natural)†	5	75	95	150
	35	75	90	150
SBR rubber	2	75	95	200
Butyl rubber	5	90	105	200
	35	85	100	200
Oil-base rubber	35	70	85	200
Polyethylene (crosslinked)†	35	90	130	250
EPR rubber†	35	90	130	250
Chlorosulfonated polyethylene‡	2	90	130	250
Polyvinyl chloride	2	60	85	150
	2	75	95	150
	2	90	105	150
Silicone rubber	5	125	150	250
Ethylene tetra fluoroethylene‡‡	2	150	200	250

*Operation at these overload temperatures shall not exceed 100 h/yr. Such 100 h overload periods shall not exceed five.
‡Cables are available in 69 kV and higher ratings.
‡‡For example, hypalon.
‡‡‡For example, tefzel.

(d) *Moisture Resistance*. Insulations such as crosslinked polyethylene, polyethylene, and EPR exhibit excellent resistance to moisture as measured by standard industry tests such as ICEA EM-60 [12]. The electrical stability of these insulations in water as measured by capacitance and power factor is impressive, and moisture is cause for little concern. The capacitance and power factor by the ICEA EM-60 [12] test of natural polyethylene and some crosslinked polyethylenes are lower than those of EPR or other elastomeric power cable insulations.

(2) *Insulations in General Use*. Insulations in general use for 2 kV and above are shown in Table 52. Solid dielectrics of both plastic and thermosetting types are being more and more commonly used while the laminated-type constructions, such as paper-lead cables, are declining in popularity in commercial building service.

The generic names given for these insulations cover a broad spectrum of actual materials, and the history of performance on any one type may not properly be related to another in the same generic family.

8.3.4 Cable Design. The selection of power cable for particular circuits or feeders develops around the following considerations.

(1) *Electrical*. Dictates conductor size, type and thickness of insulation, correct materials for low- and medium-voltage designs, consideration of dielectric strength, insulation resistance, specific inductive capacitance (dielectric constant), and power factor.

(2) *Thermal*. Compatible with ambient and overload conditions, expansion, and thermal resistance.

(3) *Mechanical*. Involves toughness and flexibility, consideration of jacket-ing or armoring, and resistance to impact, crushing, abrasion, and mositure.

(4) *Chemical*. Stability of materials on exposure to oils, flame, ozone, sunlight, acids, and alkalies.

The installation of cable in conformance with ANSI/NFPA 70-1981 [4] and state and local codes under the jurisdiction of a local electrical inspection authority requires evidence of approval for use in the intended service by a nationally recognized testing laboratory, such as Underwriters Laboratories Inc (UL) labeled cable. Some of the types listed in ANSI/NFPA 70-1981 [4] are discussed below.

8.3.4.1 Low-Voltage Cables. Low-voltage power cables are generally rated at 600 V, regardless of the voltage use, whether 120 V, 208 V, 240 V, 277 V, 480 V, or 600 V.

The selection of 600 V power cable is oriented more to physical rather than to electrical service requirements. Resistance to forces such as crush, impact, and abrasion becomes a predominant factor, although good electrical properties for wet locations are also needed.

The 600 V compounds of crosslinked polyethylene are usually filled (carbon black or mineral fillers) to further enhance the relatively good toughness of conventional polyethylene. The combination of crosslinking the polyethylene molecules through vulcanization plus fillers produces superior mechanical properties. Vulcanization eliminates polyethylene's main drawback of a relatively low melting point of 105 °C. The 600 V cable consists simply of the conductor with a single extrusion of insulation in the specified thickness.

Rubberlike insulations such as EPR and SBR have been provided with outer jackets for mechanical protection, usually of polyvinyl chloride, neoprene, or CP rubber (such as hypalon). However, the newer EPR insulations have improved physical properties that do not require an outer jacket for mechanical protection. A guide list of the more commonly used 600 V cables is provided below. Cables are classified by conductor operating temperatures and coverings with ANSI/NFPA 70-1981 [4] insulation thickness.

(1) EPR insulated, with or without jacket, type RHW for 75 °C maximum operating temperature in wet or dry locations and type RHH for 90 °C in dry locations only.

(2) Crosslinked polyethylene insulated, without jacket, type XHHW, for 75 °C maximum operating temperature in wet or dry locations and 90 °C in dry locations only.

(3) Polyvinyl chloride insulated, nylon jacketed, type THWN, for 75 °C maximum operating temperature in wet or dry locations and type THHN, for 90 °C in dry locations only.

(4) Polyvinyl chloride insulated, without jacket, type THW, for 75 °C maximum operating temperature in wet or dry locations.

(5) Metal-clad or interlocked armor cable, type MC; individual insulated conductors are usually type XHHW or RHH/RHW and the cable has the rating of the conductors used; for use in any raceway, in cable tray, as open runs of cable, direct buried, or as aerial cable on a messenger.

(6) Tray cable, type TC; multiconductor with an overall flame-retardant nonmetallic jacket; individual conductors may be any of the above and cable takes rating of insulation selected; for use in cable trays, raceways, or where supported in outdoor locations by a messenger wire.

The preceding cables are suitable for installation in conduit, duct, or other raceway, and, when specifically approved for the purpose, may be installed in cable tray, or direct buried, provided ANSI/NFPA 70-1981 [4] requirements are satisfied.

Cables (2) and (4) are usually restricted to conduit or duct. Single conductors may be furnished paralleled or multiplexed, as multiconductor cables with overall nonmetallic jacket or as preassembled aerial cable on a messenger.

Note that the temperatures listed are the maximum rated operating temperatures as specified in ANSI/NFPA 70-1981 [4].

8.3.4.2 Power-Limited Circuit Cables. When the power in the circuit is limited to the levels defined in ANSI/NFPA 70-1981 [4], Article 725 for remote-control, signalling and power-limited circuits, then the wiring method may utilize power-limited circuit cable or type PLTC, Power-limited tray cable. These cables, which are rated 300 V, include both copper conductors for electrical circuits and thermocouple alloys for thermocouple extension wire.

Similarly power-limited fire-protective signaling-circuit cable used on circuits which comply with the power limitations of ANSI/NFPA 70-1981 [4], Article 760.

8.3.4.3 Medium-Voltage Cables. Type MV medium-voltage power cables have solid extruded dielectric insulation and are rated from 2001 V to 35 000 V. These single-conductor and multiple-conductor cables are available with nominal voltage ratings of 5 kV, 8 kV, 15 kV, 25 kV, and 35 kV.

EPR and crosslinked polyethylene are the usual insulating compounds for type MV cables, however, polyethylene and butyl rubber are also authorized as insulations. The maximum operating temperatures are 90 °C for EPR and cross-linked polyethylene, 85 °C for butyl rubber, and 75 °C for polyethylene.

Type MV cables may be installed in raceways in wet or dry locations. The cable must be specifically approved for installation in cable tray, direct burial, exposure to sunlight, or for messenger supported wiring.

Multiconductor medium-voltage cables that also comply with the requirements for type MC metal-clad cables will be labeled as type MV or MC and may be installed as open runs of cable.

8.3.4.4 Shielding of Medium-Voltage Cable. For operating voltages below 2 kV, nonshielded constructions are normally used, while above 2 kV, cables are required to be shielded to comply with ANSI/NFPA 70-1981 [4] and various ICEA standards. ANSI/NFPA 70-1981 [4] does provide for the use of nonshielded cables up to 8 kV provided the conductors are listed by a nationally recognized testing laboratory and are approved for the purpose. Where nonshielded conductors are used in wet locations, the insulated conductor(s) must have an overall nonmetallic jacket or a continuous metallic sheath, or both. Refer to ANSI/NFPA 70-1981 [4] for specific insulation thicknesses for wet or dry locations.

Since shielded cable is usually more expensive than nonshielded cable, and the more complex terminations require a larger space in the terminal boxes, the nonshielded cable has been used extensively at 2400 V and 4160 V and occasionally at 7200 V. However, any of the following conditions may dictate the use of shielded cable:

(1) Personnel safety
(2) Single conductors in wet locations
(3) Direct-earth burial

(4) Where the cable surface may collect unusual amounts of conducting materials (salt, soot, conductive pulling compounds)

Shielding of an electric power cable is defined as the practice of confining the electric field of the cable to the insulation surrounding the conductor by means of conducting or semiconducting layers, or both, which are in intimate contact or bonded to the inner and outer surfaces of the insulation. In other words, the outer insulation shield confines the electric field to the space between the conductor and the shield. The inner or strand stress relief layer is at or near the conductor potential. The outer or insulation shield is designed to carry the charging currents and in many cases fault currents. The conductivity of the shield is determined by its cross-sectional area and the resistivity of the metal tapes or wires employed in conjunction with the semiconducting layer.

The stress control layer at the inner and outer insulation surfaces, by its close bonding in the insulation surface, presents a smooth surface to reduce the stress concentrations and minimize void formation. Ionization of the air in such voids can progressively damage certain insulating materials to eventual failure.

Insulation shields have several purposes.
(1) Confine the electric field within the cable
(2) Equalize voltage stress within the insulation, minimizing surface discharges
(3) Protect cable from induced potentials
(4) Limit electromagnetic or electrostatic interference (radio, TV)
(5) Reduce shock hazard (when properly grounded)

Figure 48 illustrates the electrostatic field of a shielded cable.

The voltage distribution between an

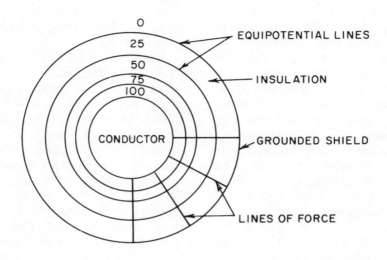

Fig 48
Electric Field of Shielded Cable

unshielded cable and a grounded plane is illustrated in Fig 49. There it is assumed that the air is the same, electrically, as the insulation, so that the cable is in a uniform dielectric above the ground plane to permit a simpler illustration of the voltage distribution and field associated with the cable.

In a shielded cable (Fig 48) the equipotential surfaces are concentric cylinders between conductor and shield. The voltage distribution follows a simple logarithmic variation, and the electrostatic field is confined entirely within the insulation. The lines of force and stress are uniform and radial, and cross the equipotential surfaces at right angles eliminating any tangential or longitudinal stresses within the insulation or on its surface.

The equipotential surfaces for the un-

shielded system (Fig 50) are cylindrical but not concentric with the conductor, and cross the cable surface at many different potentials. The tangential creepage stress to ground at points along the cable may be several times the normal recommended for creepage distance at terminations in dry locations for unshielded cable operating on 4160 V systems.

Surface tracking, burning, and destructive discharges to ground could occur under these conditions. However, properly designed nonshielded cables as described in ANSI/NFPA 70-1981 [4] limit the surface energies available which could compromise the cable from any of these effects.

Typical cables supplied for shielded and nonshielded applications are illustrated in Fig 51.

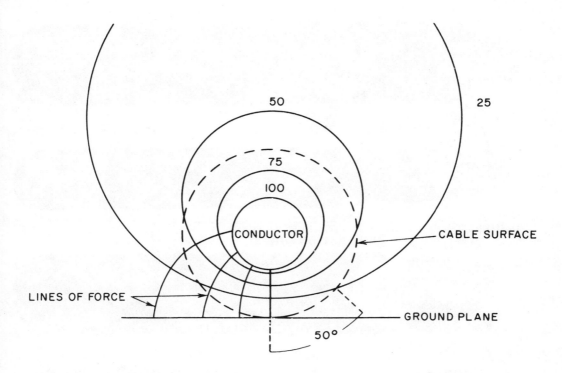

**Fig 49
Electric Field of Conductor on Ground Plane in Uniform Dielectric**

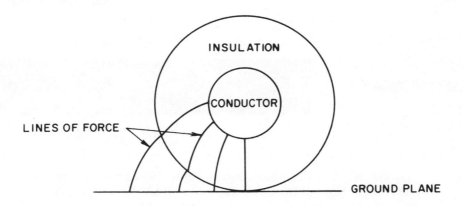

**Fig 50
Unshielded Cable on Ground Plane**

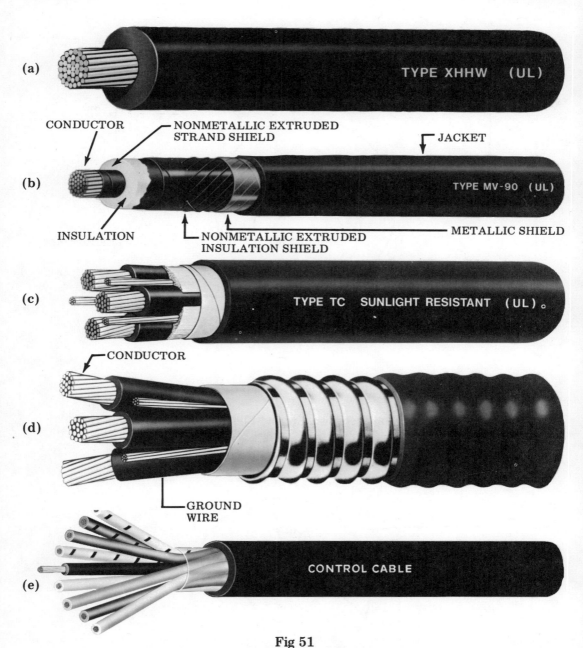

Fig 51
Commonly Used Shielded and Nonshielded Constructions
(a) Single-Conductor Cable (600 V or 5 kV Nonshielded)
(b) Type MV Medium-Voltage Single-Conductor Cable (5 kV—35 kV)
(c) Type TC Power and Control Tray Cable (600 V)
(d) Type MC Metal-Clad Power and Control Cable (600 V—35 kV)
(e) Multiconductor Control Cable (600 V)

8.4 Cable Outer Finishes. Cable outer finishes or outer coverings are used to protect the underlying cable components from the environmental and installation conditions associated with intended service. The choice of cable outer finishes for a particular application is based on the same performance categories as for insulations specifically related to electrical, thermal, mechanical, and chemical considerations. Except for highly protected and essentially room temperature operating environments, the usual conditions associated with cable operation cannot be met with a single outer finishing material. Therefore, combinations of metallic and nonmetallic finishes are required to provide the total protection needed for the installation and operation. Specific industry requirements for these materials are covered in standards of the ICEA, ASTM, and UL.

8.4.1 Nonmetallic Finishes

(1) *Extruded Jackets.* There are outer coverings, either thermoplastic or vulcanized, which may be extruded directly over insulation, or over electrical shielding systems of metal sheaths or tapes, copper braid, or semiconducting layers with copper drain wires or spiraled copper concentric wires; or over multiconductor constructions. Commonly used materials include polyvinyl chloride, nitrile butadiene/polyvinyl chloride (NBR/PVC), crosslinked polyethylene, polychloroprene (neoprene), chlorosulfonated polyethylene, and polyurethane. While the detailed characteristics may vary with individual manufacturers' compounding, these materials provide a high degree of moisture, chemical, and weathering protection, are reasonably flexible, provide some degree of electrical isolation, and are of sufficient mechanical strength to protect the insulating and shielding components from normal service and installation damage. Materials are available for service temperatures from $-55\,°C$ to $+115\,°C$ (Table 53).

(2) *Fiber Braids.* This category includes braided, wrapped or served synthetic or natural fiber materials selected by the cable manufacturer to best meet the intended service. While asbestoes fiber has been the most common material used in the past, fiberglas is now used extensively for employee health reasons. Some special industrial applications may

Table 53
Properties of Jackets and Braids

Material	Abrasion Resistance	Flexibility	Low Temperature	Heat Resistance	Fire Resistance
Neoprene	Good	Good	Good	Good	Good
Class CP rubber*	Good	Good	Fair	Excellent	Good
Crosslinked polyethylene	Good	Poor	Poor	Excellent	Poor
Polyvinyl chloride	Fair	Good	Fair	Good	Fair
Polyurethane	Excellent	Good	Good	Good	Poor
Glass braid	Fair	Good	Good	Excellent	Excellent
Nylon	Excellent	Fair	Good	Good	Fair

NOTE: Chemical resistance and barrier properties depend on the particular chemicals involved, and the question should be referred to the cable manufacturer.

*For example, hypalon.

require synthetic or cotton fibers applied in braid form. All fiber braids require saturants or coating and impregnating materials to provide some degree of moisture and solvent resistance as well as abrasive and weathering resistance.

Glass braid is used on cables to minimize flame propagation, smoking, and other hazardous or damaging products of combustion which may be evolved by some extruded jacketing materials.

8.4.2. Metallic Finishes. This category of materials is widely used where a high degree of mechanical, chemical, or short-time thermal protection of the underlying cable components may be required in the intended service. Commonly used materials are interlocked galvanized steel, aluminum, or bronze armor; extruded lead or aluminum; strip formed, welded, and corrugated aluminum or copper and spirally applied round or flat armor wires. The use of any of these materials, alone or in combination with others, will reduce flexibility of the overall cable. This characteristic must be sacrificed to obtain the other benefits.

Installation and operating conditions may involve localized compressive loadings, occasional impact from external sources, vibration and possible abrasion, heat shock from external sources, extended exposure to corrosive chemicals, and condensation.

(1) *Interlocked Armor.* The unprotected interlocked armor provides a high degree of mechanical protection without seriously reducing flexibility. While not entirely impervious to moisture or corrosive agents, interlocked armor does provide protection from thermal shock by acting as a heat sink for short-time localized exposure.

Where corrosion and moisture resistance are required in addition to mechanical protection, an overall jacket of extruded material may be used.

The use of interlocked galvanized steel armor must be avoided on single-conductor alternating-current power circuits due to its high hysteresis and eddy current losses. This effect, however, is minimized in three-conductor cables with armor overall and with aluminum armor on single-conductor cables.

Commonly used interlocked armor materials are galvanized steel, aluminum (for less weight and corrosion resistance) and marine bronze and other alloys for highly corrosive atmosphere.

(2) *Corrugated Metal Sheath.* Longitudinally welded and corrugated metal sheaths (corrugations or bellows formed perpendicular to the cable axis) have been used for many years in direct-burial communications cables, but only since 1960 has this method of cable core protection been applied to control and power cable. The sheath material may be of copper, aluminum, copper alloy, or a bimetallic composition with the choice of material selected to best meet the intended service.

The corrugated metal sheath offers mechanical protection equal to interlocked armor but at a lower weight. The aluminum or copper sheath may also be used as the equipment grounding conductor, either alone or in parallel with a grounding conductor within the cable.

The sheath is made from a metal strip which is longitudinally formed around the cable, welded into a continuous, impervious metal cylinder, and corrugated for pliability and increased radial strength. This sheath offers maximum protection from moisture and liquid or gaseous contaminants.

An extruded nonmetallic jacket must be used over the metal sheath for direct burial, embedment in concrete, or in areas which are corrosive to the metal sheath.

(3) *Lead.* Pure or alloy lead is used for industrial power cable sheaths for maximum moisture protection in underground manhole and tunnel or underground duct distribution systems subject to flooding. While not as resistant to crushing loads as interlocked armor or a corrugated metal sheath, its very high degree of corrosion and moisture resistance makes lead attractive in the above applications. Protection from installation damage can be provided by an outer jacket of extruded material.

Pure lead is subject to work hardening and should not be used in applications where flexing may be involved. Copper or antimony-bearing lead alloys are not as susceptible to work hardening as is pure lead, and may be used in applications involving limited flexing. Lead or its alloys must never be used for repeated flexing service.

One problem encountered today with the use of lead sheathed cable is in the area of splicing and terminating. Installing personnel experienced in the *art* of wiping lead sheath joints are not as numerous as they were many years ago, posing an installation problem for many potential users. However, many insulation systems do not require lead sleeves at splices and treat the lead like any other metallic sheath.

(4) *Aluminum or Copper.* Extruded aluminum or copper or die-drawn aluminum or copper sheaths are used in certain applications for weight reduction and moisture penetration protection. While more crush resistant than lead, aluminum sheaths are subject to electrolytic attack when installed underground. Under these conditions aluminum sheathed cable should be protected with an outer extruded jacket.

Mechanical splicing sleeves are available for use with aluminum sheathed cables, and sheath joints can be made by inert gas welding, provided that the underlying components can withstand the heat of welding without deterioration. Specifically designed hardware is available for terminating the sheath at junction boxes and enclosures.

(5) *Wire Armor.* A high degree of mechanical protection and longitudinal strength can be obtained with the use of spirally wrapped or braided round steel armor wire. This type of outer covering is frequently used in submarine cable and vertical riser cable for support. As noted for steel interlocked armor, this form of protection should be used only on three-conductor power cables to minimize sheath losses.

While not properly *armor*, spirally laid tinned copper wires, round or flat, are used in direct-burial cables as concentrics or neutrals. In potentially corrosive environments, these should be protected with an extruded jacket.

8.4.3 Single- and Multiconductor Constructions. The single-conductor cables are usually easier to handle and can be furnished in longer lengths as compared to multiconductor cables. The multiconductor constructions give smaller overall dimensions in comparison with the equivalent number of single-conductor cables, which can be an advantage where space is important.

Sometimes the outer finish can influence whether the cable should be supplied as a single- or multiconductor cable. For example, as mentioned previously, the use of steel interlocked or steel wire armor on alternating-current cables is practical on multiconductor constructions, but must be avoided over single-conductor cables. It is also economical to apply the more rugged finishes over multiconductor constructions rather than over each of the single-conductor cables.

8.4.4 Physical Properties of Materials for Outer Coverings. Depending on the environment and application, the selection of outer finishes to provide the degree of protection needed can be complex. For a general appraisal, Table 53 lists the relative properties of various commonly used materials.

8.5 Cable Ratings

8.5.1 Voltage Rating. The selection of the cable insulation (voltage) rating is made on the basis of the phase-to-phase voltage of the system in which the cable is to be applied, whether the system is grounded or ungrounded, and the time in which a ground fault on the system is cleared by protective equipment. It is possible to operate cables on ungrounded systems for long periods of time with one phase grounded due to a fault. This results in line-to-line voltage stress across the insulation of the two ungrounded conductors. Therefore such a cable must have greater insulation thickness than a cable used on a grounded system where it is impossible to impose full line-to-line potential on the other two unfaulted phases for an extended period of time.

Consequently 100% voltage rated cables are applicable to grounded systems provided with protection which will clear ground faults within 1 min. On ungrounded systems where the clearing time of the 100% level category cannot be met, and yet there is adequate assurance that the faulted section will be cleared within 1 h, 133% rated cables are required. On systems where the time required to de-energize a grounded section is indefinite, 173% voltage level insulation is used.

8.5.2 Conductor Selection. The selection of conductor size is based on the following considerations:

(1) Load-current criteria as related to loadings, ANSI/NFPA 70-1981 [4] requirements, thermal effects of the load current, mutual heating, losses produced by magnetic induction, and dielectric losses

(2) Emergency overload criteria

(3) Voltage-drop limitations

(4) Fault-current criteria

(5) Frequency criteria

8.5.3 Load-Current Criteria. The ANSI/ NFPA 70-1981 [4] ampacity tables for low- and medium-voltage cables must be used where the code has been adopted. These are derived from IEEE/ICEA S-135 [10] Power Cable Ampacities.

All ampacity tables show the minimum conductor size required, but conservative engineering practice, future load growth considerations, voltage drop, and short-circuit heating may make the use of larger conductors necessary.

Large groups of cables must be carefully considered, as derating due to mutual heating may be limiting. Conductor sizes over 500 kcmil — 750 kcmil require consideration of paralleling two or more smaller size cables because the current-carrying capacity per circular mil of conductor decreases for alternating-current circuits due to skin effect and proximity effect. The reduced ratio of surface to cross-sectional area of the larger size conductors is a factor in the reduced ability of the larger cable to dissipate heat. When cables are used in multiple, consideration must be given to the phase placement of the cable to minimize the effects of maldistribution of current in the cables, which will reduce ampacity. Although the material cost of cable may be less for two smaller conductors, this saving may be offset by higher installation costs.

The use of load factor in underground runs takes into account the heat capacity of the duct bank and surrounding soil which responds to the average heat losses. The temperatures in the underground section will follow the average loss, thus permitting higher short-period loadings. The load factor is the ratio of average load to peak load. Average load is usually measured on a daily basis. The peak load is usually the average of a $\frac{1}{2}$ to 1 h period of the maximum loading that occurs in 24 h.

For direct-buried cables, the average surface temperature is limited to 60 °C — 70 °C depending on soil condition, to prevent moisture migration and thermal runaway.

Cables must be derated when in proximity to other loaded cables or heat sources, or when the ambient temperature exceeds the ambient temperature on which the ampacity (current-carrying capacity) tables are based.

The normal ambient temperature of a cable installation is the temperature the cable would assume at the installed location with no load being carried on the cable. A thorough understanding of this temperature is required for a proper determination of the cable size required for a given load. For example, the ambient temperature for a cable exposed in the air isolated from other cables is the temperature of that cable before load is applied, assuming, of course, that this temperature is measured at the same time of day and with all other conditions exactly the same as they will be when the required load is being carried. It is assumed that for cables in air, the space around cable is large enough so that the heat generated by the cable can be dissipated without raising the temperature of the

room as a whole. Unless exact conditions are specified, the following ambients are commonly used for calculation of the current-carrying capacity.

(1) *Indoors.* The ANSI/NFPA 70-1981 [4] ampacity tables are based upon an ambient temperature of 30 °C for low-voltage cables. In most parts of the United States, 30 °C is too low for summer months, at least for some parts of the building. The ANSI/NFPA 70-1981 [4] type MV cable ampacity tables use 40 °C for air ambient temperature. In any specific case where the conditions are accurately known, the measured temperature should be used; otherwise, use 40 °C. Refer to ANSI/NFPA 70-1981 [4] Sections 318–10 and 318–12 for cables installed in cable tray.

Sources of heat adjacent to the cables under the most adverse condition must be taken into consideration in figuring the current-carrying capacity. This is usually done by correcting the ambient temperature for these localized hot spots. These may be caused by steam lines or heat sources adjacent to the cable, or they may be due to sections of the cable running through boiler rooms or other hot locations. Rerouting may be necessary to avoid this problem.

(2) *Outdoors.* An ambient temperature of 40 °C is commonly used as the maximum for cables installed in the shade and 50 °C for cables installed in the sun. In using these ambient temperatures, it is assumed that the maximum load occurs during the time when the ambient will be as specified. Some circuits probably do not carry their full load during the hottest part of the day or when the sun is at its brightest, so that an ambient temperature of 40 °C for outdoor cables is probably reasonably safe for such conditions. See ANSI/NFPA 70-1981 [4], NOTES to Tables

310–39 through 310–54, for the procedure to be used for outdoor installations. Refer to ANSI/NFPA 70-1981 [4] Sections 318–10 and 318–12 for cables installed in cable tray.

(3) *Underground.* The ambient temperature used for underground cables varies in different sections of the country. For the northern section, an ambient of 20 °C is commonly used. For the central part of the country, 25 °C is commonly used, while for the extreme south and southwest, an ambient of 30 °C may be necessary. The exact geological boundaries for these ambient temperatures cannot be set up, and the maximum ambient should be measured in the earth at a point away from any sources of heat at the depth at which the cable will be buried. Changes in the earth ambient will lag changes in the air ambient by several weeks.

The thermal characteristics of the medium surrounding the cable are of primary importance in determining the current-carrying capacity of the cable. The type of soil in which the cable or duct bank is buried has a major effect on the current-carrying capacity of cables. Porous soils, such as gravel and cinder fill, usually result in higher temperatures and lower ampacities than normal sandy or clay soil. The type of soil and its thermal resistivity should be known before the size of the conductor is calculated.

The moisture content of the soil has a major effect on the current-carrying capacity of cables. In dry sections of the country, cables may have to be derated or other precautions taken to compensate for the increase in thermal resistance due to the lack of moisture. On the other hand, in ground which is continuously wet or under tidewater conditions, cables may carry higher than normal currents. Shielding for even 2400 V circuits is necessary for continuously wet or alternately wet and dry conditions, for where the cable passes from a dry area to a wet area which provides *natural shielding*, there will be an abrupt voltage gradient stress, just as at the end of shielded cables terminated without a stress cone. Nonshielded cables specifically designed for this service are available.

Ampacities in the ANSI/NFPA 70-1981 [4] tables take into account the grouping of adjacent circuits. For ambient temperatures different from those shown in the tables, derating factors to be applied are shown in Tables 310–16 through 310–19, NOTES to Tables 310–16 through 310–39, and NOTES to Tables 310–39 through 310–54.

8.5.4 **Emergency Overload Criteria.** Normal loading limits of insulated wire and cable are based on many years of practical experience and represent a rate of deterioration that results in the most economical and useful life of such cable systems. The rate of deterioration is expected to result in approximately a useful life of 20 to 30 years. The life of cable insulation is about halved and the average rate of thermally caused service failures about doubled for each 5 °C—15 °C increase in normal daily load temperature. Additionally, sustained operation over and above maximum rated operating temperatures or ampacities is not a very effective or economical expedient, because the temperature rise is directly proportional to the conductor loss which increases as the square of the current. The greater voltage drop might also increase the risks to equipment and service continuity.

As a practical guide, the ICEA has established maximum emergency overload temperatures for various types of insulation. Operation at these emergency overload temperatures should not exceed 100 h/y, and such 100 h overload periods should not exceed five times during the life of the cable. Table 54 gives uprating factors for short-time overloads for various types of insulated cables. The uprating factor, when multiplied by the nominal current rating for the cable in a particular installation, will give the emergency or overload current rating for the particular insulation type.

A more detailed discussion on emergency overload and cable protection is contained in IEEE Std 242-1975 [8], Chapter 11.

8.5.5 Voltage-Drop Criteria. The supply conductor, if not of sufficient size, will cause excessive voltage drop in the circuit, and the drop will be in direct proportion to the circuit length. Proper starting and running of motors, lighting equipment, and other loads having heavy inrush currents must be considered. ANSI/NFPA 70-1981 [4] recommends that the steady-state voltage drop in power, heating, or lighting feeders be no more than 3%, and the total drop including feeders and branch circuits be no more than 5% overall.

8.5.6 Fault-Current Criteria. Under short-circuit conditions, the temperature of the conductor rises rapidly. Then, due to thermal characteristics of the insulation, sheath, surrounding materials, etc, it cools off slowly after the short-circuit condition is removed. The ICEA has recommended a transient temperature limit for each type of insulation for short-circuit duration times not in excess of 10 s.

Failure to check the conductor size for short-circuit heating could result in permanent damage to the cable insulation due to disintegration of insulation material, which may be accompanied by smoke and generation of combustible vapors. These vapors will, if sufficiently heated, ignite, possibly starting a serious fire. Less seriously, the insulation or sheath of the cable may be expanded to produce voids leading to subsequent failure. This becomes especially serious in 5 kV and higher voltage cables.

In addition to the thermal stresses, mechanical stresses are set up in the cable through expansion upon heating. As the heating is rapid, these stresses may result in undesirable cable movement. However, on modern cables, reinforcing binders and sheaths considerably reduce the effect of such stresses. Within the range of temperatures expected with coordinated selection and application, the mechanical aspects can normally be discounted except with very old or lead-sheathed cables.

During short-circuit or heavy pulsing currents, single-conductor cables will be subjected to forces tending to either attract or repel the individual conductors with respect to each other. Therefore such cables laid in trays, or racked, should be secured to prevent damage caused by such movements.

The minimum conductor size requirements for various rms short-circuit currents and clearing times are shown in Table 55. The ICEA initial and final conductor temperatures are shown for the various insulations. Table 52 gives conductor temperatures (maximum operating, maximum overload, and maximum short-circuit current) for various insulated cables.

A more detailed discussion of fault current and cable protection is contained in IEEE Std 242-1975 [8].

8.5.7 Frequency Criteria. In general, three-phase, 400 Hz power systems are designed in the same way that 60 Hz systems are designed, keeping in mind that the increased frequency will in-

Table 54
Uprating for Short-Time Overloads *

Insulation Type	Voltage Class (kV)	Conductor Operating Temperature (°C)	Conductor Overload Temperature (°C)	Uprating Factors for Ambient Temperature							
				20 °C		30 °C		40 °C		50 °C	
				Cu	Al	Cu	Al	Cu	Al	Cu	Al
Paper (solid type)	9	95	115	1.09	1.09	1.11	1.11	1.13	1.13	1.17	1.17
	29	90	110	1.10	1.10	1.12	1.12	1.15	1.15	1.19	1.19
	49	80	100	1.12	1.12	1.15	1.15	1.19	1.19	1.25	1.25
	69	65	80	1.13	1.13	1.17	1.17	1.23	1.23	1.38	1.38
Varnished cambric	5	85	100	1.09	1.08	1.10	1.10	1.13	1.13	1.17	1.17
	15	77	85	1.05	1.05	1.07	1.07	1.09	1.09	1.13	1.13
	28	70	72								
Polyethylene (natural)	35	75	95	1.13	1.13	1.17	1.17	1.22	1.22	1.30	1.30
SBR rubber	0.6	75	95	1.13	1.13	1.17	1.17	1.22	1.22	1.30	1.30
	5	90	105	1.08	1.08	1.09	1.09	1.11	1.11	1.14	1.14
Butyl RHH	15	85	100	1.09	1.08	1.10	1.10	1.13	1.13	1.17	1.17
	35	80	95	1.09	1.09	1.11	1.11	1.14	1.14	1.20	1.20
Oil-base rubber	35	70	85	1.11	1.11	1.14	1.14	1.20	1.20	1.29	1.29
Polyethylene (crosslinked)	35	90	130	1.18	1.18	1.22	1.22	1.26	1.26	1.33	1.33
Silicone rubber	5	125	150	1.08	1.08	1.09	1.09	1.10	1.10	1.12	1.11
EPR rubber	35	90	130	1.18	1.18	1.22	1.22	1.26	1.26	1.33	1.33
Chlorosulfonated polyethylene‡	0.6	75	95	1.13	1.13	1.17	1.17	1.22	1.22	1.30	1.30
Polyvinyl chloride	0.6	60	85	1.22	1.22	1.30	1.30	1.44	1.44	1.80	1.79
	0.6	75	95	1.13	1.13	1.17	1.17	1.22	1.22	1.30	1.30

*To be applied to normal rating determined for such installation conditions.
‡For example, hypalon.

Table 55

Minimum Conductor Sizes, in AWG or kcmil, for Indicated Fault Current and Clearing Times

Total RMS Current (amperes)	Polyethylene and Polyvinyl Chloride, 75–150 °C				Oil Base and SBR, 75–200 °C				Crosslinked Polyethylene and EPR, 90–250 °C			
	1/2 Cycle (0.0083 s)		10 Cycles (0.166 s)		1/2 Cycle (0.0083 s)		10 Cycles (0.166 s)		1/2 Cycle (0.0083 s)		10 Cycles (0.166 s)	
	Cu	Al	Cu	Al	Cu	Al	Cu	Al	Cu	Al	Cu	Al
5 000	10	8	4	2	10	8	4	3	12	10	4	3
15 000	6	4	2/0	4/0	6	4	1/0	3/0	6	4	1	3/0
25 000	3	2	4/0	350	4	2	3/0	250	4	3	3/0	250
50 000	1/0	2/0	400	700	1	2/0	350	500	2	1/0	300	500
75 000	2/0	4/0	600	1000	1/0	3/0	500	750	1/0	3/0	500	700
100 000	4/0	300	800	1250	3/0	250	700	1000	2/0	4/0	600	1000

crease the skin and proximity effects on the conductors, thereby increasing the effective copper resistance. For a given current, this increase in resistance results in increased heating and may require additional copper. The increased frequency will also increase the reactance, and this combined with the increased resistance will increase the voltage drop. The higher frequency will also increase the effect of magnetic materials upon cable reactance and heating. For this reason the cables should not be installed in steel or magnetic conduit or run along on magnetic structures in the building, etc.

The curves in Fig 52 show the ac/dc resistance ratio which exists on a 400 Hz system and the resulting reduction in current rating which is necessary from a heating standpoint to counteract the effect of the increased frequency.

The reactance can be taken as directly proportional to the frequency without introducing any appreciable errors. This method of determining reactance does not take into account the reduction due to proximity effect, but this change is not large and the error introduced by neglecting it is small.

The curves are applicable to any 600 V cable in the same nonmagnetic conduit, or to any interlocked armor cable with aluminum or bronze armor.

When voltage drop is the limiting factor, then paralleling smaller conductors should be considered.

8.6 Installation. There are a variety of ways to install power distribution cables in commercial installations. The engineer's responsibility is to select the method most suitable for each particular application. Each mode has characteristics which make it more suitable for certain conditions than others, that is, each mode will transmit power with a unique combination of reliability, safety, economy, and quality for any specific set of conditions. These conditions include the quantity and characteristics of the power being transmitted, the distance of transmission, and the degree of exposure to adverse mechanical and environmental conditions.

8.6.1 Layout. The first consideration in the layout of wiring systems is to keep the distance between the source and the load as short as possible. This consideration must be tempered by many other important factors to arrive at the lowest cost system that will operate within the reliability, safety, economy, and performance required. Some other factors that must be considered for various routings are the cost of additional cable and raceway versus the cost of additional supports, inherent mechanical protection provided in one alternative versus additional protection required in another, clearance for and from other facilities, and the need for future revision.

8.6.2 Open Wire. This mode was used extensively in the past. Although it has now been replaced in most applications, it is still quite often used for primary power distribution over large areas where conditions are suitable.

Open-wire construction consists of uninsulated conductors on insulators which are mounted on poles or structures. The conductor may be bare or it may have a covering for protection from corrosion or abrasion.

The attractive features of this method are its low initial cost and the fact that damage can be detected and repaired quickly. On the other hand, the uninsulated conductors are a safety hazard and are also highly susceptible to mechanical damage and electrical outage from birds, animals, lightning, etc. There

is increased hazard where crane or boom truck use may be involved. In some areas contamination on insulators and conductor corrosion can result in high maintenance costs.

Due to the large conductor spacing, open-wire circuits have a higher reactance than circuits with more closely spaced conductors which results in a higher voltage drop. This problem is reduced with higher voltage and higher power-factor circuits.

Exposed open-wire circuits are more susceptible to outages from lightning than other modes. The effects may be minimized though by the use of overhead ground wires and lightning arresters.

8.6.3 Aerial Cable. Aerial cable is usually used for incoming or service distribution between commercial buildings. The greatest gain is in replacing open wiring, where it provides greater safety and reliability and requires less space. Properly protected cables are not a safety hazard and are not easily damaged by casual contact. They are, however, open to the same objections as open wire so far as vertical clearance is concerned. Aerial cables are frequently used in place of the more expensive conduit systems, where the high degree of mechanical protection of the latter is not required. They are also generally more economical for long runs of one or two cables than are cable tray installations. It is cautioned that aerial cable having a portion of the run in conduit must be derated to the in-conduit ampacity for this condition.

Aerial cables may be either self-supporting or messenger supported. They may be attached to pole lines or structures. Self-supporting aerial cables have high tensile strength conductors for this application. Cables may be messenger supported either by spirally wrapping a steel band around the cables and the

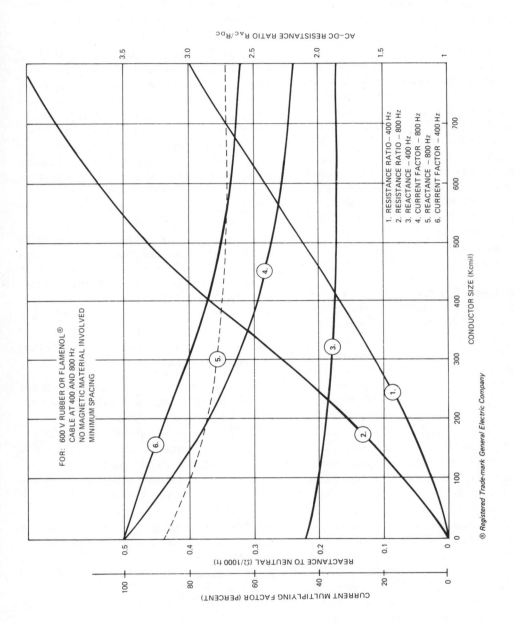

® Registered Trade-mark General Electric Company

**Fig 52
Curves Showing the ac/dc Resistance Ratio
which Exists on a 400 Hz System**

messenger or by pulling the cable into rings suspended from the messenger. The spiral wrap method is used for factory-assembled cable, while both methods are used for field assembly. A variety of spinning heads is available for application of the spiral wire banding in the field. Factory preassembled aerial cables are especially subject to installation damage from high stress at support sheaves when *pulled in.*

Self-supporting cable is suitable for only relatively short spans. Messenger-supported cable can span large distances, dependent on the weight of the cable and the tensile strength of the messenger. The supporting messenger provides high strength to withstand climatic rigors or mechanical shock. It may also serve as the grounding conductor of the power circuit.

A convenient feature available in one form of factory-assembled aerial cable makes it possible to form a slack loop to connect a circuit tap without cutting the cable conductors. This is done by reversing the direction of the spiral of the conductor cabling every $10 - 20$ ft.

Spacer cable is a type of electric supply-line construction that consists of an assembly of one or more covered conductors separated from each other and supported from a messenger by insulating spacers. This is another economical means of transmitting power overhead between buildings. Available for use in three-phase grounded or ungrounded systems at utilization voltages of 5 kV, 15 kV, 25 kV, or 35 kV, the insulated nonshielded phase conductors provide protection from accidental discharge through contact with ground-level equipment such as aerial ladders or crane booms. Uniform line electrical characteristics are obtained through the balanced geometric positioning of the con-

ductors with respect to each other by the use of plastic or ceramic spacers located at regular intervals along the line. Low terminating costs are obtained because the conductors are unshielded.

8.6.4 Direct Attachment. This is a low-cost method where adequate support surfaces are available between the source and the load. It is most useful in combination with other methods such as branch runs from cable trays and when adding new circuits to existing installations. Its use in commercial buildings is usually limited to low-energy control and telephone circuits.

This method employs multiconductor cable attached to surfaces such as structural beams and columns. A cable with metallic covering should be used where exposed to adverse mechanical conditions. Otherwise, plastic or rubber jacketed cable is satisfactory, provided this is approved by local or regional codes. For architectural reasons it is usually limited to service areas, hung ceilings, and electric shafts.

8.6.5 Cable Bus. Cable bus is used for transmitting large amounts of power over relatively short distances.

Cable bus uses insulated conductors in an enclosure which is similar to the cable tray with covers. The conductors are supported at maintained spacings by some form of nonmetallic spacer blocks. Cable buses are furnished either as components for field assembly or as completely assembled sections. The completely assembled sections are best if the run is short enough so that splices may be avoided. Multiple sections requiring joining may preferably employ the continuous conductors.

The spacing of the conductors is such that their maximum rating in air may be attained. This spacing is also close enough to provide low reactance, resulting in low-voltage drop.

8.6.6 Conduit. Among conduit systems, rigid steel affords the highest degree of mechanical protection available in aboveground conduit systems. Unfortunately, this is also a relatively high-cost system. For this reason their use is being superseded, where possible, by other types of conduit and wiring systems. Where applicable, rigid aluminum, intermediate grade steel conduit, thin-wall EMT, intermediate metal conduit, plastic, fiber, and cement ducts may be used.

Conduit systems offer some degree of flexibility in permitting replacement of existing conductors with new ones. However, in case of fire or faults it may be impossible to remove the conductors. In this case it is necessary to replace both conduit and wire at great cost and delay. Also during fires, conduits may transmit corrosive fumes into equipment where these gases can do much damage. To keep flammable gases out of such areas, seals must be installed.

With magnetic conduits, an equal number of conductors of each phase must be installed; otherwise, losses and heating will be excessive. For example, a single-conductor cable should not be used in steel conduit.

NOTE: Refer to ANSI/NFPA 70-1981 [4] for code regulations on conduit use.

Underground ducts are used where it is necessary to provide a high degree of mechanical protection. For example, when overhead conduits are subject to extreme mechanical abuse or when the cost of going underground is less than providing overhead supports. In the latter case, direct burial (without conduit) may be satisfactory under certain circumstances.

Underground ducts use rigid steel, plastic and fiber conduits encased in concrete, or precast multihole concrete duct banks with close fitting joints. Clay tile duct is also used to some extent. Where the added mechanical protection of concrete is not required, heavy wall versions of fiber conduits are direct buried as are rigid steel and plastic. High voltage, low voltage, signal and communications systems should not be run in the same manhole. Power-cable manholes should contain grounding conductors and drain provisions.

Cables used in underground conduits must be suitable for use in wet areas. Some cost savings can be realized by using flexible plastic conduits with conductors factory installed.

Where a relatively long distance between the point of service entrance into a building and the service entrance protective device is unavoidable, ANSI/NFPA 70-1981 [4], Section 230—202 (i) should be followed. The conductors must be placed under in a least 2 inches of concrete beneath the building, or they must be placed in conduit or duct and enclosed by concrete or brick not less than 2 inches thick. They are then considered outside the building.

8.6.7 Direct Burial. Cables may be buried directly in the ground where permitted by codes when the need for future maintenance along the cable run is not anticipated nor the protection of conduit required. The cables used must be suitable for this purpose, that is, resistant to moisture, crushing, soil contaminants and insect and rodent damage. Direct buried cables rated over 600 V nominal must be shielded and an exterior ground path for personnel safety must be provided in the event of accidental dig-in. Refer to ANSI/NFPA 70-1981 [4], Tables 300–5 and 710–3(b) for minimum depth requirements. The cost savings of this method over duct banks

Table 56
Wiring Methods for Hazardous Locations
(Based on ANSI/NFPA 70-1981)

Wiring Method	Class I Division 1	Class I Division 2	Class II Division 1	Class II Division 2	Class III Division 1 or 2
Threated rigid metal conduit	x	x	x	x	x
Threaded steel intermediate metal conduit*	x	x	x	x	x
Type MI mineral insulated cable	x	x	x	x	x
Type MC metal-clad cable		x		x	x
Type SNM shielded nonmetallic cable		x		x	
Type TC power and control tray cable		x			
Type MV medium-voltage cable		x			
Type PLTC power-limited tray cable		x			
Enclosed gasketed busways		x			
Dust-tight wireways				x	

*IMC must have NPT threads.

can vary from very little to a considerable amount. While this system cannot be readily added to or maintained, the current-carrying capacity for a cable of a given size is usually greater than that of cables in ducts. Buried cable must have selected back fill for suitable heat dissipation. It must be used only where chances of its being disturbed are small or it must be suitably protected if used where these chances exist. Relatively recent advances in the design and operating characteristics of cable fault location equipment and subsequent repair methods and material have diminished the maintenance problem.

8.6.8 Hazardous (Classified) Locations.
Wire and cable installed in locations where fire or explosion hazards may exist must comply with ANSI/NFPA 70-1981 [4], Articles 500 — 517. The authorized wiring methods are dependent upon the class and division of the specific area, see Table 56. The wiring method must be approved for the class and division, but is not dependent upon the group which defines the hazardous substance.

Equipment and the associated wiring system approved as intrinsically safe is permitted in any hazardous location for which it has been approved, and the requirements of ANSI/NFPA 70-1981 [4], Articles 500 — 517 are not applicable. However, the installation must prevent the passage of gases or vapors from one area to another. Intrinsically safe equipment and wiring is not capable of releasing sufficient electrical or thermal energy under normal or abnormal conditions to cause ignition of a specific hazardous atmospheric mixture in its most easily ignited concentration.

Seals must be provided in the wiring system to prevent the passage of the hazardous atmosphere along the wiring system from one division to the other or from a division to a non-hazardous location. The sealing requirements are defined in ANSI/NFPA 70-1981 [4], Articles 501-503. The use of multi-conductor cables with a gas/vapor-tight continuous outer sheath, either metallic or nonmetallic, which will not transmit gases or vapors through the cable core in excess of the allowable limits, can significantly reduce the sealing requirements in a Class 1, Division 2 hazardous location.

8.6.9 Installation Procedures. Care must be taken in the installation of raceways to make sure that no sharp edges exist to cut or abrade the cable as it is pulled in. Another important consideration is not to exceed the maximum allowable tensile strength or side-wall pressure of a cable. These forces are directly related to the force exerted on the cable when it is pulled in. This can be decreased by shortening the length of each pull and reducing the number of bends. The force required for pulling a given length can be reduced by the application of a pulling compound on cables in conduit and the use of rollers in cable trays.

If the cable is to be pulled by the conductors, the maximum tension in pounds is limited to 0.008 times the area of the conductors in circular mils. This tension may be further reduced if pulled by grips over the outside covering. A reasonable figure for most jacketed constructions would be 1000 lb per grip, but the calculated conductor tension should not be exceeded.

Side-wall pressures on most single-conductor constructions limit pulling tensions to approximately 450 lb times cable diameter, in inches, times radius of bend, in feet. Triplexed and paralleled cables would use their single-conductor diameters and a figure of 225 lb and 675 lb, respectively instead of the 450 lb factor for single-conductor construction.

For duct installations involving many bends, it is preferable to feed the cable into the end closest to the majority of the bends (in that the friction through the longer duct portion without the bends is not yet a factor) and pull from the other end. Each bend gives a multiplying factor to the tension it sees; therefore, the shorter runs to the bends will keep this increase in pulling tensions to a minimum. However, it is best to calculate pulling tensions for installation from either end of the run and install from the end offering the least tension.

The minimum bending radius for metal-taped cables is normally taken at twelve times the cable diameter, although cables with non-metallic covering can be bent to at least half this radius without disrupting the cable components. The minimum bending radius is applicable to bends of even a fraction of an inch in length, not just the average of a long length being bent.

When installing cables in wet underground locations, the cable ends must be sealed to prevent entry of moisture into the conductor strands. These seals should be left intact or remade after pulling if disrupted, until splicing, terminating, or testing is to be done. This practice is recommended to avoid unnecessary corrosion of the conductors and to safeguard against the generation of steam under overload, emergency loadings, or short-circuit conditions after the cable is placed in operation.

8.7 Connectors

8.7.1 Types Available. Connectors are classified as thermal or pressure, depending upon the method of attaching them to the conductor.

Thermal connectors use heat to make soldered, silver-soldered, brazed, welded, or cast-on terminals. Soldered connections have been used with copper conductors for many years, and their use is well understood. Aluminum connections may also be soldered satisfactorily with the proper materials and technique. However, soldered joints are not commonly used with aluminum. Shielded arc welding of aluminum terminals to aluminum cable makes a satisfactory termination for cable sizes larger than No 4/0. Torch brazing and silver soldering of copper cable connections are in use, particularly for underground connections with bare conductors such as are found in grounding mats. Thermite welding kits utilizing carbon molds are also in use for making connections with bare copper cable for ground mats and for junctions which will be below grade. These are satisfactory as long as the charge and tool are proper. The thermite welding process has also proved satisfactory for attaching connectors to insulated power cables.

Mechanical and compression pressure connectors are used for making joints in electric conductors. Mechanical type connectors obtain the pressure to attach the connector to the electric conductor from an integral screw, cone, or other mechanical parts. A mechanical connector thus applies force and distributes it suitably through the use of bolts or screws and properly designed sections. The bolt diameter and number of bolts are selected to produce the clamping and contact pressures required for the most satisfactory design. The sections are made heavy enough to carry rated current and withstand the mechanical operating conditions. These are frequently not satisfactory with aluminum, since only a portion of the strands are distorted by this connector.

Compression connectors are those in which the pressure to attach the connector to the electric conductor is applied externally, changing the size and shape of the connector and conductor.

The compression connector is basically a tube with the inside diameter slightly larger than the outer diameter of the conductor. The wall thickness of the tube is designed to carry the current, withstand the installation stresses, and withstand the mechanical stresses resulting from thermal expansion of the conductor. A joint is made by compressing the conductor and tube into another shape by means of a specially designed die and tool. The final shape may be indented, cup, hexagon, circular, or oval. All methods have in common the reduction in cross-sectional area by an amount sufficient to assure intimate and lasting contact between the connector and the conductor. Small connectors can be applied with a small hand tool. Larger connectors are applied with a hydraulic compression tool.

A properly crimped joint deforms the conductor strands sufficiently to have good electrical conductivity and mechanical strength, but not so much that the crimping action overcompresses the strands, thus weakening the joint.

Mechanical and compression connectors are available as tap connectors. Many types have an independent insulating cover. After a connection is made, the cover is assembled over the joint to insulate, and in some cases to seal against the environment.

8.7.2 Connectors for Aluminum. Aluminum conductors are different from copper in several ways, and these differences should be considered in specifying and using connectors for aluminum conductors (see Table 50). The normal oxide coating on aluminum is of relatively high electrical resistance. Aluminum has a coefficient of thermal expansion higher than that of copper. The ultimate and the yield strength properties and the resistance to creep of aluminum are different from the corresponding properties of copper. Corrosion is possible under some conditions because aluminum is anodic to other commonly used metals, including copper, when electrolytes even from humid air are present.

(1) *Mechancial Properties and Resistance to Creep.* Creep has been defined as the continued deformation of the material under stress. The effect of excessive creep resulting from the use of an inadequate connector which applies excessive stress, could be relaxation of contact pressure within the connector, and a resulting deterioration and failure of the electric connection. In mechanical connectors for aluminum, as for copper, proper design can limit residual unit-bearing loads to reasonable values, with a resulting minimum plastic deformation and creep subsequent to that initially experienced on installation. Connectors for aluminum wire can accommodate a range of conductor sizes, provided that the design takes into account the residual pressure on both minimum and maximum conductors.

(2) *Oxide Film.* The surface oxide film on aluminum, though very thin and quite brittle, has a high electrical resistance and therefore must be removed or penetrated to ensure a satisfactory electric joint. This film can be removed by abrading with a wire brush, steel wool, emery cloth, or similar abrasive tool or material. A plated surface, whether on the connector or bus, should never be abraded. It can be cleaned with a solvent or other means which will not remove the plating.

Some aluminum fittings are factory filled with a connection aid compound, usually containing particles which aid in obtaining low contact resistance. These compounds act to seal connections against oxidation and corrosion by preventing air and moisture from reaching contact surfaces. Connection to the inner strands of a conductor requires deformation of these strands in the presence of the sealing compound to prevent the formation of an oxide film.

(3) *Thermal Expansion.* The linear coefficient of thermal expansion of aluminum is greater than that of copper and is important in the design of connectors on aluminum conductors. Unless provided for in the design of the connector, the use of metals with coefficients of expansion less than that of aluminum can result in high stresses in the aluminum during heat cycles, causing additional plastic deformation and significant creep. Stresses can be quite high, not only because of the differences of coefficients of expansion, but also because the connector may operate at an appreciably lower temperature than the conductor. This condition will be aggravated by the use of bolts which are of a dissimilar metal or have different thermal expansion characteristics from those of the terminal.

(4) *Corrosion.* Direct corrosion from chemical agents affects aluminum no more severely than it does copper and, in most cases, less. However, since aluminum is more anodic than other common conductor metals, the opportunity exists for galvanic corrosion in the presence of moisture and a more cathodic metal. For this to occur, a wetted path must exist

between external surfaces of the two metals in contact to set up an electric cell through the electrolyte (moisture), resulting in erosion of the more anodic of the two, in this instance, the aluminum.

Galvanic corrosion can be minimized by the proper use of a joint compound to keep moisture away from the points of contact between dissimilar metals. The use of relatively large aluminum anodic areas and masses minimizes the effects of galvanic corrosion.

Plated aluminum connectors must be protected by taping or other sealing means.

(5) *Types of Connections Recommended for Aluminum.* UL has listed connectors approved for use on aluminum. Such connectors have successfully withstood UL performance tests which have recently been increased in severity. Both mechanical and compression-type connectors are available. The most satisfactory connectors are specifically designed for aluminum conductors to prevent any possible troubles from creep, the presence of oxide film, and the differences of coefficients of expansion between aluminum and other metals. These connectors are usually satisfactory for use on copper conductors in noncorrosive locations. The connection of an aluminum connector to a copper or aluminum pad is similar to the connection of bus bars. When both the pad and the connector are plated and the connection is made indoors, few precautions are necessary. The contact surfaces must be clean; if not, a solvent must be used. Abrasive cleaners are undesirable since the plating may be removed. In normal application, steel, aluminum, or copper alloy bolts, nuts, and flat washers may be used. A light film of a joint compound is acceptable, but not mandatory. When

either of the contact surfaces is not plated, the bare surface should be cleaned by wire brushing and then coated with a joint compound. Belleville washers are suggested for heavy-duty applications where cold flow or creep may occur, or where bare contact surfaces are involved. Flat washers should be used wherever Belleville washers or other load-concentrating elements are employed. The flat washer must be located between the aluminum lug, pad, or bolt and the outside edge of the Belleville washer with the neck or crown of the Belleville against the bolting nut to obtain satisfactory operation. In outdoor or corrosive atmosphere, the above applies with the additional requirement that the joint be protected. An unplated aluminum-to-aluminum connection can be protected by the liberal use of compound.

In an aluminum-to-copper connection, a large aluminum volume compared to the copper is important as is the placement of the aluminum above the copper. Again, coating with a joint compound is the minimum protection; painting with a zinc chromate primer or thoroughly sealing with a mastic or tape is even more desirable. Plated aluminum should be completely sealed against the elements.

(6) *Welded Aluminum Terminals.* For aluminum cables in 250 kcmil and larger sizes carrying high currents, excellent terminations can be made by welding special terminals to the cable. This is best done by the inert-gas shielded metal arc method. The use of inert gas eliminates the need for any flux to be used in making the weld. The welding type terminal is shorter than a compression terminal because the barrel for holding the cable can be very short. It has the advantage of requiring less room in junction or terminal boxes of equipment. Another

advantage is the reduced resistance of the connection. Each strand of the cable is bonded to the terminal, resulting in a continuous metal path for the current from every strand of the cable to the terminal.

Welding of these terminals to the cables may also be done with the tungsten electrode type of welding equipment with alternating-current power. The tungsten arc method is slower, but for small work gives somewhat better control.

The tongues or pads of the welding-type terminals, such as the large compression type, are available with bolt holes to conform to the standards of the National Electrical Manufacturers Association (NEMA) for terminals to be used on equipment.

(7) *Procedure for Connecting Aluminum Conductors (Fig 53)*

(a) When cutting cable, avoid nicking the strands. Nicking makes the cable subject to easy breakage [Fig 53(a)].

(b) Contact surfaces must be cleaned. The abrasion of contact surfaces is helpful even with new surfaces, and is essential with weathered surfaces. Do not abrade plated surfaces [Fig 53(b)].

(c) Apply joint compound to conductor if the connector does not already have it [Fig 53(c)].

(d) Use only connectors specifically tested and approved for use on aluminum conductors.

(e) For mechanical-type connectors tighten the connector with screwdriver or wrench to required torque. Remove excess compound [Fig 53(d)].

The cube terminal connector has the disadvantage of the mounting screw being located under the conductor, requiring conductor removal to tighten or check the mounting screw.

NOTE: This type of aluminum connector should be used only where permitted by local codes.

(a)

(b)

(c)

(d)

(e)

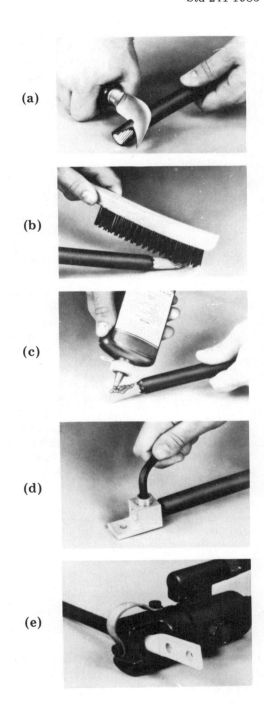

**Fig 53
Procedures for Connecting Aluminum
Conductors**

(f) For compression-type connectors crimp the connector using proper tool and die. Remove excess compound [Fig 53(e)].

(g) Always use a joint compound compatible with the insulation and as recommended by the manufacturer. The oxide-film penetrating or removing properties of some compounds aid in obtaining high initial conductivity. The corrosion inhibiting and sealing properties of some compounds help ensure the maintenance of continued high conductivity and prevention of corrosion.

(h) In making an aluminum-to-copper connection which is exposed to moisture, place the aluminum conductor above the copper. This prevents soluble copper salts from reaching the aluminum conductor, which could result in corrosion. If there is no exposure to moisture, the relative position of the two metals is not important.

(i) When using insulated conductors outdoors, extend the conductor insulation or covering as close to the connector as possible to minimize weathering of the joint. Outdoors, whenever possible, joints should be completely protected by tape or other means. If outdoor joints are to be covered or protected, the protection should completely exclude moisture, as the retention of moisture could lead to severe corrosion.

8.7.3 Connectors for Cables of Various Voltages. Standard mechanical or compression-type connectors are recommended for all primary voltages provided the bus is uninsulated. Welded connectors may also be used for conductors sized in circular mils. Up to 600 V, standard connector designs present no problem for insulated or uninsulated conductors.

Mechanical connectors should not be used over 600 V. The standard compression-type connectors are recommended for use on insulated conductors up to 5 kV. Above 5 kV, dielectric stress considerations make it desirable to use tapered-end compression connectors or semiconducting tape construction to give the same effect.

8.7.4 Performance Requirements. Electric connectors for commercial buildings are designed to meet the requirements of ANSI/NFPA 70-1981 [4]. They are evaluated on the basis of their ability to pass secureness, heating, heat-cycling, and pull-out tests as outlined in ANSI/ UL 486-1975 [5] and UL 486A-1981, [12]. This standard has been recently revised to incorporate more stringent requirements for aluminum terminating devices. The reader is cautioned to use only those lugs meeting the current UL standard.

electrical and mechanical operating requirements. Electrically, the connectors must carry the current without exceeding the temperature rise of the conductors being joined. Joint resistance not appreciably higher than that of an equal length of conductor being joined is recommended to assure long life and satisfactory operation of the joint. In addition, the connector must be able to withstand momentary overloads or short-circuit currents to the same degree as the conductor itself. Mechanically, a connector must be able to withstand the effects of the environment within which it is operating. If outdoors, it must stand up against temperature extremes, wind, vibration, rain, ice, sleet, chemical attack, etc. If used indoors, any vibration from rotating machinery, corrosion caused by plating or manufacturing processes, high temperatures from furnaces, etc, must not materially affect the performance of the joint.

8.8 Terminations

8.8.1 Purpose. A termination for an insulated power cable must provide certain basic electrical and mechanical functions. The following list is included in these essential requirements.

(1) Electrically connect the insulated cable conductor to electric equipment, bus, or uninsulated conductor.

(2) Physically protect and support the end of the cable conductor, insulation, shielding system, and overall jacket, sheath, or armor of the cable.

(3) Effectively control electrical stresses to provide both internal and external dielectric strength to meet desired insulation levels for the cable system.

The current-carrying requirements are the controlling factors in the selection of the proper type and size of connector or lug to be used. Variations in these components are related, in turn, to the base material used to make up the conductor within the cable, the type of termination used, and the requirements of the electric system.

The physical protection offered by the termination will vary considerably, depending on the requirements of the cable system, the environment, and the type of termination used. The termination must provide an insulating cover at the cable end to protect the cable components (conductor, insulation, and shielding system) from damage by any contaminants which may be present, including gases, moisture, and weathering.

Shielded high-voltage insulated cables are subject to unusual electrical stresses where the cable shield system is ended just short of the point of termination. The creepage distance that must be provided between the end of the cable shield, which is at ground potential, and the cable conductor, which is at line potential will vary with the magnitude of

the voltage, the type of terminating device used, and, to some degree, the kind of cable used. The net result is the introduction of both radial and longitudinal voltage gradients which impose dielectric stress of varying magnitude at the end of the cable. The termination provides a means of reducing and controlling these stresses within the working limits of the cable insulation and materials used to make up the terminating device itself.

8.8.2 Definition. The definitions for cable terminations are contained in IEEE Std 48-1975 [7].

A Class 1 high-voltage cable termination, or more simply, a Class 1 termination, provides:

(1) Some form of electric stress control for the cable-insulation shield terminus

(2) Complete external leakage insulation between the high-voltage conductor(s) and ground

(3) A seal to prevent the entrance of the external environment into the cable and to maintain the pressure, if any, within the cable system. This classification encompasses what was formerly referred to as a pothead.

A Class 2 termination is one that provides only (1) and (2): some form of electric stress control for the cable insulation shield terminus and complete external leakage insulation, but no seal against external elements. Terminations falling into this classification would be, for example, stress cones with rain shields or special outdoor insulation added to give complete leakage insulation, and the more recently introduced slip-on terminations for cables having extruded insulation when not providing a seal as in Class 1.

A Class 3 termination is one that provides only (1): some form of electric stress control for the cable-insulation shield terminus. This class of termination would be for use primarily indoors. Typically, this would include hand-wrapped stress cones (tapes or pennants), and the slip-on stress cones.

Refer to IEEE Std 48-1975 [7].

8.8.3 Cable Terminations. The requirements of the installation dictate that the termination must be designed for a specific end use. The least imposing is an indoor installation such as within a building or inside a protective housing. Here the termination is subjected to a minimum exposure to the elements.

Outdoor installations expose the termination to the elements and require that features be included in its makeup to withstand this exposure. In some areas the air can be expected to carry a high percentage of gaseous contaminants, liquid or solid particles which may be conducting, either alone or in the presence of moisture. These environments impose an even greater demand on the termination for protection of the cable end from the damaging contaminants and for the termination itself to withstand exposure to these contaminants. The termination may be required to perform its intended function while partially or fully immersed in a liquid or gaseous dielectric. These exposures impose upon the termination the necessity of complete compatibility between the liquids and exposed parts of the termination, including any gasket sealing material. Generally speaking, cork gaskets, have been used in the past, but the more recent materials such as tetrafluroethylene (for example, teflon) and silicone provide superior gasketing characteristics. The gaseous dielectrics may be nitrogen or any of the electronegative gases, such as sulfur hexafluoride, that are used for the dielectric property of the gas to fill electric equipment.

8.8.3.1 Nonshielded Cable. Cables usually have a copper or aluminum conductor with a rubber or plastic type insulating system without a shield. Terminations for these cables generally consist of a lug and may be taped. The lug is fastened to the cable by one of the several methods described in 8.7 and tape is applied over the lower portion of the barrel of the lug and down onto the cable insulation. Tapes used for this purpose are selected on the basis of compatibility with the cable insulation and suitability for application in the environmental exposure anticipated.

8.8.3.2 Shielded Cable. Cables above 600 V may have a copper or aluminum conductor with either an extruded solid-type insulation such as rubber, polyethylene, etc, or a laminated insulating system such as oil-impregnated paper tapes, varnished-cloth tapes, etc. A shielding system must be used on solid dielectric cables rated 2 kV and higher unless the cable is specifically listed or approved for nonshielded use (see 8.3.4.3).

At the point of termination, the cutback of the cable shielding to provide necessary creepage distances between the conductor and shielding introduces a longitudinal stress over the surface of the exposed cable insulation. The resultant combination of radial and longitudinal electrical stress at the termination of the cable end results in maximum stress occurring at the point. However, these stresses can be controlled and reduced to values within safe working limits of the materials used to make up the termination. The most common method of reducing these stresses is to gradually increase the total thickness of insulation at the termination by adding insulating tapes to form a cone. The cable shield-

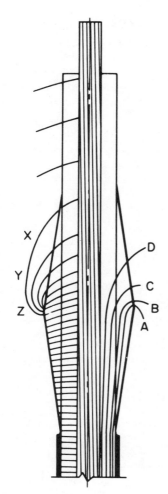

X, Y, Z — Electric Stress Lines;
A, B, C, D — Equipotential Lines

Fig 54
Stress-Relief Cone

ing is carried up the cone surface and terminated at a point approximately $\frac{1}{8}$ in behind the largest diameter of the cone. This construction, illustrated in Fig 54, is commonly referred to as a stress-relief cone.

It is advisable to consult individual cable manufacturers for their recommendations in terminating and splicing shielded cables.

8.8.3.3. Class 1 Terminator. A Class 1 terminator (Fig 55) is a hermetically sealed device used to enclose and protect a cable end. It consists of a metallic body with one or more porcelain insulators. The body is arranged to accept a variety of optional cable entrance sealing fittings, while the porcelain bushings, in turn, are designed to accommodate a number of optional cable conductor and aerial connections. These parts are field assembled to the prepared cable ends (taped stress-relief cones required for shielded cables), and the assembled unit is filled with an insulating compound. Considerable skill is required for proper installation of Class 1 terminations, particularly in filling and cooling out, to avoid shrinkage and void formation in the fill material.

Class 1 terminators are available in ratings of 5 kV and above for either single-conductor or three-conductor installation and for indoor, outdoor, or liquid-immersed application. Mounting variations include bracket, plate, and flanged types.

Class 1 terminators offer a greater degree of protection to the cable elements (conductor, insulation, and shielding system) than the taped terminations and are generally preferred for applications involving exposure to a high degree of environmental contamination. Equipment applications where there is an exposure to liquid dielectrics usually employ Class 1 terminators since these units may be directly attached to the equipment.

Both cable construction and type of application must be considered in the selection of a Class 1 terminator. Voltage rating, desired basic impulse insula-

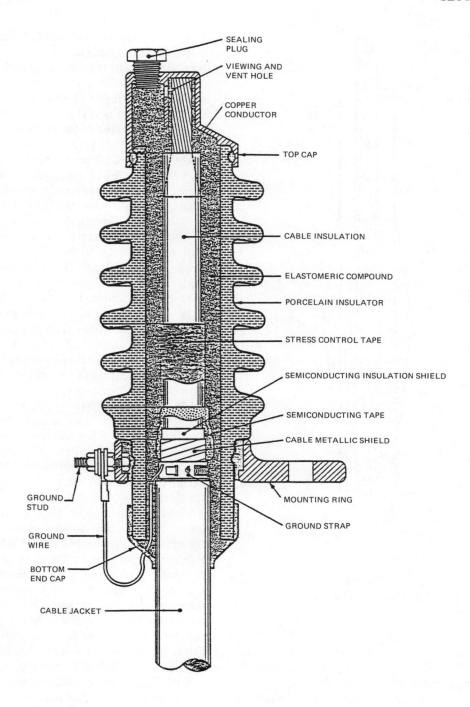

**Fig 55
Typical Class 1 Terminator**

tion level, conductor size, and current requirements are also basic considerations in the selection of a Class 1 terminator. Cable construction is the controlling factor in the selection of proper entrance sealing fittings, stress-relief cone materials, and filling compound. Application, in turn, is the prime consideration for selecting the type of Class 1 terminator, method of mounting, and desired terminal connectors.

Class 1 terminators can be subdivided into a number of groups or types related to cable system requirements and construction of the devices. Cable system requirements may be arranged in two general groups, nonpressurized and pressurized. Most power cable systems used in commercial buildings are of the nonpressurized type using solid dielectric insulated cables. The two most commonly used Class 1 terminator types for these solid-type cables are the capnut and solder seal. The capnut Class 1 terminator is made up of cast metal parts (iron, aluminum, or bronze) with gasketed joints between metal parts and the porcelains. The metal parts for the solder seal Class 1 terminator are copper spinnings which are solder-bonded directly to the porcelain insulator, thus eliminating several or all gasketed joints. Arrangements other than terminal up, cable down, may be expected to have higher failure rates.

Class 1 terminators are filled with a high-strength dielectric and designed to be electrically stronger internally than externally. However, they require great care and skill in installation.

Dielectrics in common use for filling Class 1 terminators include asphaltic based materials, resins, and oils. The asphaltic and some resin materials must be heated to liquify them, then poured into the Class 1 terminator, and allowed to cool. Special techniques are employed to control the rate of cooling so that voids which may result from the shrinkage of the compound with cooling can be forced to occur in areas of little or no electrical stress.

8.8.3.4 Class 2 Terminator. Advances in the art of terminating single-conductor cables include several types of units designed to reduce the required cable end preparation and eliminate the *hot fill with compound step* associated with Class 1 terminators. One type applies elastomeric materials directly to the cable end. This type is offered both with and without a metal-porcelain housing and is applicable only on solid dielectric cables (Fig 56). The other type consists of a metal-porcelain housing filled with a gelatin-like substance designed to be partially displaced as the terminator is installed on the cable. This latter unit may be used on any non-pressure-type cable.

Advantages of the preassembled terminators include simplified installation procedures and reduced installation time. Accordingly, they can be installed by less skilled workers, yet they offer a high degree of consistency to the overall quality and integrity of the installed system.

Preassembled terminators are available in ratings of 15 kV and above for most types of application. The porcelain-housed units include flanged mounting arrangements for equipment mounting and liquid-immersed applications. Although ratings below 15 kV are not offered, it is often desirable to use the 15 kV units on 5 kV systems when terminating shielded cables.

Selection of Class 2 preassembled terminators is essentially the same as for Class 1 terminators, with the exception that those units using solid elastomeric materials must be sized, with close tolerance, to the cable diameters to provide proper fit.

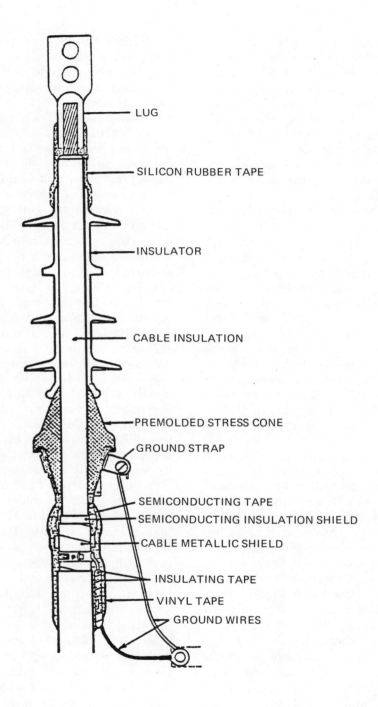

LUG

SILICON RUBBER TAPE

INSULATOR

CABLE INSULATION

PREMOLDED STRESS CONE

GROUND STRAP

SEMICONDUCTING TAPE
SEMICONDUCTING INSULATION SHIELD

CABLE METALLIC SHIELD

INSULATING TAPE

VINYL TAPE

GROUND WIRES

Fig 56
Typical Class 2 Terminator

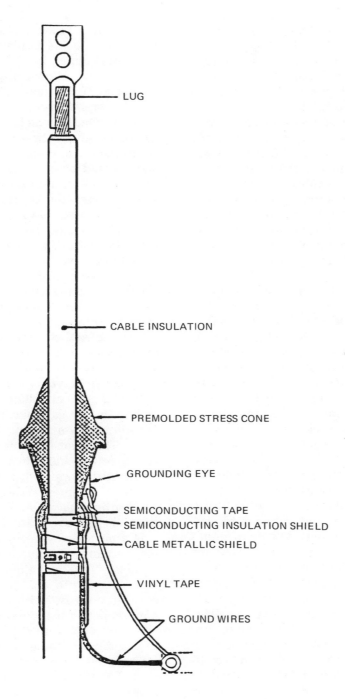

Fig 57
Typical Class 3 Terminator

8.8.3.5 Class 3 Terminator. Indoor applications or equipment-type installations where the apparatus housing provides weather protection for the cable termination may be made up using a preformed stress-relief cone (Fig 57). The most common preformed stress cone is a two-part elastomeric assembly consisting of a semiconducting lower section formed in the shape of a stress-relief cone and an insulating upper section. In addition, some types are housed in a hard plasticlike-insulating protective housing. Both types are applicable for use on shilded cables with extruded solid-type insulation, such as rubber, polyethylene, etc.

Taped terminations (Fig 55) may be used either indoors or outdoors and on shielded or unshielded cables. Generally, taped terminations are used at 15 kV and below; however, there are instances where taped terminations are used on some cable types to 69 kV. On unshielded cables the termination is made up with only a lug and tape, either indoors or outdoors. Terminations of shielded cables generally require the use of a stress-relief cone and cover tapes in addition to the lug. The size and location of the stress cone is controlled primarily by the operating voltage and location of the termination, that is, indoors or outdoors.

Creepage of 1 in per kilovolt of nominal system voltage is commonly used for indoor application, and from 1 to 2 inches or more per kilovolt of system voltage is allowed for outdoor installations. Additional creepage may be gained by using an insulator of neoprene, rubber, plastic, or porcelain for outdoor installations. Insulating tapes for the stress-relief cone are selected to be compatible with the cable insulation, and tinned copper braid or semiconducting tape is used as a conducting material for the cone. Cover tapes are applied over the stress cone, cable insulation, and up into the connecting lug. These cover tapes may be vinyl, silicone rubber, tetrafluoroethylene, (for example, teflon) or other suitable taping materials.

The insulators which are generally used on the outdoor taped terminations are positioned in the creepage path between the lug and stress cone. On upright termination the rain hood is usually placed directly over the stress-relief cone, and its primary function is to keep some portion of the cable insulation along the creepage path dry at all times. Some installations have been made with two or more insulators in the creepage path. Where unusual surface contamination is expected, Class 1 terminators should be used.

8.8.4 Sheathed and Armored Cable Connectors. Outer coverings for these cables may be nonmetallic, such as neoprene, polyethylene, or polyvinyl chloride; or metallic, such as lead, aluminum, or galvanized steel, or both, depending upon the installation environment. These latter two metallic coverings are generally furnished in a helical or corrugated form with sections joined either by an interlocking arrangement or by continuous weld. The terminations available for use with these cables provide a means of securing the outer covering and may include conductor terminations. The techniques for applying them vary with the type of cable, its construction, its voltage rating, and the requirements for the installation.

The outer covering of multiconductor cables must be secured at the point of termination using cable connectors approved both for the cable and the installation conditions. Type MC metal-clad cables with a continuously welded and corrugated sheath or an interlocking tape

280

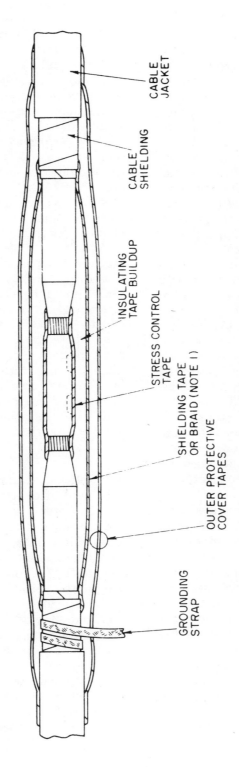

NOTES: (1) Heavy braid jumper or perforated strip should be used across splice to carry possible ground-fault current. Stress control tape should cover strands completely, lapping slightly onto insulation taper.
(2) Consult individual cable supplier for recommended installation procedures and materials.

Fig 58
Typical-Taped Splice in Shielded Cable

armor require, in addition to cable terminators, an arrangement to secure and ground the armor. Fittings available for this purpose are generally called armored cable connectors. These armored cable connectors provide mechanical termination and electrically ground the armor. This is particularly important on the continuous corrugated aluminum sheath since the sheath is the grounding conductor. In addition the connector may provide a water-tight seal for the cable entrance to a box, compartment, pothead, or other piece of electric equipment.

These connectors are sized to fit the cable armor and are designed for use on the cable alone, with brackets, or with locking nuts or adaptors for application to other pieces of equipment.

8.8.5 Separable Insulated Connectors. These are two-part devices used in conjunction with high-voltage electric apparatus. A bushing assembly is attached to the high-voltage apparatus (transformer, switch, or fusing device, etc), and a molded plug-in connector is used to terminate the insulated cable and connect the cable system to the bushing. The dead-front feature is obtained by fully shielding the plug-in connector assembly.

Two types of separable insulated connectors for application at 15 kV and 25 kV are available, one load break and the other nonload break. Both are essentially of a molded construction design for use on solid dielectric insulated cables (rubber, cross-linked polyethylene, etc) and are suitable for submersible application. The connector section of the device assumes an elbow-type (90 °) configuration to facilitate installation, improve separation, and save space. See ANSI/IEEE Std 386-1977 [3].

Electric apparatus may be furnished with a *universal bushing wall* only for future installation of bushings for either the load break or nonload break deadfront assemblies. Shielded-elbow connectors may be furnished with a *voltage detection tap* to provide a means of determining whether or not the circuit is energized.

8.8.6 Performance Requirements. Design test criteria have been established for terminators and are listed in IEEE Std 48-1975 [7] which outlines shorttime alternating-current 60 Hz and impulse withstand requirements. Also listed in this design standard are maximum direct-current field-proof test voltages. Individual types may safely withstand higher test voltages, and the manufacturer should be contacted for such information. All devices employed to terminate insulated power cables should meet these basic requirements. Additional performance requirements may include thermal load cycle capabilities of the current-carrying components, environmental performance of completed units, and long-time overvoltage withstand capabilities of the device.

8.9 Splicing Devices and Techniques. Splicing devices are subjected to a somewhat different set of voltage gradients and dielectric stress from that of a cable termination. In a splice, as in the cable itself, the highest stresses are around the conductor and connector area and at the end of the shield. Splicing design must recognize this fundamental consideration and provide the means to control these stresses to values within the working limits of the materials used to make up the splice.

In addition, on shielded cables the splice is in the direct line of the cable system and must be capable of handling any ground currents or fault currents that may pass through the cable shielding.

The connectors used to join the cable conductors together must be electrically capable of carrying full-rated load, emergency overload, and fault currents without overheating as well as being mechanically strong enough to prevent accidental conductor pullout or separation.

Finally, the splice housing or protective cover must provide adequate protection to the splice, giving full consideration to the nature of the application and its environmental exposure.

(1) *600 V and Below.* An insulating tape is applied over the conductor connection to electrically and physically seal the joint. The same taping technique is employed in the higher voltages, but with more refinement to cable-end preparation and tape applications.

Insulated connectors are used where several relatively large cables must be joined together. These terminators, called *moles* or *crabs* are, fundamentally, insulated buses with provision for making a number of tap connections which can be very easily taped or covered with an insulating sleeve. Connectors of this type enable a completely insulated multiple connection to be made without the skilled labor normally required for careful *crotch* taping or the expense of special junction boxes. One widely used type is a preinsulated multiple-outlet joint in which the cable connections are made mechanically by compression cones and clamping nuts. Another type is a more compact preinsulated multiple joint in which the cable connections are made by standard compression tooling which indents the conductor to the tubular cable sockets. Also available are tap connectors accomodating a range of conductor sizes and having an independent insulating cover. After the connection is

made, the cover is snapped closed to insulate the joint.

Insulated connectors lend themselves particularly well to underground services and industrial wiring where a large number of multiple-connection joints must be made.

(2) *Over 600 V.* Splicing of unshielded cables up to 8 kV consists of assembling a connector, usually soldered or pressed onto the cable conductors, and applying insulating tapes to build up an insulation wall to a thickness of 1½ to 2 times that of the factory-applied insulation on the cable. Care must be exercised in applying the connector and insulating tapes to the cables, but it is not as critical with unshielded cables as with shielded cables.

Aluminum conductor cables require a waterproof joint to prevent moisture entry into the stranding of the aluminum conductors. Splices on solid dielectric cables are made with uncured tapes which will fuse together after application and provide a waterproof assembly. It is necessary, however, to use a moistureproof adhesive between the cable insulation and the first layer of insulating tapes. Additional protection may be obtained through the use of a moistureproof cover over the insulated splice. This cover may consist of additional moistureproof tapes and paint or a sealed weatherproof housing of some form.

8.9.1 Taped Splices (Fig 58). Taped splices for shielded cables have been used quite successfully for many years. Basic considerations are essentially the same as for unshielded cables. Insulating tapes are selected not only on the basis of dielectric properties but also for compatability with the cable insulation. The characteristics of the insulating tapes must also be suitable for the application of the splice. This latter consideration gives attention to details such as providing a moisture seal for splices subjected to water immersion or direct burial, thermal stability of tapes for splices subjected to high-ambient and operating temperatures, and ease of handling for applications of tapes on wye- or tee-type splices.

Connector surfaces must be smooth and free from any sharp protrusions or edges. The connector ends are tapered, and indentations or distortion caused by pressing tools are filled and shaped to provide a round smooth surface. Semiconducting tapes are recommended for covering the connector and exposed conductor stranding to provide a uniform surface over which insulating tapes can be applied. Cables with a solid-type insulation are tapered and those with a tape-type insulation are stepped to provide a gradual transition between conductor-connector diameter and cable-insulation diameter prior to the application of insulating tapes. This is done to control the voltage gradients and resultant voltage stress to values within the working limits of the insulating materials. The splice should not be overinsulated to provide additional protection since this could restrict heat dissipation at the splice area and risk splice failure.

A tinned copper braid is used to provide the shielding function over the splice area. Grounding straps are applied to at least one end of the splice for grounding purposes, and a heavy braid jumper is applied across the splice to carry available ground-fault current.

Final cover tapes or weather barriers are applied over the built up splice to seal it against moisture entry. A splice on a cable with lead sheath is generally housed in a lead sleeve which is solder wiped to the cable sheath at each end of the splice. These lead sleeves are filled with compound in much the same manner as potheads.

Hand-taped splices may be made between lengths of dissimilar cables if proper precautions are taken to ensure the integrity of the insulating system of each cable and the tapes used are compatible with both cables. One example would be a splice between a rubber-insulated cable and an oil-impregnated paper-insulated cable. Such a splice must have an oil barrier to prevent the oil impregnant in the paper cable from coming in contact with the insulation on the rubber cable. In addition, the assembled splice must be made completely moistureproof. This requirement is usually accomplished by housing the splice in a lead sleeve with wiped joints at both ends. A close-fitting lead nipple is placed on the rubber cable and is taped or epoxy sealed to the jacket of the rubber cable. The solder wipe is made to this lead tube.

Three-way wye and tee splices and the several other special hand-taped splices that can be made all require special design considerations. In addition, a high degree of skill on the part of the installer is a prime requirement for proper makeup and service reliability.

8.9.2 Preassembled Splices. Similar to the preassembled terminators are several types of factory-made splices. The most elementary is an elastomeric unit consisting of a molded housing sized to fit the cables involved, a connector for joining the conductors, and tape seals for sealing the ends of the molded housing to the cable jacket. Other versions of elastomeric units include an overall protective metallic housing which completely encloses the splice. These preassembled elastomeric-type splices are available in two-way-, three-way-tee and multiple configurations for application to 35 kV and can be used on most cables having an extruded solid-type insulation.

The preassembled splice provides a waterproof seal to the cable jacket and is suitable for submersible, direct-burial, and other applications where the splice housing must provide protection for the splice to the same degree that the cable jacket provides protection to the cable insulation and shielding system. An advantage of these preassembled splices is the reduced time to complete the splice after cable-end preparation. However, the solid elastomeric materials used for the splice must be sized, with close tolerance, to the cable diameters to provide proper fit.

8.10 Grounding of Cable Systems. For safety and reliable operation, the shields and metallic sheaths of power cables must be grounded. Without such grounding, shields would operate at a potential considerably above ground. Thus they would be hazardous to touch, and would cause rapid degradation of the jacket or other material intervening between shield and ground. This is due to the capacitive charging current of the cable insulation which is of the order of 1 mA/ft of conductor length. This current normally flows, at power frequency, between the conductor and the earth electrode of the cable, normally the shield. In addition, the shield or metallic sheath provides a fault return path in the event of insulation failure, permitting rapid operation of the protection devices.

The grounding conductor, and its attachment to the shield or metallic sheath, normally at a termination or splice, needs to have an ampacity no lower than that of the shield. In the case of a lead sheath, the ampacity must be ample to carry the available fault current until it is interrupted without overheating. Attachment to shield or sheath is frequently by means of solder, which has a low melting point; thus an adequate area of attachment is required.

There is much disagreement as to whether the cable shield should be grounded at both ends or at only one end. If grounded at only one end, any possible fault current must traverse the length from the fault to the grounded end, imposing high current on the usually very light shield conductor. Such a current could readily damage or destroy the shield, and require replacement of the entire cable rather than only the faulted section. With both ends grounded, the fault current would divide and flow to both ends, reducing the duty on the shield, with consequently less chance of damage. There are modifications of both systems. In one, single-ended grounding may be attained by insulating the shields at each splice or sectionalizing point, and grounding only the source end of each section. This limits possible shield damage to only the faulted section. Multiple grounding, rather than just two-end grounding, is simply the grounding of the cable shield or sheath at all access points, such as manholes or pull boxes. This also limits possible shield damage to only the faulted section.

8.10.1 Sheath Losses. Currents are induced in the multigrounded shields and sheaths of cables by the current flow in the power conductor. These currents increase with the separation of the power conductors, and increase with decreasing shield or sheath resistance. With three-conductor cables this sheath current is negligible, but with the single-conductor cables separated in direct-burial or separate ducts it can be appreciable. For example, with three single-conductor 500 kmcil cables, flat, on 8 in centers, with twenty spiral No 16 copper-shield wires, the ampacity is reduced by approximately 20% by this shield current. With single-conductor lead-sheathed cables in separate ducts, this current is important enough that single-end grounding is obligatory. As an alternate, the shields are insulated at each splice (at approximately 500 ft intervals), and cross-bonded to provide sheath transposition. This neutralizes the sheath currents, but still provides double-ended grounding. Of course, these sheaths and bonding jumpers must be insulated; their voltage differential from ground may be in the 30 V–50 V range. For details on calculating sheath losses in cable systems, consult [1].

Difficulties may arise from current attempting to flow via the cable shield, unrelated to cable insulation failures. To prevent this, all points served by a multiple-grounded shield cable need to be interconnected with an ample grounding system. (Insulation between shield sections at splices of single-end grounded shield systems needs enough dielectric strength to withstand possible abnormal voltages as well.) This system requires interconnecting grounding conductors of suitably low impedance so that fault lightning and stray currents will follow this path rather than the cable shield. Cable-shield ground connections must be made to this system, which must also connect to the grounded element of the source supplying the energy to the cable. Duct runs, or direct-burial routes, generally include a heavy grounding conductor to ensure such interconnection.

For further details, see IEEE Std 142-1982 [2], Section 6, and ANSI C2-1981 [1], Part 3, Rules for the Installation and Maintenance of Electric Supply Stations and Equipment.

8.11 Protection from Transient Overvoltage. Cables up through 35 kV used in commercial building service have insulation strengths well above that of essentially all other types of electric equipment of similar voltage ratings. This is to compensate for installation handling and possibly a higher deterioration rate than insulation which is exposed to less severe ambient conditions. This high insulation strength may or may not exist in splices or terminations, depending on their design and construction. Except for deteriorated points in the cable itself, the splices or terminations are most affected by overvoltages of lightning and switching transients. The terminations of cable systems not provided with surge protection may flash over due to switching transients. In this event, the cable proper would be subjected to possible wave reflections of even higher levels, possibly damaging the cable insulation; however, this is unlikely in this medium-voltage class.

Like other electric equipment, the means employed for protection from these overvoltages is usually surge arresters. These may be for protection of associated equipment as well as the cable. Distribution or intermediate type arresters are used, applied at the junctions of open-wire lines and cables, and at terminals where switches may be open. Surge arresters are not required at intermediate positions along the cable run in contrast to open-wire lines.

It is recommended that surge arresters be connected between the conductor and the cable shielding system with short leads to maximize the effectiveness of the arrester. Similarly recommended is the direct connection of the shields and arrester ground wires to a substantial grounding system to prevent surge current propagation through the shield.

Aerial messenger-supported fully-insulated cables and spacer-type cables are subject to direct-lightning strokes, and a number of such cases are on record. The incidence rate is, however, rather low, and in most cases no protection is provided. Where, for reliability, such incidents must be guarded against, a grounded shield wire similar to that used for bare aerial circuits should be installed on the poles a few feet above the cable. Down-pole grounding conductors need to be carried past the cable messenger with a lateral offset of approximately 18 in to guard against side flashes consequent to direct strokes. Metal bayonets, where used to support the grounded shielding wire, should also be kept no less than 18 in clear of the cables or messengers.

8.12 Testing

8.12.1 Application and Utility. Testing, particularly of elastomeric and plastic (solid) insulations, is a useful method of checking the ability of a cable to withstand service conditions for a reasonable future period. Failure to pass the test will either cause in-test breakdown of the cable or otherwise indicate the need for its immediate replacement.

Whether or not to routinely test cables is a decision each user has to make. The following factors must be taken into consideration.

(1) If there is no alternate source for the load supplied, testing must be done when the load equipment is not in operation.

(2) The costs of possible service outages due to cable failures must be weighed against the cost of testing. With solid-type insulation, in-service cable failures may be reduced approximately 90% by direct-current maintenance testing.

(3) Personnel with adequate technical capability must be available to do the testing and evaluate observations and results.

The procedures outlined here are intended as a guide, and many variations are possible. At the same time, variations made without sound technical basis can negate the usefulness of the test, or even damage equipment.

With solid dielectric-cable types (elastomeric and plastic), the principal failure mechanism results from progressive degradation due to alternating-current corona cutting during service at locations of either manufacturing defects, installation damage, or accessory workmanship shortcomings. Initial tests reveal only gross damage, improper splicing or terminating, or cable imperfections. Subsequent use on alternating-current usually causes progressive enlargement of such defects proportional to their severity.

Oil-paper (laminated) cable with lead sheath fails, usually from water entrance at a perforation in the sheath, generally within three to six months after the perforation occurs. Periodic testing, unless very frequent, is therefore likely to miss many of these cases, making the method less effective with this type of cable.

Testing is not useful in detecting possible failure from moisture-induced tracking across termination surfaces, since this develops principally during periods of precipitation, condensation, or leakage failure of the enclosure or housing. However, terminals should be examined regularly for signs of tracking and the condition corrected whenever detected.

8.12.2 Alternating Current Versus Direct Current. Cable insulation can, without damage, sustain application of direct-current potential equal to the system basic impulse insulation level for very long periods. In contrast, most cable insulations will sustain degradation from alternating-current overpotential, proportional to a high power of overvoltage and to time (and frequency) of the application. Hence it is desirable to utilize direct current for any testing that will be repetitive. While the manufacturers use alternating current for the original *factory* test, it is almost universal practice to employ direct current for any subsequent testing. All discussion of field testing hereafter applies to direct-current high-voltage testing.

8.12.3 Factory Tests. All cable is tested by the manufacturer before shipment, normally with alternating voltage for a 5 min period. Unshielded cable is immersed in water (ground) for this test, shielded cable is tested using the shield as the ground return. Test voltages are specified by the manufacturer, by the applicable specification of the ICEA, or by other specifications such as that of the Association of Edison Illuminating Companies[36] (AEIC). In addition, a test may be made using direct voltage of two to three times the rms value used in the alternating-current test. On cable rated 3000 V and above, corona tests also may be made.

8.12.4 Field Tests. As well as having no deteriorating effect on good insulation, direct-current high voltage is most convenient for field testing since the test power sources or test sets are relatively light and portable.

Voltages for such testing should fulfill both of the following requirements. Voltages must be:

(1) Not high enough to damage sound cable or component insulation

(2) High enough to indicate incipient failure of unsound insulation which may fail in service before the next scheduled test

[36] 51 East 42nd Street, New York, NY 10017.

Test voltages and intervals require coordination to attain suitable performance. One large industrial company with cable-testing experience of over 20 years has reached over 90% reduction of cable-system service failures through use of ICEA specified voltages. These are applied at installation, after approximately three years of service, and every five to six years thereafter. The majority of test failures occur at the first two tests; test (or service) failures after eight years of satisfactory service are less frequent. The importance of uninterrupted service should also influence the test frequency for specific cables. Tables 57 and 58 specify cable-field test voltages.

The AEIC has specified test values for 1968 and later cables of approximately 20% higher than the ICEA values.

ANSI/IEEE Std 400-1980 [9] specifies much higher voltage than either ICEA or AEIC. These much more severe test voltages, as shown in Table 59, are intended to reduce cable failures during operation by overstressing the cable during shutdown testing and causing the cable to fail at this time. These test voltages should not be used without the concurrence of the cable manufacturer, otherwise the cable warranty will be voided.

Cables to be tested must have their ends free of equipment and clear from ground. All conductors not under test must be grounded. Since equipment to which cable is customarily connected may not withstand the test voltages allowable for cable, either the cable must be disconnected from this equipment, or the test voltage must be limited to levels which the equipment can tolerate. The latter constitutes a relatively mild test on the cable condition, and the predominant leakage current measured is

Table 57
ICEA Specified Direct-Cable Test Voltages (kV)
Pre-1968 Cable

Insulation Type	Grounding	5 kV	Maintenance Test Rated Cable Voltage		
			15 kV	25 kV	35 kV
Elastomeric:	Grounded	27	47	—	—
butyl, oil base, EDM, EPDM	Ungrounded	—	67	—	—
Polyethylene,	Grounded	22	40	67	88
including cross-linked polyethylene	Ungrounded	—	52	—	—

Table 58
ICEA Specified Direct-Cable Test Voltages (kV)
1968 and Later Cable† (CB)

Insulation Type	Insulation Level (%)	Rated Cable Voltage							
		5 kV		15 kV		25 kV		35 kV	
		1	2	1	2	1	2	1	2
Elastomeric:	100	25	19	55	51	80	60	—	—
butyl and oil base	133	25	19	65	49	—	—	—	—
Elastomeric:	100	25	19	55	41	80	60	100	75
EPM	133	25	19	65	49	100	75	—	—
Polyethylene,	100	25	19	55	41	80	60	100	75
including cross-linked polyethylene	133	25	19	65	49	100	75	—	—

NOTE: Columns 1 — Installation tests, made after installation, before service; columns 2 — maintenance tests, made after cable has been in service.

†These test values are lower than for pre-1968 cables, because the insulation is thinner. Hence the alternating-current test voltage is lower. The direct-current test voltage is specified as three times the alternating-current test voltage, so it is also lower than for older cables.

Table 59
Specified Direct-Current Cable
Test Voltages (kV)
Installation and Maintenance
ANSI/IEEE Std 400-1980

System Voltage (kV)	BIL (kV)	Test Voltage (kV)	
		100% Insulation Level	133% Insulation Level
2.5	60	40	50
5	75	50	65
8.7	95	65	85
15	110	75	100
23	150	105	140
28	170	120	
34.5	200	140	

NOTE: These test voltages should not be used without the cable manufacturers' concurrence as the cable warranty will be voided.

likely to be that of the attached equipment. In essence, this tests the equipment, not the cable.

In field testing, in contrast to the go-no-go nature of factory testing, the leakage current of the cable system must be closely watched and recorded for signs of approaching failure. The test voltage may be raised continuously and slowly from zero to the maximum value, or it may be raised in steps, pausing for 1 min or more at each step. Potential differences between steps are of the order of the alternating-current rms rated voltage of the cable. As the voltage is raised, current will flow at a relatively high rate to charge the capacitance, and to a much lesser extent to supply the dielectric absorption characteristics of the cable, as well as to supply the leakage current. The capacitance charging current subsides within a second or so, the absorption current subsides much more slowly and would continue to decrease for 10 min or more, ultimately leaving only the leakage current flowing.

At each step, and for the 5–15 min duration of the maximum voltage, the current meter (normally a microammeter) is closely watched. If, except when the voltage is being increased, the current starts to increase, slowly at first, then more rapidly, the last remnants of insulation at a weak point are failing, and total failure will occur shortly thereafter unless the voltage is reduced. This is characteristic of approximately 80% of all elastomeric-type insulation test failures.

In contrast to this *avalanche* current increase to failure, sudden failure (flashover) can occur if the insulation is already completely (or nearly) punctured. In the latter case voltage increases until it reaches the sparkover potential of the air gap length, then flashover occurs. Polyethylene cables exhibit the latter characteristic for all failure modes. Conducting leakage paths, such as at terminations or through the body of the insulation, exhibit a constant leakage resistance independent of time or voltage.

One advantage of step testing is that a 1 min *absorption-stabilized* current may be read at the end of each voltage step. The calculated resistance of these steps may be compared as the test progresses to higher voltage. At any step where the calculated leakage resistance decreases markedly (approximately 50% of that of the next lower voltage level), the cable could be near failure and the test should be discontinued short of failure as it may be desirable to retain the cable in serviceable condition until a replacement can be made ready. On any test in which the cable will not withstand the prescribed test voltage for the full test period (usually 5 min) without current increase, the cable is considered to have failed the test and is subject to replacement as soon as possible.

The polarization index is the ratio of the current after 1 min to the current after 5 min of maximum-voltage test, and on good cable it will be between 1.25 and 2. Anything less than 1.0 should be considered a failure, and between 1.0 and 1.25 only a marginal pass.

After completion of the 5 min maximum-voltage test step, the supply voltage-control dial should be returned to zero and the charge in the cable allowed to drain off through the leakage of the test set and voltmeter circuits. If this requires too long a time, a bleeder resistor of 1 MΩ per 10 kV of test potential can be added to the drainage path, discharging the circuit in a few seconds. After the remaining potential drops below 10% of the original value, the cable conductor may be solidly grounded. All conductors should be left grounded when not on test during the testing of other conductors and for at least 30 min after the removal of a direct-current test potential. They may be touched only while the ground is connected to them; otherwise the release of absorption current by the dielectric may again raise their potential to a dangerous level.

8.12.5 Procedure. Load is removed from the cables either by diverting the load to an alternate supply, or by shutdown of the load served. The cables are de-energized by switching; they are tested to ensure voltage removal, then grounded and disconnected from their attached switching equipment. (In case they are left connected, lower test potentials are required.) Surge arresters, potential transformers, and capacitors must be disconnected.

All conductors and shields must be grounded. The test set is checked for operation, and after its power is turned off, the test lead is attached to the conductor to be tested. At this time (and not before) the ground should be removed from that conductor, and the bag or jar (see 8.12.6) applied over all of its terminals, covering all uninsulated parts at both ends of the run. The test voltage is then slowly applied, either continuously or in steps as outlined in 8.12.4. At completion of the maximum voltage test duration, the charge is drained off, the conductor grounded, and the test lead removed for connection to the next conductor. This procedure is repeated for each conductor to be tested. Grounds should be left on each tested conductor for no less than 30 min.

8.12.6 Direct-Current Corona and Its Suppression. Starting at approximately 10—15 kV and increasing at a high power of the incremental voltage, the air surrounding all bare conductor portions of the cable circuit becomes ionized from the test potential on the conductor and draws current from the conductor. This ionizing current indication is not separable from that of the normal leakage current, and reduces the apparent leakage resistance value of the cable. Wind and other air currents tend to blow the ionized air away from the terminals, dissipating the space charge, and allowing ionization of the new air, thus increasing the *direct corona current* as this may be called.

Enclosing the bare portions of both end terminations in plastic or glass jars, or plastic bags, prevents the escape of this ionized air, thus it becomes a captive space charge. Once formed, it requires no further current, so the *direct corona current* disappears. With this treatment, testing up to approximately 100 kV is possible. Above 100 kV, larger bags or a small bag inside a large one are required. In order to be effective, the bags must be blown up so that no part of the bag touches the conductor.

An alternative method to minimize corona is to completely tape all bare conductor surfaces with standard electrical insulating tape. This method is superior to the bag method for corona suppression, but it requires more time to adequately tape all exposed ends.

8.12.7 Voltage Fluctuations. The very large capacitance of the cable circuit makes the microammeter extremely sensitive to even minor variations in 120 V, 60 Hz supply to the test set. Normally it is possible only to read average-current values or the near-steady-current values. A low-harmonic-content constant-voltage transformer improves this condition moderately. Complete isolation and stability are attainable only by use of a storage battery and 120 V, 60 Hz inverter to supply the test set.

8.12.8 Resistance Evaluation. High-voltage cable exhibits extremely high insulation resistance, frequently many thousands of megohms. While insulation resistance alone is not a primary indication of the condition of the cable insulation, the comparison of the insulation resistances of the three-phase conductors is useful. On circuits under 1000 ft long, a ratio in excess of 5:1 between any two conductors is indicative of some questionable condition. On longer circuits, a ratio of 3:1 should be regarded as maximum. Comparison of insulation resistance values with previous tests may be informative; but insulation resistance varies inversely with temperature, with winter insulation resistance measurements being much higher than those obtained under summer conditions. An abnormally low-insulation resistance is frequently indicative of a faulty splice, termination, or a weak spot in the insulation. (Higher than standard test voltages have been found practical to locate these by causing a test failure where the standard voltage will not cause breakdown. Fault location methods may be used to locate the failure.)

8.12.9 Megohmmeter Test. Since the insulation resistance of a sound high-voltage cable circuit is generally in the order of thousands to hundreds of thousands of megohms, a megohmmeter test will reveal only grossly deteriorated insulation conditions of high-voltage cable. For low-voltage cable, however, the megohmmeter tester is quite useful, and is probably the only practicable test. Sound 600 V cable insulation will normally withstand 20 000 V or higher direct current. Thus a 1000 V or 2500 V megohmmeter is preferable to the lower 500 V testers for such cable testing.

For this low-voltage class, temperature-corrected comparisons of insulation resistances with other phases of the same circuit, with previous readings on the same conductor, and with other similar circuits are useful criteria for adequacy. Continued reduction in the insulation resistance of a cable over a period of several tests is indicative of degrading insulation; however, a megohmmeter will rarely initiate final breakdown of such insulation.

8.13 Locating Cable Faults. In commercial buildings a wide variety of cable faults can occur. The problem may be in a communication circuit or in a power circuit, either in the low- or high-voltage class. Circuit interruption may have resulted, or operation may continue with some objectionable characteristic. Regardless of the class of equipment involved or the type of fault, the one common problem is to determine the location of the fault so that repairs can be made.

The vast majority of cable faults encountered in a commercial-building power system occur between conductor and ground. Most fault-locating techniques are made with the cirucit de-energized. In ungrounded or high-resistance-grounded low-voltage systems, however, the occurrence of a single line-to-ground fault will not result in automatic circuit interruption, and therefore the process of locating the fault may be carried out by special procedures with the circuit energized.

8.13.1 Influence of Ground-Fault Resistance. Once a line-to-ground fault has occurred, the resistance of the fault path can range from almost zero up to millions of ohms. The fault resistance has a bearing on the method used to locate the failure. In general a low-resistance fault can be located more readily than one of high resistance. In some cases the fault resistance can be reduced by the application of voltage sufficiently high to cause the fault to break down with sufficient current to cause the insulation to carbonize. The equipment required to do this is quite large and expensive, and its success is dependent to a large degree on the type of insulation involved. Large users indicate that this method is useful with paper and elastomeric cables, but generally of little use with plastic types.

The fault resistance which exists after the occurrence of the original fault depends on the type of cable insulation and construction, the location of the fault, and the cause of the failure. A fault which is immersed in water will generally exhibit a variable fault resistance and will not consistently arc over at a constant voltage. Damp faults behave in a similar manner until the moisture has been vaporized. In contrast, a dry fault will normally be much more stable and consequently can more readily be located.

For failures which have occurred in service, the type of system grounding and available fault current, as well as the speed of relay protection, will be influencing factors. Because of the greater carbonization and conductor vaporization, a fault resulting from an in-service failure can generally be expected to be of a lower resistance than one resulting from over-potential testing.

8.13.2 Equipment and Methods. A wide variety of commercially available equipment and a number of different approaches can be used to locate cable faults. The safety considerations outlined in 8.12 should be observed.

The method used to locate a cable fault depends on:

(1) Nature of fault
(2) Type and voltage rating of cable
(3) Value of rapid location of faults
(4) Frequency of faults
(5) Experience and capability of personnel

(1) *Physical Evidence of the Fault.* Observation of a flash, sound, or smoke accompanying the discharge of current through the faulted insulation will usually locate a fault. This is more probable with an overhead circuit than with underground construction. The discharge may be from the original fault or may be intentionally caused by the application of test voltages. The burned or disrupted appearance of the cable will also serve to indicate the faulted section.

(2) *Megohmmeter Instrument Test.* When the fault resistance is sufficiently low that it can be detected with a megohmmeter, the cable can be sectionalized and each section tested to determine which contains the fault. This procedure may require that the cable be opened in a number of locations before

the fault is isolated to one replaceable section. This could, therefore, involve considerable time and expense, and might result in additional splices. Since splices are often the weakest part of a cable circuit, this method of fault locating may introduce additional failures at a subsequent time.

(3) *Conductor-Resistance Measurement.* This method consists of measuring the resistance of the conductor from the test location to the point of fault by using either the Varley loop or the Murray loop test. Once the resistance of the conductor to the point of fault has been measured, it can be translated into distance by using handbook values of resistance per unit length of the size and type of conductor involved, correcting for temperature as required. Both of these methods give good results which are independent of fault resistance, provided the fault resistance is low enough that sufficient current for readable galvanometer deflection can be produced with the available test voltage. Normally a low-voltage bridge is used for this resistance measurement. For distribution systems using cables insulated with organic materials, relatively low resistance faults are normally encountered. The conductor-resistance measurement method has its major application on such systems. Loop tests on large conductor sizes may not be sensitive enough to narrow down the location of the fault.

High-voltage bridges are available for higher-resistance faults but have the disadvantage of increased cost and size as well as requiring a high-voltage direct-current power supply. High-voltage bridges are generally capable of locating faults with a resistance to ground of up to 1 MΩ or 2 MΩ, while a low-voltage bridge is limited to the application where the resistance is several kilohms or less.

(4) *Capacitor Discharge.* This method consists of applying a high-voltage high-current impulse to the faulted cable. A high-voltage capacitor is charged by a relatively low-current capacity source such as that used for high-potential testing. The capacitor is then discharged across an air gap or by a timed-closing contact into the cable. The repeated discharging of the capacitor provides a periodic pulsing of the faulted cable. The maximum-impulse voltage should not exceed 50% of the allowable dc cable-test voltage since voltage doubling can occur at open-circuit ends. Where the cable is accessible, or the fault is located at an accessible position, the fault may be located simply by sound. Where the cable is not accessible, such as in duct or directly buried, the discharge at the fault may not be audible. In such cases, detectors are available to trace the signal to the point of fault. The detector generally consists of a magnetic pickup coil, an amplifier, and a meter to display the relative magnitude and direction of the signal. The direction indication changes as the detector passes beyond the fault. Acoustic detectors are also employed, particularly in situations where no appreciable magnetic field external to the cable is generated by the tracing signal.

In applications where relatively high-resistance faults are anticipated, such as with solid dielectric cables or through compound in splices and terminations, the impulse method is the most practical method presently available and is the one most commonly used.

(5) *Tone Signal.* A fixed-frequency signal, generally in the audio frequency range, is imposed on the faulted cable. The cable route is then traced by means of a detector which consists of a pickup coil, receiver, and head set or visual dis-

play, to the point where the signal leaves the conductor and enters the ground return path. This class of equipment has its primary application in the low-voltage field and is frequently used for fault location on energized ungrounded-type circuits. On systems over 600 V the use of a tone signal for fault location is generally unsatisfactory because of the relatively large capacitance of the cable circuit.

(6) *Radar System.* A short-duration low-energy pulse is imposed on the faulted cable and the time required for propagation to and return from the point of fault is monitored on an oscilloscope. The time is then translated into distance in order to locate the point of fault. Although this type of equipment has been available for a number of years, its major application in the power field has been on long-distance high-voltage lines. In older test equipment the propagation time is such that it cannot be displayed with good resolution for relatively short cables encountered in commercial-building systems. However, recent equipment advances have largely overcome this problem. The major limitation to this method is the inability to adequately determine the difference between faults and splices on multitapped circuits. An important feature of this method is that it will locate an *open* in an otherwise unfaulted circuit.

8.13.3 Selection. The methods listed represent some of the means available to industrial-plant power-system operators to locate cable faults. They range from very simple to relatively complex. Some require no equipment, others require equipment which is inexpensive and can be used for other purposes, while still others require special equipment. As the complexity of the means used to locate a fault increases, so does the cost of the equipment, and also the training and experience required for those who are to use it.

In determining which approach is most practical for any particular facility, the size of the installation and the amount of circuit redundancy which it contains must be considered. The importance of minimizing the outage time of any particular circuit must be evaluated. The cable installation and maintenance practices and the number and time of anticipated faults will determine the expenditure for test equipment which can be justified. Equipment which requires considerable experience and operator interpretation of accurate results may be satisfactory for an application with frequent cable faults but ineffective where the number of faults is so small that adequate experience cannot be obtained. Because of these factors, most commercial-building managers employ firms which offer the service of cable-fault locating. Such firms are established in large cities and cover a large area with mobile test equipment.

While the capacitor discharge method is most widely used, no single method of cable-fault locating can be considered to be most suitable for all applications. The final decision on which method or methods to use must depend upon evaluation of the merits and disadvantages of each in relation to the particular circumstances of the plant in question.

8.14 Cable Specification. Once the correct cable has been determined, it can be described in a cable specification. Cable specifications generally start with the conductor and progress radially through the insulation and coverings. The following is a check list which can be used in preparing a cable requirement:

(1) Number of conductors in cable and phase identification required

(2) Conductor size (AWG, kcmil) and material

(3) Insulation type (rubber, polyvinyl chloride, polyethylene, EPR, etc)

(4) Voltage rating

(5) Shielding system; applicable to cable used on systems 8 kV and above, and may be required on 2.1 kV to 8 kV

(6) Outer finishes

(7) Installation (cable tray, direct burial, wet location, exposure to sunlight or oil, etc)

(8) Applicable UL Listing

(9) Test voltage and partial discharge voltage

An alternative method of specifying cable is to furnish the ampacity of the circuit (amperes), the voltage (phase-to-phase, phase-to-ground, grounded, or ungrounded), and the frequency along with any other pertinent system data. Also required is the method of installation anticipated and the installation conditions (ambient temperature, load factor, etc). For either method, the total number of lineal feet of conductors required, the quantity desired shipped in one length, the pulling eyes, and whether it is desired to have several single-conductor cables paralleled on a reel should also be given.

8.15 Busway. Busways have become increasingly popular in commercial buildings as the size of the electrical load has increased. Power circuits over 600 A—800 A are usually more economical and require less space with busway than with conduit and wire. When other factors are considered (number of tapoffs, future tapoffs and rearrangement) busways of any size may be more economical.

Rated 600 V and below, busways in commercial buildings are used to transmit power from:

(1) The electric utility service, usually a vault or transformer, to the service entrance switchboard

(2) The service entrance switchboard to electrical rooms, closet, or individual load

(3) To supply small loads, such as lighting and office machines

In extremely large buildings, if voltages above 600 V are available, it may prove economical to distribute power at the higher voltage to secondary unit or spot network substations at 2 or 3 locations in the building. In such cases busway is usually used to interconnect main switchboards in network or secondary-selective configuration.

8.16 Types. In commercial buildings four types of busways are generally used which, complete with fittings and accessories, provide a unified and continuous system of enclosed conductors.

8.17 Feeder Busway. Feeder busways are designed to transmit large blocks of power from one point to another with provision for tapping off power by bolted connections at a joint or other factory built location. It has a very low and balanced circuit reactance to minimize voltage regulation and phase imbalance at the utilization equipment.

In commercial buildings feeder busway is frequently used between the source of power, such as a utility vault or transformer, and the service entrance switchboard. It may be used from the service-entrance switchboard to supply large blocks of power, such as motor-control centers. Busway is usually made in 10 ft sections with all possible combinations of elbows, tees, and crosses available, since it must conform to the building. Feed and tap fittings to other electrical equipment, such as switchboards, transformers, motor-control centers, etc are available. Bolted tapoffs using fusible switches, molded case circuit breakers and low-voltage power circuit breakers with ratings to 4000 A are available.

Available current ratings range from 600 A to 5000 A, 600 V ac in single and three-phase ratings with 50 and 100% neutral conductor. A ground bus is available with all ratings and types. Available short-circuit ratings are 50 000—200 000 A, symmetrical rms (see 8.2.2).

The voltage drop of low-impedance feeder busway with the entire load at the end of the run ranges from 1 V to 3 V per 100 ft line-to-line, depending on the type of construction and the current rating (see Tables 62 and 63).

Feeder busway is available in indoor and weatherproof (outdoor) designs. Weatherproof busway is designed to shed. Where busway is located adjacent to the fire sprinkler system or in other applications where the busway may be subjected to water or other liquids, even temporarily during construction, weatherproof busway or other means such as a shield should be used to protect the busway.

When busway passes through the floor as in a vertical riser, curbing should be used around the hole in the floor to prevent the busway from becoming a drainpipe for any water in the building.

8.18 Plug-in Busway. Plug-in busway is used to supply power to utilization equipment. It serves as an elongated switchboard or panelboard with covered plug-in openings provided at closely spaced intervals (every 1 or 2 ft) to accommodate plug-in devices.

Most plug-in busway is totally enclosed with current ratings from 100 A to 4000 A. Plug-in and feeder busway sec-

tions of the same manufacturer, above 600 A, have compatible joints so they are interchangeable in a run. All fittings are of the feeder design and plug-in straight lengths 10 ft long are inserted in a run when a tap is desired.

Plug-in tapoffs *(bus plugs)* are usually fusible switches (up to 600 A) or molded case circuit breakers (up to 800 A). Bolted tapoffs with feeder busway are used when larger capacity is required.

A ground bar is often added to provide a higher integrity and lower resistance ground path than furnished by the standard busway housing. A neutral bar with 50 to 100% of the capacity of phase bars may be provided for single-phase loads, such as lighting.

Short-circuit ratings vary from 14 000 to 200 000 A, symmetrical rms (see Table 61). The voltage drop ranges from 1 V to 3 V per 100 ft line-to-line, for evenly distributed loads. If the entire load is concentrated at the end of the run, these values double (see 8.22.3 and Tables 62 and 63).

8.19 Lighting (Small Plug-in) Busway. Lighting busway by Underwriters Laboratories' standard is rated a maximum of 60 A, 300 V to ground with two, three, or four conductors. It may be used on 480Y/277 and 208Y/120 V systems and is, specifically, designed and approved to supply and support fluorescent and high-intensity discharge lighting fixtures. It is also suitable for supplying power to small loads, such as air conditioners and office machines. When supplying power to lighting the busway it is usually protected by a 20 A circuit breaker, which in accordance with ANSI/NFPA 70-1981 [4] protects individual fixtures. If the busway is protected by a device larger than 20 A then

Table 60
Typical Busway Parameters
Line to Neutral Milliohms per 100 ft, 25 °C

Current Rating (amperes)	Feeder Busway				Plug-In Busway			
	Aluminum		Copper		Aluminum		Copper	
	Resistance R	Reactance X	Resistance R	Reactance X	Resistance R	Reactance X	Resistance R	Reactance X
100	—	—	—	—	29.1	5.0	—	—
225	—	—	—	—	9.54	3.94	5.24	3.94
400	—	—	—	—	3.78	4.33	2.73	2.76
600	3.33	1.35	2.68	1.68	3.58	3.80	2.26	4.33
800	2.21	0.97	2.06	1.35	2.40	2.52	2.10	3.80
1000	1.66	0.65	1.35	0.97	1.59	1.59	1.42	2.52
1350	1.10	0.45	0.96	0.61	1.04	1.06	0.88	1.44
1600	1.02	0.45	0.83	0.53	1.10	1.24	0.72	1.17
2000	0.78	0	0.68	0.49	0.80	0.86	0.66	1.24
2500	0.55	0.23	0.51	0.32	0.51	0.51	0.44	0.86
3000	0.49	0.20	0.41	0.27	0.50	0.62	0.37	0.66
4000	0.36	0.15	0.30	0.21	0.33	0.39	0.30	0.62
5000	—	—	0.23	0.15	—	—	0.20	0.39
Lighting Busway								
30	—	—	—	—	—	—	79.0	3.0
60	—	—	—	—	—	—	51.0	3.0

NOTE: Resistance values increase as temperature increases. Reactance values are not affected by temperature. The above values are based on conductor temperatures of 25 °C (normal room temperature) since short circuits may occur when busway is initially energized or lightly loaded. To calculate voltage drop when fully loaded (75 °C) multiply resistance of copper and aluminum by 1.19.

Table 61
Typical Busway Short-Circuit Ratings

Current Rating (Amperes)	Symmetrical Amperes in Thousands			
	Feeder		Plug-in	
	Aluminum Bus Bars	Copper Bus Bars	Aluminum Bus Bars	Copper Bus Bars
100			14	14
225			20	20
400			55	22
600	70	60	55	50
800	80	65	60	50
1000	90	75	65	60
1200	90	85	70	65
1350	90	90	70	65
1600	160	100	125	70
2000	160	150	135	120
2500	170	155	140	130
3010	175	160	175	135
4000	185	165	185	165
5000		180	185	180

Table 62
Voltage Drop Values for Three-Phase Busways with Aluminum Bus Bars in Volts per 100 ft, Line to Line at Rated Current

Current Rating (amperes)	Load Power Factor, Percent (Lagging)									
	30	40	50	60	70	80	85	90	95	100
Totally Enclosed Feeder Busway Concentrated Loading										
600	2.38	2.67	2.95	3.20	3.42	3.61	3.68	3.73	3.73	3.46
800	2.20	2.46	2.70	2.91	3.10	3.26	3.31	3.34	3.33	3.06
1000	1.94	2.18	2.41	2.63	2.82	2.98	3.04	3.08	3.08	2.88
1200	1.88	2.12	2.34	2.54	2.72	2.87	2.93	2.97	2.97	2.76
1350	1.78	1.99	2.20	2.39	2.55	2.69	2.74	2.77	2.77	2.57
1600	2.04	2.27	2.49	2.69	2.87	3.01	3.06	3.09	3.07	2.83
2000	1.83	2.07	2.28	2.48	2.66	2.81	2.86	2.90	2.90	2.70
2500	1.66	1.87	2.05	2.23	2.38	2.50	2.55	2.58	2.57	2.38
3000	1.76	1.97	2.17	2.36	2.52	2.66	2.71	2.74	2.74	2.55
4000	1.74	1.95	2.15	2.33	2.49	2.62	2.67	2.70	2.69	2.49
Totally Enclosed Plug-In Busway Distributed Loading										
225	1.29	1.45	1.59	1.73	1.85	1.95	1.98	2.00	2.00	1.86
400	1.83	1.90	1.96	1.99	1.99	1.95	1.91	1.83	1.71	1.31
600	2.44	2.56	2.64	2.70	2.71	2.68	2.62	2.54	2.39	1.86
800	2.17	2.27	2.35	2.40	2.41	2.38	2.34	2.26	2.13	1.67
1000	1.73	1.82	1.88	1.93	1.95	1.93	1.90	1.84	1.74	1.38
1200	1.59	1.66	1.72	1.77	1.78	1.76	1.73	1.68	1.58	1.25
1350	1.55	1.62	1.68	1.72	1.74	1.72	1.69	1.64	1.54	1.22
1600	2.10	2.19	2.25	2.29	2.30	2.25	2.20	2.12	1.99	1.53
2000	1.84	1.92	1.99	2.03	2.04	2.00	1.96	1.90	1.78	1.39
2500	1.51	1.58	1.62	1.65	1.66	1.63	1.59	1.53	1.44	1.11
3000	1.93	2.00	2.05	2.07	2.06	2.01	1.96	1.87	1.74	1.30
4000	1.63	1.70	1.74	1.77	1.77	1.78	1.69	1.62	1.51	1.15

each fixture must have overcurrent protection.

Lighting busway has a complete line of fittings and power takeoff devices to form a continuous, enclosed electric distribution system. These include tap boxes, fusible and circuit breaker plugs. Fluorescent fixtures may be suspended from the busway or they are ordered with plugs and hangers attached for close coupling of the fixture to the busway. Lighting busway must be supported at 5 ft intervals. Auxiliary-supporting means called *strength beams* are available which require support at maximum intervals of 16 ft. The busway may be recessed in or surface mounted to dropped ceilings.

8.20 Busway Construction. Modern busways use insulated conductors which are placed in intimate thermal contact with each other and their steel or aluminum enclosure. This provides several benefits. The intimate thermal contact dissipates heat by conduction and current densities are achieved for totally enclosed busway that are comparable to older ventilated designs. The close spacing of conductors results in a compact busway and in low and balanced reactance. Totally enclosed busways of this type have the same cur-

Table 63
Voltage Drop Values for Three-Phase Busways with Copper Bus Bars in Volts per 100 ft Line to Line at Rated Current

Current Rating (amperes)	Load Power Factor, Percent (Lagging)									
	30	40	50	60	70	80	85	90	95	100
Totally Enclosed Feeder Busway, Concentrated Loading										
600	2.50	2.71	2.90	3.07	3.20	3.28	3.29	3.27	3.19	2.79
800	2.64	2.86	3.05	3.21	3.33	3.41	3.41	3.38	3.30	2.85
1000	2.30	2.48	2.62	2.75	2.84	2.88	2.87	2.84	2.75	2.34
1200	1.91	2.07	2.21	2.33	2.42	2.47	2.48	2.46	2.40	2.08
1350	2.03	2.21	2.36	2.49	2.59	2.65	2.66	2.64	2.58	2.24
1600	2.09	2.27	2.42	2.56	2.66	2.72	2.73	2.71	2.64	2.30
2000	2.33	2.50	2.65	2.77	2.86	2.90	2.90	2.86	2.77	2.36
2500	1.98	2.15	2.30	2.43	2.54	2.60	2.61	2.59	2.53	2.21
3000	1.98	2.14	2.28	2.40	2.49	2.55	2.55	2.53	2.46	2.13
4000	1.95	2.10	2.24	2.36	2.44	2.49	2.50	2.47	2.41	2.08
5000	1.84	1.99	2.12	2.23	2.32	2.37	2.38	2.36	2.30	1.99
Totally Enclosed Plug-in Busway, Distributed Loading										
225	1.04	1.11	1.18	1.23	1.27	1.28	1.27	1.26	1.21	1.02
400	1.20	1.26	1.30	1.33	1.35	1.33	1.31	1.27	1.20	0.95
600	2.50	2.53	2.54	2.51	2.43	2.29	2.19	2.04	1.82	1.18
800	2.95	3.00	3.01	2.98	2.90	2.75	2.63	2.46	2.21	1.46
1000	2.45	2.49	2.51	2.49	2.42	2.30	2.20	2.06	1.85	1.23
1200	2.15	2.19	2.21	2.20	2.15	2.04	1.96	1.85	1.67	1.14
1350	1.92	1.96	1.97	1.97	1.92	1.84	1.76	1.66	1.51	1.03
1600	1.85	1.89	1.91	1.90	1.86	1.77	1.70	1.61	1.46	1.00
2000	2.39	2.43	2.43	2.41	2.34	2.21	2.11	1.97	1.76	1.15
2500	2.10	2.13	2.15	2.13	2.07	1.97	1.89	1.77	1.59	1.06
3000	1.93	1.96	1.97	1.95	1.90	1.80	1.72	1.62	1.45	0.96
4000	2.36	2.39	2.38	2.34	2.26	2.12	2.02	1.87	1.66	1.04
5000	1.87	1.90	1.90	1.87	1.81	1.71	1.63	1.52	1.35	0.87
Lighting Busway, Single Phase, Distributed Loading										
30	0.84	1.11	1.38	1.65	1.89	2.13	2.40	2.51	2.20	2.75
60	1.08	1.38	1.62	1.98	2.22	2.46	2.70	2.88	3.00	3.00

NOTE: Voltage drop values are based on bus-bar resistance at 75 °C (room ambient temperature 25 °C plus average conductor temperature at full load of 50 °C rise).

rent rating regardless of mounting position. A stack of one bus bar per phase up to approximately 7 inches wide is used up to the 1600 A rating. Higher ratings will use two (3000 A) or three stacks (5000 A). Each stack will contain all three phases and neutral to minimize circuit reactance.

Early busway designs required multiple nuts, bolts, and washers to electrically join adjacent sections. The most recent designs use a single bolt for each stack (with bars up to 7 in wide). All hardware is captive to the busway section when shipped from the factory. Installation labor is greatly reduced with corresponding savings in installation costs.

Busway are available with either copper or aluminum conductors. Compared to copper, aluminum has lower electrical conductivity, less mechanical strength, and upon exposure to the atmosphere quickly forms an insulating film on the surface. For equal current-carrying ability aluminum is lighter in weight and less costly.

For these reasons aluminum conductors have tin or silver electro-plated contact surfaces and at electrical joints use Belleville springs and bolting practices which accommodate aluminum mechanical properties. Copper busway is physically smaller in cross section while aluminum busway is lighter in weight and lower in cost for a given current capacity. Copper plug-in busway is more tolerant of cycling loads such as welding.

Busway is usually made in 10 ft sections. Since the busway must conform to the building, all possible combinations of elbows, tees, and crosses are available. Feed and tap fittings to other electric equipment such as switchboards, transformers, motor-control centers, etc, is provided. Standard busway current ratings are 20 A—5000 A for single-phase and three-phase service. Neutral conductors may be supplied if required. Newer designs of busway including plug-in devices can incorporate a ground bar if specified.

8.21 Standards. Busways are designed to conform to:

1. ANSI/NFPA 70-1981 [4]
2. ANSI/UL 857-1976[6]
3. NEMA BU 1-1978 [11]

UL and NEMA are primarily manufacturing and testing standards. The NEMA standard is generally an extension of the UL standard to areas that UL does not cover. The most important areas are busway parameters (R, X, Z) and short-circuit testing and rating.

The National Electrical Code, ANSI/NFPA 70-1981 [4] is the most impor-

tant standard for busway installation. Some of its most important requirements are:

(1) Busway may be installed only where located in the open and visible. Installation behind panels is permitted if access is provided and the following conditions are met.

(a) Overcurrent devices are not installed on the busway, other than for an individual fixture.

(b) The space behind the panels is not used for air handling purposes.

(c) The busway is totally enclosed, nonventilating type.

(d) Busway is so installed that the joints between sections and fittings are accessible for maintenance purposes.

(2) Busway may not be installed where subject to severe physical damage, corrosive vapors or in hoistways.

(3) When specifically approved for the purpose, busway may be installed in a hazardous location, or outdoors, or in wet or damp locations.

(4) Busway must be supported at intervals not to exceed 5 ft - 0 in unless otherwise approved. Where specifically approved for the purpose, horizontal busway may be supported at intervals up to 10 ft–0 in and vertical busway may be supported at intervals up to 16 ft-0 in.

(5) Busway must be totally enclosed where passing through floors and for a minimum distance of 6 ft–0 in above the floor to provide adequate protection from physical damage. It may extend through walls if joints are outside walls.

(6) State and local electrical codes may have specific requirements over and above Underwriters Laboratories and the National Electrical Code. Appropriate code authorities and manufacturers should be contacted to ensure requirements are met.

8.22 Selection and Application of Busway. To properly apply busway in an electrical power distribution system, some of the more important items to consider are listed below.

8.22.1 Current-Carrying Capacity. Busways should be rated on a temperature rise basis to provide safe operation, long life, and reliable service.

Conductor size (cross-sectional area) must not be used as the sole criterion for specifying busway. Busway may have seemingly adequate cross-sectional area and yet have a dangerously high temperature rise. The ANSI/UL 857-1976 [6] requirement of 55 °C, a 40 °C ambient temperature, should be used to specify the maximum temperature rise permitted. Larger cross-section areas can be used to provide lower-voltage drop and temperature rise.

In well-designed busway, the gradual thermal degradation of the insulation is the predominant limiting factor on service life. Life is cut approximately in half for every 10 °C of total *hot spot* operating temperature above the design value. Consequently if the ambient temperature exceeds 40 °C or a total temperature in excess of 95 °C is expected, then the busway manufacturer should be consulted.

8.22.2 Short-Circuit Current Rating. The bus bars in busways may be subject to electromagnetic forces of considerable magnitude by a short-circuit current. The force generated per unit length of bus bar is directly proportional to the square of the short-circuit current and is inversely proportional to the spacing between bus bars. Short-circuit ratings are generally assigned and tested in accordance with NEMA BU1-1978 [11].

The ratings are based on:

(1) The use of an adequately rated protective device ahead of the busway that will clear the short circuit in 3 cycles

(2) Application in a system with power factor not less than those listed in Table 64.

If the system on which the busway is to be applied has a lower power factor (larger X/R ratio) or a protective device with a longer clearing time, the short-circuit rating may have to be reduced. The busway manufacturer should be consulted.

The required short-circuit current rating should be determined by calculating the available short-circuit current and X/R ratio at the point where the input end of the busway is to be connected. The short-circuit rating of the busway must equal or exceed the available short-circuit current.

When properly protected by current limiting devices (reactors, fuses, and circuit breakers) busway may be applied in a system having an available short-circuit current in excess of the busway short-circuit rating. Critical items are busway short-circuit rating, available short-circuit current, and let-through peak amperes. See Table 65 for maximum fuse size. (See Section 9.8.9 for further explanation.)

Table 64
Short-Circuit Test Power Factor

Busway Rating (symmetrical rms amperes)	Power Factor	X/R Ratio
10 000 or less	0.50	1.7
10 001—20 000	0.30	3.2
Above 20 000	0.20	4.9

Table 65
Maximum Fuse Rating for Busway Short-Circuit Protection

Busway short-circuit rating amperes, symmetrical	Maximum Current Limiting Fuse Rating Class J, L				
	Available RMS Symmetrical Short-Circuit Current				
	25 000	50 000	75 000	100 000	200 000
5000	200	100	100	100	60
7500	400	200	200	200	100
14 000	600	600	400	400	400
22 000	1200	1000	600	600	600
35 000		1600	1200	1200	1000
42 000		2000	1600	1600	1200
60 000			2500	2500	2000
65 000			3000	2500	2000
70 000			3000	2500	2500
85 000				4000	3000
105 000					4000
140 000					4000
175 000					4000

Busway may be used in systems where the available short-circuit current exceeds the busway short-circuit rating if current limiting fuses protect the busway and do not exceed the sizes listed in Table 65.

Short-circuit ratings are dependent on many factors such as bus bar centerline spacing, size and strength of bus bars, and mechanical supports. See Table 61 for typical ratings. Short-circuit ratings should include ability of ground-return path (housing and ground bar if provided) to carry rated short-circuit current. Failure of the ground-return path to adequately carry this current can result in arcing and spitting at joints with attendant fire hazard. The ground-fault current can also be reduced to the point that the overcurrent protective device does not operate.

8.22.3 Voltage Drop.
Voltage drop (line to neutral) in busways may be calculated by the following formulas:

Exact formulas for concentrated loads at end of line:

With e_R known

$$e_D = \sqrt{(e_R \cos \phi + IR)^2 + (e_R \sin \phi + IX)^2} - e_R \qquad \text{(Eq 1)}$$

With e_S known

$$e_D = e_S + IR \cos \phi + IX \sin \phi - \sqrt{e_3^2 - (IX \cos \phi - IR \sin \phi^3} \qquad \text{(Eq 2)}$$

where

$$e_R = e_S \cdot \frac{Z_L}{Z_S}\,^*$$

$$e_D = e_S - e_R \quad \text{numerically} \qquad \text{(Eq 3)}$$

*NOTE: Multiply line-to-neutral voltage drop by $\sqrt{3}$ to obtain line-to-line voltage drop in three-phase systems. For single-phase systems, multiply the line-to-neutral voltage drop by two to obtain line-to-line voltage drop.

Approximate formulas for concentrated loads at end of line:

$$I(R \cos \phi + X \sin \phi) \qquad \text{(Eq 4)}$$

Percent volts drop =

$$\frac{\text{kVA} (R \cos \phi + X \sin \phi)}{10 \text{ kV}^2} \qquad \text{(Eq 5)}$$

Approximate formula for distributed load on a line:

Percent volts drop =

$$\frac{\text{kVA} (R \cos \phi + X \sin \phi) L}{10 \text{ kV}^2} \left(1 - \frac{L_1}{2L}\right) \qquad \text{(Eq 6)}$$

Nomenclature:

e_D = line-to-neutral voltage drop, in volts

e_S = line-to-neutral voltage at sending end

e_R = line-to-neutral voltage at receiving end

ϕ = angle whose cosine is the load power factor

R = resistance for the circuit in ohms per phase

X = reactance of the circuit in ohms per phase

I = load current in amperes

Z_L = load impedance in ohms

Z_S = circuit impedance in ohms, plus load impedance in ohms, added vectorially

kVA = three-phase kVA for three-phase circuits and single-phase kVA for single-phase circuits

kV = line-to-line kV

L_1 = distance from source to desired point

L = total length of line.

The foregoing formulas for concentrated loads may be verified by a trigonometric analysis of Fig 59. From this figure, it can be seen that the approximate formulas are sufficiently accurate for practical purposes. In practical cases the angle between e_R and e_S will be small (much smaller than in Fig 59 which has been exaggerated for illustrative purposes). The error in the approximate formulas diminishes as the angle between e_R and e_S decreases and is zero if that angle is zero. This latter condition will exist when the X/R ratio (or power factor) of the load is equal to the X/R ratio (or power factor) of the circuit through which the load current is flowing.

In actual practice, loads may be concentrated at various locations along the feeders, uniformly distributed along the feeder, or any combination of the same. A comparison of the approximate formulas for concentrated end loading and uniform loading will show that a uniformly loaded line will exhibit one-half the voltage drop as that due to the same total load concentrated at the end of the line. This aspect of the approximate formula is mathematically exact and entails no approximation. Therefore, in calculations of composite loading involving approximately uniformly loaded sections and concentrated loads, the uniformly loaded sections may be treated as end-loaded sections having one-half normal voltage drop of the same total load. Thus, the load can be divided into a number of concentrated loads distributed at various distances along the line. The voltage drop in each section may then be calculated for the load which it carries.

Three-phase voltage drops may be determined with reasonable accuracy by the use of Tables 62 and 63. These are typical values for the particular types of busway shown. The voltage drops will be different for other types of busway and

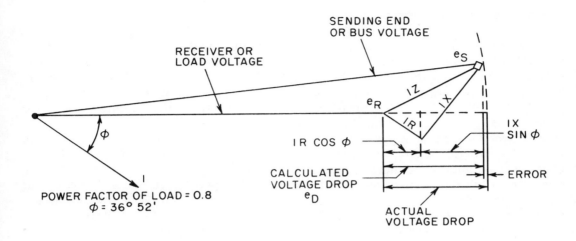

Fig 59
Diagram Illustrating Voltage Drop and Indicating Error
when Approximate Voltage-Drop Formulas 4, 5, and 6 Are Used

will vary slightly by manufacturer within each type. The voltage drop shown is three-phase line-to-line per 100 ft at rated load on a concentrated loading basis for feeder and trolley busway. Plug-in busway is shown for distributed loading. Lighting busway values are single-phase distributed loading. For other loading and distance use the following:

voltage drop =

$$\left(\text{Table } V_D\right) \bullet \left(\frac{\text{actual load}}{\text{rated load}}\right) \bullet \left(\frac{\text{actual distance (ft)}}{100 \text{ ft}}\right)$$

The voltage drop for a single-phase load connected to a three-phase busway is 15.5% higher than the values shown in the table. Typical values of resistance and reactance are shown in Table 60. Resistance is shown at normal room temperature (25 °C). This value should be used in calculating short-circuit current available in systems since short circuits can occur when busway is lightly loaded or initially energized. To calculate voltage drop when fully loaded (75 °C) resistance of copper and aluminum should be multiplied by 1.19.

8.22.4 Thermal Expansion. As load is increased, bus bars expand as the temperature increases. The lengthwise expansion between no-load and full load will range from $\frac{1}{2}$ to 1 in per 100 ft. The amount of expansion will depend on the total load, the size and location of tap-

305

offs, and the size and duration of varying loads. To accommodate the expansion, the busway should be mounted using hangers which permit the busway to move. In horizontal runs this can be accomplished by suspending hangers on drop rods which provide sufficient movement. In vertical riser runs spring hangers may be necessary. These support the busway yet permit movement relative to the building to accommodate thermal expansion. In addition, expansion lengths which expand and contract may be necessary in both horizontal and vertical runs. To determine if expansion lengths are required and to locate their position, the method of busway support, location of power takeoffs, the degree of movement at each end of the run that is permissible, and the orientation of the busway must be known. The manufacturer can determine the number and location of expansion lengths required, if any.

8.22.5 Building Expansion Joints. Busway when crossing a building expansion joint must include provisions for accommodating movement of the building structure. Fittings providing for 6 in of movement are available.

8.22.6 Layout. Busway must be tailored to the building in which it is installed. Once the basic engineering work has been completed and the busway type, ampere rating, number of poles, etc determined, a layout should be made for all but the simplest straight runs. The initial step in the layout is to identify and locate the building structure (walls, ceilings, columns, etc) and other equipment which is in the busway route. A layout of the busway to conform to this route is made. Although the preliminary layout (drawings for approval) can be made from architectural drawings, it is essential that field measurements be taken to verify building and busway dimensions prior to release of busway for manufacture. Where dimensions are critical, it is recommended that a section be held for field check of dimensions and manufactured after remainder of run has been installed. Manufacturers will provide quick delivery on limited numbers of these *field check* sections.

Another important consideration when laying out busway is coordination with other trades. Since there is a finite time lapse between job measurement and actual installation, other trades may use the busway *clear area* if coordination is lacking. Again, standard components can help since they are more readily available (sometimes from stock). By reducing the time between final measurement and installation, in addition to proper coordination, the chances of interference from other trades can be reduced to a minimum.

Finally, terminations are a significant part of busway layout considerations. For ratings of 600 A and above, direct bussed connections to the switchboard, motor-control center, etc can reduce installation time and problems. For ratings up to 600 A, direct bussed terminations are generally not practical or economical. These lower ampere ratings of busway are usually fed by short cable runs.

8.22.7 Installation. Busway installs quickly and easily. When compared with other distribution methods, the reduced installation time for busway can result in direct dollar savings on installation costs. In order to ensure maximum safety, reliability, and long life from a busway system, proper installation is a must. The guide lines below can serve as an outline from which to develop a complete installation procedure and time table.

8.22.8 Prior to Installation

(1) Manufacturer supplies installation drawings on all but the simplest of bus-

way layouts. Study these drawings carefully. Where drawings are not supplied, make your own.

(2) Verify actual components on hand against those shown on installation drawing to be sure there are no missing items. Drawings identify components by catalog number and location in the installation. Catalog numbers appear on the section nameplate and carton label. Location on the installation (item number) will also be on each section.

(3) During storage (prior to installation) all components, even the weatherproof type, should be stored in a clean, dry area and protected from physical damage.

(4) Read manufacturer's instructions for installation of individual components. If you are still in doubt, ask for more information; don't guess!

(5) Finally, preposition hanger supports (drop rods, etc) and hangers, if of the type which can be prepositioned. You are now ready to begin the actual installation of busway components.

(6) Perform electrical testing of individual components prior to installation. (See 8.22.11.) Identification of defective pieces prior to installation will save considerable time and money.

8.22.9 During Installation

(1) Almost all busway components are built with two dissimilar ends which are commonly called the *bolt end* and *slot end*. Refer to the installation drawing to properly orient the bolt and slot end of each component. This is important because it is not possible to properly connect two slot ends or bolt ends.

(2) Lift individual components into position and attach to hangers. It is generally best to begin this process at the end of the busway run which is most rigidly fixed (for example, the switchboards).

(3) Pay particular attention to *top* labels and other orientation marks where applicable.

(4) As each new component is installed in position, tighten the joint bolt to proper torque per manufacturer's instructions. Also install any additional joint hardware which may be required.

(5) Finally, on plug-in busway installations, attach plug-in units in accordance with manufacturer's instructions and proceed with wiring.

(6) Outdoor busway may require removal of *weep hole* screws and addition of joint shields. Pay particular attention to installation instructions to ensure all steps are followed.

8.22.10 After Installation

(1) Be sure to recheck all steps to ensure that you have not forgotten something. Be particularly sure that all joint bolts have been properly tightened.

(2) If spring hangers are used in vertical risers start at the top of the building and check adjustment in accordance with manufacturer's instruction.

At this point, the busway installation should be almost complete. Before energizing, however, the complete installation should be properly tested.

8.22.11 Field Testing. The completely installed busway run should be electrically tested prior to being energized. The testing procedure should first verify that the proper phase relationships exist between the busway and associated equipment. This phasing and continuity test can be performed in the same manner as similar tests on other pieces of electrical equipment on the job.

All busway installations should be tested with a megohmmeter or high-potential tester to be sure that excessive leakage paths between phases or ground, or both, do not exist. Megohmmeter values depend on the busway construction, type of insulation, size and length of busway, and atmospheric conditions. Acceptable values for a particular busway should be obtained from the manufacturer.

If a megohmmeter is used, it should be rated at 1000 V direct current. Normal high-potential test voltages are twice rated voltage plus 1000 V for 1 min. Since this may be above the corona, starting voltage of some busway frequent testing is undesirable. 120% of twice rated voltage plus 1000 V for 1 s is a satisfactory alternative.

8.23 References

[1] ANSI C2-1981, National Electrical Safety Code.[37]

[2] IEEE Std 142-1982, Recommended Practice for Grounding of Industrial and Commercial Power Systems.

[3] ANSI/IEEE Std 386-1977, Separable Insulated Connectors for Power Distribution Systems Above 600 V.

[4] ANSI/NFPA 70-1981, National Electrical Code.

[5] ANSI/UL 486-1975, Safety Standard for Wire Connectors and Soldering Lugs.

[6] ANSI/UL 857-1976, Safety Standard for Busways and Associated Fittings.

[7] IEEE Std 48-1975, Test Procedures and Requirements for High-Voltage AC Cable Terminations.

[8] IEEE Std 242-1975, Recommended Practice for Protection and Coordination of Industrial and Commercial Power Systems.

[9] ANSI/IEEE Std 400-1980, Guide for Making High-Direct Voltage Tests on Power Cable Systems in the Field.

[10] IEEE/ICEA S-135, Power Cable Ampacities.[38]

[11] NEMA BU 1-1978, Busways, Instructions for Safe Handling Installation, Operation and Maintenance of Busway and Associated Fittings Rated 600 V or Less.[39]

[12] UL 486A-1981, Wire Connectors and Soldering Lugs for Use with Copper Conductors.[40]

8.24 Bibliography

AEIC No 5-75, Specifications for Polyethlene and Crosslinked Polyethylene-Insulated Shielded Power Cables Rated 5 kV Through 69 kV.[41]

AEIC No 6-75, Specifications for Ethylene Propylene Rubber-Insulated Shielded Power Cables Rated 5 kV Through 69 kV.

ANSI/UL 44-1981, Safety Standard for Rubber-Insulated Wires and Cables.

ANSI/UL 83-1980, Safety Standard for Thermoplastic-Insulated Wires and Cables.

---

[37] ANSI standards are available from the Sales department of American National Standards Institute, 1430 Broadway, New York, NY 10018.

[38] ICEA publications are available from Insulated Cable Engineers Association, Inc, PO Box P, South Yarmouth, MA 02664.

[39] NEMA publications are available from the National Electrical Manufacturers Association (NEMA), 2101 L Street, NW, Washington, DC 20037.

[40] UL standards are available from Underwriters Laboratories Inc, Publications Stock, 333 Pfingsten Rd, Northbrook, IL 60062.

[41] AEIC. This document is available from the Publication Department of the Association of Edison Illuminating Companies, 51 East 42nd Street, New York, NY 10017.

ANSI/UL 493-1978, Safety Standard for Thermoplastic-Insulated Underground Feeder and Branch-Circuit Cables.

ANSI/UL 854-1979, Safety Standard for Service-Entrance Cables.

NEMA/W3-1980, Rubber-Insulated Wire and Cable for the Transmission and Distribution of Electrical Energy (ICEA S-19-81, 6th ed).

NEMA WC5-1973 (R 1979), Thermoplastic-Insulated Wire and Cable for the Transmission and Distribution of Electrical Energy (ICEA S-61-402, 3rd ed).

NEMA WC7-1971 (R 1976), Cross-Linked - Thermosetting-Polyethylene-Insulated Wire and Cable for the Transmission and Distribution of Electrical Energy (ICEA S-66-524).

NEMA WC8-1976, Ethylene-Propylene-Rubber-Insulated Wire and Cable for the Transmission and Distribution of Electrical Energy (ICEA S-68-516).

IEEE Std 404-1977, Standard for Power Cable Joints.

IEEE Std 525-1978, Guide for Selection and Installation of Control and Low-Voltage Cable Systems in Substations.

IEEE Std 592-1977, Standard for Exposed Semiconducting Shields on Premolded High-Voltage Cable Joints and Separable Insulated Connectors.

Underground Systems Reference Book. New York: Association of Illuminating Companies, 1957, chap 10.

9. System Protection and Coordination

9.1 General Discussions. Electric power systems in commercial and institutional buildings must be designed to serve loads in a safe and reliable manner. One of the major considerations in the design of a power system is adequate control of phase-to-ground, phase-to-phase, and three-phase short-circuit faults. Uncontrolled short circuits can cause service outages with accompanying lost time and associated inconvenience, interruption of essential facilities or vital services, extensive equipment damage, personnel injury or fatality, and possible fire damage.

Electric power systems should be as fault free as possible through careful system and equipment design and should be properly installed and maintained. However, even with these precautions, faults do occur. Some precipitating causes are loose connections, voltage surges, deterioration of insulation, accumulation of moisture, vermin or rodents, dust, seepage from concrete, contaminants, the intrusion of metallic or conducting objects such as fish tapes, tools, jack hammers, or construction equipment and undetermined phenomena.

When a short circuit occurs on a power system, undesirable things happen.

(1) Arcing and burning can occur at the fault location.

(2) Increased current will flow from the various sources to the fault location. All components carrying the fault currents are subject to increased thermal and mechanical stress. This mechanical stress varies as a function of the current squared and the thermal stress varies as a function both current squared and of the duration of current flow $(I^2 t)$.

(3) Voltages decrease throughout the system for the duration of the fault; voltage drops in proportion to the magnitude of the current; maximum voltage drop will occur at the fault location (to zero voltage for bolted fault).

(4) Enclosures which are in contact with live conductors can be subjected to elevated voltages and can increase hazard of electric shock.

The fault should be quickly removed from the power system to minimize the effects of these undersirable conditions including arcing and burning. This is the job of the circuit protective devices, the circuit breakers and fuses. The protective

311

device must have the ability to interrupt the maximum short-circuit current which can flow for a bolted fault at the device location.

All conductive components must have the capability to carry the short-circuit current until it is successfully interrupted. Equipment grounding must be adequate to limit voltage on faulted enclosures to safe values.

The bolted-fault value of short-circuit current results when the fault offers no impedance to the flow of short-circuit current and the magnitude of current is limited only by the impedance of the circuit elements. This condition results in a maximum short-circuit current and is frequently referred to as the available short-circuit current. Bolted short-circuits are very rare, however, and the fault usually involves arcing and burning. Under these conditions fault currents may be very much lower than bolted-fault values and may present special problems of detection and isolation.

When the fault involves ground, as it very often will, the protective enclosure may experience elevated potential which can increase the exposure of personnel to shock hazard. The likelihood of injury and death increases as a function of shock voltage and duration. It is important to maintain adequate equipment grounds to minimize exposure voltage and to detect and isolate the fault rapidly to reduce the duration of exposure.

For a simple example, consider Fig 60(a). The impedance which determines the flow of load current is the 20 Ω impedance of the motor. If a short circuit occurs at F, the only impedance limiting the flow of short-circuit current is the transformer impedance (0.1 Ω compared with 20 Ω for the motor). Therefore, the short-circuit current is 1000 A, or 200 times as great as the load current. Consequently, the circuit protective device must have the ability to interrupt 1000 A.

If the load grows and a larger transformer is substituted for the original unit, then the short-circuit at F_1 [Fig 60(b)] becomes limited by 0.01 Ω, the impedance of the larger transformer. Although the load current is still 5 A, the short-circuit current increases to 10 000 A which the circuit protective device must be able to interrupt.

9.1.1 Single-Pole and Multipole Interrupters. Circuit breakers and fuses can be designed for single- or multipole use. The protective function of circuit breakers automatically actuates the switching function. Fused switches including safety switches, service protectors, and bolted-pressure switches have in general separate protective and switching functions. If one pole of a multipole circuit breaker is actuated by the protective sensors, all poles are usually opened simultaneously. This same feature can be specified for service protectors, bolted-pressure switches, and certain other types of fused switches.

There are many instances where the opening of all phases (poles) because of a fault on one phase is undesirable. In commercial buildings, particularly in public areas where lighting is connected phase to neutral, opening of a single phase would inactivate only one third of the lighting, permitting almost normal operation. Opening of all three phases could, even with emergency lighting, make the area unusable.

Single or unbalanced phase-voltage conditions resulting from loss of voltage on one line conductor of a multiphase or three-wire single-phase system may arise from failure of the utility supply, system

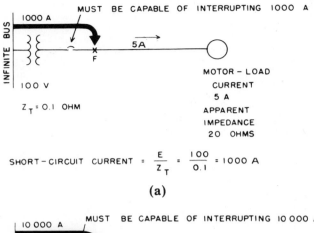

$$\text{SHORT-CIRCUIT CURRENT} = \frac{E}{Z_T} = \frac{100}{0.1} = 1000 \text{ A}$$

(a)

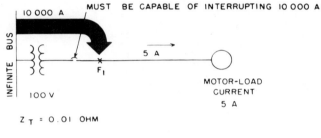

$$\text{SHORT-CIRCUIT CURRENT} = \frac{E}{Z_T} = \frac{100}{0.01} = 10\,000 \text{ A}$$

(b)

NOTE: Values were chosen for illustrative purposes only.

**Fig 60
Short Circuit on Load Side of Main Switch**

defects, or operation of single-pole interrupters. Under such conditions, some portions of the system may be subjected to undervoltage, unbalanced voltages, backfeeds through loads including voltages from rotating equipment, or prolonged faults. The designer should evaluate the extent of protection required to provide an effective system including undervoltage protection, ground-fault protection, and their relationship to the type of circuit-operating device.

9.1.2 Sources of Short-Circuit Currents. When determining the magnitude of short-circuit currents, it is extremely important that all sources of short circuit be considered and that the impedance characteristics of these sources be known.

There are four basic sources of short-circuit current:

(1) Generators in local system
(2) Synchronous motors
(3) Induction motors

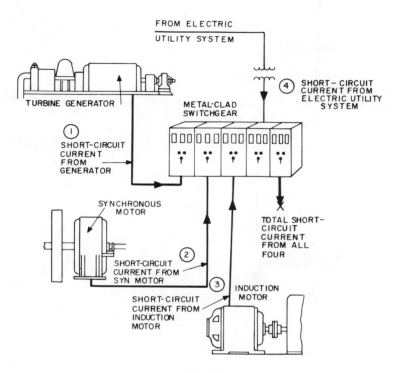

Fig 61
Total Short-Circuit Current Equals Sum of Sources

(4) Electric utility systems (remote generation)

All these can feed current into a fault (Fig 61).

9.1.3 Rotating-Machine Reactance. The impedance of a rotating machine consists primarily of reactance and is not one simple value as for a transformer or a piece of cable. It is complex and variable with time. For example, if a short circuit is applied to the terminals of a generator, the short-circuit current starts out at a high value and decays to a steady-state value after some time has elapsed from its inception. Since the field excitation voltage and speed have remained relatively constant within the short interval of time after inception of the fault, the reactance of the machine may be assumed to have changed with time, to explain the change in the current value.

Expression of such a variable reactance at any instant requires a complicated formula involving time as one of the variables. Therefore, for the sake of simplification, three values of reactance are assigned to rotating machines, for example motors and generators, for the purpose of calculating short-circuit currents at specified times. These values are called the subtransient reactance, transient reactance, and synchronous reactance. They are described as follows.

(1) The subtransient reactance X_d'' is the apparent reactance of the stator winding at the instant the short circuit occurs, and it determines the current flow during the first few cycles after the short circuit.

(2) The transient reactance X_d' determines the current following the period when the subtransient reactance is the controlling value; it is effective up to one half second or longer, depending upon the design of the machine.

(3) The synchronous reactance X_d is the reactance that determines the current flow when a steady-state condition is reached; it is not effective until several seconds after the short circuit occurs; consequently, it is not generally used in short-circuit current calculations.

A synchronous motor has the same kinds of reactances as a generator but usually has different values. Induction motors have no field coils, but the rotor bars act like the amortisseur winding in a generator. Therefore, induction motors

are said to have subtransient reactance only.

9.1.4 Utility Source. The available short-circuit current from the utility must be obtained from the serving utility. If the data furnished is at the primary voltage, it must be modified by the transformer impedance and voltage ratio.

9.1.5 Symmetrical and Asymmetrical Currents. The word *symmetrical* describes the shape of the alternating-current waves about the zero axis. If the envelopes of the peaks of the current waves are symmetrical around the zero axis, they are called symmetrical current envelopes (Fig 62). If the envelopes are not symmetrical around the zero axis, they are called asymmetrical current envelopes (Fig 63). The envelope is a line drawn through the peaks of the waves. The magnitude of the direct-current component of an asymmetrical current at any instant is the value of the offset between the axis of symmetry of the asymmetrical current and the zero axis (see Figs 68 and 69).

Most short-circuit currents are asymmetrical during the first few cycles after

Fig 62
Symmetrical Alternating-Current Wave

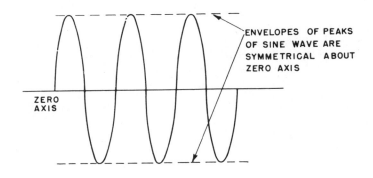

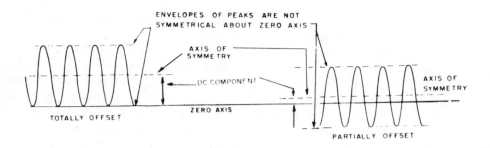

Fig 63
Asymmetrical Alternating-Current Wave

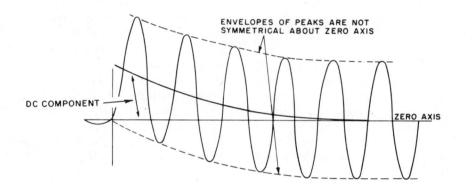

Fig 64
Typical Short Circuit

the short circuit occurs. The asymmetrical current is at a maximum during the first cycle after the short circuit occurs and in a few cycles gradually becomes symmetrical. An oscillogram of a typical short-circuit current is shown in Fig 64 (see Figs 68 and 69).

9.1.6 Why Short-Circuit Currents Are Asymmetrical. In ordinary power systems, the applied or generated voltage wave shapes are sinusoidal. When a short circuit occurs, approximately sinusoidal currents result. The following discussion assumes sinusoidal wave voltages and currents.

The power factor of a current circuit is determined by the series resistance and reactance of the circuit (from the fault back to and including the source or sources of the short-circuit currents). For example, in Fig 65 the reactance equals 19%, the resistance 1.4%, and the short-circuit power factor 7.3%, determined by the formula

$$\text{power factor} = \frac{R}{\sqrt{R^2 + X^2}}$$

The relationship of the resistance and re-

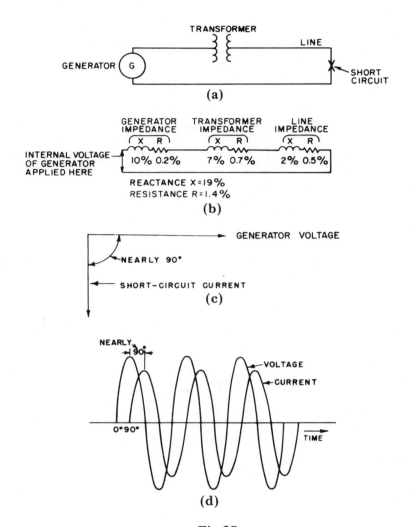

Fig 65
Phase Relations of Voltage and Short-Circuit Currents.
(a) Circuit Diagram. (b) Impedance Diagram. (c) Vector Diagram.
(d) Sine Waves Corresponding to Vector Diagram (c) for Circuit (a).

actance of a circuit is sometimes expressed in terms of the X/R ratio. For example, the X/R ratio of the circuit shown in Fig 65 is 13.6.

In high-voltage power circuits, the resistance of the circuit back to and including the power source is low compared with the reactance of the circuit. Therefore, the short-circuit current lags the source voltage by almost 90° (see Fig 65). Low-voltage power circuits (below 600 V) tend to have a larger percentage of resistance, and the current will lag behind the voltage by less than 90°.

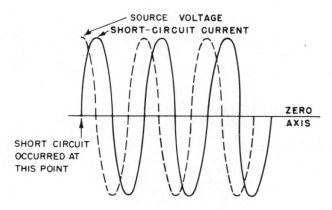

Fig 66
Symmetrical Current and Voltage in a Zero
Power-Factor Circuit

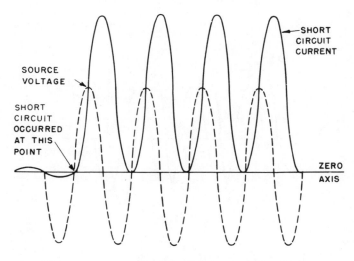

Fig 67
Asymmetrical Current and Voltage in a Zero
Power-Factor Circuit

If a short circuit occurs at the peak of the voltage wave in a circuit containing only reactance, the short-circuit current will start at zero and trace a sine wave which will be symmetrical about the zero axis (Fig 66). If a short circuit occurs at the zero point of the voltage wave, the current will start at zero, but cannot follow a sine wave symmetrically about the zero axis because the current must lag behind the voltage by 90°. This can happen only if the current is displaced from the zero axis as shown in Fig 67.

The two cases shown in Figs 66 and 67 are extremes. One shows a totally symmetrical current and the other a completely asymmetrical current. If the fault occurs at any point between zero voltage and peak voltage, the current will be asymmetrical to a degree dependent upon the point at which the short circuit occurs on the voltage wave.

To produce maximum asymmetry, when a circuit contains resistance, the short circuit must always occur at the zero point on the voltage wave. However, the point on the voltage wave at which the short circuit must occur to produce a symmetrical short-circuit current wave depends on the ratio of reactance to resistance (X/R ratio). The actual point on the voltage wave at which a short circuit must be initiated to produce a symmetrical current is the angle whose tangent equals the X/R ratio of the circuit.

For example, when X/R = 6.6 (15% pf) the angle on the voltage wave = arctan 6.6 = 81.384°; when X/R = 3.0 the angle on the voltage wave = arctan 3.0 = 71.565°.

9.1.7 The Direct-Current Component of Asymmetrical Short-Circuit Currents.

Asymmetrical currents are analyzed in terms of two components, a symmetrical current and a direct-current component, as shown in Fig 68. As previously discussed, the symmetrical component is at a maximum at the inception of the short circuit and decays to a steady-state value due to the apparent change in machine reactance. In all practical circuits, that is, those containing resistance, the direct-current component will also decay (to zero) as the energy represented by the direct-current component is dissipated as I^2R loss in the resistance of the circuit. Fig 69 illustrates the decay of the direct-current component.

The rate of decay of the direct-current

Fig 68
Components of Current Shown in Fig 67

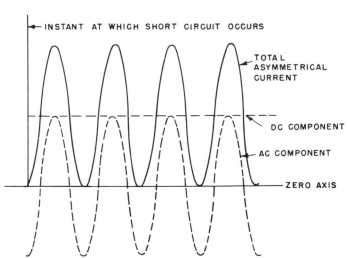

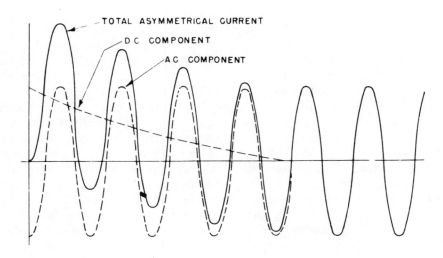

Fig 69
Decay of Direct-Current Component and Effect of
Asymmetry of Current

component is a function of the resistance and reactance of the circuit. In practical low-voltage circuits the direct-current component decays to zero in from one to six cycles.

9.1.8 Total Short-Circuit Current. The total symmetrical short-circuit current usually has several sources as illustrated in Fig 70. The first source is the utility. The second is local generation. Synchronous motors, if any, are a third source. Induction motors, a fourth source, are located in every building. Because rotating-machine currents usually decay with time due to reduction of flux in the machine after a short-circuit, the total short-circuit current decays with time (Fig 70). Considering only the symmetrical part of the short-circuit current, the magnitude is highest at the first half-cycle after a short circuit and is of lower value a few cycles later. Note that the induction-motor component will usually disappear almost entirely after one or two

cycles, except for very large motors where it may be present longer than four cycles.

The magnitude during the first few cycles is further increased by the direct-current component. This component also decays with time, accentuating the difference in magnitude of a short-circuit current at the first cycle after short circuit and a few cycles later. The maximum asymmetrical current is available on only one phase of a three-phase system due to a three-phase fault.

9.2 Short-Circuit Current Calculations. The calculation of the precise value of an asymmetrical current at a given time after the inception of a fault is rather complex. Consequently, simplified methods have been developed which yield short-circuit currents required to match the assigned ratings of various system protective devices and equipment.

The value of the symmetrical short-cir-

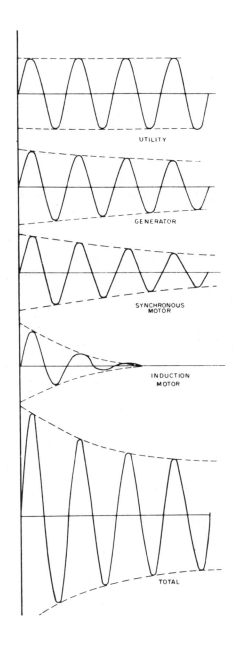

**Fig 70
Symmetrical Short-Circuit
Currents from Four Sources
Combined into Total**

cuit current is determined through the use of the proper impedance in the basic equation

$$I = E/Z$$

where E is the system driving voltage and Z (or X) is the proper system impedance (or reactance) of the power system back to and including the source or sources of the short-circuit current. The value of the proper impedance is determined with regard to the basis of rating for the device or equipment under consideration.

9.2.1 Types of Power-System Faults. Faults or short circuits can occur on a three-phase power system in several ways. The protective device or equipment must have the ability to interrupt or withstand the fault current and conductive components must have the ability to withstand the resulting mechanical and thermal stresses for any type of fault which can occur. The basic types of faults will be described, but it should be noted that the basic fault calculation for the selection of equipment is the three-phase bolted fault.

9.2.2. Three-Phase Bolted Fault. A three-phase bolted fault describes the condition where the three phase conductors are physically held together with zero impedance between them, just as if they were bolted together. This type of fault condition is not the most frequent in occurrence; however, it generally results in maximum short-circuit values and for this reason is the basic fault calculation in commercial power systems.

9.2.3 Line-to-Line Bolted Faults. In most three-phase power systems the line-to-line bolted fault currents are approximately 87% of the three-phase bolted fault currents. A detailed calculation is seldom required.

9.2.4 Line-to-Ground Bolted Fault. In solidly grounded systems the line-to-ground bolted fault-current value is usually about equal to the three-phase bolted fault-current value for the location being examined. Under certain conditions, such as a bolted line-to-ground fault at the secondary terminals of the Δ—Y connected transformer, the line-to-ground bolted fault-current value can theoretically exceed the three-phase bolted current value (however, tests show that the ground fault current in practical systems is less than the bolt three-phase fault current). Most often the ground-fault current will be significantly lower than the three-phase bolted fault current due to the relatively high impedance of the ground return circuit (that is, conduit, busway enclosure, grounding conductor, etc).

In resistance-grounded high-voltage systems the resistor is generally selected to limit the ground-fault current to a value ranging between 1 A and 2000 A. Line-to-ground fault magnitudes on these systems are limited primarily by the resistor itself, and a complicated line-to-ground short-circuit current calculation is generally not required.

9.2.5 Arcing Faults. Power-system faults may also be arcing in nature. Arcing faults can display a much lower level of short-circuit current than a bolted fault at the same location. These lower levels of current are due in part to the impedance of the arc. While system components must be capable of interrupting and withstanding the thermal and mechanical stresses of bolted short-circuit currents, arcing faults usually present different problems. Arcing faults may be difficult to detect because of the smaller currents. Sustained arcs can present safety hazards to people and also cause extensive damage due to the burning and welding effect of the arc as well as from the conductive products of ionization. Table 66 gives multipliers which can be applied to bolted-fault currents at point of fault to estimate approximate values of arcing-fault current. Maximum arcing-fault currents can approach bolted-fault values.

9.2.6 Single-Phase Transformer Faults. Single-phase transformers, particularly those with low impedance such as distribution types, may have lower per unit impedances on a line-to-neutral or line-to-ground basis than on a line-to-line basis. This may result in a higher neutral or ground-fault current than line-to-line fault current under bolted fault conditions. The transformer impedance under either type of fault can be identified by the manufacturer.

9.3 Selection of Equipment. In order to provide for personnel safety, to mini-

Table 66
Approximate Minimum Values of
Arcing-Fault Currents, in Per Unit of Bolted Values

| Type of Fault | Nominal System Voltage | | |
	570 V	480 V	208 V
Three phase	0.94	0.89	0.12
Single phase, line to line	0.85	0.74	0.02
Single phase, line to ground	0.40	0.38	0
Three phase, one transformer primary fuse open	0.88	0.80	0

mize equipment damage, and to maintain a high degree of service continuity, equipment must be selected to detect faults quickly and accurately and remove them in the shortest possible period of time.

Most protective devices employ the detection of current for operation. Fuses and certain types of circuit breakers are inherently current sensitive. A wide variety of protective relays is available to detect abnormal conditions of voltage, frequency, or real or reactive power. Relays can be used to determine current or power direction, and differential relays can be used to compare current magnitude and direction at two or more locations. Relays are only detecting devices and must be used in conjunction with circuit breakers or motorized switches to remove the detected faults.

The circuit protective devices must be selected to successfully detect and interrupt the fault condition rapidly enough so that any circuit element is not subjected to conditions beyond its rating. Proper selection is dependent upon a knowledge of the magnitudes of short-circuit current that can be expected for the various types of faults that may be experienced. Short-circuit calculations are the method by which these values are predicted.

9.3.1 Equipment Ratings. In order to provide for personnel safety and to minimize equipment damage, it is absolutely essential to use equipment with short-circuit ratings equal to or greater than the available short-circuit current to which the equipment can be subjected. ANSI/NFPA 70-1981 [7],[42] Section 110-9, states that devices intended to

break current shall have an interrupting capacity sufficient for the voltage employed and for the current which must be interrupted.

For any given location there may be a choice of one of several types of protective devices. Selection of a specific device then depends on other factors such as protection characteristics, economics, component protection, maintainability, user preference, etc.

Equipment can be applied at a location where the available short-circuit current is higher than the short-circuit rating of the device, provided that current-limiting fuses or circuit breakers "upstream" from the device limit the "let-through" current to a level the "downstream" equipment can withstand.

The calculated available short-circuit current may be found on the line side of the device. For a fault on the load side of the device, the actual current that the device does interrupt may be less than the available current due to the impedance of the device, the impedance of the arc on contact parting, and the ability of the device to limit current as in the case of a current-limiting fuse or circuit breaker. The basic concept is that the device must have the ability, when applied at a location with a given available short-circuit current, to satisfactorily interrupt a fault at its load terminals.

It is also necessary to identify the short-circuit rating of circuit conducting components such as busway, bus structures within switchgear and panelboards, and insulated conductors. The short-circuit rating refers to the ability of the equipment to withstand the available short-circuit current at the location where it is to be connected. Short-circuit ratings are somewhat imprecise for many noninterrupting circuit components.

---

[42] The numbers in brackets correspond to those in the references at the end of this Section.

9.4 Basis of Short-Circuit Current Calculations. The basis of rating of interrupting devices and the time after the inception of the fault at which the devices operate, determines the type of short-circuit calculation required.

Most equipment is rated on a symmetrical basis. However, unusual circuit configurations may result in a device being applied within its symmetrical rating but being subjected to asymmetrical currents beyond its capabilities. To avoid this possibility all short-circuit calculations should include a consideration of the X/R ratio at the point of fault. The X/R ratio establishes the power factor of the short-circuit current. Data is available to allow estimating asymmetrical multiplying factors based upon the X/R ratio and circuit power factor.

Since the short-circuit current may change during the time following the inception of a fault, the speed of operation and the basis of rating of the devices establish the circuit impedances to be used in the basic equation $I = E/Z$.

9.4.1 Total Current Basis of Rating. ANSI C37.6-1971 [2] Preferred Ratings for AC High-Voltage Circuit Breakers Rated on a Total Current Basis, lists such high-voltage circuit breakers. ANSI C37.5-1953 previously described the calculation of short-circuit duties to apply these circuit breakers. It was superseded by ANSI C37.5-1969, Methods for Determining Values of a Sinusoidal Current Wave, a Normal-Frequency Recovery Voltage, and a Guide for Calculation of Fault Currents for Application of AC High-Voltage Circuit Breakers Rated on a Total Current Basis, which describes a revised calculation for obtaining short-circuit duties to apply to total-current-rated circuit breakers.

The first-cycle duty (momentary) was determined by ANSI C37.5-1953 as follows. First a symmetrical short-circuit current value was calculated using subtransient reactance X_d'' for all sources of short-circuit current in the equivalent circuit of the power system. Next, multiplying factors were applied to this calculated symmetrical value to determine asymmetrical short-circuit duty. In the revised calculation procedure in ANSI C37.5-1969, the first-cycle duty (momentary) calculation is very similar. Differences occur in modified reactance values for medium and small induction motors.

The interrupting duty according to ANSI C37.5-1953 was determined using an equivalent circuit with subtransient reactance X_d'' for synchronous generators, transient reactance X_d' for synchronous motors, and ignoring the contribution of induction motors. The short-circuit interrupting current calculated from the circuit was then multiplied by a factor that depends on the circuit breaker rated interrupting time and on power-system operating conditions.

The contact-parting-time short-circuit (interrupting) duty calculated by the ANSI C37.5-1969 method uses subtransient reactance X_d'' for synchronous generators, 1.5 times subtransient reactance $1.5X_d''$ for synchronous motors, and modified subtransient reactances for induction motors which are divided into three categories, each with a different reactance multiplier in the power-system reactance network equivalent circuit. The circuit is then reduced to an equivalent X (reactance) value and an E/X symmetrical short-circuit current calculated. Then a multiplying factor obtained from curves in ANSI C37.5-1969 is applied to obtain the total short-circuit duty to be compared with the capability of a total current rated circuit breaker. The multiplying factor depends on the circuit-breaker contact parting time, the fault point X/R ratio, and the proximity of generation. ANSI/IEEE C37.5-1979 [5] describes the fault point X/R ratio calculation utilizing a resistance network corresponding to the reactance network.

Low-voltage protective devices and equipment, including power circuit breakers, molded case circuit breakers, motor control centers, motor controllers, fuses and busway are rated on the basis of maximum available symmetrical current at some specified power factor (X/R ratio). Their short-circuit ratings are based on current during the first cycle. Therefore, subtransient reactance X_d'' is used for all sources of short-circuit current.

9.4.2 Symmetrical Current Basis of Rating. ANSI C37.06-1980 [1] lists such high-voltage circuit breakers. The rated symmetrical short-circuit current listed for a circuit breaker in the standard applies only at rated maximum voltage. The short-circuit capability at an actual lower operating voltage is higher and is found by applying the voltage ratio to the rated short-circuit current.

The calculation method used to apply symmetrically rated circuit breakers is described in ANSI/IEEE C37.010-1980 [4]. The first-cycle duty calculation by this standard is exactly the same as in ANSI/IEEE C37.5-1979 [5]. The result is an asymmetrical first-cycle duty that is compared with the asymmetrical closing and latching capabilities of the symmetrically rated circuit breaker.

The contact-parting-time short-circuit (interrupting) duty calculation, as described by ANSI/IEEE C37.010-1979 [4], uses the same reactance network as the calculation described in ANSI/IEEE C37.5-1979 [5] and the same E/X calculated current value. A different multiplying factor is applied to E/X to establish the duty to be compared with the symmetrical short-circuit interrupting capability of a symmetrically rated circuit breaker.

As long as the X/R ratio for each network element or the fault point X/R ratio is 15 or less, the multiplying factor is 1.0. When the X/R ratio is 15 or less, the asymmetrical short-circuit duty never exceeds the symmetrical short-circuit duty by a margin greater than that by which the circuit breaker's asymmetrical short-circuit capability, as required by the standards, exceeds its symmetrical short-circuit capability.

When the X/R ratio exceeds 15, the multiplier usually exceeds 1.0. Multiplying factors are determined from curves in ANSI/IEEE C37.010-1979 [4] and depend on the contact parting (interrupting) time of the circuit breaker, the fault point X/R ratio, and the proximity of generation to the point of fault. The X/R ratio calculation of ANSI/IEEE C37.010-1979 [4] is the same as in ANSI/IEEE C37.5-1979 [5].

9.4.3 Comparison of Duty-Calculation Methods. The newer calculation methods, ANSI/IEEE C37.5-1979 [5] (for total current basis rated circuit breakers) and ANSI/IEEE C37.010-1979 [4] (for symmetrical current basis rated circuit breakers), differ from ANSI C37.5-1953 principally in data collection (not only reactance values, but also X/R ratios or resistance values are needed for system components) and in the treatment of reactances.

The first-cycle (momentary) duty calculated by the newer methods will not generally be greatly different from that calculated by the earlier method. The interrupting duty calculated by the newer methods often is higher because

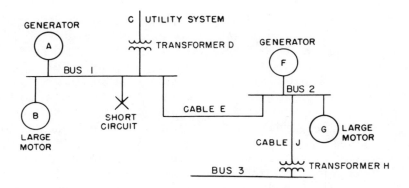

Fig 71
Typical System One-Line Diagram

of the increased motor contributions recognized.

For a further description of these procedures see IEEE Std 141-1976 [9].

9.5 Details of Short-Circuit Current Calculations. The general nature of alternating-current short-circuit currents has been discussed and it was determined that the basic equation for the calculation of short-circuit currents is $I = E/Z$, where E is the system driving voltage and Z (or X) is the proper impedance (or reactance) of the power system back to and including the source(s) of short-circuit current. Furthermore, the proper value of impedance depends on the basis of short-circuit current rating for the device or equipment under consideration.

In this section the details of short-circuit current calculations will be presented. Much of the work of such a study involves the representation of the proper system impedances from the point of fault back to and including the source(s) of short-circuit current.

9.5.1 Step-by-Step Procedure. The following steps identify the basic considerations in making short-circuit current calculations. In the simpler systems, several steps may be combined; for example, a combined one-line and impedance diagram may be used.

(1) Prepare a system one-line diagram, which is fundamental to short-circuit analysis. It should include all significant equipment and components and show their interconnections. Figure 71 illustrates a typical system one-line diagram.

(2) Decide on fault locations and type of short-circuit current calculations required, based on the type of equipment being applied. Consider the variations of system operating conditions required to display the most severe duties. Assign bus numbers of suitable identification to the fault locations.

(3) Prepare an impedance diagram. For systems above 600 V, two diagrams are usually required to calculate interrupting and momentary duty for high-voltage circuit breakers. Determine the type of short-circuit current rating required for various kinds of equipment as well as the machine reactances to use in the impedance diagram. Select suitable kilovoltampere and voltage bases for the study when the per-unit system is used.

In order to develop accurate fault currents it is necessary to know the subtransient and transient reactances of synchronous machines and the subtransient reactances of induction machines. In calculating the short-circuit currents of low-voltage systems, a realistic approximation involving a mix of synchronous and induction machines assumes a contribution at the machine terminals under bolted conditions of four times rated full-load current. This implies a machine reactance of 25% for purposes of approximation.

(4) For the designated fault locations and system conditions, resolve the impedance network and calculate the required symmetrical currents (E/Z or E/X). When calculations are made on a computer, submit impedance data in proper form as required by the specific program.

9.5.2 System Conditions for Most Severe Duty. Sometimes several of the intended or possible system conditions must be investigated to reveal the most severe duties for various components. Severe duties are those that are most likely to tax the capabilities of components.

Future growth and change in the system can modify short-circuit currents. For example, the initial utility available short-circuit duty for an in-building system being investigated may be 150 MVA. But future growth plans may call for an increase in available duty to 750 MVA several years later. This increase could substantially raise the short-circuit duties on the in-building equipment. Therefore the increase must be included in present calculations so that adequate in-building equipment can be selected. In a similar manner, future in-building expansions very often will raise short-circuit duties in various parts of

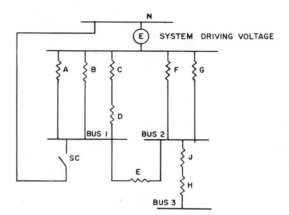

Fig 72
Equivalent Impedance Diagram for System of Fig 71

the power system, so that future expansions must also be considered.

The most severe duty usually occurs when the maximum concentration of machinery is in operation and all interconnections are closed. To determine the conditions that will most likely influence the critical duty, the following questions should be answered.

(1) Which machines and circuits are to be considered in actual operation?

(2) Which switching units are to be open, which closed?

(3) What future expansions or system changes will affect in-building short-circuit currents?

9.5.3 Preparing Impedance Diagrams. The impedance diagram displays the interconnected circuit impedances that control the magnitude of short-circuit currents. The diagram is derived from the system one-line diagram, showing an impedance for every system component that exerts a significant effect on short-circuit current magnitude. Not only must the impedance be interconnected to reproduce actual circuit conditions, but it will be helpful to preserve the same arrangement pattern used in the one-line diagram (Fig 72.)

9.5.4 Component Impedance Values. Component impedance values as collected may be expressed in terms of any of the following units:

(1) Ohms per phase (actually line-to-neutral single-phase impedance)

(2) Percent on rated kilovoltamperes or a reference kilovoltampere base

(3) Per unit on a reference kilovoltampere base

In formulating the impedance diagram, all impedance values must be expressed in the same units, either in ohms per phase or per unit on a reference kilovoltampere base. (Percent is a form of per unit: percent = per unit $\cdot$ 100.)

9.5.5. Neglecting Resistance. All system components have an impedance Z consisting of resistance R and inductive reactance X, where

$$Z = \sqrt{R^2 + X^2}$$

Many system components such as rotating machines, transformers, and reactors have high values of reactance compared to resistance. When the system impedance consists mainly of such components, the magnitude of a short-circuit current as derived by the basic equation $I = E/Z$ is primarily determined by the reactance, so that the resistance can practically be neglected in the calculation. This allows a much simpler calculation because then $I = E/X$.

Conductors (cables, buses, and open-wire lines), however, have a significant resistance compared to their reactance, so that when the system impedance contains considerable conductor impedance, the resistance may have an effect on the magnitude of the short-circuit current and should be included in the calculation.

Many descriptions of calculating methods appear to use Z or X interchangeably. The proper concept is that whenever the resistance does not significantly affect the calculated short-circuit current, a network of reactances alone can be used to represent the system impedance. When the ratio of reactance to resistance (X/R ratio) of the system impedance is greater than 4, negligible errors (less than 3%) will result from neglecting resistance. Neglecting R introduces some error but always increases the calculated current.

On systems above 600 V, circuit X/R ratios usually are greater than 4 and resistance can generally be neglected in short-circuit current calculations. However, on systems below 600 V, the circuit X/R ratio at locations remote from the supply transformer can be low and the resistance of circuit conductors should be included in the short-circuit current calculation. Because of their high X/R ratios, rotating machines, transformers, and reactors are generally represented by reactance only, regardless of the system voltage, an exception being transformers with impedance less than 4%. Figure 73 summarizes the locations in a system where resistance is generally used in the short-circuit current calculation.

9.5.6 Combining of Impedances. An impedance Z containing resistance R and reactance X is a complex quantity analyzed as a vector. It is frequently expressed in the form $R + jX$ and is illustrated in Fig 74.

When combining impedances in series, impedances Z cannot be added directly. The resistance R and reactance X must be added together separately, and then Z can be computed, $Z = \sqrt{R^2 + X^2}$. Figure 75 illustrates the addition of impedance in series.

When combining several impedances in

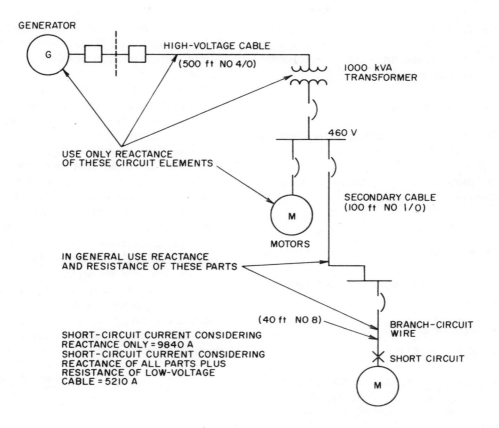

Fig 73
Locations in System where Reactance and Resistance Are
Generally Used for Short-Circuit-Current Calculations

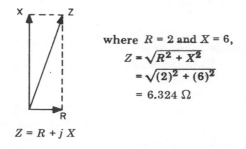

where $R = 2$ and $X = 6$,

$$Z = \sqrt{R^2 + X^2}$$
$$= \sqrt{(2)^2 + (6)^2}$$
$$= 6.324 \ \Omega$$

$Z = R + jX$

Fig 74
Impedance Vectors

parallel, the equivalent impedance is found by taking the reciprocal of the sum of the several impedance reciprocals, using the expression

$$1/Z_T = 1/Z_1 + 1/Z_2 + \ldots$$

The following formulas are used to find impedance reciprocals where $Z = R + jX$ and $1/Z(=Y) = G + jB$. Components of $1/Z$ found from components of Z are

$$G = R/(R^2 + X^2)$$
$$-B = X/(R^2 + X^2)$$

Components of Z found from components of $1/Z$ are

$$R = G/(G^2 + B^2)$$
$$X = -B/(G^2 + B^2)$$

Figure 76 illustrates summing reciprocals to find an equivalent impedance, using a table for recording the calculation steps.

9.5.7 Use of Per Unit, Percent or Ohms. Short-circuit current calculations can be made with impedances represented in per unit, percent or ohms. All representation will yield identical results. A single system should be used throughout any calculation and at the outset this decision must be made.

In general, if the system being studied has several different voltage levels or is a high-voltage system, per-unit impedance representation will often provide the easier more straight-forward calculation. The per-unit system is ideal for studying multivoltage systems. Also data for most of the components included in high-voltage networks (machines, transformers, and utility systems) are given in per-unit or percent values, and further conversion is simple.

Percent impedance representation is only a variation of the per-unit system: Percent impedance = per-unit impedance (100)

For commercial buildings short-circuit calculations, the percent and per-unit methods are equivalent and both have the same advantages.

Where few or no voltage transformations are involved and for low-voltage systems where many conductors are included in the impedance network, representation of system elements in ohms may provide the easier, more straight-forward calculations.

Characteristic impedance data is given for system components in Tables 67 through 73. Such data is commonly given as a percent based on the equipment kVA rating or as an ohmic value. The conversion equations for the three systems are:

$$\text{Per-unit impedance} = \frac{\text{percent impedance}}{100}$$

Per-unit impedance
(on chosen kVA base)

$$= \frac{\text{ohms} \cdot \text{kVA base}}{1000 \cdot \text{kV}^2}$$

where

$$\begin{aligned}
\text{ohms} &= \text{line-to-neutral values} \\
&\quad \text{(single conductor)} \\
\text{kVA base} &= \text{the three-phase base kVA} \\
\text{kV} &= \text{line-to-line voltage}
\end{aligned}$$

9.5.8 Per-Unit Representations. The per-unit system is a method of expressing numbers in a form that allows them to be easily compared. Impedances of circuit components are, therefore, a ratio on a prechosen base number, the chosen kVA base. The kVA base chosen may be

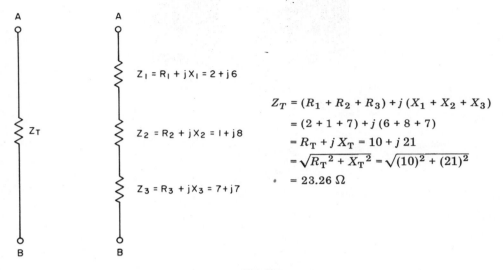

$$Z_T = (R_1 + R_2 + R_3) + j(X_1 + X_2 + X_3)$$
$$= (2 + 1 + 7) + j(6 + 8 + 7)$$
$$= R_T + jX_T = 10 + j21$$
$$= \sqrt{R_T{}^2 + X_T{}^2} = \sqrt{(10)^2 + (21)^2}$$
$$= 23.26\ \Omega$$

Fig 75
How Series Impedances Are Added

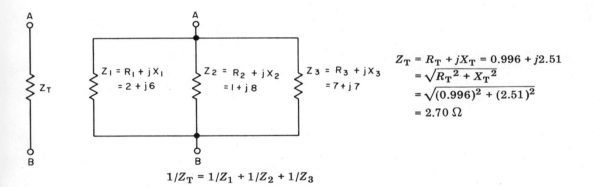

$$Z_T = R_T + jX_T = 0.996 + j2.51$$
$$= \sqrt{R_T{}^2 + X_T{}^2}$$
$$= \sqrt{(0.996)^2 + (2.51)^2}$$
$$= 2.70\ \Omega$$

$$1/Z_T = 1/Z_1 + 1/Z_2 + 1/Z_3$$

Element	R	X	G	$-B$	Sum of Squares*
		Z		$1/Z$	
Z_1	2.0	6.0	→ 0.05	0.15	40.0
Z_2	1.0	8.0	→ 0.0154	0.1231	65.0
Z_3	7.0	7.0	→ 0.0714	0.0714	98.0
Z_T	0.996	2.51	← 0.1368	0.3445	0.1374

*$R^2 + X^2$ when finding $1/Z$ from Z and $G^2 + B^2$ when finding Z from $1/Z$.

Fig 76
Combining Impedances in Parallel

Table 67
Transformer Impedance Data

Transformer Rating (kVA)	X/R	R (%)	X (%)	Z (%)
150	3.24	1.23	4.0	4.19
225	3.35	1.19	4.0	4.17
300	3.50	1.14	4.0	4.16
500	3.85	1.04	4.0	4.12
750	5.45	0.94	5.1	5.19
1000	5.70	0.89	5.1	5.19
1500	6.15	0.83	5.1	5.18
2000	6.63	0.77	5.1	5.17
150	1.5	1.111	1.665	2.0
225	1.5	1.111	1.665	2.0
300	1.5	1.111	1.665	2.0
500	1.5	1.111	1.665	2.0

From NEMA AB1-1969 [12] Table A-1.

NOTES: (1) These values are for three-phase liquid-filled self-cooled transformers.

(2) Due to the trend toward lower impedance transformers for better voltage regulation, the actual transformer impedance may deviate from the NEMA standard. Therefore, for actual values obtain nameplate impedance from owner or manufacturer. The percent X and percent R values are desirable for calculation.

Table 68
Approximate Reactances of Current Transformers, in Ohms per Phase*

Primary Current Rating (amperes)	Voltage Ratings		
	600—5000 V	7500 V	15 000 V
100—200	0.0022	0.0040	
250—400	0.0005	0.0008	0.0002
500—800	0.00019	0.00031	0.00007
1000—4000	0.00007	0.00007	0.00007

*For actual values, refer to manufacturers' data.

Table 69
Disconnecting-Switch Approximate Reactance Data, in Ohms*

Switch Size (amperes)	Reactance (ohms)
200	0.0001
400	0.00008
600	0.00008
800	0.00007
1200	0.00007
1600	0.00005
2000	0.00005
3000	0.00004
4000	0.00004

SINGLE POLE

NOTE: The reactance of disconnecting switches for low-voltage circuits (600 V and below) is in the order of magnitude of 0.00008—0.00005 Ω/pole at 60 Hz for switches rated 400—4000 A, respectively.

*For actual values, refer to manufacturers' data.

Table 70
Circuit Breaker Reactance Data

(a) Reactance of Low-Voltage Power Circuit Breakers			(b) Typical Molded-Case Circuit-Breaker Impedances*		
Circuit-Breaker Interrupting Rating (amperes)	Circuit-Breaker Rating (amperes)	Reactance (ohms)	Molded-Case Circuit-Breaker Rating (amperes)	Resistance (ohms)	Reactance (ohms)
15 000 and 25 000	15—35	0.04	20	0.00700	Negligible
	50—100	0.004	40	0.00240	Negligible
	125—225	0.001	100	0.00200	0.00070
	250—600	0.0002	225	0.00035	0.00020
50 000	200—800	0.0002	400	0.00031	0.00039
	1000—1600	0.00007	600	0.00007	0.00017
75 000	2000—3000	0.00008			
100 000	4000	0.00008			

NOTES: (1) Due to the method of rating low-voltage power circuit breakers, the reactance of the circuit breaker which is to interrupt the fault is not included in calculating the fault current.
(2) Above 600 A the reactances of molded-case circuit breakers are similar to those given in (a).

*For actual values, refer to manufacturers' data.

Table 71
Copper-Cable Impedance Data, in Ohms per 1000 ft at 75 °C*

(a) Three Single Conductors

AWG or kcmil	In Magnetic Duct						In Nonmagnetic Duct					
	600 V and 5 kV Nonshielded			5 kV Shielded and 15 kV			600 V and 5 kV Nonshielded			5 kV Shielded and 15 kV		
	R	X	Z	R	X	Z	R	X	Z	R	X	Z
8	0.811	0.0754	0.814	0.811	0.0860	0.816	0.811	0.0603	0.813	0.811	0.0688	0.814
8 (solid)	0.786	0.0754	0.790	0.786	0.0860	0.791	0.786	0.0603	0.788	0.786	0.0688	0.789
6	0.510	0.0685	0.515	0.510	0.0796	0.516	0.510	0.0548	0.513	0.510	0.0636	0.514
6 (solid)	0.496	0.0685	0.501	0.496	0.0796	0.502	0.496	0.0548	0.499	0.496	0.0636	0.500
4	0.321	0.0632	0.327	0.321	0.0742	0.329	0.321	0.0506	0.325	0.321	0.0594	0.326
4 (solid)	0.312	0.0632	0.318	0.312	0.0742	0.321	0.312	0.0506	0.316	0.312	0.0594	0.318
2	0.202	0.0585	0.210	0.202	0.0685	0.214	0.202	0.0467	0.207	0.202	0.0547	0.209
1	0.160	0.0570	0.170	0.160	0.0675	0.174	0.160	0.0456	0.166	0.160	0.0540	0.169
1/0	0.128	0.0540	0.139	0.128	0.0635	0.143	0.127	0.0432	0.134	0.128	0.0507	0.138
2/0	0.102	0.0533	0.115	0.103	0.0630	0.121	0.101	0.0426	0.110	0.102	0.0504	0.114
3/0	0.0805	0.0519	0.0958	0.0814	0.0605	0.101	0.0766	0.0415	0.0871	0.0805	0.0484	0.0939
4/0	0.0640	0.0497	0.0810	0.0650	0.0583	0.0929	0.0633	0.0398	0.0748	0.0640	0.0466	0.0792
250	0.0552	0.0495	0.0742	0.0557	0.0570	0.0797	0.0541	0.0396	0.0670	0.0547	0.0456	0.0712
300	0.0464	0.0493	0.0677	0.0473	0.0564	0.0736	0.0451	0.0394	0.0599	0.0460	0.0451	0.0644
350	0.0378	0.0491	0.0617	0.0386	0.0562	0.0681	0.0368	0.0393	0.0536	0.0375	0.0450	0.0586
400	0.0356	0.0490	0.0606	0.0362	0.0548	0.0657	0.0342	0.0392	0.0520	0.0348	0.0438	0.0559
450	0.0322	0.0480	0.0578	0.0328	0.0538	0.0630	0.0304	0.0384	0.0490	0.0312	0.0430	0.0531
500	0.0294	0.0466	0.0551	0.0300	0.0526	0.0505	0.0276	0.0373	0.0464	0.0284	0.0421	0.0508
600	0.0257	0.0463	0.0530	0.0264	0.0516	0.0580	0.0237	0.0371	0.0440	0.0246	0.0412	0.0479
750	0.0216	0.0445	0.0495	0.0223	0.0497	0.0545	0.0194	0.0356	0.0405	0.0203	0.0396	0.0445

*Resistance values (R_L) at lower copper temperatures (T_L) are obtained by using the formula $R_L = \dfrac{R_{75} \, (234.5 + T_L)}{309.5}$

Table 71 (Continued)
Copper-Cable Impedance Data, in Ohms per 1000 ft at 75 °C*

(b) Three-Conductor Cable

AWG or kcmil	In Magnetic Duct and Steel Interlocked Armor						In Nonmagnetic Duct and Aluminum Interlocked Armor					
	600 V and 5 kV Nonshielded			5 kV Shielded and 15 kV			600 V and 5 kV Nonshielded			5 kV Shielded and 15 kV		
	R	X	Z	R	X	Z	R	X	Z	R	X	Z
8	0.811	0.0577	0.813	0.811	0.0658	0.814	0.811	0.0503	0.812	0.811	0.0574	0.813
8 (solid)	0.786	0.0577	0.788	0.786	0.0658	0.789	0.786	0.0503	0.787	0.786	0.0574	0.788
6	0.510	0.0525	0.513	0.510	0.0610	0.514	0.510	0.0457	0.512	0.510	0.0531	0.513
6 (solid)	0.496	0.0525	0.499	0.496	0.0610	0.500	0.496	0.0457	0.498	0.496	0.0531	0.499
4	0.321	0.0483	0.325	0.321	0.0568	0.326	0.321	0.0422	0.324	0.321	0.0495	0.325
4 (solid)	0.312	0.0483	0.316	0.312	0.0508	0.317	0.312	0.0422	0.315	0.312	0.0495	0.316
2	0.202	0.0448	0.207	0.202	0.0524	0.209	0.202	0.0390	0.206	0.202	0.0457	0.207
1	0.160	0.0436	0.166	0.160	0.0516	0.168	0.160	0.0380	0.164	0.160	0.0450	0.166
1/0	0.128	0.0414	0.135	0.128	0.0486	0.137	0.127	0.0360	0.132	0.128	0.0423	0.135
2/0	0.102	0.0407	0.110	0.103	0.0482	0.114	0.101	0.0355	-0.107	0.102	0.0420	0.110
3/0	0.0805	0.0397	0.0898	0.0814	0.0463	0.0936	0.0766	0.0346	0.0841	0.0805	0.0403	0.090
4/0	0.0640	0.0381	0.0745	0.0650	0.046	0.0788	0.0633	0.0332	0.0715	0.0640	0.0389	0.0749
250	0.0552	0.0379	0.0670	0.0557	0.0436	0.0707	0.0541	0.0330	0.0634	0.0547	0.0380	0.0666
300	0.0464	0.0377	0.0598	0.0473	0.0431	0.0640	0.0451	0.0329	0.0559	0.0460	0.0376	0.0596
350	0.0378	0.0373	0.0539	0.0386	0.0427	0.0576	0.0368	0.0328	0.0492	0.0375	0.0375	0.0530
400	0.0356	0.0371	0.0514	0.0362	0.0415	0.0551	0.0342	0.0327	0.0475	0.0348	0.0366	0.0505
450	0.0322	0.0361	0.0484	0.0328	0.0404	0.0520	0.0304	0.0320	0.0441	0.0312	0.0359	0.0476
500	0.0294	0.0349	0.0456	0.0300	0.0394	0.0495	0.0276	0.0311	0.0416	0.0284	0.0351	0.0453
600	0.0257	0.0343	0.0429	0.0264	0.0382	0.0464	0.0237	0.0309	0.0389	0.0246	0.0344	0.0422
750	0.0216	0.0326	0.0391	0.0223	0.0364	0.0427	0.0197	0.0297	0.0355	0.0203	0.0332	0.0389

NOTE: Resistance based on tinned copper at 60 Hz. 600 V and 5 kV nonshielded cable based on varnished cambric insulation. 5 kV shielded and 15 kV cable based on neoprene insulation.

*Resistance values (R_L) at lower copper temperatures (T_L) are obtained by using the formula $R_L = R_{75} \dfrac{(234.5 + T_L)}{3.09.5}$

Table 72
Aluminum-Cable Impedance Data, in Approximate Ohms per 1000 ft at 90 °C*
(Cross-Linked Polyethylene Insulated Cable)

(a) Three Single Conductors

AWG or kcmil	In Magnetic Duct						In Nonmagnetic Duct					
	600 V and 5 kV Nonshielded			5 kV Shielded and 15 kV			600 V and 5 kV Nonshielded			5 kV Shielded and 15 kV		
	R	X	Z	R	X	Z	R	X	Z	R	X	Z
6	0.847	0.053	0.849	—	—	—	0.847	0.042	0.848	—	—	—
4	0.532	0.050	0.534	0.532	0.068	0.536	0.532	0.040	0.534	0.532	0.054	0.535
2	0.335	0.046	0.338	0.335	0.063	0.341	0.335	0.037	0.337	0.335	0.050	0.339
1	0.265	0.048	0.269	0.265	0.059	0.271	0.265	0.035	0.267	0.265	0.047	0.269
1/0	0.210	0.043	0.214	0.210	0.056	0.217	0.210	0.034	0.213	0.210	0.045	0.215
2/0	0.167	0.041	0.172	0.167	0.055	0.176	0.167	0.033	0.170	0.167	0.044	0.173
3/0	0.133	0.040	0.139	0.132	0.053	0.142	0.133	0.037	0.137	0.132	0.042	0.139
4/0	0.106	0.039	0.113	0.105	0.051	0.117	0.105	0.031	0.109	0.105	0.041	0.113
250	0.0896	0.0384	0.0975	0.0892	0.0495	0.102	0.0894	0.0307	0.0945	0.0891	0.0396	0.0975
300	0.0750	0.0375	0.0839	0.0746	0.0479	0.0887	0.0746	0.0300	0.0804	0.0744	0.0383	0.0837
350	0.0644	0.0369	0.0742	0.0640	0.0468	0.0793	0.0640	0.0245	0.0705	0.0638	0.0374	0.0740
400	0.0568	0.0364	0.0675	0.0563	0.0459	0.0726	0.0563	0.0291	0.0634	0.0560	0.0367	0.0700
500	0.0459	0.0355	0.0580	0.0453	0.0444	0.0634	0.0453	0.0284	0.0535	0.0450	0.0355	0.0573
600	0.0388	0.0359	0.0529	0.0381	0.0431	0.0575	0.0381	0.0287	0.0477	0.0377	0.0345	0.0511
700	0.0338	0.0350	0.0487	0.0332	0.0423	0.0538	0.0330	0.0280	0.0433	0.0326	0.0338	0.0470
750	0.0318	0.0341	0.0466	0.0310	0.0419	0.0521	0.0309	0.0273	0.0412	0.0304	0.0335	0.0452
1000	0.0252	0.0341	0.0424	0.0243	0.0414	0.0480	0.0239	0.0273	0.0363	0.0234	0.0331	0.0405

*Resistance values (R_L) at lower aluminum temperatures (T_L) are obtained by the formula $R_L = \dfrac{R_{90}\,(228.1 + T_L)}{318.1}$

Table 72 (*Continued*)
Aluminum-Cable Impedance Data, in Approximate Ohms per 1000 ft at 90 °C*
(Cross-Linked Polyethylene Insulated Cable)

(b) *Three-Conductor Cable*

AWG or kcmil	In Magnetic Duct						In Nonmagnetic Duct					
	600 V and 5kV Nonshielded			5 kV Shielded and 15 kV			600 V and 5 kV Nonshielded			5 kV Shielded and 15 kV		
	R	X	Z	R	X	Z	R	X	Z	R	X	Z
6	0.847	0.053	0.849	—	—	—	0.847	0.042	0.848	—	—	—
4	0.532	0.050	0.534	—	—	—	0.532	0.040	0.534	—	—	—
2	0.335	0.046	0.338	0.335	0.056	0.340	0.335	0.037	0.337	0.335	0.045	0.338
1	0.265	0.048	0.269	0.265	0.053	0.270	0.265	0.035	0.267	0.265	0.042	0.268
1/0	0.210	0.043	0.214	0.210	0.050	0.216	0.210	0.034	0.213	0.210	0.040	0.214
2/0	0.167	0.041	0.172	0.167	0.049	0.174	0.167	0.033	0.170	0.167	0.039	0.171
3/0	0.133	0.040	0.139	0.133	0.048	0.141	0.133	0.037	0.137	0.133	0.038	0.138
4/0	0.106	0.039	0.113	0.105	0.045	0.114	0.105	0.031	0.109	0.105	0.036	0.111
250	0.0896	0.0384	0.0975	0.0895	0.0436	0.100	0.0894	0.0307	0.0945	0.0893	0.0349	0.0959
300	0.0750	0.0375	0.0839	0.0748	0.0424	0.0860	0.0746	0.0300	0.0804	0.0745	0.0340	0.0819
350	0.0644	0.0369	0.0742	0.0643	0.0418	0.0767	0.0640	0.0245	0.0705	0.0640	0.0334	0.0722
400	0.0568	0.0364	0.0675	0.0564	0.0411	0.0700	0.0563	0.0291	0.0634	0.0561	0.0329	0.0650
500	0.0459	0.0355	0.0580	0.0457	0.0399	0.0607	0.0453	0.0284	0.0535	0.0452	0.0319	0.0553
600	0.0388	0.0359	0.0529	0.0386	0.0390	0.0549	0.0381	0.0287	0.0477	0.0380	0.0312	0.0492
700	0.0338	0.0350	0.0487	0.0335	0.0381	0.0507	0.0330	0.0280	0.0433	0.0328	0.0305	0.0448
750	0.0318	0.0341	0.0466	0.0315	0.0379	0.0493	0.0309	0.0273	0.0412	0.0307	0.0303	0.0431
1000	0.0252	0.0341	0.0424	0.0248	0.0368	0.0444	0.0239	0.0273	0.0363	0.0237	0.0294	0.0378

*Resistance values (R_L) at lower aluminum temperatures (T_L) are obtained by the formula $R_L = R_{90} \dfrac{(228.1 + T_L)}{318.1}$

From Kaiser Aluminum Electrical Products Division.

Table 73
Busway Impedance Data, in Ohms per 1000 ft, Line to Neutral, 60 Hz

(a) Plug-In Busway

Current Rating (amperes)	Copper Bus Bars			Aluminum Bus Bars		
	R	X	Z	R	X	Z
225	0.0836	0.0800	0.1157	0.1090	0.0720	0.1313
400	0.0437	0.0232	0.0495	0.0550	0.0222	0.0592
600	0.0350	0.0179	0.0393	0.0304	0.0121	0.0327
800	0.0218	0.0136	0.0257	0.0243	0.0154	0.0288
1000	0.0145	0.0135	0.0198			

(b) Low-Impedance Feeder Busway

Current Rating (amperes)	R	X	Z
800	0.0219	0.0085	0.0235
1000	0.0190	0.0050	0.0196
1350	0.0126	0.0044	0.0134
1600	0.0116	0.0035	0.0121
2000	0.0075	0.0031	0.0081
2500	0.0057	0.0025	0.0062
3000	0.0055	0.0017	0.0058
4000	0.0037	0.0016	0.0040

(c) Current-Limiting Busway

Current Rating (amperes)	R	X	Z	X/R
1000	0.013	0.063	0.064	4.85
1350	0.012	0.061	0.062	5.08
1600	0.009	0.056	0.057	6.22
2000	0.007	0.052	0.052	7.45
2500	0.006	0.049	0.049	8.15
3000	0.005	0.046	0.046	9.20
4000	0.004	0.042	0.042	10.50

From *Actual Specifying Engineer* (Oct. 1965).

the kilovoltampere rating of one of the predominant pieces of system equipment such as a generator or transformer. However, an arbitrary number such as 10 000 kVA may be selected as the kVA base. The number selected should be one that will result in component impedances that are not excessively large or small and can be easily handled in the calculations.

Component impedance may be given on bases other than the chosen kVA base. The conversion equation for one kVA base to another is:

Per-unit impedance on new base =

$$\text{Per unit on old base} \cdot \frac{\text{new kVA base}}{\text{old kVA base}}$$

The procedure for making short-circuit calculations using the per-unit system is as given in 9.5.1 and involves converting the impedance of each circuit element to a per-unit value on the common kVA base. The network is resolved to the point of fault to obtain a total per-unit fault impedance.

The fault kVA is calculated by the following equation

Fault kVA =

$$\frac{\text{kVA base}}{\text{total per-unit fault impedance}}$$

The short-circuit current (for three-phase systems) can be calculated from the following equation

Short-circuit current =

$$\frac{\text{fault kVA}}{\sqrt{3} \cdot \text{rated kV at fault}}$$

where
rated kV = line-to-line voltage.

9.5.9 The Electric Utility System.

The electric utility system is usually represented by a single equivalent reactance referred to the user's point of connection which is equivalent to the available short-circuit current from the utility for a fault at the user's point of connection. The per-unit reactance of the utility is, therefore, 1.0 based on the available short-circuit current from the utility. This value is obtained from the utility and may be expressed in several ways.

(1) Three-phase short-circuit current in kilovoltamperes available at a given voltage

(2) Three-phase short-circuit current in amperes available at a given voltage

(3) Percent or per-unit reactance on a specified kilovoltampere and voltage base

(4) Reactance in ohms per phase (sometimes $R + jX$) referred to a given voltage

Examples
(1) Conversion to per unit on a 10 000 kVA base (kVA_b).

(a) Available three-phase short-circuit kilovoltamperes = 500 000 kVA (500 MVA):

$$X_{pu} = \frac{\text{kVA}_b}{\text{kVA}_{sc}} = \frac{10\ 000}{500\ 000} = 0.02$$

(b) Available three-phase short-circuit current = 20 940 A at 13.8 kV:

$$X_{pu} = \frac{\text{kVA}_b}{\sqrt{3}\ I_{sc}\ \text{kV}} = \frac{10\ 000}{\sqrt{3}\ (20\ 940)\ (13.8)} = 0.02$$

(c) Equivalent utility reactance = 0.02 pu on a 100 000 kVA base:

$$X_{pu} = X_{pu_{old}}\left(\frac{\text{kVA}_b}{\text{kVA}_{old}}\right) = 0.2\left(\frac{10\ 000}{100\ 000}\right) = 0.02$$

(d) Equivalent utility reactance = 0.38 Ω/phase at 13.8 kV:

$$X_{pu} = X\left(\frac{\text{kVA}_b}{\text{kV}^2\ (1000)}\right) = 0.38\left(\frac{10\ 000}{(13.8)^2\ 1000}\right)$$
$$= 0.02$$

(2) Conversion to ohms per phase at 480 V.

(a) Available three-phase short-circuit kilovoltamperes = 62 280 kVA:

$$X = \frac{kV^2 \ (1000)}{kVA} = \frac{(0.48)^2 \ 1000}{62 \ 280}$$

= 0.0037 Ω/phase at 480 V

(b) Available three-phase short-circuit current = 75 000 A at 480 V:

$$X = \frac{V_{L-N}}{I_{sc}} = \frac{277}{75 \ 000} = 0.0037 \ \Omega/\text{phase at } 480 \ V$$

(c) Equivalent utility reactance = 0.1605 pu on a 10 000 kVA base:

$$X = X_{pu} \left(\frac{kV^2 \ (1000)}{kVA} \right)$$

$$= 0.1605 \left(\frac{(0.48)^2 \ 1000}{10 \ 000} \right)$$

= 0.0037 Ω/phase at 480 V

(d) The X/R ratio for this available short-circuit current equals 4, and

$$R = \frac{X}{X/R} = \frac{0.0037}{4}$$

= 0.000925 Ω/phase at 480 V

9.5.10 Transformers. Transformers reactance (impedance) will most commonly be expressed as a percent value ($\%X_T$ or $\%Z_T$) on the transformer rated kilovoltamperes. (Impedance values are usually expressed on the self-cooled kilovoltampere rating at rated temperature rise.)

Example. A 500 kVA transformer with an impedance of 5 percent on its kilovolt-ampere rating (assume impedance is all reactance).

(1) Conversion to per unit on a 10 000 kVA base (kVA$_b$):

$$X_{pu} = \frac{\%X_T}{100} \left(\frac{kVA_b}{\text{transformer kVA}} \right)$$

$$= \frac{5}{100} \left(\frac{10 \ 000}{500} \right) = 1.0$$

(2) Conversion to ohms per phase at 480 V:

$$X = \frac{\%X_T}{100} \left(\frac{kV^2 \ (1000)}{\text{transformer kVA}} \right)$$

$$= \frac{5}{100} \left(\frac{(0.48)^2 \ 1000}{500} \right)$$

= 0.023 Ω/phase at 480 V

9.5.11 Busways, Cables, and Conductors. The resistance and reactance of busway, cables, and conductors will most frequently be available in terms of ohms per-phase per-unit length.

Example. 250 ft of a three-conductor 500 kcmil cable (600 V) installed in steel conduit on a 480 V system.

Conversion to per unit on a 10 000 kVA base (kVA$_b$):

$$R = 0.0294 \ \Omega/1000 \text{ ft}$$
$$R = 0.00735 \ \Omega/250 \text{ ft}$$
$$X = 0.0349 \ \Omega/1000 \text{ ft}$$
$$X = 0.00872 \ \Omega/250 \text{ ft}$$

$$R_{pu} = R \left(\frac{kVA_b}{kV^2 \ (1000)} \right)$$

$$= 0.00735 \left(\frac{10 \ 000}{(0.48)^2 \ 1000} \right) = 0.319$$

$$X_{pu} = X \left(\frac{kVA_b}{kV^2 \ (1000)} \right)$$

$$= 0.00872 \left(\frac{10 \ 000}{(0.48)^2 \ 1000} \right) = 0.378$$

$$Z_{pu} = 0.319 + j0.378$$

For high-voltage cables (above 600 V) the resistance of cables can generally be omitted; in fact, for short high-voltage cable runs (less than 1000 ft) the entire impedance of the cable can be omitted with negligible error in calculating short-circuit current magnitudes. Sometimes it is beneficial to include resistances of high-voltage elements because they may significantly affect the X/R ratio needed for high-voltage interrupting duty calculations.

9.5.12 Rotating Machines. Machine reactances are usually expressed in terms of percent reactance $\%X_m$, or per-unit reactance X_{pu} on the normal rated kilovoltamperes of the machine. Either the subtransient reactance X'' or the transient reactance X' should be selected, depending on the type of short-circuit current calculation required. Motor rated kilovoltamperes can be estimated, given the motor horsepower, as follows:

Type of Machine	Rated kVA
All	$\dfrac{V_{rated}\ I_{rated}\ \sqrt{3}}{1000}$ (exact)
Induction motors and 0.8 power-factor synchronous motors	rated hp (approximate)
1.0 power-factor synchronous motors	0.8 rated hp (approximate)

9.5.13 Motors Rated 600 V or Less. In systems of 600 V or less, the large motors (that is, motors of several hundred horsepower) are usually few in number and represent only a small portion of the total connected horsepower. These large motors can be represented individually, or they can be combined with the smaller motors, representing the complete group as one equivalent motor in the impedance diagram. Small motors are turned OFF and ON frequently, so it is practically impossible to predict which ones will be on the line when a short circuit occurs. Therefore, all small motors are generally assumed to be running and all considered as one large motor.

Where more accurate data are not available, the following procedures may be used in representing the combined re-

actance of a group of miscellaneous motors.

(1) In all 208 V systems and 480 V commercial-building systems, a substantial portion of the load consists of lighting; so assume that the running motors are grouped at the transformer secondary bus and have a reactance of 25% on a kilovoltampere base equal to 50% of the transformer kVA rating.

(2) Groups of small induction motors as served by a motor-control center can be represented by considering the group to have a reactance of 25% on a kilovoltampere rating equal to the connected motor horsepower.

Examples
(1) Conversion to per unit on a 10 000 kVA base (kVA_b). A 500 hp 0.8 pf synchronous motor has a subtransient reactance X_d'' of 15%

$$X''_{pu} = \frac{\%X_d''}{100}\left(\frac{kVA_b}{\text{motor kVA}}\right)$$

$$= \frac{15}{100}\left(\frac{10\ 000}{500}\right) = 3.0$$

(2) Conversion to ohms per phase at 480 V. A motor-control center has induction motors with a connected horsepower totaling 420 hp. Assume a group of motors to have a reactance of 25% on a kilovoltampere rating of 420 kVA:

$$X = \frac{\%X}{100}\left(\frac{kV^2\ (1000)}{\text{motor kVA}}\right)$$

$$= \frac{25}{100}\left(\frac{(0.48)^2\ 1000}{420}\right)$$

$$= 0.137\ \Omega/\text{phase at 480 V}$$

9.5.14 Other Circuit Impedances. There are other circuit impedances such as those associated with circuit breakers, current transformers, bus structures, and connections which for ease of calculation are usually neglected in short-circuit-current calculations. Accuracy of the calculation is not generally affected because the effects of the impedances are small and omitting them provides conservative (higher) short-circuit currents. However, on low-voltage systems and particularly at 208Y/120 V, there are cases where their inclusion can result in a lower calculated short-circuit current indicating the use of lower rated circuit components. The system designer may want to include these impedances in such cases.

9.5.15 Shunt-Connected Impedances. In addition to the components already mentioned, every system includes other components or loads that are represented in a diagram as shunt-connected impedances. Examples are lights, furnaces, and capacitors. A technically accurate solution requires that these impedances be included in the equivalent circuit used in calculating a short-circuit current, but practical considerations allow the general practice of omitting them. Such impedances are relatively high values and their omission will not significantly affect the calculated results.

9.5.16 System Driving Voltage. The system driving voltage E in the basic equation can be represented by the use of a single overall driving voltage as illustrated in Fig 72, rather than the array of individual unequal generated voltages acting within individual rotating machines. This single driving voltage is equal to the prefault voltage at the point of fault connection. The equivalent circuit is a valid transformation accomplished by Thevenin's theorem and permits an accurate determination of the short-circuit current for the assigned values of system impedance. The prefault voltage referred to is ordinarily taken as system nominal voltage at the point of fault, as this calculation leads to the full value of short-circuit current that may be produced by the probable maximum operating voltage.

In making a short-circuit current calculation on three-phase balanced systems, a single-phase representation of a three-phase system is utilized so that all impedances are expressed in ohms per phase, and the system driving voltage E is expressed in line-to-neutral volts. Line-to-neutral voltage is equal to line-to-line voltage divided by $\sqrt{3}$.

When using the per-unit system, if the system per-unit impedances are established on voltage bases equal to system nominal voltages, the per-unit driving voltage is equal to 1.0. In the per-unit system, both line-to-line voltage and line-to-neutral voltage have equal values, that is, both would have values of 1.0.

When system impedance values are expressed in ohms per phase rather than per unit, the system driving voltage is equal to the system line-to-neutral voltage, that is, 277 V for a 480Y/277 V system.

9.5.17 Determination of Short-Circuit Currents. After the impedance diagram has been prepared, the short-circuit currents can be determined. This can be accomplished by longhand calculation, network analyzer, or digital computer techniques.

Simple radial systems, such as those used in most low-voltage systems, can be easily resolved by longhand calculations, though digital computers can yield significant time savings, particularly when short-circuit duties at many system locations are required and when resistance is being included in the calculation.

9.5.18 Longhand Solution. A longhand solution requires the combining of impedances in series and parallel from the source driving voltage to the location of the fault being calculated to determine the simple equivalent network impedance. The calculation to derive the symmetrical short-circuit current is

$I = E/Z$ (or E/X) where E is the system driving voltage and Z (or X) is the single equivalent network impedance (or reactance).

When calculations are made using per unit, the following formulas apply:

(1) Symmetrical three-phase short-circuit current, in per unit:

$$I_{pu} = \frac{E_{pu}}{Z_{pu}}$$

(2) Symmetrical three-phase short-circuit current, in amperes

$$I = I_b (I_{pu}) = * \frac{I_b}{Z_{pu}}$$

(3) Symmetrical three-phase short-circuit per-unit kilovoltamperes

$$kVA_{pu} = E_{pu} (I_{pu})$$

(4) Symmetrical three-phase short-circuit kilovoltamperes

$$kVA = kVA_b (kVA_{pu})$$

$$= * \frac{kVA_b}{Z_{pu}}$$

where

I_{pu} = per-unit current
Z_{pu} = equivalent network per-unit impedance
E_{pu} = per-unit voltage
kVA_{pu} = per-unit kilovoltamperes
I_b = base current in amperes
kVA_b = base kilovoltamperes
* = a simplified equality that applies only where E_{pu} = 1.0

When calculations are made using ohms, the symmetrical three-phase short-circuit current, in amperes, is

$$I = \frac{E_{L-N}}{Z}$$

where

E_{L-N} = line-to-neutral voltage
Z = equivalent network impedance, in ohms per phase

A new combination of impedances to determine the single equivalent network impedance is required for each fault location.

The longhand solution for a radial system is fairly simple. For systems containing loops, simultaneous equations may be necessary, though $\Delta-Y$ network transformations can usually be used to combine impedances. Some of the newer electronic calculators can be excellent time savers in making longhand calculations. An example of a longhand solution is included.

Example: The Building Power System

Step 1: System One-Line Diagram. Figure 77 is a one-line diagram of a building power system served from a utility spot network. The diagram includes:

(1) Utility short-circuit duty at the network bus

(2) Conductor type and length

(3) Kilovoltamperes and impedance of 30 kVA and 150 kVA transformers

(4) Lumped connected horsepower of induction motors

Step 2: Type and Locations of Short Circuits. Short-circuit currents are required at all buses where protective devices will be located (buses 1 – 18). Symmetrical short-circuit currents are required since all devices are rated 480 V and below, and three-phase bolted fault values are also required. The most severe duty will occur with all circuit breakers closed, with a maximum short-circuit duty of 55 600 A, three phase, symmetrical, from the utility spot network.

Step 3: System Impedance Diagrams. The impedance diagram for this system is shown in Fig 78. Since most buses are at the 480 V level, the example uses sys-

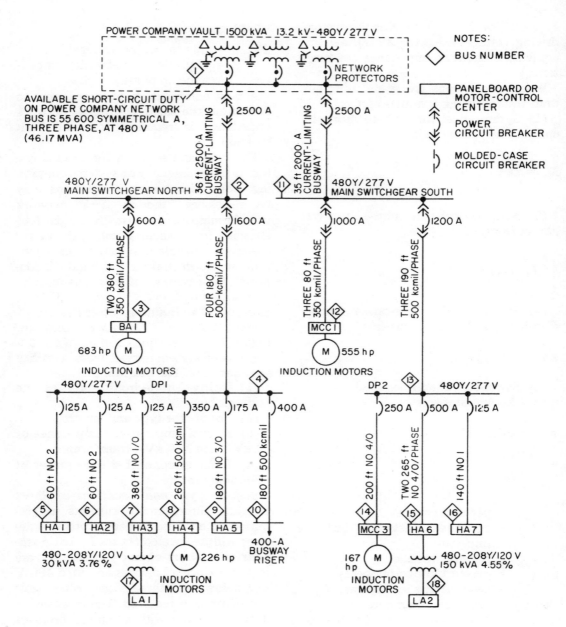

Fig 77
One-Line Diagram for Typical Building Served from
480Y/277 V Network

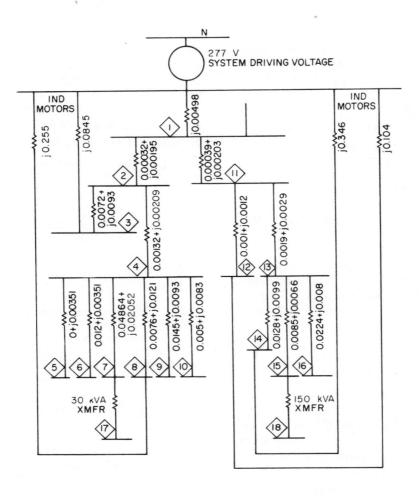

NOTE: All impedances listed are in ohms per phase.

Fig 78
Impedance Diagram for Building System of Fig 77

tem impedances in ohms rather than per unit. All impedances are given in ohms per phase. The impedance values as shown on the diagram are derived by calculations of the following types:

(1) Utility spot network. Available short-circuit duty at 480 V = 55 600 A, symmetrical rms

equivalent $X = \dfrac{E_{L-N}}{I_{SC}} = \dfrac{277V}{55\,600}$

$= 0.00498 \; \Omega/\text{phase}$

It is assumed, conservatively, that $R = 0$.

(2) Motors (typical). 555 hp of induc-

tion motors connected to motor-control center MCC 1 (bus 12). Assume a reactance of 25% on kilovoltampere base equal to the motor horsepower

$$X = \frac{\%X_{\mathrm{m}}}{100}\left(\frac{kV^2 \ (1000)}{\mathrm{motor \ kVA}}\right)$$

$$= \frac{25}{100}\left(\frac{(0.48)^2 \ 1000}{555}\right)$$

$$= 0.104 \ \Omega/\mathrm{phase}$$

It is assumed, conservatively, that $R = 0$.

(3) Conductors (typical). Feeder to panel BA1 380 ft of two 350 kcmil cables per phase (bus 2 to bus 3). Impedance of 350 kcmil cable is $0.0378 + j0.0491 \ \Omega/1000$ ft

$$Z = \frac{380 \ (0.0378 + j0.0491)}{(1000) \ 2 \ (\mathrm{cables/phase})}$$

$$= 0.0072 + j0.0093 \ \Omega/\mathrm{phase}$$

Step 4: Determination of Fault Duties. The impedances of Fig 78 are resolved into a single equivalent impedance for each fault location. In order to reduce the calculation time required, a simplified procedure is illustrated which uses a tabular form similar to that of Fig 76 for recording calculation results.

Many of the circuits in the example radial system have impedances in series, and equivalents are determined by summing resistance and reactance components separately. Where the utility and induction motor sources of short-circuit current act together, impedance paralleling is necessary, and this is done by summing reciprocals.

A record is kept of the steps used to calculate the short-circuit currents, because many of the impedance combinations found initially are used repeatedly

and can simply be copied. For recording purposes, utility and motor sources of short-circuit current are identified in this example with their bus number preceded by the letter S.

Step 5: Calculations. The following impedance combinations are used in finding the short-circuit current at bus 1.

(1) Z_a, utility system S1, as previously determined:

$$Z_a \ (S1) = 0 + j0.00498 \ \Omega$$

(2) Z_b, 683 hp of induction motors and feeders to bus 2 (S3-2):

$$\begin{aligned} \mathrm{motors \ (S3)} &= \ \ \ 0 \ \ \ \ + j0.0845 \\ \mathrm{feeder, \ bus \ 3 \ to \ 2 \ (3\text{-}2)} &= \underline{0.0072 + j0.0093} \\ \mathrm{total, \ } Z_b \ \mathrm{(S3\text{-}2)} &= 0.0072 + j0.0938 \ \Omega \end{aligned}$$

(3) Z_c, 226 hp of induction motors and feeders to bus 2 (S8-2):

$$\begin{aligned} \mathrm{motors \ (S8)} &= \ \ \ 0 \ \ \ \ \ + j0.255 \\ \mathrm{feeder, \ bus \ 8 \ to \ 4 \ (8\text{-}4)} &= 0.0076 \ \ + j0.0121 \\ \mathrm{feeder, \ bus \ 4 \ to \ 2 \ (4\text{-}2)} &= \underline{0.00132 + j0.00209} \\ \mathrm{total, \ } Z_c \ \mathrm{(S8\text{-}2)} &= 0.00892 + j0.2692 \ \Omega \end{aligned}$$

(4) Z_d, parallel combination of Z_b and Z_c (S3, 8-2):

(a) Components of $1/Z_b$ are

$$\begin{aligned} G_b &= R_b/(R_b{}^2 + X_b{}^2) \\ &= 0.0072/((0.0072)^2 + (0.0938)^2) \\ &= 0.0072/0.00885 = 0.814 \ \mathrm{S} \\ -B_b &= X_b/(R_b{}^2 + X_b{}^2) \\ &= 0.0938/0.00885 = 10.60 \ \mathrm{S} \end{aligned}$$

$$1/Z_b = 0.814 - j10.60 \ \mathrm{S}$$

(b) Components of $1/Z_c$ are

$$\begin{aligned} G_c &= 0.00892/((0.00892)^2 + (0.2692)^2) \\ &= 0.00892/0.07255 = 0.123 \ \mathrm{S} \\ -B_c &= 0.2692/0.07255 = 3.71 \ \mathrm{S} \end{aligned}$$

$$1/Z_c = 0.123 - j3.71 \ \mathrm{S}$$

$$1/Z_d = 1/Z_b + 1/Z_c = 0.937 - j14.31 \ \mathrm{S}$$

(c) Components of Z_d are

$$\begin{aligned} R_d &= G_d/(G_d{}^2 + B_d{}^2) \\ &= 0.937/((0.937)^2 + (14.31)^2) \\ &= 0.937/205.7 = 0.00456 \ \Omega \\ X_d &= -B_d/(G_d{}^2 + B_d{}^2) \\ &= 14.31/205.7 = 0.0696 \ \Omega \end{aligned}$$

$$Z_d \ (S3,8\text{-}2) = 0.00456 + j0.0696 \ \Omega$$

(5) Z_e, impedance Z_d and feeder to bus 1 (S3,8-1):

$$Z_d \text{ (S3,8-2)} = 0.00456 + j0.0696$$
$$\text{feeder, bus 2 to 1 (2-1)} = \underline{0.00032 + j0.00195}$$
$$\text{total, } Z_e \text{ (S3,8-1)} = 0.00488 + j0.07155 \ \Omega$$

(6) Z_f, 555 hp of induction motors and feeders to bus 11 (S12-11):

$$\text{motors (S12)} = \quad 0 \quad + j0.104$$
$$\begin{array}{l}\text{feeder, bus 12 to 11} \\ \quad\quad (12\text{-}11) = \underline{0.001 + j0.0012}\end{array}$$
$$\text{total, } Z_f \text{ (S12-11)} = 0.001 + j0.1052 \ \Omega$$

(7) Z_g, 167 hp of induction motors and feeders to bus 11 (S14-11):

$$\text{motors (S14)} = \quad 0 \quad + j0.346$$
$$\begin{array}{l}\text{feeder, bus 14 to 13} \\ \quad\quad (14\text{-}13) = 0.0128 + j0.0099 \\ \text{feeder, bus 13 to 11} \\ \quad\quad (13\text{-}11) = \underline{0.0019 + j0.0029}\end{array}$$
$$\text{total, } Z_g \text{ (S14-11)} = 0.0147 + j0.3588 \ \Omega$$

(8) Z_h, parallel combination of Z_f and Z_g (S12,14-11), using the method detailed when calculating Z_d:

$$\begin{aligned} 1/Z_f &= 0.0904 \quad - j \ 9.505 \\ 1/Z_g &= \underline{0.1140 \quad - j \ 2.782} \\ 1/Z_h &= 0.2044 \quad - j12.287 \ \text{S} \end{aligned}$$

$$Z_h(\text{S12,14-11}) = 0.00135 + j0.0814 \ \Omega$$

(9) Z_i, impedance Z_h and feeder to bus 1 (S12,14-1):

$$Z_h \text{ (S12,14-11)} = 0.00135 + j0.0814$$
$$\begin{array}{l}\text{feeder, bus 11 to 1} \\ \quad\quad (11\text{-}1) = \underline{0.00039 + j0.00203}\end{array}$$
$$\text{total, } Z_i \text{ (S12,14-1)} = 0.00174 + j0.0834 \ \Omega$$

(10) Z_j, parallel combination of Z_a, Z_e, and Z_i (S1,3,8,12,14-1), total equivalent impedance for bus 1 short circuit:

$$\begin{aligned} 1/Z_a &= \quad 0 \quad - j200.8 \\ 1/Z_e &= 0.949 - j \ 13.91 \\ 1/Z_i &= \underline{0.250 - j \ 11.99} \\ 1/Z_j &= 1.199 - j226.7 \ \text{S} \end{aligned}$$

$$Z_j \text{ (S1,3,8,12,14-1)} = 0.00002 + j0.00441 \ \Omega$$
$$|Z_j| = \sqrt{(0.00002)^2 + (0.00441)^2}$$
$$= 0.00441 \ \Omega$$

The short-circuit current at bus 1 is

$$I_1 = E_{\text{L}-\text{N}}/Z = 277/0.00441 = 62\,810 \ \text{A, symmetrical rms}$$

For the short-circuit current at bus 2, similar impedance reduction calculations are recorded in an abbreviated table as follows:

Element*		Z			$1/Z$		Sum of Squares†
		R	X		G	$-B$	
S12,14-1	(Z_i)	0.00174	0.0834	→	0.250	11.99	0.00696
S1	(Z_a)	0	0.00498	→	0	200.8	
S1,12,14-1		0.00001	0.00470	←	0.250	212.79	45 280
1-2		0.00032	0.00195				
S1,12,14-2		0.00033	0.00665	→	7.444	150.00	4.433×10^{-5}
S3-2	(Z_b)	0.0072	0.0938	→	0.814	10.6	0.00885
S8-2	(Z_c)	0.00892	0.2692	→	0.123	3.71	0.07255
S1,3,8,12,14-2		0.00031	0.00607	←	8.381	164.31	27 070

*Source, branch, or combination.
†$R^2 + X^2$ when finding $1/Z$ from Z, $G^2 + B^2$ when finding Z from $1/Z$.

The total equivalent impedance is

$$Z = \sqrt{(0.00031)^2 + (0.00607)^2} = 0.00608 \, \Omega$$

and the short-circuit current at bus 2 is

$$I_2 = 277/0.00608 = 45\,560 \ \text{A, symmetrical rms}$$

For the bus 3 short-circuit current the abbreviated table is as follows:

Element	Z R	Z X		1/Z G	1/Z −B	Sum of Squares
S1,12,14-2	0.00033	0.00665	→	7.444	150.00	4.433×10^{-5}
S8-2	0.00892	0.2692	→	0.123	3.71	0.0725
S1,8,12,14-2	0.00032	0.00651	←	7.567	153.71	23 680
2-3	0.0072	0.0093				
S1,8,12,14-3	0.00752	0.01581	→	24.535	51.58	3.065×10^{-4}
S3	0	0.0845	→	0	11.83	
S1,3,8,12,14-3	0.00531	0.01372	←	24.535	63.41	4623

The total equivalent impedance is

$$Z = \sqrt{(0.00531)^2 + (0.01372)^2} = 0.147\,\Omega$$

and the short-circuit current at bus 3 is

$$I_3 = 277/0.0147 = 18\ 840 \text{ A, symmetrical rms}$$

For the bus 4 short-circuit current the values are as follows:

Element	Z R	Z X		1/Z G	1/Z −B	Sum of Squares
S1,12,14-2	0.00033	0.00665	→	7.444	150.00	4.433×10^{-5}
S3-2	0.0072	0.0938	→	0.814	10.60	0.00885
S1,3,12,14-2	0.00032	0.00621	←	8.258	160.60	25 860
2-4	0.00132	0.00209				
S1,3,12,14-4	0.00164	0.00830	→	22.91	115.95	7.158×10^{-5}
S8	0	0.255				
8-4	0.0076	0.0121				
S8-4	0.0076	0.2671	→	0.11	3.74	0.0714
S1,3,8,12,14-4	0.00155	0.00806	←	23.02	119.69	14 860

The total equivalent impedance is

$$Z = \sqrt{(0.00155)^2 + (0.00806)^2} = 0.00821\,\Omega$$

and the short-circuit current at bus 4 is

$$I_4 = 277/0.00821 = 33\ 740 \text{ A, symmetrical rms}$$

For the bus 5 short-circuit current the values are as follows:

Element	R	X
S1,3,8,12,14-4	0.00155	0.00806
4-5	0	0.00351
S1,3,8,12,14-5	0.00155	0.01157

The total equivalent impedance is

$$Z = \sqrt{(0.00155)^2 + (0.01157)^2} = 0.01167\,\Omega$$

and the short-circuit current at bus 5 is

$$I_5 = 277/0.01167 = 23\ 740 \text{ A, symmetrical rms}$$

For the bus 6 short-circuit current the following tabulation is used:

Element	R	X
S1,3,8,12,14-4	0.00155	0.00806
4-6	0.012	0.00351
S1,3,8,12,14-6	0.01355	0.01157

The total equivalent impedance is

$$Z = \sqrt{(0.01355)^2 + (0.01157)^2} = 0.0178\,\Omega$$

and the short-circuit current at bus 6 is

$$I_6 = 277/0.0178 = 15\ 560\ \text{A, symmetrical rms}$$

For the bus 7 short-circuit current the values are as follows:

Element	R	X
S1,3,8,12,14-4	0.00155	0.00806
4-7	0.04864	0.02052
S1,3,8,12,14-7	0.05019	0.02858

The total equivalent impedance is

$$Z = \sqrt{(0.05019)^2 + (0.02858)^2} = 0.0578\,\Omega$$

and the short-circuit current at bus 7 is

$$I_7 = 277/0.0578 = 4790\ \text{A, symmetrical rms}$$

For the bus 8 short-circuit current the calculations are tabulated as follows:

| Element | Z | | 1/Z | | Sum of Squares |
	R	X	G	$-B$	
S1,3,12,14-4	0.00164	0.00830			
4-8	0.0076	0.0121			
S1,3,12,14-8	0.00924	0.02040	→ 18.42	40.67	5.015×10^{-4}
S8	0	0.255	→ 0	3.92	
S1,3,8,12,14-8	0.00791	0.01916	← 18.42	44.59	2328

The total equivalent impedance is

$$Z = \sqrt{(0.00791)^2 + (0.01916)^2} = 0.0207\,\Omega$$

and the short-circuit current at bus 8 is

$$I_8 = 277/0.0207 = 13\ 380\ \text{A, symmetrical rms}$$

Similar calculations for short-circuit currents at the remaining buses have the following results:

Bus	Symmetrical RMS Short-Circuit Current (amperes)
9	11 720
10	15 730
11	44 760
12	37 750
13	30 040
14	12 030
15	14 690
16	9 280

For the bus 17 short-circuit current the transformer reactance in ohms per phase at 208 V is found as follows:

$$X_T = \frac{\%X_T}{100}\left(\frac{kV^2\,(1000)}{\text{transformer kVA}}\right)$$

$$= \frac{3.76}{100}\left(\frac{(0.208)^2\,1000}{30}\right)$$

$$= 0.05422\ \Omega/\text{phase at 208 V}$$

The equivalent impedance to bus 7 was previously calculated to be $0.0502 + j0.0286$ Ω/phase at 480 V. Multiplying by $(208/480)^2$ converts this to an equivalent impedance at 208 V of $0.00943 + j0.00537\ \Omega$/phase. The total equivalent impedance Z is the sum of

$$\begin{array}{l} 0\ \ \ \ \ \ \ \ \ \ \ + j0.05422 \\ \underline{0.00943 + j0.00537} \\ Z = 0.00943 + j0.05959 \end{array}$$

$$|Z| = \sqrt{(0.00943)^2 + (0.05959)^2} = 0.0603\,\Omega$$

The short-circuit current at bus 17 is

$$I_{17} = 120/0.0603 = 1990\ \text{A, symmetrical rms}$$

For the bus 18 short-circuit current at 208 V, per-unit calculations are illustrated. The calculation base is chosen equal to the transformer rating of 150 kVA. At a base voltage equal to the low-voltage rating of 208 V the base current is $150/(\sqrt{3} \times 0.208) = 416$ A.

The available short-circuit current at 480 V transformer primary terminals is assumed to be all reactive, accepting for the sake of simplicity, less accuracy than a complex calculation would provide. The short-circuit kilovoltamperes at the 480 V primary bus 15 are $\sqrt{3}(0.48)(14\ 690) = 12\ 210$ kVA. The per unit reactance of the system to bus 15 is $X = 150/12\ 210 = 0.0123$ pu. The per-unit reactance of the transformer is $X = 4.55\%/100 = 0.0455$ pu. The total equivalent reactance is

$$X_{pu} = 0.0578 \text{ pu}$$

The per-unit short-circuit current at bus 17 is

$$I_{pu} = E_{pu}/X_{pu} = 1.0/0.0578 = 17.3 \text{ pu}$$

and the short-circuit current at bus 17 is

$$I_{17} = I_b \cdot I_{pu} = 416 \cdot 17.3$$
$$= 7197 \text{ A, symmetrical rms}$$

Comments on Example. This example illustrates the use of current-limiting busway to reduce the short-circuit currents at the main switchboards (buses 2 and 11). If conventional busway had been used, the short-circuit current at these buses would have been significantly higher.

Identical feeders feed buses 5 and 6 from bus 4. The calculation includes $R + jX$ for bus 6, but ignores feeder R and uses only jX for bus 5. The major difference between short-circuit currents for buses 5 and 6 illustrates why resistance should not be ignored in low-voltage fault calculations.

9.5.19 Determination of Line-to-Ground Fault Currents. The technique of symmetrical components will allow us to express the bolted line-to-ground fault current as follows

$$I_{L-G} = \frac{3E}{Z_1 + Z_2 + Z_0 + 3R_0}$$

where

E = line-to-neutral voltage
Z_1 = positive-sequence impedance
Z_2 = negative-sequence impedance
Z_0 = zero-sequence impedance
R_0 = resistance of neutral grounding resistor, if any

In the case of a solidly grounded system, $3R_0 = 0$, and assuming that Z_2 is approximately equal to Z_1, the expression becomes

$$I_{L-G} = \frac{3E}{2Z_1 + Z_0}$$

From this expression we can derive the following

$$I_{L-G} = \frac{E}{Z_1} \left(\frac{3}{Z_0/Z_1 + 2} \right)$$

This expression shows the line-to-ground fault current as a function of the three-phase bolted fault current (E/Z_1) and the ratio of the zero-sequence impedance and the positive-sequence impedance.

In the strictest sense, the quantity above can only be treated as a scalar if Z_0 and Z_1 have the same phase angles. It has been shown, however, that useful results can be obtained using Z_0 and Z_1 as scalars. (See Kaufman [20] for details.)

Practical circuit values of the Z_0/Z_1 ratio may range from 1 to 50, depending on the construction of the ground-return circuit. Some typical values of the Z_0/Z_1 ratio are 2 for aluminum conduit (with or without internal ground conductor), 4 to 14 for steel conduit (with internal ground conductor the ratio will generally not exceed 4) depending on the size of conduit, and 15 to 30 for cable in magnetic armor [17], [20].

The type of ground-return circuit must be known to calculate the bolted line-to-ground fault currents. A selection must then be made to determine the points where ground-fault current levels are required. Generally a good indication will be those locations where the lower levels of three-phase bolted fault currents were found.

The selection and coordination of ground-fault protective device settings must consider the minimum arcing ground-fault currents. This type of fault can be particularly destructive. With a knowledge of the levels of ground-fault current for a system, settings can be selected which will avoid excessive equipment damage.

Ground-fault calculations will be performed for the system shown in Fig 77. The following locations were selected to be calculated:

Bus Number	Equipment Name
2	Main switchboard—north
11	Main switchboard—south
4	DP1
13	DP2
16	Panel HA 7
7	Panel HA 3

These calculations will be performed utilizing the following procedure for the series elements in the ground circuit:

Element Z_1	Z_0/Z_1 Ratio	Element Z_0
( )	$\times$ ( ) =	( )
( )	$\times$ ( ) =	( )
( )	$\times$ ( ) =	( )
$\Sigma(Z_1)$		$\Sigma(Z_0)$

$$\text{Effective overall } Z_0/Z_1 = \frac{\Sigma(Z_0)}{\Sigma(Z_1)}$$

The effective overall Z_0/Z_1 ratio is then substituted in the expression

$$I_{L-G} = \frac{E}{Z_1}\left(\frac{3}{Z_0/Z_1 + 2}\right)$$

which is solved for the bolted line-to-ground fault current. The minimum arcing ground-fault current value is then obtained by utilizing the 0.38 multiplier from Table 66.

For these calculations, representative values of Z_0/Z_1 ratios were taken from [1] and [2]. The cable conductors are run in metallic conduit, and a representative Z_0/Z_1 ratio of 10 is used. A Z_0/Z_1 ratio of 5 is used with the current-limiting busway. A good assumption is that the Z_0/Z_1 ratio for the utility source is approximately 1.

Calculations

Bus 2, main switchboard — north:

	Z_1	Z_0/Z_1	Z_0
Utility	$0.00498 \times$	1	$= 0.00498$
Current-limiting busway	$0.00197 \times$	5	$= 0.00985$
	0.00695		0.01483

$$\text{Overall } Z_0/Z_1 = \frac{0.01483}{0.00695} = 2.13$$

$$I_{\text{L}-\text{G}} = \frac{E}{Z_1}\left(\frac{3}{Z_0/Z_1 + 2}\right) = \frac{277}{0.00695}\left(\frac{3}{2.13 + 2}\right) = 28\ 951 \text{ A, symmetrical}$$

Minimum arcing ground-fault-current value:

$$0.38 \times 28\ 951 \text{ A} = 11\ 000 \text{ A, symmetrical}$$

Bus 4, DP1:

	Z_1	Z_0/Z_1	Z_0
Source (bus 2)	$0.00695 \times$	2.13	$= 0.0148$
180 ft of four-conductor 500-kcmil cable per phase	0.00247×10		$= 0.0247$
	0.00942		0.0395

$$\text{Overall } Z_0/Z_1 = \frac{0.0395}{0.00942} = 4.19$$

$$I_{\text{L}-\text{G}} = \frac{277}{0.00942}\left(\frac{3}{4.19 + 2}\right) = 14\ 250 \text{ A, symmetrical}$$

Minimum arcing ground-fault-current value:

$$0.38 \times 14\ 250 \text{ A} = 5415 \text{ A, symmetrical}$$

Bus 7, panel HA3:

	Z_1	Z_0/Z_1	Z_0
Source (bus 4)	$0.00942 \times$	4.19	$= 0.03946$
380 ft of No 1/0 cable	0.05279×10		$= 0.52790$
	0.6221		0.56736

$$\text{Overall } Z_0/Z_1 = \frac{0.56736}{0.06221} = 9.12$$

$$I_{\text{L}-\text{G}} = \frac{277}{0.06221}\left(\frac{3}{9.12 + 2}\right) = 1200 \text{ A, symmetrical}$$

Minimum arcing ground-fault-current value:

$$0.38 \times 1200 \text{ A} = 456 \text{ A, symmetrical}$$

Bus 11, main switchboard — south:

	Z_1	Z_0/Z_1		Z_0
Utility	$0.00498 \times$	1	=	0.00498
Current-limiting busway	$0.00206 \times$	5	=	0.01030
	0.00704			0.01528

$$\text{Overall } Z_0/Z_1 = \frac{0.01528}{0.00704} = 2.17$$

$$I_{L-G} = \frac{277}{0.00704}\left(\frac{3}{2.17 + 2}\right) = 28\ 305 \text{ A, symmetrical}$$

Minimum arcing ground-fault-current value:

$$0.38 \times 28\ 305 \text{ A} = 10\ 755 \text{ A, symmetrical}$$

Bus 13, DP2:

	Z_1	Z_0/Z_1		Z_0
Source (bus 11)	$0.00704 \times$	2.17	=	0.01527
190 ft of 500-kcmil cable	$0.00346 \times$	10	=	0.03460
	0.01050			0.04987

$$\text{Overall } Z_0/Z_1 = \frac{0.04987}{0.01050} = 4.75$$

$$I_{L-G} = \frac{277}{0.0105}\left(\frac{3}{4.75 + 2}\right) = 11\ 725 \text{ A, symmetrical}$$

Minimum arcing ground-fault-current value:

$$0.38 \times 11\ 725 \text{ A} = 4455 \text{ A, symmetrical}$$

Bus 16, panel HA7:

	Z_1	Z_0/Z_1		Z_0
Source (bus 13)	$0.01050 \times$	4.75	=	0.04987
140 ft of No 1 cable	$0.02378 \times$	10	=	0.23780
	0.03428			0.28767

$$\text{Overall } Z_0/Z_1 = \frac{0.28767}{0.03428} = 8.39$$

$$I_{L-G} = \frac{277}{0.03428}\left(\frac{3}{8.39 + 2}\right) = 2333 \text{ A, symmetrical}$$

Minimum arcing ground-fault-current value:

$$0.38 \times 2333 \text{ A} = 885 \text{ A, symmetrical}$$

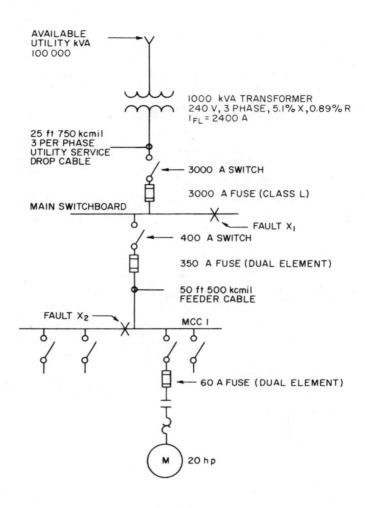

AVAILABLE
UTILITY kVA
100 000

1000 kVA TRANSFORMER
240 V, 3 PHASE, 5.1% X, 0.89% R
I_{FL} = 2400 A

25 ft 750 kcmil
3 PER PHASE
UTILITY SERVICE
DROP CABLE

3000 A SWITCH

3000 A FUSE (CLASS L)

MAIN SWITCHBOARD

FAULT X_1

400 A SWITCH

350 A FUSE (DUAL ELEMENT)

50 ft 500 kcmil
FEEDER CABLE

FAULT X_2

MCC I

60 A FUSE (DUAL ELEMENT)

M   20 hp

Fig 79
System A

9.5.20 Simplified Short-Circuit-Current Calculations.

Consider the following system, supplied by a 1000 kVA three-phase transformer having a full-load current of 2400 A at 240 V (Fig 79).

To start, the available short-circuit kilovoltamperes must be obtained from the local utility company. The utility estimates that their system can deliver a short-circuit current of 100 000 kVA (100 MVA) at the transformer. Since the X/R ratio of the utility system is usually quite high, only the reactance need be considered.

With this available short-circuit current fault information, the necessary calcula-

Table 74
Asymmetrical Factors

Short-Circuit Power Factor (%)	Short-Circuit X/R	Ratio to Symmetrical RMS Current		
		Maximum Single-Phase Instantaneous Peak Current M_p	Maximum Single-Phase RMS Current at Half-Cycle M_m	Average Three-Phase RMS Current at Half-Cycle M_a
0	∞	2.828	1.732	1.394
1	100.00	2.785	1.696	1.374
2	49.993	2.743	1.665	1.355
3	33.322	2.702	1.630	1.336
4	24.979	2.663	1.598	1.318
5	19.974	2.625	1.568	1.301
6	16.623	2.589	1.540	1.285
7	14.251	2.554	1.511	1.270
8	12.460	2.520	1.485	1.256
9	11.066	2.487	1.460	1.241
10	9.950	2.455	1.436	1.229
11	9.0354	2.424	1.413	1.216
12	8.2733	2.394	1.391	1.204
13	7.6271	2.364	1.372	1.193
14	7.0721	2.336	1.350	1.182
15	6.5912	2.309	1.330	1.171
16	6.1695	2.282	1.312	1.161
17	5.7947	2.256	1.294	1.152
18	5.4649	2.231	1.277	1.143
19	5.1672	2.207	1.262	1.135
20	4.8990	2.183	1.247	1.127
21	4.6557	2.160	1.232	1.119
22	4.4341	2.138	1.218	1.112
23	4.2313	2.11	1.205	1.105
24	4.0450	2.095	1.192	1.099
25	3.8730	2.074	1.181	1.093
26	3.7138	2.054	1.170	1.087
27	3.5661	2.034	1.159	1.081
28	3.4286	2.015	1.149	1.075
29	3.3001	1.996	1.139	1.070
30	3.1798	1.978	1.130	1.066
31	3.0669	1.960	1.121	1.062
32	2.9608	1.943	1.113	1.057
33	2.8606	1.926	1.105	1.053
34	2.7660	1.910	1.098	1.049
35	2.6764	1.894	1.091	1.046
36	2.5916	1.878	1.084	1.043
37	2.5109	1.863	1.078	1.039
38	2.4341	1.848	1.073	1.036
39	2.3611	1.833	1.068	1.033
40	2.2913	1.819	1.062	1.031
41	2.2246	1.805	1.057	1.028
42	2.1608	1.791	1.053	1.026
43	2.0996	1.778	1.049	1.024
44	2.0409	1.765	1.045	1.022
45	1.9845	1.753	1.041	1.020
46	1.9303	1.740	1.038	1.019
47	1.8780	1.728	1.034	1.017
48	1.8277	1.716	1.031	1.016
49	1.7791	1.705	1.029	1.014
50	1.7321	1.694	1.026	1.013
55	1.5185	1.641	1.015	1.008
60	1.3333	1.594	1.009	1.004
65	1.1691	1.553	1.004	1.002
70	1.0202	1.517	1.002	1.001
75	0.8819	1.486	1.0008	1.0004
80	0.7500	1.460	1.0002	1.00005
85	0.6198	1.439	1.00004	1.00002
100	0.0000	1.414	1.00000	1.00000

From NEMA AB 1-1969 [12] pt 3, pp 4, 5.

tions can be made to determine the fault current at any point in the electrical system. Tables 67–73 list the impedance and reactance data and Table 74 the asymmetrical factors.

An ohmic method, a per-unit method, or a percent method can be used for calculating short-circuit current. The ohmic and per-unit methods will be examined.

To determine the fault current at any point in the system, a one-line diagram must first be drawn, showing all the sources of short-circuit current feeding into the fault, as well as the impedances furnished by the circuit components.

The impedances may be represented by ohms, percent ohms, or per-unit ohms. The ohmic method, and then the per-unit method, is used in this study of three-phase faults on typical distribution systems. An arbitrary kilovoltampere base and the system line-to-line voltage are selected as base values for the per-unit method.

To make the study, the system components, including those of the utility system, are represented as impedances in Fig 79.

(1) *Ohmic Method (Figs 80–83)*. Most circuit component impedances are given in ohms, except utility and transformer impedances which are found by the following formulas.

NOTES: (1) For simplicity of calculation, all ohmic values are single-phase distance one way, later compensated for in the three-phase short-circuit formulas by the factor $\sqrt{3}$ (see Step 7).

(2) The transformer and utility ohms are referred to the secondary kilovolts by squaring the secondary voltage.

Step 1:

utility X (ohms)
$$= \frac{1000 \ (\text{secondary kV})^2}{\text{utility short-circuit kVA}}$$

NOTE: Only X is considered in this procedure since utility X/R ratios are usually quite high. For more accurate results obtain R of utility source.

Step 2:

transformer X (ohms)
$$= \frac{(10)(\%X)(\text{secondary kV})^2}{\text{transformer kVA}}$$

transformer R (ohms)
$$= \frac{(10)(\%R)(\text{secondary kV})^2}{\text{transformer kVA}}$$

Step 3: The impedance (in ohms) given for current transformers, large switches, and large circuit breakers is essentially all X.

Step 4: Cable and bus X (ohms); cable and bus R (ohms).

Step 5: Total all X and all R in the system to the point of fault.

Step 6: Determine the impedance (in ohms) of the system by

$$Z_T = \sqrt{R_T{}^2 + X_T{}^2}$$

Step 7: Calculate the symmetrical rms short-circuit current at the point of fault:

$$I_{\text{SC sym rms}} = \frac{\text{secondary line voltage}}{\sqrt{3}Z_T}$$

Step 8: Determine the motor load. Add up the full-load motor currents. The full-load motor current in the system is generally a percentage of the transformer full-load current, depending upon the type of load. (The generally accepted procedure assumes 50% motor load when both motor and lighting loads are considered, such as supplied by four-wire 208Y/120 V and 480Y/277 V three-phase systems.)

Step 9: The short-circuit current that the motor load can contribute is an asym-

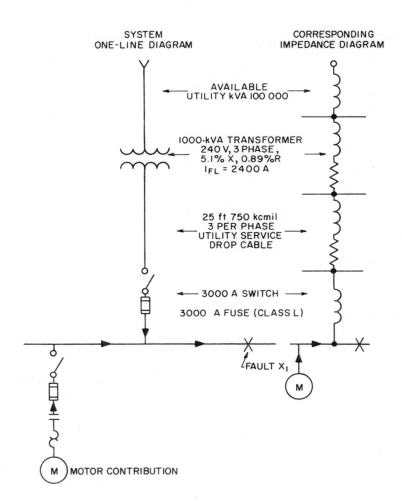

Fig 80
One-Line Diagram to Fault X_1, System A

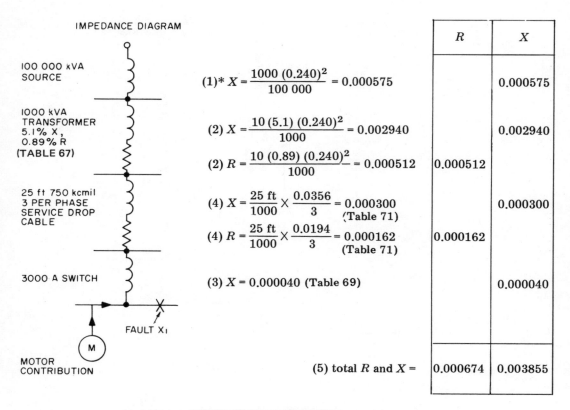

IMPEDANCE DIAGRAM

100 000 kVA
SOURCE

$(1)^* \ X = \dfrac{1000 \,(0.240)^2}{100\,000} = 0.000575$

1000 kVA
TRANSFORMER
5.1% X,
0.89% R
(TABLE 67)

$(2) \ X = \dfrac{10\,(5.1)\,(0.240)^2}{1000} = 0.002940$

$(2) \ R = \dfrac{10\,(0.89)\,(0.240)^2}{1000} = 0.000512$

25 ft 750 kcmil
3 PER PHASE
SERVICE DROP
CABLE

$(4) \ X = \dfrac{25 \text{ ft}}{1000} \times \dfrac{0.0356}{3} = 0.000300$ (Table 71)

$(4) \ R = \dfrac{25 \text{ ft}}{1000} \times \dfrac{0.0194}{3} = 0.000162$ (Table 71)

3000 A SWITCH

$(3) \ X = 0.000040$ (Table 69)

FAULT X_1

MOTOR
CONTRIBUTION

(5) total R and $X =$

	R	X
(1)		0.000575
(2)		0.002940
(2)	0.000512	
(4)		0.000300
(4)	0.000162	
(3)		0.000040
(5) total	0.000674	0.003855

(6) total $Z = \sqrt{(0.000674)^2 + (0.003855)^2} = 0.00392 \ \Omega/\text{phase}$

$(7) \ I_{sc \ sym \ rms} = \dfrac{240}{\sqrt{3}\,(0.00392)} = 35\,300 \text{ A}$

$(8,9)$ asymmetrical motor contribution (100%) = 5 × 2400 = 12 000 A

(10) symmetrical motor contribution $= \dfrac{12\,000}{1.25} = 9600 \text{ A}$

(11) total $I_{sc \ sym \ rms}$ (fault X_1) = 35 300 + 9600 = 44 900 A

$(12) \ X/R = \dfrac{0.003855}{0.000674} = 5.72$

(13) asymmetrical factor $M_m = 1.290$ (Table 74)

$(14) \ I_{sc \ asym \ rms} = 1.290 \times 35\,300 = 45\,500 \text{ A}$

(15) total $I_{sc \ asym \ rms}$ (fault X_1) = 45 500 + 12 000 = 57 500 A

Fig 81
Three-Phase Short-Circuit Current Calculation, Ohmic Method, Fault X_1

*Steps as listed in text.

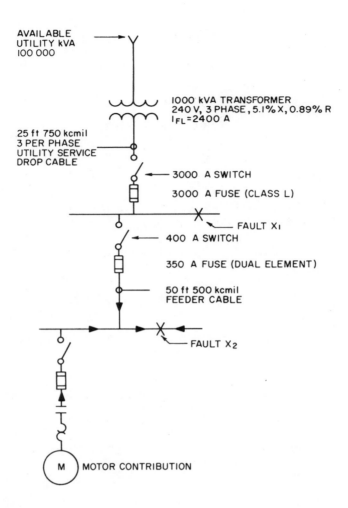

**Fig 82
One-Line Diagram to Fault X_2, System A**

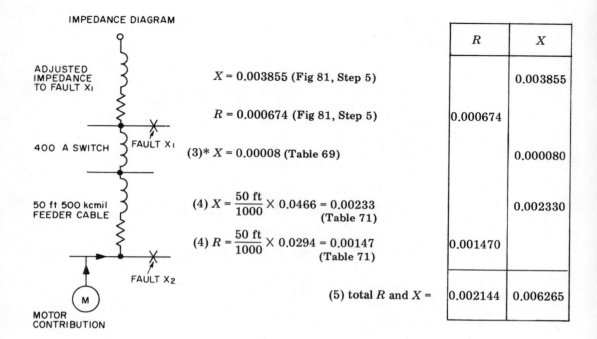

$X = 0.003855$ (Fig 81, Step 5)

$R = 0.000674$ (Fig 81, Step 5)

(3)* $X = 0.00008$ (Table 69)

(4) $X = \dfrac{50 \text{ ft}}{1000} \times 0.0466 = 0.00233$
(Table 71)

(4) $R = \dfrac{50 \text{ ft}}{1000} \times 0.0294 = 0.00147$
(Table 71)

(5) total R and $X =$

	R	X
		0.003855
	0.000674	
		0.000080
		0.002330
	0.001470	
	0.002144	0.006265

(6) total $Z = \sqrt{(0.002144)^2 + (0.006265)^2} = 0.00662 \; \Omega/\text{phase}$

(7) $I_{\text{sc sym rms}} = \dfrac{240}{\sqrt{3}\,(0.00662)} = 20\,930$ A

(8,9) asymmetrical motor contribution (100%) $= 5 \times 2400 = 12\,000$ A

(10) symmetrical motor contribution $= \dfrac{12\,000}{1.25} = 9600$ A

NOTE: Actual motor contribution will be somewhat smaller than calculated due to the impedance of the feeder cable.

(11) total $I_{\text{sc sym rms}}$ (fault X_2) $= 20\,930 + 9600 = 30\,530$ A

(12) $X/R = \dfrac{0.006265}{0.002144} = 2.92$

(13) asymmetrical factor $M_{\text{m}} = 1.112$ (Table 74)

(14) $I_{\text{sc asym rms}} = 1.112 \times 20\,930 = 23\,310$ A

(15) total $I_{\text{sc asym rms}}$ (fault X_2) $= 23\,310 + 12\,000 = 35\,310$ A

**Fig 83
Three-Phase Short-Circuit Current Calculation
Ohmic Method, Fault X_2**

*Steps as listed in text.

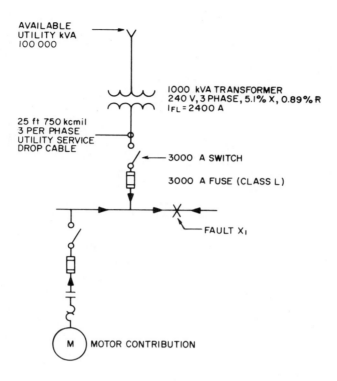

AVAILABLE
UTILITY kVA
100 000

1000 kVA TRANSFORMER
240 V, 3 PHASE, 5.1% X, 0.89% R
I_{FL} = 2400 A

25 ft 750 kcmil
3 PER PHASE
UTILITY SERVICE
DROP CABLE

3000 A SWITCH

3000 A FUSE (CLASS L)

FAULT X_1

M   MOTOR CONTRIBUTION

Fig 84
One-Line Diagram to Fault X_1, System A

metrical current usually approximated as being equal to five times the full load current of the motors. As a close approximation with a margin of safety use

asymmetrical motor contribution

= 5 · full-load motor current

NOTE: A more exact determinaton depends upon the subtransient reactances of the motors in question and the associated circuit impedances. A less conservative method involves the total motor circuit impedance to a common bus (sometimes referred to as a zero-reactance bus).

Step 10: The symmetrical motor contribution can be approximated by using the average asymmetry factor associated with the motors in the system. This asymmetry factor varies according to motor design and in this text may be chosen as 1.25 for approximate calculation purposes. To solve for the symmetrical motor contribution,

symmetrical motor contribution

$$= \frac{\text{asymmetrical motor contribution}}{1.25}$$

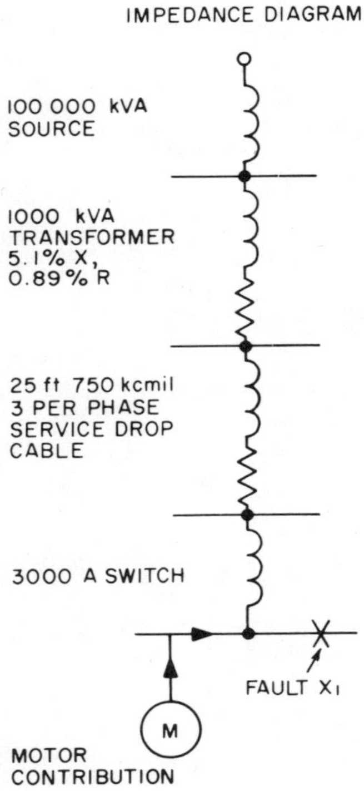

10 000 kVA Base

IMPEDANCE DIAGRAM

	pu R	pu X

100 000 kVA
SOURCE

$(1)^* \text{ pu } X = \dfrac{10\ 000}{100\ 000} = 0.1000$ — pu X = 0.1000

1000 kVA
TRANSFORMER
5.1% X,
0.89% R

$(2) \text{ pu } X = \dfrac{(5.1)\ (10\ 000)}{(100)\ (1000)} = 0.5100$ — pu X = 0.5100

$(2) \text{ pu } R = \dfrac{(0.89)\ (10\ 000)}{(100)\ (1000)} = 0.0890$ — pu R = 0.0890

25 ft 750 kcmil
3 PER PHASE
SERVICE DROP
CABLE

$(3) \text{ pu } X = \dfrac{(0.0003)\dagger\ (10\ 000)}{(1000)\ (0.240)^2} = 0.0520$ — pu X = 0.0520

$(4) \text{ pu } R = \dfrac{(0.00016)\dagger\ (10\ 000)}{(1000)\ (0.240)^2} = 0.0281$ — pu R = 0.0281

3000 A SWITCH

$(3) \text{ pu } X = \dfrac{(0.00004)\ (10\ 000)}{(1000)\ (0.240)^2} = 0.0069$ — pu X = 0.0069

FAULT X_1

MOTOR
CONTRIBUTION

(5) total pu R and pu X = pu R = 0.1171 pu X = 0.6689

(6) total pu $Z = \sqrt{(0.1171)^2 + (0.6689)^2} = 0.6800$

$(7) \ I_{sc\ sym\ rms} = \dfrac{10\ 000}{\sqrt{3}\ (0.240)\ (0.68)} = 35\ 300 \text{ A}$

(8,9) asymmetrical motor contribution (100%) = 5 × 2400 = 12 000 A

$(10) \text{ symmetrical motor contribution} = \dfrac{12\ 000}{1.25} = 9600 \text{ A}$

(11) total $I_{sc\ sym\ rms}$ (fault X_1) = 35 300 + 9600 = 44 900 A

$(12) \ X/R = \dfrac{0.6689}{0.1171} = 5.72$

(13) asymmetrical factor M_m = 1.290 (Table 74)

(14) $I_{sc\ asym\ rms}$ = 1.290 × 35 300 = 45 500 A

(15) total $I_{sc\ asym\ rms}$ (fault X_1) = 45 500 + 12 000 = 57 500 A

Fig 85
Three-Phase Short-Circuit Current Calculation,
Per-Unit Method, Fault X_1

*Steps are listed in text.
†See Fig 81 for determination of these values.

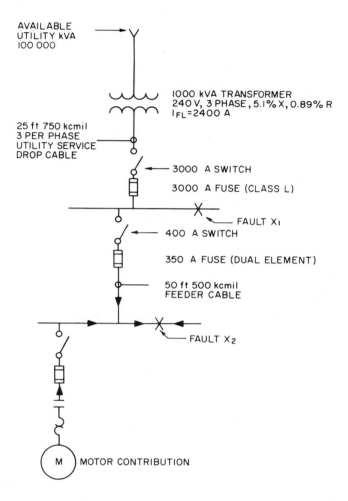

AVAILABLE
UTILITY kVA
100 000

1000 kVA TRANSFORMER
240 V, 3 PHASE, 5.1% X, 0.89% R
I_{FL} = 2400 A

25 ft 750 kcmil
3 PER PHASE
UTILITY SERVICE
DROP CABLE

3000 A SWITCH

3000 A FUSE (CLASS L)

FAULT X_1

400 A SWITCH

350 A FUSE (DUAL ELEMENT)

50 ft 500 kcmil
FEEDER CABLE

FAULT X_2

M   MOTOR CONTRIBUTION

Fig 86
One-Line Diagram to Fault X_2, System A

Step 11: The total symmetrical rms short-circuit current is calculated as

$$\text{total } I_{\text{sc sym rms}} = I_{\text{sc sym rms}}$$
$$+ \text{ symmetrical motor contribution}$$

NOTE: Arithmetic addition results in conservative values of fault current. More accurate values involve vectorial addition of the currents.

Step 12: Determine the X/R ratio of the system to the point of fault:

$$X/R = \frac{\text{total } X \text{ (ohms)}}{\text{total } R \text{ (ohms)}}$$

Step 13: The asymmetrical factor corresponding to the X/R ratio in Step 12 is found in Table 74, column M_m. This multiplier will provide the worst case

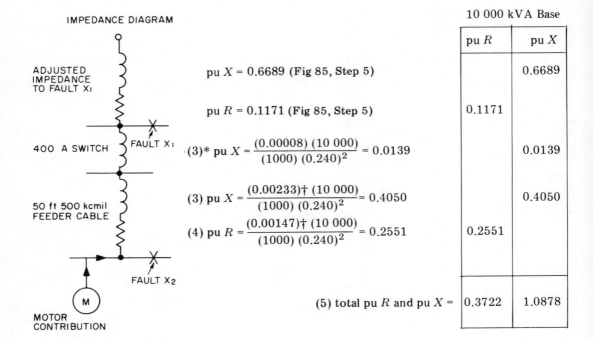

IMPEDANCE DIAGRAM

10 000 kVA Base

	pu R	pu X
		0.6689
	0.1171	
		0.0139
		0.4050
	0.2551	
(5) total pu R and pu X =	0.3722	1.0878

ADJUSTED IMPEDANCE TO FAULT X₁

pu $X = 0.6689$ (Fig 85, Step 5)

pu $R = 0.1171$ (Fig 85, Step 5)

400 A SWITCH

FAULT X₁

(3)* pu $X = \dfrac{(0.00008)\,(10\,000)}{(1000)\,(0.240)^2} = 0.0139$

50 ft 500 kcmil FEEDER CABLE

(3) pu $X = \dfrac{(0.00233)\dagger\,(10\,000)}{(1000)\,(0.240)^2} = 0.4050$

(4) pu $R = \dfrac{(0.00147)\dagger\,(10\,000)}{(1000)\,(0.240)^2} = 0.2551$

FAULT X₂

MOTOR CONTRIBUTION

(6) total pu $Z = \sqrt{(0.3722)^2 + (1.0878)^2} = 1.150$

(7) $I_{sc\ sym\ rms} = \dfrac{10\,000}{\sqrt{3}\,(0.240)\,(1.15)} = 20\,930$ A

(8,9) asymmetrical motor contribution (100%) = $5 \times 2400 = 12\,000$ A

(10) symmetrical motor contribution = $\dfrac{12\,000}{1.25} = 9600$ A

NOTE: Actual motor contribution will be somewhat smaller than calculated due to the impedance of the feeder cable.

(11) total $I_{sc\ sym\ rms}$ (fault X₂) = $20\,930 + 9600 = 30\,530$ A

(12) $X/R = \dfrac{1.0878}{0.3722} = 2.92$

(13) asymmetrical factor $M_m = 1.112$ (Table 74)

(14) $I_{sc\ asym\ rms} = 1.112 \times 20\,930 = 23\,310$ A

(15) total $I_{sc\ asym\ rms}$ (fault X₂) = $23\,310 + 12\,000 = 35\,310$ A

Fig 87
Three-Phase Short-Circuit Current Calculation,
Per-Unit Method, Fault X₂

*Steps are listed in text.
†See Fig 83 for determination of these values.

asymmetry occurring in the first half-cycle. Where the average three-phase multiplier is desired, use column M_a.

Step 14: Calculate the asymmetrical rms short-circuit current:

$$I_{sc\ asym\ rms} = I_{sc\ sym\ rms}$$
$$\cdot \text{asymmetrical factor}$$

Step 15: The total asymmetrical rms short-circuit current is calculated as

$$\text{total } I_{sc\ asym\ rms} = I_{sc\ asym\ rms}$$
$$+ \text{asymmetrical motor}$$
$$\text{contribution}$$

The resistances of the circuit components must be referred to the same voltage. If there is more than one voltage transformation in the system, the ohmic method becomes more complicated. It is then recommended that the per-unit method be used for ease in calculation.

(2) *Per-Unit Method (Figs 84–87).* The per-unit method is generally used for calculating short-circuit currents where the electrical system is more complex than the simple example given.

After establishing a one-line diagram of the system, the following calculations are made.

NOTES: (1) As in the ohmic method, all ohmic values are single-phase distance one way, later compensated for in the three-phase short-circuit formula by the factor $\sqrt{3}$ (see Step 7).

(2) The base kilovoltamperes used throughout this section will be 10 000 kVA.

Step 1:

$$\text{utility pu } X = \frac{\text{base kVA}}{\text{utility short-circuit kVA}}$$

NOTE: Only pu X is considered in the procedure since the utility X/R ratio is usually quite high. For more accurate results obtain pu R of utility source.

Step 2:

$$\text{transformer pu } X = \frac{(\%X)\,(\text{base kVA})}{100\,(\text{transformer kVA})}$$

$$\text{transformer pu } R = \frac{(\%R)\,(\text{base kVA})}{100\,(\text{transformer kVA})}$$

Step 3:

$$\text{component pu } X = \frac{(\text{ohm } X)\,(\text{base kVA})}{1000\,(\text{kV})^2}$$

Step 4:

$$\text{component pu } R = \frac{(\text{ohm } R)\,(\text{base kVA})}{1000\,(\text{kV})^2}$$

NOTE: The reactance and resistance are calculated for the system components, such as cable, switches, current transformer, and bus.

Step 5: Next, total all pu X and pu R in the system to the point of fault.

Step 6: Determine the pu impedance of the system by

$$\text{pu } Z_T = \sqrt{(\text{pu } R_T)^2 + (\text{pu } X_T)^2}$$

Step 7: Calculate the symmetrical rms short-circuit current at the point of fault:

$$I_{sc\ sym\ rms} = \frac{\text{base kVA}}{\sqrt{3}\,(\text{kV})\,(\text{pu } Z_T)}$$

Step 8: Determine the motor load. Add up the full-load motor currents. (Whenever motor and lighting loads are considered, such as supplied by four-wire 208Y/120 V and 480Y/277 V three-phase systems, the generally accepted procedure is to assume 50% motor load based on the full-load current rating of the transformer.)

Step 9: The short-circuit current that the motor load can contribute is an asymmetrical current usually approximated as being equal to five times the full load current of the motors. As a close approximation with a margin of safety use

asymmetrical motor contribution
$$= 5 \cdot \text{full-load motor current}$$

NOTE: A more exact determination depends upon the subtransient reactances of the motors in question and the associated circuit impedance. A less conservative method involves the total motor circuit impedance to a common bus (sometimes referred to as a zero-reactance bus).

Step 10: The symmetrical motor contribution can be approximated by using the average asymmetry factor associated with the motors in the system. This asymmetry factor varies according to motor design and in this text may be chosen as 1.25 for approximate calculation purposes. To solve for the symmetrical motor contribution,

symmetrical motor contribution
$$= \frac{\text{asymmetrical motor contribution}}{1.25}$$

Step 11: The total symmetrical rms short-circuit current is calculated as

$$\text{total } I_{\text{sc sym rms}} = I_{\text{sc sym rms}}$$
$$+ \text{symmetrical motor}$$
$$\text{contribution}$$

NOTE: Arithmetic addition results in conservative values of fault current. More accurate values involve vectorial addition of the currents.

Step 12: Determine the X/R ratio of the system to the point of fault:

$$X/R = \frac{\text{pu } X_T}{\text{pu } R_T}$$

Step 13: From Table 74, column M_m, obtain the asymmetrical factor corresponding to the X/R ratio determined in Step 12. This multiplier will provide the

worst case asymmetry occurring in the first half-cycle. Where the average three-phase multiplier is desired, use column M_a.

Step 14: The asymmetrical rms short-circuit current can be calculated as

$$I_{\text{sc asym rms}} = I_{\text{sc asym rms}}$$
$$\cdot \text{asymmetrical factor}$$

Step 15: The total asymmetrical rms short-circuit current is calculated as

$$\text{total } I_{\text{sc asym rms}} = I_{\text{sc asym rms}}$$
$$+ \text{asymmetrical motor contribution}$$

(3) *Simplified Per-Unit Method, Use of Constants (Figs 84 and 88).* The per-unit method lends itself rather well to the use of constants based on base kilovoltamperes and a base voltage. Tables 75–78 are convenient when used in conjunction with per-unit short-circuit current calculations. They are derived using 10 000 kVA base.

Table 75
Utility Short-Circuit kVA and Per-Unit Impedance

$$\text{pu}X = \frac{\text{base kVA}}{\text{utility short-circuit kVA}}$$

Utility Short-Circuit kVA	puX
25 000	0.400
50 000	0.200
75 000	0.133
100 000	0.100
200 000	0.050
300 000	0.033
400 000	0.025
500 000	0.020
1 000 000	0.010
Infinite	0

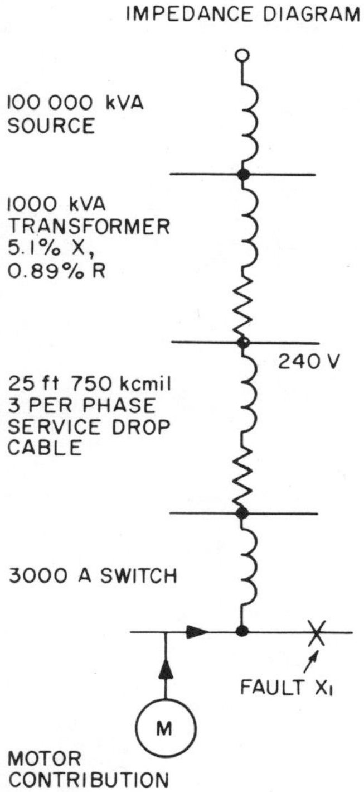

IMPEDANCE DIAGRAM

10 000 kVA Base

	pu R	pu X
		0.1000
		0.5100
	·0.0890	
		0.0520
	0.0281	
		0.0069·
total pu R and pu X =	0.1171	0.6689

100 000 kVA
SOURCE

pu X = 0.1000 (Table 75)

1000 kVA
TRANSFORMER
5.1% X,
0.89% R

pu X = (5.1) (0.1000) = 0.5100
(Table 76)

pu R = (0.89) (0.1000) = 0.0890
(Table 76)

240 V

25 ft 750 kcmil
3 PER PHASE
SERVICE DROP
CABLE

pu X = (0.0003)*(173.61) = 0.0520
(Table 77)

pu R = (0.00016)*(173.61) = 0.0281
(Table 77)

3000 A SWITCH

pu X = (0.00004) (173.61) = 0.0069
(Table 77)

FAULT X_1

M

MOTOR
CONTRIBUTION

total pu $Z = \sqrt{(0.1171)^2 + (0.6689)^2} = 0.6800$

$I_{\text{sc sym rms}} = \dfrac{24\,039}{0.6800} = 35\,300$ A (Table 78)

asymmetrical motor contribution (100%) = 5 × 2400 = 12 000 A

symmetrical motor contribution = $\dfrac{12\,000}{1.25} = 9600$ A

total $I_{\text{sc sym rms}}$ (fault X_1) = 35 300 + 9600 = 44 900 A

Fig 88
Three-Phase Short-Circuit Current Calculation, Simplified
Per-Unit Method, Use of Constants, Fault X_1

*See Fig 81 for determination of these values.

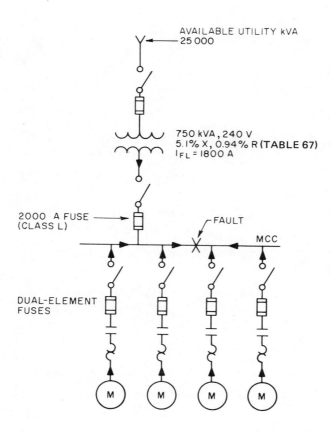

AVAILABLE UTILITY kVA
25 000

750 kVA, 240 V
5.1% X, 0.94% R (TABLE 67)
I_{FL} = 1800 A

2000 A FUSE
(CLASS L)

FAULT

MCC

DUAL-ELEMENT
FUSES

Fig 89
System B

9.5.21 Effect of Low Available Utility Kilovoltamperes (Figs 89 and 90). Even when utility fault currents are held down to a low level, it is not always safe to specify protective devices with limited interrupting capacity. Overnight the available fault kilovoltamperes which the utility can deliver might be doubled or tripled. Since the destructive thermal and magnetic forces vary as the square of the current, any increase in fault level could result in a disastrous situation. The protective device selected should be one that takes system growth into consideration.

Figure 89 points out that despite very limited utility short-circuit kilovoltamperes, there is considerable short-circuit current available, and any future increase in the utility system will result in even more fault current.

9.5.22 General Discussion of Short-Circuit Current Calculations

(1) *Motor Contribution.* Synchronous and induction motors will feed additional short-circuit current to a fault at their terminals at a value approximately equal to their locked rotor rating. For this reason they can be represented in equivalent circuits by their locked rotor im-

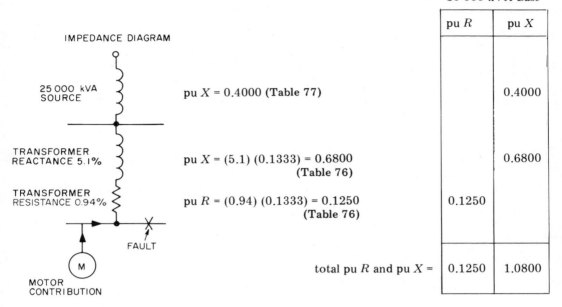

10 000 kVA Base

	pu R	pu X
pu X = 0.4000 (Table 77)		0.4000
pu X = (5.1) (0.1333) = 0.6800 (Table 76)		0.6800
pu R = (0.94) (0.1333) = 0.1250 (Table 76)	0.1250	
total pu R and pu X =	0.1250	1.0800

total pu $Z = \sqrt{(0.1250)^2 + (1.0800)^2} = 1.0850$

$I_{\text{sc sym rms}} = \dfrac{24\ 039}{1.0850} = 22\ 150$ A (Table 78)

asymmetrical motor contribution (100%) = 5 × 1800 = 9000 A

symmetrical motor contribution $= \dfrac{9000}{1.25} = 7200$ A

total $I_{\text{sc sym rms}}$ = 22 150 + 7200 = 29 350 A

$X/R = \dfrac{1.08}{0.125} = 8.64$

asymmetrical factor M_{m} = 1.40 (Table 74)

$I_{\text{sc asym rms}}$ = 1.40 × 22 150 = 31 000 A

total $I_{\text{sc asym rms}}$ = 9000 + 31 000 = 40 000 A

NOTE: Properly selected current-limiting fuses in the motor-control center clears short-circuit current in less than one quarter-cycle and coordinates selectively with the KRP-C 2000 A main fuses.

Fig 90
Three-Phase Short-Circuit Current Calculation, Per-Unit
Method, System B

Table 76
Transformer Impedance Multipliers

$$\text{pu}X = \frac{(\%X)\,(\text{base kVA})}{100\,(\text{transformer kVA})}$$
$$= \%Z\,(\text{Multiplier})$$

Transformer kVA	Multiplier
150	0.6666
225	0.4444
300	0.3333
500	0.2000
750	0.1333
1000	0.1000
1500	0.0666
2000	0.0500
2500	0.0400

NOTE: Example—750 kVA transformer, 5% X, 1% R:
$$\text{pu}X = 5 \times 0.1333 = 0.6665$$
$$\text{pu}R = 1 \times 0.1333 = 0.1333$$

Table 77
Component Impedance Multipliers
(Based on System Voltage)

$$\text{pu}X = \frac{(\text{ohms } X)\,(\text{base kVA})}{1000\,(\text{kV})^2}$$

System Voltage	Multiplier
208 V	231.14
220 V	206.50
240 V	173.61
440 V	51.60
460 V	47.26
480 V	43.40
550 V	33.10
2400 V	1.736
4160 V	0.579
12.47 kV	0.0649
13.2 kV	0.0574
13.8 kV	0.0525

NOTE: Examples
(1) 1000 ft 500 kcmil cable, $R = 0.0294\ \Omega$, $X = 0.0466\ \Omega$, 480 V system:
$$\text{pu}X = 0.0466 \times 43.4 = 2.060$$
$$\text{pu}R = 0.0294 \times 43.4 = 1.277$$
(2) 1800 A current transformer, $X = 0.00007\ \Omega$, 208 V system:
$$\text{pu}X = 0.00007 \times 231.14 = 0.0162$$

Table 78
Symmetrical RMS Short-Circuit-Current
Formulas
(Based on System Voltage)

$$I_{\text{sc}} = \frac{\text{base kVA}}{\sqrt{3}\,(\text{kV})\,(\text{total pu}Z)}$$

Line-to-Line System Voltage	Symmetrical RMS Short-Circuit Current (amperes)
208 V	27 758 : total pu$Z$
220 V	26 280 : total pu$Z$
240 V	24 039 : total pu$Z$
440 V	13 120 : total pu$Z$
460 V	12 551 : total pu$Z$
480 V	12 019 : total pu$Z$
550 V	10 500 : total pu$Z$
2400 V	2406 : total pu$Z$
4160 V	1389 : total pu$Z$
12.47 kV	463 : total pu$Z$
13.2 kV	438 : total pu$Z$
13.8 kV	419 : total pu$Z$

NOTE: Example—Total pu$Z$ to fault is 0.825 and system voltage is 240 V. Find symmetrical rms short-circuit current:
$$I_{\text{sc}} = \frac{24\ 039}{0.825} = 29\ 180\ \text{A rms}$$

pedances fed by line voltage. In the preceding example the locked rotor current rating is assumed to be five times the motor full-load current. This is a conservative figure and on the safe side. Actual contribution is normally somewhat less.

(2) *Limiting Fault Current.* The asymmetrical short-circuit current will continue to flow for several cycles depending upon the X/R ratio of the system. The asymmetrical fault current will eventually decay to the final symmetrical value of current which was calculated in the examples. Since the asymmetrical current is always greater than the symmetrical current, we find that the largest amount of destructive energy flows during the first several cycles after the fault is initiated. The amount of destructive energy is proportional to the square of the current and the time the fault persists. Therefore it is very important to limit the current to the smallest value possible.

9.6 Methods of Reducing Available Short-Circuit Current.

The available short-circuit current on a distribution system decreases from the source to the load because the circuit impedance increases. The rate of current decrease or impedance increase is a function of circuit design. With design and insertion of impedance in the circuit between the power source and the building protective equipment, the short-circuit values throughout the building may be appreciably decreased and, at some points, may be lowered enough to permit lower-rated, less expensive equipment to be used.

9.6.1 Effect of Distribution Circuit Lengths on Short-Circuit Current.

When design considerations of voltage regulation, space, and economics permit, the circuit impedance at any point may be increased by the proper selection of cables, busways, and principally the choice of the circuit arrangement.

The circuit length, physically, depends on the location of the service-entrance switch, load or distribution centers, riser shafts and riser tapoffs, and is affected by the type of circuit and method of installation.

Where the available short-circuit current is high, a small increase in impedance of service-entrance feeders and parts of the network system, such as can be obtained by using increased spacing between phase conductors, is very effective for reducing the maximum fault currents.

9.6.2 Current-Limiting Reactors.

Reactors are useful devices for reducing the interrupting duty imposed on protective equipment. Where standard-rated circuit breakers can be used, it is generally not economical to substitute a reactor and a lower rated circuit breaker. However, where a reactor can be used to reduce the ratings of several circuit breakers or to reduce interrupting duty to within the capacity of standard circuit breakers, the installation may be economically justified.

When installing reactors, consideration must be given to power loss, space, and voltage drop. If they are to be installed in combination power and lighting circuits, lamp flicker problems as well as motor starting torque requirements must be investigated.

9.6.3 Current-Limiting Busways.

Current-limiting busways are another means of reducing short-circuit currents. They are available in ratings of approximately 1000 A—4000 A. Typical reactance values are 0.006 Ω per 100 ft (line to neutral) for the 1000 A rating, 0.005 Ω for the 2000 A, 0.0015 Ω for the 3000 A, and 0.001 Ω for the 4000 A rating. At 0.70 to 0.90 lagging power factor, the voltage drop in the busway ranges from 1 V to 2 V per 10 ft (line to line) at full current rating. On large network systems with short-circuit currents up to 200 000 A symmetrical, the short run from the network bus to the switchboard is often sufficient to reduce the short-circuit currents to 100 000 A or less. As an example, the length of busway required to reduce a 180 000 A duty to 100 000 A is about 40 ft at 480 V or 20 ft at 240 V.

The impedance of the current-limiting busways is constant during all types of faults such as low-level arcing faults.

A disadvantage of the current-limiting busway is that inherently there is some voltage drop in the busway. In every case where it is decided to use the current-limiting busway, the voltage drop should be calculated. Although to obtain the desired reduction in short-circuit current the voltage drop in the current-limiting busway is generally small, there are some applications on which the voltage drop is so high that the busway is not recommended and current-limiting fuses should be used instead (current limiting fuse application discussed later in this section). Another possibility, when the voltage drop in the current-limiting busway is too high, is to break the load into smaller parts. Dividing the load among four feeders reduces the voltage drop to 25% of its former value, provided four current-limiting busways, each with the same impedance as the single busway, are utilized. To make a given reduction in short-circuit current, the percent voltage drop is the same on a 480 V system as on a 240 V system, but the length of busway required at 480 V is twice that at 240 V, and thus costs twice as much. On those current-limiting busway applications where the voltage drop is

satisfactory, the power loss in the busway is not a significant item.

When all factors are considered, including cost, it may be advisable to use current-limiting busways or current-limiting reactors on some applications and combinations of current-limiting fuses and other equipment on other applications. Sometimes both are applicable on the same job. For example, current-limiting busway might be used from the transformer to the switchboard to reduce the duty to 75 000 A; current-limiting fuses might be used in the switchboard in combination with circuit breakers rated less than 75 000 A interrupting; current-limiting feeders might supply some equipment such as a motor-control center to reduce the short-circuit current to 25 000 A; the first 20 or 30 ft of some busway feeders might be changed to current-limiting busways; and then combinations of current-limiting fuses and molded-case circuit breakers or other equipment might be used in other places.

9.6.4 Examples of Reducing Available Short-Circuit Current. Figure 91 illustrates the interrupting duty of protective equipment that might be required in various parts of a building for both a higher- and lower-voltage system. Cable conductors may be used in some circuits and low- or high-impedance busway in others.

The building wiring may be designed to connect the service switch or circuit breaker directly to the bus takeoff or, if it is not feasible because of structural limitations or for other reasons, it may be located some distance away as shown in Fig 91. Examples of Fig 91(a) and (c) assume that the service-entrance switch or circuit breaker is directly at the end of the network bus takeoff. Figure 91(b) assumes that 37 ft of high-impedance

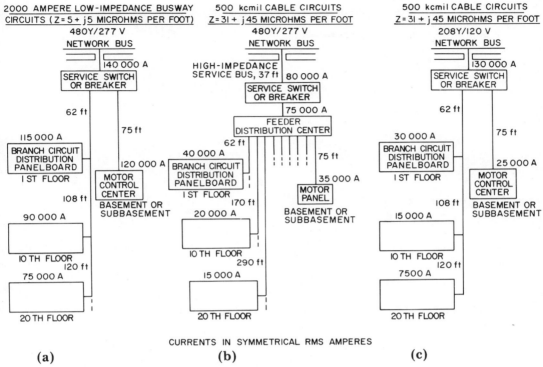

CURRENTS IN SYMMETRICAL RMS AMPERES

(a)   (b)   (c)

NOTE: Supply source six 2000 kVA 7 percent Z transformers for 460Y/265 V
and six 1000 kVA 4 percent Z transformers for 208Y/120 V.

**Fig 91
Typical Available Short-Circuit Currents on Large Office
Building Distribution Systems**

busway is between the network bus and the service switch. This demonstrates the rapid reduction of available short-circuit current by lengthening the circuit when the available current value is high. In this case, the reduction is from 140 000 A to 80 000 A. Generally it is advisable to have the service switch or circuit breaker as near the bus takeoff as possible. The main interrupting device should be rated for the full available short-circuit current at the point of entrance.

Progressing away from the service switch, the available fault current decreases, but not necessarily at a rate desirable for the protective equipment.

The effect of circuit length is illustrated in Fig 92 which depicts the initial rapid decrease in available fault current with increase in length of circuit and shows the diminishing rate of improvement as the circuit lengthens. The reduction of short-circuit current by lengthening the circuit is more effective at the higher

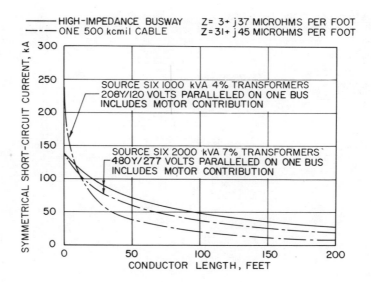

Fig 92
Available Current Versus Conductor Length

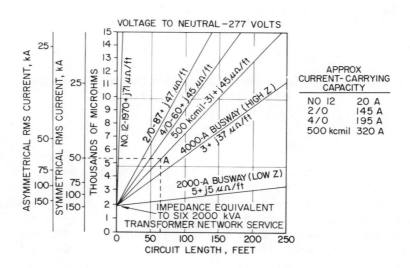

Fig 93
Distribution Circuit Lengths to Limit Available
Short-Circuit Current

current values than at the lower current values where the relative improvement is much less and probably would not justify a lengthening of the circuit.

Illustrated in a different manner, Fig 93 shows the variation of impedance and available short-circuit values with circuit length for various sizes of cables and types of bus design. It includes an initial impedance of $200 + j2000\ \mu\Omega$, approximately equivalent to the impedance value of a network installation of six 2000 kVA 7% transformers.

The purpose of Fig 93 is to give an idea of the approximate length of circuit required to reduce the available fault current to a desired value. The curves for the low-impedance busway and the No 12 cable are included to show extremes.

As an example, assume that it is desired to reduce an available short-circuit current from a value of 140 000 A symmetrical rms to a value of 50 000 A using single-conductor 500 kcmil cable. Referring to Fig 93 and following the dotted line from the 50 000 A value on the ordinate scale over to point A, the required circuit length on the abscissa, is found to be approximately 70 ft.

Figure 94 illustrates general examples of the application of the methods and devices outlined above for controlling and limiting short-circuit currents in large building distribution circuits.

There are many other combinations of circuit elements that can be used in the layout of building wiring. The actual design in specific instances depends on the type and magnitude of load, the service supply installation, building structure, local code requirements, reliability of service required, economic considerations, and the engineer's evaluation of these factors.

9.7 Coordination. The major objectives of the electrical power system designer is to design a system such that faults will be removed in the shortest period of time possible, while maintaining a high degree of service continuity. The area of outage should be restricted as far as practical. The goals of maximum protection and maximum service continuity can most nearly be realized by proper selection and adjustment of high-speed protective devices. In order to properly select and adjust protective devices, a protective-device coordination study must first be performed. Only through a coordination study can the capabilities of today's modern protective devices be fully utilized.

9.7.1 Coordination of Protective Devices. For a great many power systems, the optimum degree of protective-device coordination consists of selective coordination wherein only the protective device nearest the fault opens to remove a short circuit, and the other upstream protective devices remain closed.

On all power systems the protective device must be selected and set to open before the thermal and mechanical limitations of the components protected are exceeded.

9.7.2 Preliminary Steps in Coordination Study. Protective-device coordination which balances protection against the needs of service continuity is achieved and maintained only as a result of following a multistep procedure through to completion. If a short-circuit study has been performed as described earlier in this section many of the preliminary steps to coordination may already have been taken.

Initially a one-line diagram must be made of the system to be coordinated (Fig 95). The diagram is used as a base on which to record pertinent data and information regarding relays, circuit

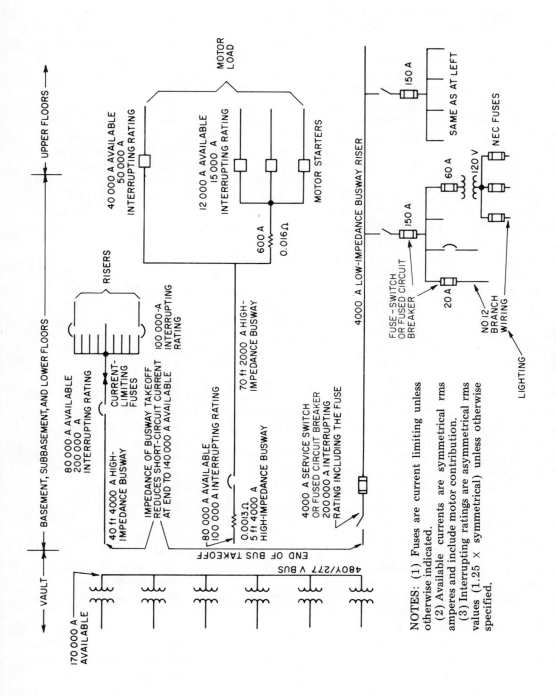

NOTES: (1) Fuses are current limiting unless otherwise indicated.

(2) Available currents are symmetrical rms amperes and include motor contribution.

(3) Interrupting ratings are asymmetrical rms values (1.25 × symmetrical) unless otherwise specified.

Fig 94

Some Possible Arrangements to Limit and Control Available Short-Circuit Current

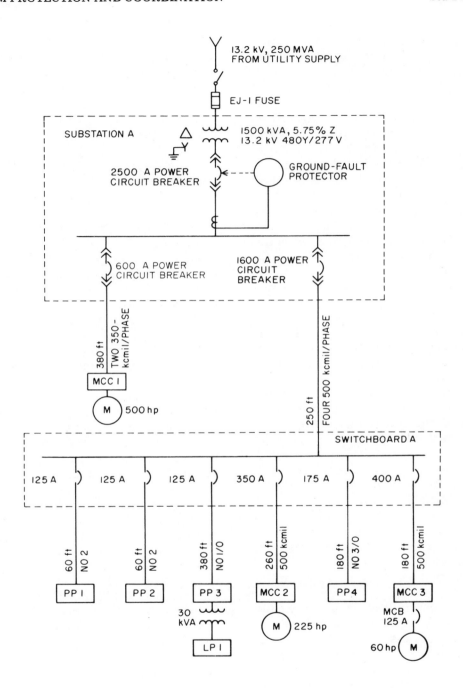

**Fig 95
Typical Distribution One-Line Diagram (Coordination Example)**

breakers, fuses, current transformers, and operating equipment, while at the same time it provides a convenient representation of the relationship of circuit protective devices with one another. The next step is to record all applicable impedances and ratings. Using these values a short-circuit study is then made to determine maximum and minimum short-circuit currents available at any particular point in the system.

Available fault current values can be noted on the system one-line diagram, and on the partial single-line diagrams used in coordination studies; for example see Fig 94.

A further step is to ascertain the maximum load currents which will exist under normal operating conditions in each of the power-system circuits, the transformer magnetizing in-rush currents, and the starting currents, and accelerating times of large motors. These values will determine the maximum currents which circuit protective devices must carry without operating. The upper boundary of current sensitivity will be determined by the smallest values resulting from the following considerations:

(1) Maximum available short-circuit current obtained by calculation.

(2) Requirements of applicable codes and standards for the protection of equipment such as cables, motors, and transformers.

(3) Thermal and mechanical limitations of equipment.

As a last preliminary step, the characteristic time-current curves of all the protective devices to be coordinated must be obtained. These should be plotted on standard log — log coordination paper to facilitate the coordination study (Figs 96 and 97).

9.7.3 Mechanics of Achieving Coordination. The process of achieving coordination among protective devices in series is essentially one of selecting individual units to match particular circuit or equipment protection requirements, and of plotting the time-current characteristic curves of these devices on a single overlay sheet of log — log coordination paper.[43]

The achievement of coordination is a trial-and-error routine in which the various time — current characteristic curves of the series array of devices are matched one against another on the graph plot.

When selecting protective devices one must recognize ANSI and NEC requirements and adhere to the limiting factors of coordination such as load current, short-circuit current, and motor starting. The protective devices selected must operate within these boundaries, while providing selective coordination where possible.

Selective coordination is usually obtained in low-voltage systems when the log — log plot of time-current characteristics displays a clear space between the characteristics of the protective devices operating in series, that is, no overlap should exist between any two time-current characteristics if full selective coordination is to be obtained. Allowance must be made for relay overswing and for relay- and fuse-curve accuracy. Quite often the coordination study will stop at a point short of complete selective coordination because a compromise must be made between the competing objectives of maximum protection and maximum service continuity as in Fig 84.

9.7.4 Coordination Example with Explanation. Let us examine the coordina-

---

[43] Similar to K&E paper 53599.

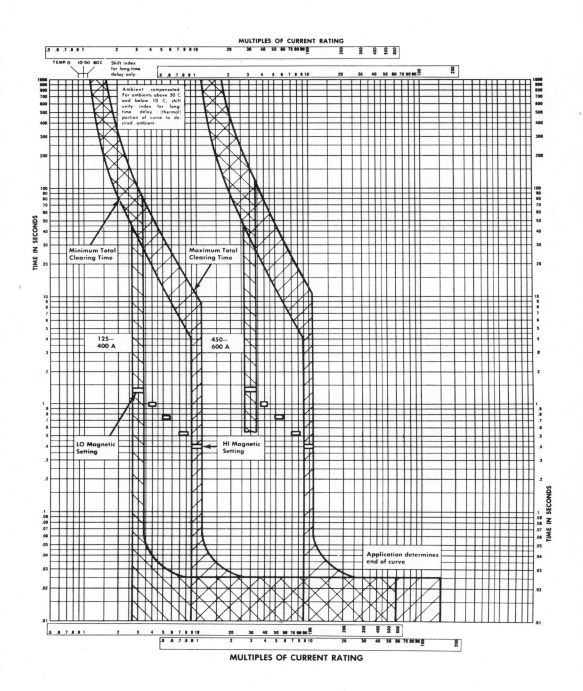

Fig 96
Time-Current Curves for 125 — 600 A Molded-Case Circuit Breakers

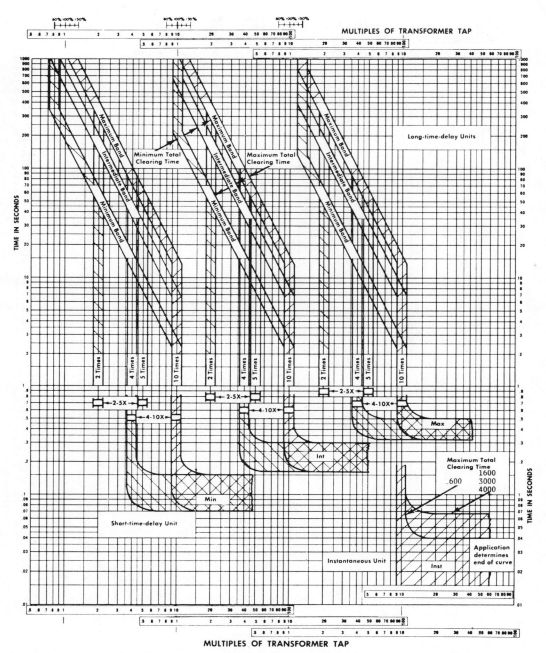

NOTE: Taps provided above the continuous current rating do not provide thermal self-protection

Fig 97
Time-Current Curves for 600 A—4000 A Power Circuit Breakers

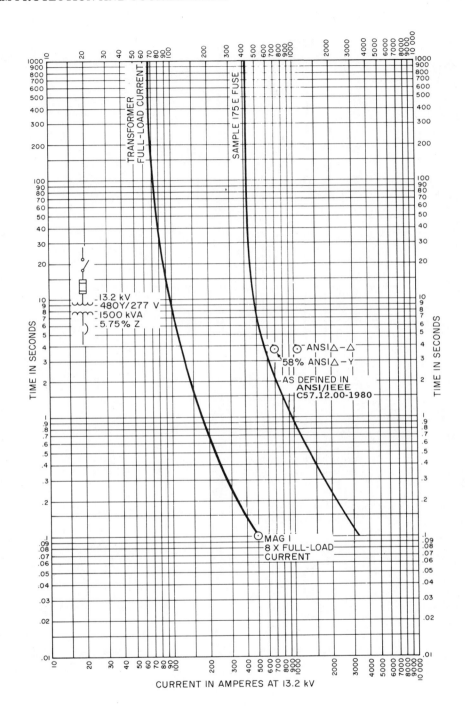

Fig 98
Transformer Protection Zone

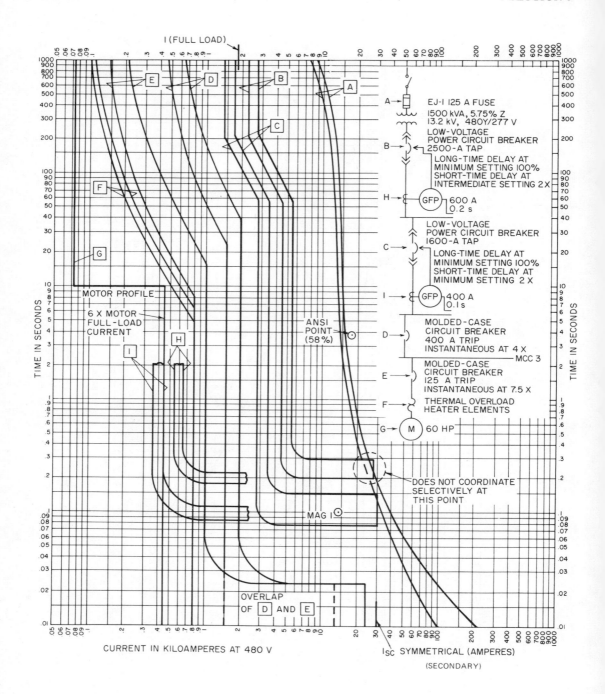

**Fig 99
Coordination of Example System**

tion of a portion of the power system shown in the one-line diagram of Fig 95.

The first level of coordination and protection to be considered is that at the transformer primary. When selecting the primary protection, four factors must be taken into account:

(1) Transformer full-load current
(2) ANSI/NFPA 70-1981 [7] Section 450-3
(3) ANSI C57.12 series [11]
(4) Magnetizing inrush current of the transformer Mag I

In considering these factors one can plot a band in which the transformer protective device must operate (Fig 98).

To fully utilize the transformer, the protective device must carry transformer full-load current. Furthermore, the ANSI/NFPA 70-1981 [7] limits the maximum fuse size or device setting that can be utilized. For the transformer shown (1500 kVA), a maximum fuse rating of three times or relay pickup value of six times transformer full-load current is based on using a main secondary protective device. See ANSI/NFPA 70-1981 [7] Section 450-3 for complete information.

Whatever primary device is used, it must be capable of withstanding transformer magnetizing inrush current. This point is usually selected as eight to twelve times transformer full-load current for a period of 0.1 s.

Finally the transformer withstand point established by ANSI/IEEE C57.12.00-1980 [11] often referred to as the ANSI point, must be considered. To establish this point for transformers with 4% impedance or less, use 25 times transformer full-load current for 2 s. For transformers with impedances of 7% or greater, use 14.3 times transformer full-load current for 5 s. For transformer impedances between 4 and 7%, use

$$I_{ANSI} = \frac{\text{transformer full-load current}}{\% Z/100}$$

and

$$t_{ANSI} = \% Z - 2$$

EXAMPLE: For 1500 kVA, 13.2 kV, 480Y/277 V, 5.75%Z,

$$I_{ANSI} = \frac{66}{5.75/100}$$

$$= 1148 \text{ A at } 31.2 \text{ kV}$$

$$t_{ANSI} = 5.75 - 2 = 3.75 \text{ s}$$

When plotting the ANSI point for a $\Delta-Y$ connected transformer, the current value calculated from ANSI/IEEE C57.12.00-1980 [11] must be derated to 58% of the calculated value (Fig 98). This represents a single-phase line-to-ground fault at the transformer low-voltage terminals as seen by the transformer primary protective device.

Since the major portion of the example system is at 480 V, let us calculate the transformer protection points at this voltage level to facilitate the plotting of curves on one overlay sheet.

(1) Full-load current at $480 \text{ V} \cong 1800 \text{ A}$ (assume fuse protection curve at 1000 S to be twice NEC permitted fuse rating for type EJ fuse).

(2) NEC point = 6 · full-load current $\cong 10\ 800$ at 480 V
(3) ANSI point for 1500 kVA, 5.75%Z,

$$I = \frac{\text{full-load current}}{\% Z/100} = \frac{1800}{0.0575} = 31\ 300 \text{ A}$$

Since the transformer is $\Delta-Y$ connected one must use the above-mentioned 58% derating factor. Therefore,

$I_{ANSI} = (0.58)(31\ 300) = 18\ 150$ A

$t_{ANSI} = 5.75 - 2 = 3.75$ s

(4) Magnetizing inrush current Mag $I = 8 \cdot$ full-load current:

Mag $I = (8)(1800) = 14\ 400$ A for $t = 0.1$ s

These calculated values are now plotted on the graph paper to be used for coordination (Fig 99).

Once the transformer parameters are plotted, rather than setting or selecting a primary main protective device, select the settings for the largest downstream load device. By starting at the load device first, the lower boundary of coordination is established.

In the example under consideration, the largest load device is a 125 A molded-case circuit breaker feeding a 60 hp motor. Using the coordination graph paper with the transformer parameters plotted on it, sketch in the motor-current characteristics. If motor data are not available, it is usually assumed that locked rotor current is equal to six times motor full-load current, and that motor acceleration time is 10 s. Using these values of current and time, the straight-line characteristic is obtained. Motor thermal overload protection must be shown. Once the motor-starter and overload characteristics are plotted, the feeder device setting can be determined.

It is normally recommended that the instantaneous element of the motor feeder be set at twice locked rotor current. In this example, since locked rotor current is equal to 450 A, the instantaneous elements of the 125 A molded-case circuit breaker should be set at 900 A, or approximately 7.5 times the circuit-breaker trip rating. By overlaying the previously drawn curves on the circuit-breaker curve (Fig 96), its characteristics

are drawn in with an instantaneous setting as calculated (7.5 $\times$).

Once this device is drawn in, the lower limit of coordination is established. The remaining steps consist of overlaying the curves already drawn over the characteristic curves of each series upstream device in sequence, and selecting minimum settings for protection while obtaining coordination.

The next upstream device in series is a 400 A molded-case circuit breaker. Overlay the curves already drawn on its characteristic curve, and select a setting which coordinates with the 125 A circuit breaker as set. The minimum instantaneous setting to coordinate will be 4 $\times$. The 400 A circuit breaker characteristic is then drawn in using the 4 $\times$ instantaneous setting. Most adjustable molded-case circuit breakers have only one adjustment, the instantaneous element.

It should be observed from the coordination between the 125 A and the 400 A circuit breakers that selective coordination between series instantaneous protective devices is seldom possible.

If coordination of *instantaneous* operating devices is required, the use of fuses in one or more protective device may enable such coordination if adequate ratios between upstream and downstream fuses are maintained. See Table 79.

The next upstream device in the example system is a 1600 A low-voltage power circuit breaker equipped with a static trip device with long-time and short-time adjustments. For the long-time portion of its curve (Fig 97) one has a choice of a minimum, intermediate, or maximum long-time-delay band. For the short-time portion there is a choice of 2—5 times or 4—10 times the device pickup current value and a time-delay adjustment of minimum, intermediate, or maximum.

To select the settings for this device, again overlay the paper with the previously drawn curves on the characteristics curves of the device to be set. When this is done it is found that a minimum long-time-delay band and a short-time setting of 2 × with a minimum delay band coordinate well with downstream devices, while providing maximum protection with minimum settings.

The final low-voltage device to be set in the series array being studied is the main secondary circuit breaker. This device is a 2500 A low-voltage power circuit breaker equipped with a static trip device with long-time and short-time adjustments. The same choice of bands and ranges exists for this device as for the previously set 1600 A power circuit breaker.

Again overlay the coordination paper with the downstream device curves on the characteristic curve of the device to be set. When this is done, it is found that a minimum long-time-delay band, a short-time setting of 2 ×, and a short-time-delay band coordinate well with the downstream device and give room between the overall curve and the ANSI point to fit the primary protective device curve.

The end point of the short-time-delay band is cut off at maximum available short-circuit current from the 1500 kVA transformer, in this case 28 500 A.

Now that one complete series array of low-voltage protective devices has been set, one can select the primary device. In this example, an EJ type fuse is to be used as the primary protective device. A good point to start in selecting the fuse rating is 1.5 times the transformer full-load current, in this case 100 A. When the coordination plot is placed over the 100 A fuse characteristic, it is noted that very poor coordination with the main

secondary device is obtained. The next size fuse rating is 125 A. The 125 A fuse meets all transformer protection requirements and gives good coordination; thus it is used in this case.

To complete the system coordination study for phase overcurrent devices, one merely needs to select the settings of the remaining series protective devices. These settings are selected in the same manner as those described previously. Once the settings are all selected, to carry out the coordination one merely needs to set the devices as determined by the study.

Calculations similar to those in 9.5.19 were performed, and it was determined that the ground-fault protection on the main circuit breaker should be set at 600 A pickup current and 0.2 s time. The ground-fault protection on the 600 A and 1600 A power circuit breakers was set at 400 A and 0.1 s time.

These settings were selected as being the most consistent with our goals of protection and service continuity. The 400 A 0.1 s pickup on the feeder circuit breakers gives good fast protection, yet it will be selective with small downstream protective devices (25 A and 20 A molded-case circuit breakers).

The ground-fault protective device at the main circuit breaker is set higher both in time and current to provide selective coordination with the ground-fault protective devices on the feeder circuit breakers.

9.8 Fuses. The methods of system coordination which have been described in previous paragraphs cover general approaches. Characteristics of fuses such as current-limiting ability, $I^2 t$ coordination, and related material are detailed in the following sections. By observing the principles previously enunciated and

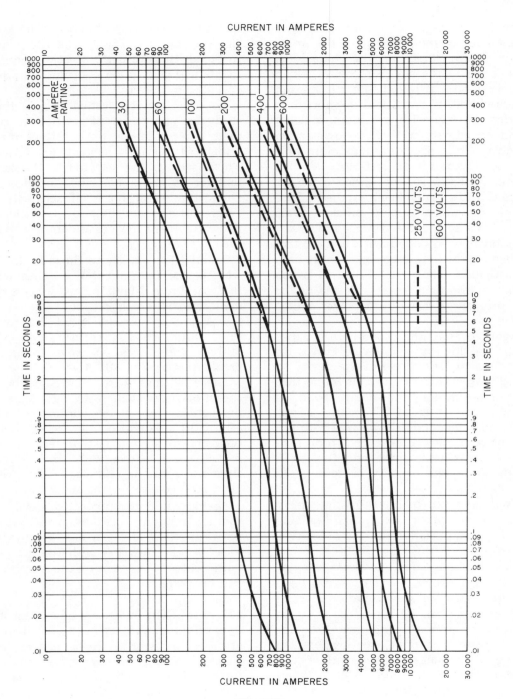

Fig 100
Typical Total Clearing Time versus Current Curves for
Type K5 Fuses

those in the following paragraphs, an effectively coordinated system involving fuses can be developed.

9.8.1 Fuse Coordination. Fuse time-current curves plotted on standard log—log coordination paper are available from fuse manufacturers (Fig 100). There are usually two sets of time-current curves for each fuse. One curve shows the minimum melt characteristic of the fuse and the second shows the maximum total clearing time. In cases where only average melting curves are available, the manufacturers' recommendations to derive minimum melt and maximum total clearing times from these curves should be adhered to.

When coordinating fuses, the maximum total clearing time characteristic of the downstream fuse must fall below the minimum melt characteristic of the next upstream fuse.

Figure 101 illustrates how fuses selectively coordinate with one another for

any value of short-circuit current. Note that for selectivity, the total clearing energy of the fuse B must be less than the melting energy of fuse A.

9.8.2 Fuse Selectivity Ratio Tables. The results of the phenomena displayed in Fig 101 for various types of fuses tested at rated voltage are presented in the form of ratio tables by various fuse manufacturers. Table 79 shows one manufacturer's selectivity schedule for various combinations of fuses.

An example of using Table 79 is found in Fig 102 where a 1200 A Class L fuse is to be selectively coordinated with a 400 A Class K5 time-delay type fuse.

Table 79 may be used as a simple check for selectivity, regardless of the short-circuit current involved. Where closer fuse sizing is desired than indicated, check with the fuse manufacturers as the ratios may be reduced for lower values of short-circuit current. A coordination study may be desired (not required if ratios are adhered to) and can be accomplished by plotting fuse time-current characteristic curves on standard NEMA log—log graph paper. Since fuse ratios for high-voltage fuses to low-voltage fuses are not available, it is recommended that the fuses in question be plotted on log—log graph paper.

9.8.3 Fuse Time-Current Characteristic Curves. Time-current characteristic fuse curves are available in the form of melting and total clearing time curves on transparent paper which is easily adapted to tracing. A typical example of coordinating high-voltage and low-voltage fuses using graphic analysis is shown in Fig 103. Note that the total clearing time curve of the 1200 A fuse is plotted against the minimum melting curve of the 125E 5 KV fuse. The curves are referred to low voltage (240 V) for the study of secondary faults.

Fig 101
Selectivity of Fuses. Total Clearing Energy of Fuse B Must Be Less than Melting Energy of Fuse A

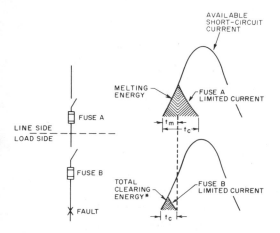

*Indicates, but does not equal, energy.

Table 79
Typical Selectivity Schedule

Line Side	Class L Time-Delay Fuse 601—6000A	Class L Fuse 601—6000A	Load Size Class K1 Fuse 0—600A	Class J Fuse 0—600A	Class K5 Dual-Element Fuse 0—600A
Class L time-delay fuse 601—6000A	2:1	2:1	2:1	2:1	3:1
Class L fuse 601—6000A	2:1	2:1	2:1	2:1	5:1
Class K1 fuse 0—600A			3:1	3:1	4:1
Class J fuse 0—600A			3:1	3:1	4:1
Class K5 dual-element fuse 0—600A			1.5:1	1.5:1	2:1

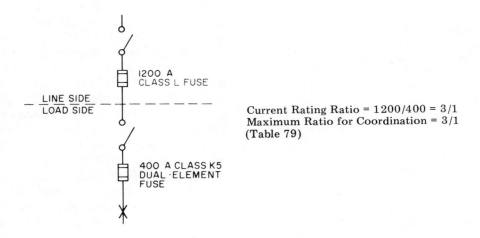

Current Rating Ratio = 1200/400 = 3/1
Maximum Ratio for Coordination = 3/1
(Table 79)

Fig 102
Typical Application Example of Table 79. Selective Coordination is Apparent as Fuses Meet Coordination-Ratio Requirements

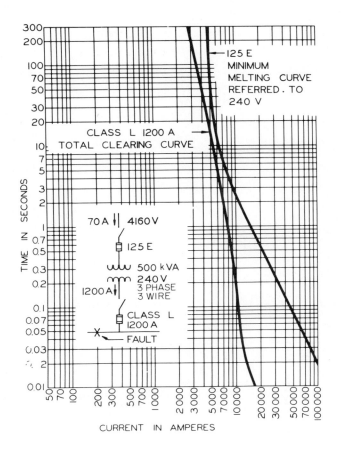

Fig 103
Coordination Study of Primary and Secondary Fuses Showing
Selective System

Care should be taken when coordinating the high-and low-side protection of a Δ—Y transformer. For a line-to-line fault on the Y side, one phase of the Δ will see 16% more per unit current than the low side line. For a phase-to-ground fault on the secondary, the primary fuses will see only 58% of phase-to-phase fault currents.

9.8.4 $I^2 t$ Values for Coordination. Depending on the class of fuses considered for application, there may be times when fuse $I^2 t$ values are required. Then one merely needs compare the total clearing $I^2 t$ of the downstream fuse with the minimum melt $I^2 t$ of the next upstream fuse. When the downstream fuse's clearing $I^2 t$ is less than the upstream fuse's

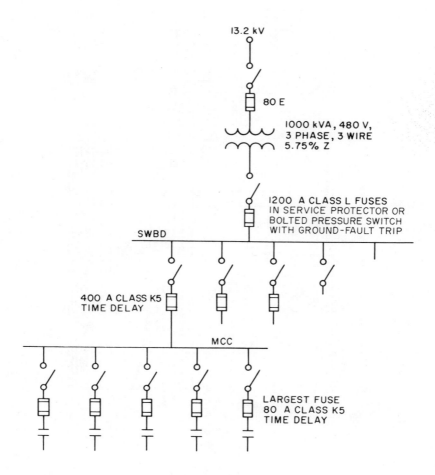

13.2 kV

80 E

1000 kVA, 480 V,
3 PHASE, 3 WIRE
5.75% Z

1200 A CLASS L FUSES
IN SERVICE PROTECTOR OR
BOLTED PRESSURE SWITCH
WITH GROUND-FAULT TRIP

SWBD

400 A CLASS K5
TIME DELAY

MCC

LARGEST FUSE
80 A CLASS K5
TIME DELAY

Fig 104
Fusible Unit Substation

minimum melt $I^2 t$, the fuses coordinate. All data used must be supplied by the manufacturer and will apply only to his fuse types.

9.8.5 Coordinating Fusible Unit Substation. The fusible unit substation shown in Fig 104 illustrates a 1000 kVA transformer supplied at 13.2 kV, serving a 480 V three-phase three-wire switchboard. The primary 13.2 kV fuses are 80E power fuses and the 480 V second-ary main fuses are 1200 A Class L fuses. The largest feeder is 400 A and is protected by 400 A Class K5 time-delay fuses. This feeder serves a 400 A motor-control center.

The first step in coordinating this system is to follow the four factors on transformer protection given in 9.7.4. Then the minimum melting curve for the 80E power fuses is traced. This curve is referred to 480 V for a study of se-

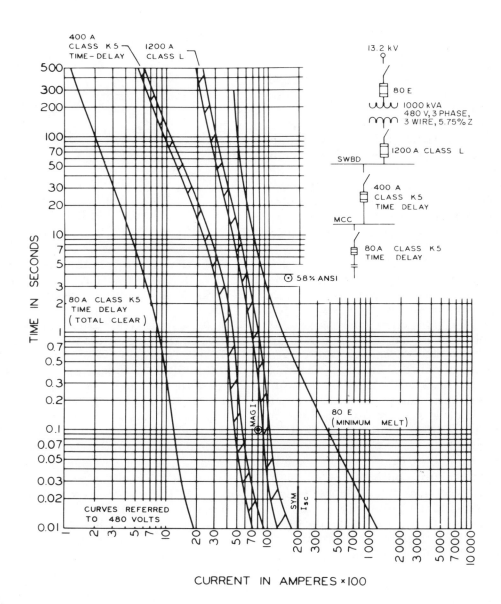

Fig 105
Completed Coordination Study of Low-Voltage Fusible Substation
Shows Complete Selectivity. Class L and Class K5 Fuse
Curves Are Typical of One Manufacturer's Fuses

condary faults. The total clearing time curve for Class L 1200 A fuses is then traced on the graph to study the coordination between primary and secondary fuses.

The next step is to trace the minimum melting curve of the 1200 A Class L fuses and the total clearing time curve of the 400 A fuses. Noting complete coordination between the main and feeder fuses, the last step is to follow the above procedure to study the largest motor-control center fuses and the 400 A feeder fuses. See Fig 105 for the completed coordination study.

The other procedure, which is quite often used to check coordination between low-voltage fuses, is to use a ratio chart that eliminates curve tracing (see 9.8.2). If a ratio chart analysis is used, the only curves that must be drawn are the primary and secondary fuse curves, as explained above. The 400 A Class K5 time-delay fuses can be installed in multipole switches equipped with ground-fault trip devices where full coordination with the upstream fault protection is to be obtained.

9.8.6 Summary. Coordination is a multistep procedure consisting of comparison and selection of protective devices and their ratings. The engineer doing a coordination study must make decisions concerning protection and selectivity.

9.8.7 Fuse Current-Limiting Characteristics. Due to the speed of response to short-circuit currents, fuses have the ability to cut off the current before it reaches dangerous proportions. Figure 106 illustrates the current-limiting ability of fuses. The available short-circuit current flows if there are no protective devices or if there is a delay as a result of operation of a mechanical inertia-type device.

The large loop is one loop of a sine

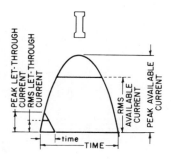

**Fig 106
Current Limitation**

wave. This represents the first half-cycle of fault current available on a standard flows if no protective device is in the circuit. The current starts at zero, in the circuit. The current starts at zero, rises to the peak of the loop, and returns to zero in a half-cycle of time (TIME). On a 60 Hz system this happens 120 times each second. The peak of the wave represents the peak available current. See 9.1.6 for effects of asymmetry.

The effective value of the half-cycle of current, which is the value read on an ammeter, is one half the square root of two (0.707) times the peak current. This is called the root-mean-square value and is not the same as the average value.

The small triangular wave in Fig 106 represents the performance of a current-limiting fuse on a fault current much higher than its rating. The fault current starts to rise, but melts the fuse element before the full available current can get through the fuse. The current through the fuse returns to zero and the total elapsed time is represented by TIME. The peak of the triangular wave represents the peak current which the fuse lets through. This current can also be expressed in equivalent or apparent rms amperes (that is the rms value of a

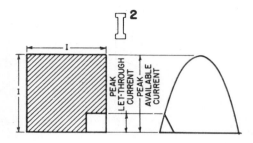

Fig 107
Mechanical (Electromagnetic) Force

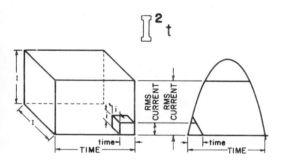

Fig 108
Heating Effect (Thermal Energy)

symmetrical sinusoidal current which has the same peak current as the fuse let-through current).

It should be noted that current-limiting fuses limit both the fault current and the fault time.

Figure 107 shows I^2, which is a measure of the mechanical force caused by peak short-circuit current because force is proportional to the square of this current. This is the electromagnetic force that mechanically stresses and can damage improperly designed bus structures, cable supports, etc. It immediately becomes apparent that squaring the peak available current can create a much larger square than squaring the peak let-

through current of the current-limiting fuse. The difference in the size of the two squares is the difference between having and not having a current-limiting fuse in the circuit.

Figure 108 shows $I^2 t$ which is a measure of the thermal energy of a fault with and without a current-limiting fuse. In the case of $I^2 t$ we must use the rms current instead of the peak current as in the case of mechanical forces (I^2). The difference in size between the large and small cube-like figures represents the difference in energy between having and not having a current-limiting fuse in a circuit involving a high-magnitude available fault current. In extreme cases where effects of $I^2 t$ heating cannot be limited, points of failure typically involve pigtails and heater coils of motor starters and possible welding of contact in circuit making devices.

For available fault currents greater than the "threshold current" of the fuse, a current-limiting fuse will limit the let-through peak current I_p to a value less than the available fault current and will clear the fault in less than one half-cycle, letting through only a portion of the available short-circuit energy. The degree of current limitation is usually represented in the form of peak let-through current charts.

Downstream equipment must be capable of withstanding voltage surges developed by a rapid drop in current or high di/dt.

9.8.7.1 Peak Let-Through Current Charts. Peak let-through charts, sometimes referred to as current-limiting effect curves, are useful from the standpoint of determining the degree of short-circuit protection that a fuse provides to the equipment beyond it. These charts plot fuse instantaneous peak let-through current as a function of available sym-

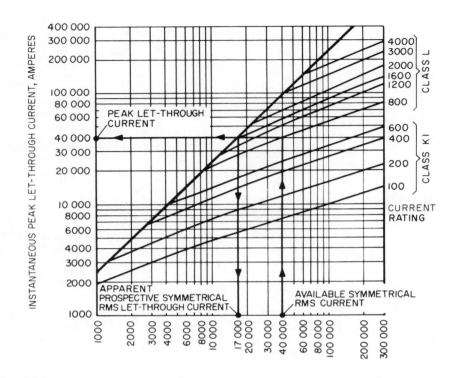

AVAILABLE SYMMETRICAL RMS SHORT-CIRCUIT CURRENT, AMPERES

**Fig 109
Peak Let-Through Current as a Function of Available
Symmetrical RMS Fault Current**

metrical rms current as shown in Fig 109, a typical manufacturers' curve. The straight line running from the lower left to the upper right shows a 2.4 (some manufacturers show 2.3 depending upon the power factor, or X/R ratio, of the test circuit) relationship between the instantaneous peak current which could occur without a current-limiting device in the circuit and the available symmetrical rms current. The following data can be determined from the let-through charts:

(1) Peak-current let-through magnetic effect

(2) Apparent symmetrical rms let-through current heating effect

These date may then be compared to short-circuit ratings of static circuit elements such as wire and bus.

An example showing the application of the let-through charts is represented in Fig 110 where the component is protected by an 800 A current-limiting fuse, and fuse let-through current values are desired with 40 000 A symmetrical rms available at the line side of the component. Using the let-through chart of Fig 109 we can enter at 40 000 A symmetrical rms available and read the

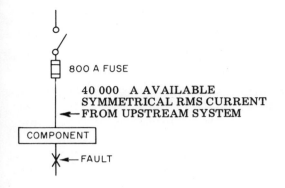

**Fig 110
Application Example of Fuse
Let-Through Charts**

following fuse let-through values:

(1) Peak let-through current 40 000 A

(2) Apparent symmetrical rms let-through current 17 000 A

This procedure will yield a value of symmetrical rms let-through current which can be compared with the rating of a downstream component, if the latter has been given a withstand time rating of one half-cycle or longer under a test power factor of 12%. When this method is used and the results are marginal it is important that the manufacturer of the equipment (particularly in the case of molded-case breakers) be consulted.

Knowing the short-circuit withstand capability of the component under consideration, a comparison can be made to establish short-circuit protection between maximum clearing $I^2 t$ and peak let-through current I_p.

$I^2 t$ is a measure of the energy which a fuse lets through while clearing a fault. Every piece of electrical equipment is limited in its capability to withstand electrical destruction. Where equipment is given an $I^2 t$ withstand rating, maxi-

mum clearing $I^2 t$ values for fuses are available from manufacturers.

Magnetic forces can be substantial under short-circuit conditions and should also be examined. These forces vary with the square of the peak current $I_p{}^2$ and can be reduced considerably when current-limiting fuses are used. Some types of electrical equipment should be examined from the standpoint of peak-current withstand as well as $I^2 t$ withstand.

9.8.8 Application of Fuses. Fuses which have 100 000 A or 200 000 A symmetrical rms interrupting rating and are sized according to ANSI/NFPA 70-1981 [7] requirements, may provide adequate protection (both overload and short-circuit) of system components as well as providing increased interrupting capacity to handle future system growth. These fuses will also prevent unnecessary outages by isolating a faulted circuit if they are selected according to selectivity ratio charts previously presented.

An examination of fuse let-through charts for current-limiting fuses will reveal adequacy of bus-bracing requirements, and wire protection when withstand ratings are known.

Time-delay fuses are most effectively applied in transformer and motor circuits because they can be sized close to the full-load rating without opening under transient conditions.

9.8.9 Bus-Bracing Requirements. Reduced bus-bracing requirements may be attained with current limiting fuses. Figure 111 shows and 800 A motor-control center protected by 800 A Class L fuses. The maximum available fault current to the motor-control center (taking into consideration future growth) is 40 000 A symmetrical rms. To this available fault current from the upstream power system must be added the maximum fault contribution from the motors served from this motor control center (for example, with maximum motors in operation, drawing rated full-load current of 700 A the local motor contribution to the fault is 700 A ∘ 4 = 2800 A). If a non-current-limiting device were used ahead of the motor-control center, the bracing requirement would be a minimum of 42 800 A symmetrical rms. Since current-limiting fuses are used, however, a substantial reduction in bracing may be possible. Entering the let-through chart of Fig 109 at 40 000 A we see that the apparent symmetrical rms let-through current for the 800 A fuse is 17 000 A. Thus, after adding the local motor fault current contribution of 2800 A the total maximum available fault current at the motor-control center main bus is 19 800 A. This would allow the standard bracing (ANSI/NEMA ICS-2-1978 [6] Part ICS 2-322, p 3) for an 800 A bus of 22 000 A symmetrical rms to be used.

Depending on the available fault current (considering future growth), other types and sizes of bus structures may be specified with reduced bracing.

9.8.10 Circuit-Breaker Protection. Circuit breakers may be applied in circuits where the available short-circuit current exceeds the interrupting rating of the circuit breakers when protected by current-limiting fuses properly applied in accordance with information from the circuit breaker manufacturer. Circuit-breaker installations of several years ago may not meet present-day short-circuit current requirements because of changes to the electrical system. These types of installations may also be protected from excessive short-circuit currents by applying current-limiting fuses.

Reference should be made to circuit breaker manufacturers' literature for recommended circuit-breaker fuse protection charts which are the results of extensive testing.

9.8.11 Wire and Cable Protection. Sizing fuses for conductor protection according to the NEC will assure short-circuit as well as overload protection of conductors. Where non-current-limiting devices are used, short-circuit protection for small conductors may not be available, and reference should be made to Insulated Power Cable Engineers Association wire-damage charts [4] for short-circuit withstand capabilities of cooper and aluminum cable.

Small conductors are protected from short-circuit currents by current limiting fuses even though the fuse rating may be 300—400 percent of the conductor rating as allowed by the NEC for motor branch circuit protection.

9.8.12 Motor-Starter Short-Circuit Protection. Underwriters' Laboratories, Inc. (UL) tests motor starters under short-circuit conditions. This short-circuit test

**Fig 111
Example for Determining Bracing
Requirements for 800 A
Motor-Current Center**

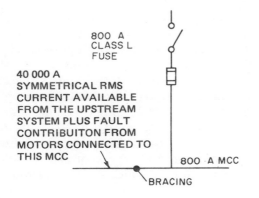

800 A
CLASS L
FUSE

40 000 A
SYMMETRICAL RMS
CURRENT AVAILABLE
FROM THE UPSTREAM
SYSTEM PLUS FAULT
CONTRIBUITON FROM
MOTORS CONNECTED TO
THIS MCC

800 A MCC

BRACING

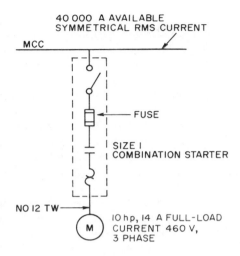

**Fig 112
Selection of Fuses to Provide Short-
Circuit Portection and Backup
Protection for Motor Starters**

may be used to establish a withstand rating for starters. Starters of 50 hp and less are tested with 5000 A available short-circuit current, starters over 50 hp with 10 000 A [5].

When applying starters in systems with high available fault currents, current-limiting fuses can reduce the let-through energy to a value less than that established by the UL test procedures described.

Figure 112 is a typical one-line diagram of a motor circuit where the available short-circuit current has been calculated to be 40 000 A symmetrical rms at the motor-control center and the fuse is to be so selected that short-circuit protection as well as backup motor-running protection is provided. If a Class K5 dual-element time-delay fuse sized at 125% of motor full-load current (17½ A fuse) is chosen, the 40 000 A

symmetrical rms will be limited by the fuse to let-through current of less than 2000 A apparent symmetrical rms, and the fault will be cleared in less than one half-cycle. Since the apparent rms let-through current and clearing time are substantially less than the short-circuit withstand values established by the UL test for size 1 starters, this starter is considered to be protected from short-circuit damage. The apparent symmetrical rms let-through current can be determined from fuse manufacturers' let-through charts for 17½ A time-delay fuses.

9.8.13 Transformer Fuse Protection. Distribution transformers with low-voltage secondaries may be protected by fusing primary and secondary connections in accordance with ANSI/NFPA 70-1981 [7] Section 450-3. Figure 113 shows low-voltage fuses for a 1000 kVA transformer to provide overload protection.

**Fig 113
Typical Low-Voltage Distribution
Transformer Secondary Protection**

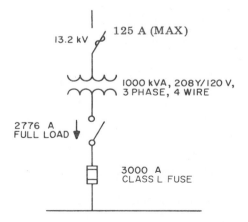

Lighting transformers are quite frequently used in low-voltage electrical distribution systems to transform 480 V to 208Y/120 V. These transformers can be protected by using time-delay fuses sized at 100—125% of the primary full-load current. Some consideration should be given to the magnetizing inrush current since for dry-type transformers this current may be as high as 20—25 times rating. These inrush currents can easily be checked against the time-delay-fuse melting curve at 0.1 s (usually taken as the maximum duration of inrush current). Where dry-type and oil-filled transformers have inrush currents of about 12 times rating lasting for 0.1 s, time-delay fuses may be sized at 100 — 125%. Figure 114 shows a 225 kVA

Fig 114
Typical Protection for 225 kVA
Lighting Transformer

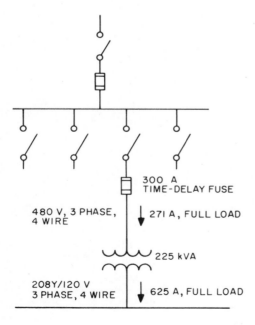

lighting transformer properly protected with time-delay fuses.

ANSI/NFPA 70-1981 [7], Section 450-3 covers overcurrent protection for transformers. It may be provided by protective devices in both primary and secondary circuits. This section of ANSI/ NFPA 70-1981 [7] does, however, spell out those conditions under which protection in the primary only is required. It also spells out the conditions under which secondary protection backed up with primary protection of the transformer is adequate. See ANSI/NFPA 70-1981 [7], Table 450-3(a)(2). With Δ—Y transformations and under line-to-ground secondary fault conditions the affected primary overcurrent devices will see only 58% of the comparable secondary short-circuit current.

9.8.14 Motor-Overcurrent Protection. Single-phase and three-phase motors can be protected by the use of time-delay fuses in motor running protection sized according to ANSI/NFPA 70-1981 [7]. These sizes vary from 100 to 125% of motor full-load current, depending on service factor and temperature rise. Where overload relays are used in motor starters, a larger size time-delay fuse may be used to coordinate with the overload relays.

Combination fused motor starters which employ overload relays sized for motor running protection (100—115%) should incorporate time-delay fuses sized at 125% or the next larger standard size to serve as backup protection. A combination motor starter with backup fuses will provide all-around protection. Figure 115 illustrates the protection for a motor cirucit.

Three-phase motor single-phasing protection will be provided by time-delay fuses that are sized at approximately 125% motor full-load current. Loss of one phase will result in an increase to

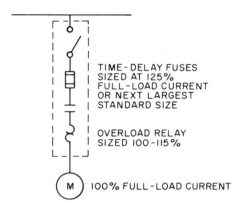

TIME-DELAY FUSES
SIZED AT 125%
FULL-LOAD CURRENT
OR NEXT LARGEST
STANDARD SIZE

OVERLOAD RELAY
SIZED 100-115%

M 100% FULL-LOAD CURRENT

**Fig 115
Protection for Typical Motor Circuit**

173—200% of the line current to the motor. This will be sensed by the motor fuses as they are sized at 125% provided the motor is operating near full load, and the single-phasing current will open the fuses before damage to the windings results.

9.8.15 Fuse Device Maintenance. Modern silver-sand and copper-link fuses require little if any maintenance. A visual inspection of the fuse-holding members is recommended occasionally to ensure adequate pressure between contact-making parts, and overheating because of bad connection is easily detected. Present-day fuse characteristics do not change with age, hence no maintenance is required for those fuses in storage.

9.9 Current-Limiting Circuit Breakers. Current-limiting characteristics can be obtained in the following ways when using circuit breakers:

(1) Auxiliary current-limiting fuses are internally mounted in molded case breakers. These are usually special-purpose fuses designed for breaker application.

(2) Current-limiting fuses are used with low-voltage power circuit breakers. The fuses are usually mounted on the circuit-breaker drawout assembly, or on switchboards or switchgear.

(3) A relatively new design of non-fused current-limiting circuit breakers utilizes very fast tripping speeds so that the potential high-magnitude fault current is limited during the first half-cycle of fault current, just as achieved by current-limiting fuses. The unit operates as a conventional circuit breaker for overload and lower-level fault currents.

Current-limiting circuit breakers, including those incorporating current-limiting fuses, are intended for applications needing the overload-overcurrent and switching functions of the circuit breaker in systems where available fault current exceeds the rated fault-current capabilities of the circuit breaker, or other components of the power distribution system. If the current-limiting element used in conjunction with the circuit breaker is properly selected, the current limiter operates only in the event of a low-impedance fault (in a power system of high-fault current capacity) to provide protection against high peak fault current for the circuit components, including the circuit breaker and the downstream circuit components. The conventional elements of the circuit breaker will clear the overloads and lower-level-magnitude fault currents, which are the most frequent causes of operation in low-voltage systems. The current limiting element handles the relatively infrequent high-magnitude fault currents, and operation of any fuse will trip the circuit breaker. This isolates the faulted portion of the system and will preclude the possibility of single-phase operation of any motors downstream from the device.

A description of the phenomenon of current limiting, including characteristics and advantages, is presented in the preceding sections of this text.

9.10 Ground-Fault Protection. ANSI/ NFPA 70-1981 [7], Sections 230-95 and 517-14 requires knowledge of the levels of ground-fault currents to properly set and coordinate ground-fault protective devices. ANSI/NFPA 70-1981 [7], Section 230-95, states that ground-fault protection of equipment shall be provided for grounded-wye electrical services of more than 150 V to ground, but not exceeding 600 V phase-to-phase for any service-disconnecting means rated 1000 A or more. The ground-fault protection may consist of overcurrent devices, or combinations of over-current devices and current transformers, or other equivalent protective equipment which shall operate to cause the service-disconnecting means to open all ungrounded conductors of the faulted circuit. The maximum setting of the ground-fault protection shall be 1200 A and the maximum time delay is 1 s for ground faults equal to or greater than 3000 A.

In ANSI/NFPA 70-1981 [7], Section 230-95, NOTES explain that it may be desirable to include ground-fault protection for service disconnecting means rated less than 1000 A and also that additional installations of ground-fault protective equipment will be needed on feeders and branch circuits where maximum continuity of electrical service is necessary. For health-care facilities, when ground-fault protection is provided on the service disconnecting means, ANSI/NFPA 70-1981, Section 517-14 requires the additional step of ground-fault protection in the next level of feeder downstream toward the load.

Figure 116 shows methods of detecting the ground-fault current. If a transformer (Fig 116(a)) is the source of supply and its ground return current can be measured, a simple current transformer may be used to detect the flow of ground-fault current back to the neutral connection of the transformer windings. This method can also be used if the power system includes the neutral conductor (that is, loads may be connected line-to-neutral) provided the current transformer is located between the power transformer ground connection and the neutral conductor connection and also that the neutral conductor remains an insulated-isolated conductor (that is, no additional neutral conductor ground connections are made downstream).

Figure 116(b) and (c) depends on the principle that the phasor sum of all currents flowing from and returning to a source of power is zero. If there is any current flow through ground, then this current, when added to that flowing through the line and neutral conductors, must equal zero. Therefore the unbalanced current or flux through the current transformers or sensor must equal that of the ground-fault current. The sensor usually consists of a single window type current transformer with an opening large enough to accept all of the phase and neutral conductors and is designed to handle only a limited-burden ground-fault relay specially matched to it.

Figure 117 illustrates a typical one-line diagram with ground-fault relaying. The term *relay* includes electronic or solid-state relays as well as electromechanical devices. These relays may be specified with various pickup levels of current and

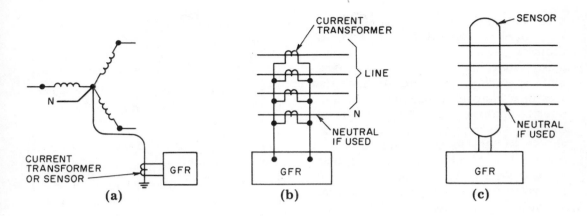

Fig 116
Detecting Ground-Fault Current. (a) With Current Transformer, or Sensor
(b) With Current Transformer. (c) With Current Sensors

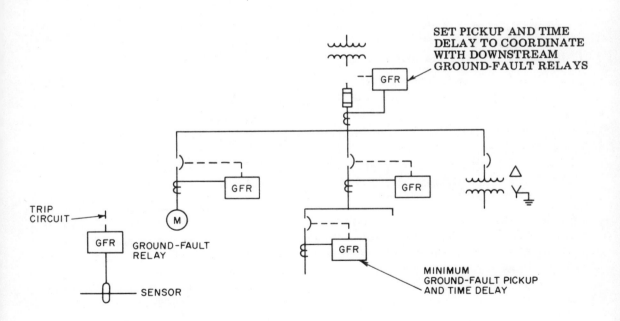

Fig 117
Typical Ground-Fault Relaying

with various time-delay setting ranges. Full coordination with the line or phase protective relaying is desirable. The simplest system involves time-delay and current selectivity. A more sophisticated system utilizes blocking signals or zone interlocking from the downstream device to delay the tripping of the upstream device to give the former a chance to clear the fault. A number of systems providing this kind of protection are available for protecting secondary unit substations, double-ended substations, networks, and other sources, and information concerning such protection can be obtained from the switch or circuit-breaker manufacturer.

9.11 References. The following references were used in preparing this section.

9.11.1 Standards

[1] ANSI C37.06-1979, Preferred Ratings and Related Required Capabilities for AC High-Voltage Circuit Breakers Rated on a Symmetrical Current Basis. [44]

[2] ANSI C37.6-1971 (1976), Preferred Ratings for AC High-Voltage Circuit Breakers Rated on a Total Current Basis.

[3] ANSI C37.16-1980, Recommendations for Low-Voltage Power Circuit Breakers and AC Power Circuit Protectors, Preferred Ratings and Related Requirements and Applications.

[4] ANSI/IEEE C37.010-1979, Application Guide for AC High-Voltage Circuit Breakers Rated on a Symmetrical Current Basis.

[5] ANSI/IEEE C37.5-1979. Guide for Calculation of Fault Currents for Application of AC High-Voltage Circuit Breakers Rated on a Total Current Basis.

[6] ANSI/NEMA ICS 2-1978 Devices, Controllers, and Assemblies for Industrial Control.

[7] ANSI/NFPA 70-1981, National Electrical Code. [45]

[8] ICEA P 32-382 (R 1969), Short-Circuit Characteristics of Insulated Cable. [45]

[9] IEEE Std 141-1976, Recommended Practice for Electric Power Distribution for Industrial Plants.

[10] IEEE Std 242-1975, Recommended Practice for Protection and Coordination of Industrial and Commercial Power Systems.

[11] ANSI/IEEE C57.12.00-1980, Standard General Requirements for Liquid Immersed Distribution, Power, and Regulating Transformers.

[12] NEMA AB 1-1969, Molded-Case circuit breakers. [46]

[13] NEMA BU 1-1969, Busways.

[14] NEMA SG 3-1971, Low-Voltage Power Circuit Breakers.

9.11.2 References

[15] BEEMAN, D.L., Ed *Industrial Power Systems Handbook.* New York: McGraw-Hill, 1955.

---

[44] ANSI documents are available from the Sales Department, American National Standards Institute, 1430 Broadway, New York, NY 10018.

[45] The National Electrical Code is published by the National Fire Protection Association, Batterymarch Park, Quincy, MA 02269. Copies are available from the Sales department of American National Standards Institute, 1430 Broadway, New York, NY 10018.

[46] NEMA publications are available from the National Electrical Manufacturers Association (NEMA), 2101 L. Street, N.W., Washington, DC 20037.

[16] FREUND, ARTHUR, *Overcurrent Protection, Electrical Construction and Maintenance.* New York: McGraw Hill, 1980.

[17] CIENGER, J.A., DAVIDSON, O.C., and BRENDEL, R.W. *Determination of Ground-Fault Current on Common AC Grounded Neutral Systems in Standard Steel or Aluminum Conduit. AIEE Transactions,* pt II, vol 79, May 1960, pp 84-90.

[18] HUENING, W.C., Jr *Interpretation of New American National Standards for Power Circuit Breaker Application. IEEE Transcations on Industry and General Applications,* vol IGA-5, Sept/Oct 1969, pp 501-523.

[19] *Industrial Control Equipment.* Underwriters Laboratories, Inc, Bull 508, para 121, 131, 144.

[20] KAUFMAN, R.H. *Let's Be More Specific About Equipment Grounding. Proceedings of the American Power Conference,* 1972.

[21] REICHENSTEIN, HERMANN W., *Applying Low-voltage Fuses, Electrical Construction and Maintenance.* New York: McGraw Hill, 1979.

10. Lighting

10.1 General Discussion. The era of electric lighting began a little more than a century ago with the invention of the incandescent lamp. Prior to that time daylight was the principal illuminant in commercial buildings, with flame sources occasionally used to allow for earlier starting times or somewhat longer operations late in the day after daylight had faded.

Electric lighting has proved to be a high-technology industry, with the manufacturers devoting considerable effort to research and development. Consequently, in recent decades a succession of new, more efficient light sources, auxiliary equipments, and luminaires have been introduced. Research in basic seeing factors has also been pursued for many years and a succession of developments has provided greater knowledge of many of the fundamental aspects of quality and quantity of lighting. Some of these developments make it possible to provide for visual task performance using considerably less lighting energy than in the past.

Today, energy conservation, energy cost, and energy availability, both present and future, must guide decisions on every energy-using subsystem of a building. Lighting, as an energy user, has had considerable attention. This section will include ways to reduce energy requirements for lighting, yet provide adequately for the well-being and needs of the occupants and the objectives of the owners.

Since there is much documentation elsewhere of lighting technology and design, reference will be made to appropriate sources of such information. Application techniques and controls that save energy and costs will be stressed in the material presented here.

10.1.1 Lighting Objectives. Owner objectives for lighting may vary over a broad scale depending on whether fast and accurate visual performance in a business-like environment is desired or whether the creation of mood and atmosphere in a space is of paramount importance. Lighting has great flexibility in this regard and designers can vary its distribution and color, use its effect on room surfaces and objects to achieve dramatic, sparkling, somber, relaxing, or attention-compelling effects, as desired.

In recent years the psychology of lighting has had some in-depth study and some guides are now available to aid designers and application engineers in using light to create the attributes in an environment that will result in the appropriate subjective reactions of the occupants of a space [2], [3], [4], [5].[47]

The desired objectives for lighting must be accomplished through an energy-efficient design.

10.1.2 Lighting Regulations. As of this writing, regulations affecting lighting use are in a fluid state. In 1976, the Federal Energy Agency (forerunner of the Department of Energy) asked the states to adopt *mandatory lighting efficiency standards* at least as stringent as ANSI/ASHRAE 90-80 [1], Section 9. IES/EMS-1 [7], Section 9 was developed by the Illuminating Engineering Society of North America and many states have adopted or referenced this document for the lighting of new buildings. Some states have developed other approaches and certain cities also have or are developing codes for lighting. Regulations for lighting in existing buildings have been or are being developed by states and cities. The National Bureau of Standards and an American National Standards committee, organized jointly by IEEE and IES, have developed another approach. When the document is ready for publication it will be processed as a consensus standard.

Most of the regulations now in effect specify a limit to the lighting power

which may be connected in a building. Since these are rather stringent limits, it behooves designers to use efficient light sources, luminaires and techniques to get the lighting desired. Equally as important, it requires appropriate controls to turn off the lighting when it is not needed.

For existing buildings, a limit on fixed lighting load expressed in watts per square foot is being applied in some legal jurisdictions. Another approach used is to make an audit of the lighting energy consumed as of some appropriate base date and mandate an arbitrary percentage cut from that figure.

The Federal Department of Energy currently has under development *performance standards* for new buildings. These standards provide for an annual energy budget (Btu/ft^2 per year) for a building on which all building subsystems must draw — lighting, heating, cooling, ventilation, hot water, etc. Budgets will vary depending on the type of building and the climate where it is located. The owner/designer will have a choice as to how to allot this energy budget among the subsystems of a particular building.

The General Services Administration has issued a circular [6] which specifies certain illuminance values (footcandles) for work stations and other areas of federal agency buildings.

As of this date, the best advice this text can provide its users is to maintain an awareness of the regulations which may affect lighting power or energy use in buildings which are their responsibility. In the meantime, they would be well advised to use the most efficient light sources, luminaires, and techniques for lighting the buildings coming under their purview.

[47] The numbers in brackets correspond to those in the references at the end of this Section.

10.2 Lighting Terms. Listed here are common lighting terms with commonly applied definitions.

ballast. An electrical device used with one or more discharge lamps to supply the appropriate voltage to a lamp for starting, to control the lamp(s) current while operating and, usually, to provide for power factor correction. Ballasts may be magnetic core and coil type, electronic, or resistive.

brightness. The subjective attribute of any light sensation, including the whole scale of qualities of being bright, light, brilliant, dim, or dark. Brightness has been used in the past in referring to the measurable *photometric brightness*. The preferable term for the latter is *luminance*, reserving *brightness* for the subjective sensation.

contrast. This indicates the degree of difference in light reflectance of the details of a task compared with its background. It includes both specular and diffuse components of reflection.

coefficient of utilization (CU). For a specific room, the ratio of the average lumens delivered by a luminaire to a horizontal work plane to the lumens generated by the luminaire's lamps alone. The work plane is usually (but not necessarily) considered to be 30 in above the floor.

efficacy. See **lumens per watt.**

equivalent sphere illumination (ESI). A measure of the effectiveness with which a practical lighting system renders a task visible compared with the visibility of the same task lit inside a sphere of uniform luminance.

fixture. See **luminaire.**

footcandle. A unit of illuminance (light incident upon a surface) equal to 1 lm/ft^2. In international terms the unit of illuminance is lux (1 footcandle = 10.76 lux).

footlambert. The unit of luminance, defined as 1 lumen uniformly emitted by an area of 1 ft^2. In the international system, the unit of luminance is candela per square meter.

glare. The undesirable sensation produced by luminance within the visual field. This may cause annoyance (discomfort glare) or temporary loss in visual performance (disabling glare).

high-intensity discharge (HID). A group of lamps filled with various gasses generically known as mercury, metal halide, and high-pressure sodium.

illuminance. The unit density of light flux (lumens per unit area) incident on a surface. In the British system one lm/ft^2 equals one footcandle; in the metric system one lm/fm^2 equals one lumen.

lamp. A generic term for a man-made source of light.

lumens. The international unit of luminous flux or time rate of flow of light.

lumens per watt (efficacy). The ratio of lumens generated by a lamp to the watts consumed by the lamp. This term has, traditionally, not included the ballast watts for discharge lamps, due to the many types of ballasts available.

luminaire. A complete lighting unit consisting of parts designed to position a lamp (or lamps), to connect them to the power supply and to distribute the light.

luminaire efficiency. The ratio of lumens emitted by a luminaire to those generated by the lamp (or lamps) used therein.

luminance. The light emanating from a

light source or the light reflected from a surface. The metric unit of measurement is cd/cm^2.

lux. The metric measure of illuminance equal to 1 lumen uniformly incident upon 1 m^2. (1 lux = 0.0929 FC.)

reflectance. The ratio of the light reflected by a surface to the light incident. An approximation of a diffuse surface's reflectance may be obtained with a light meter. Surface specularity will greatly affect reflectance measurements.

task/ambient lighting. A concept involving a component of light directed to tasks from appropriate locations by luminaires located close to the task for energy-efficiency.

Ambient lighting is also provided to fill in otherwise unlighted areas, reduce contrasts in the environment, and supply additional light on the tasks.

visual comfort probability (VCP). A rating of a lighting system expressed as a percent of people who if seated at the center of the rear of a room will find it visually acceptable with relation to the perceived glare.

visual task. The work that requires illumination in order to be accomplished.

veiling reflections. Reflected light from tasks which reduce visibility because it is reflected specularly from shiny details of a task, thereby brightening those details and reducing contrast with the background.

work plane. The plane in which visual tasks are located. For offices and schools it is usually considered to be a horizontal plane 30 in above the floor; however, it can be any plane (vertical, sloping, or horizontal) and at any height.

10.3 Illumination Quality. Some of the factors involved in the quality of light are:

(1) Providing illumination without discomfort causing glare

(2) Providing the light so that veiling reflections in task details are minimized

(3) Using a high color rendering source where the appearance of people, food, appointments of a space, etc are critical, or where the task itself has colors which must be discriminated

(4) Selecting sources and luminaires which will provide *sparkle* and *modelling* on certain types of merchandise

(5) Using sources, equipments, and techniques that will help provide the desired atmosphere in a space.

10.3.1 Visual Comfort. In offices, schools, libraries, drafting rooms, and similar spaces it is desirable to provide illumination without annoyance or discomfort due to luminaire or window brightness. Reference to visual comfort probability (VCP) data, available from luminaire manufacturers, is helpful in the selection of luminaires which will not present discomfort. The Illuminating Engineering Society of North America (IES) indicates that values above 70 VCP will generally result in satisfactory conditions; however, lower values may be satisfactory for many circumstances since the center, rear-of-room reference condition has the lowest visual comfort probability (VCP) in the room.

A great many types of shielding materials are available including a variety of lenses, polarizers and louvers. One material may differ significantly from another in its brightness properties, so it is necessary to have a manufacturer's visual comfort probability (VCP) data to properly evaluate each material being considered.

Visual comfort probability (VCP) data may also be helpful in selecting lighting for a store. If a bright, stimulating store atmosphere is desired, a luminaire with a lower VCP than desirable for an office may help. On the other hand, a store in which a subdued, relaxing atmosphere is desired should probably have a luminaire with a high VCP specified.

Windows with a direct view of the sun, clouds, sky, or bright buildings are sometimes a source of discomfort. For this reason windows should have shades, blinds, draperies, low transmission glass, or other suitable shielding to reduce the brightness in the field of view. Workers should not normally face windows, in performing their work, but should have the windows at their sides or backs.

10.3.2 Veiling Reflections. Such reflections reduce task visibility by lowering the contrast between details of the task (for example, a specular reflection from a graphite pencil stroke) and its background. They occur when a light source and the eye of a worker are at the mirror-angle of reflection with the specular detail of a visual task. They are often difficult to eliminate by shifting the viewing angle since luminaires (or windows) which produce the effects are of substantial area and there are frequently many of them to be reflected.

The most important single factor in minimizing veiling reflection effects is geometry. If the sources which light the task can be positioned out of the mirror-angle of reflection with respect to the task and worker's eyes, task visibility will be greatest. This is frequently practical in private offices where the desk location is known. It is also possible with built-in work station lighting, if lights are located to illuminate the task from both sides. Unfortunately in many work stations the sources are under a shelf or cabinet directly in front of the

tasks, which is usually the worst possible location.

In a general office or drafting room, it is best to position desks or drafting boards between rows of ceiling luminaires, with workers facing parallel with the rows so that more of the light on the tasks comes from the sides and not from luminaires on the ceiling immediately forward of the desk.

Certain lighting distributions, such as polarizing lighting panels, bat-wing lenses, bat-wing reflectors on luminaires, and indirect lighting may also reduce the effect of veiling reflections.

Indirect lighting (without accompanying direct-task lighting) works best in large rooms with low furniture and little or no use of high screens or room dividers. The veiling effects are least when the ceiling is uniformly lighted and the tasks can be lit by a large area of ceiling. This begins to approach a reference condition known as *sphere lighting*. However, even in large, open-plan spaces, if work station, furniture stands five or six feet high, or many screens are used to partition the space, or both, the utilization of the indirect lighting is greatly reduced and the effect of veiling reflections is substantially increased.

Progress has been made in predicting and evaluating the effects of veiling reflections. The term *equivalent sphere illumination* (ESI) has resulted from this new technology. See 10.4.1 (ESI).

10.3.3 Room Finishes. The reflectance of room surfaces is an important factor in the efficient utilization of light and, therefore, the efficient utilization of lighting energy. It is also important to visual comfort because luminances should be within certain well-established limits (ratios) in areas where demanding visual tasks are performed.

Table 80
Recommended Surface Reflectances for Offices

Surface	Center-point Tolerances	Equivalent Range (%)
Ceiling finishes*	0.80 + 15%	80—92
Walls	0.50 ± 20%	40—60
Furniture, machines, and equipment	0.35 ± 25%	26—44
Floors	0.30 ± 30%	21—39

*Reflectances for finish only. Overall average reflectance of textured acoustic materials may be somewhat lower.

For best utilization of light, the ceiling should be painted white. The walls, floor, and equipment finishes should be within the recommended reflectance ranges of Table 80.

To get even higher utilization of light, proposals are sometimes made to employ finishes on walls, floors, and desks whose reflectances are even higher than those of Table 80. Specifiers are cautioned against such experiments, as the recommended reflectance values have been well established over several decades of practice. Lighter finishes could create legitimate complaints of glare and upset the brightness relationships necessary for visual comfort.

The lighting engineer should include a specification for room reflectance as part of his design or ensure that he is consulted by those having the responsibility for color specifications.

Certain portions of walls, trim surfaces, or room appointments may have higher or lower reflectance than the limits of the ranges of Table 80 if these areas are thought of as accents and restricted to no more than 10% of the total visual field.

In stores, restaurants, theaters, and similar commercial areas where there is less need for balanced brightnesses, departures from the recommended reflectances may be made if done with discretion. The lighting engineer must be aware of such departures and consider them in his computations, or the lighting result may be quite different from that anticipated.

10.3.4 Color. Color is a complex subject involving both physical parameters that can be expresses in mathematical terms and psychological factors which relate to individual interpretations of color.

Certain colors seem to be warm in character while others are considered to be cool. Light sources have such characteristics and their color may sometimes be a factor in source selection in order to complement a warm, cool, or neutral color scheme. Warmth or coolness in color scheme and light source may also be a factor in the preception of temperature by occupants of space. This could have energy implications for space heating or cooling, winter and summer.

Certain light sources may have high efficacy of light production with fair or poor color rendition. Others may have excellent color rendition with only moderate efficacy. In recent years, phosphor developments have resulted in fluorescent lamps with excellent color and good to excellent efficacy. These factors must be weighed, along with many others in light source specification, for particular applications.

There are two terms which can provide useful color information about lighting. One is chromaticity, or apparent color temperature, sometimes called correlated color temperature; the other is color rendering index, symbolized as R in color literature.

Chromaticity is the measure of the *warmth* or *coolness* of a light source expressed in the Kelvin temperature scale. It describes the appearance of the theoretical *black body* of physics, a perfect absorber and emitter of radiation, if it were heated to incandescence. At the first phase of incandescence, the object is a ruddy red. At higher temperatures the color changes from a range of warm, yellowish white colors to white, and then to cool blue-white colors at still higher temperatures.

Some of the general service incandescent lamps and warm white fluorescent colors have 3000 K correlated color temperature. Cool white fluorescent lamps have a chromaticity of 4200 K. Chromaticities of sun and skylight vary over a broad range through the day.

Chromaticity provides no information about how well a light source will render various object colors. Daylight has excellent color rendition, though the appearance of colors will vary with time of day, season, latitude, weather, and other atmospheric conditions. An incandescent lamp emits relatively small amounts of blue and green light relative to red, so it tends to mute or "gray" cool object colors such as blue. Some of the discharge lamps are regarded as *high color rendering* types, others are not as good.

A measure of how well a light source renders colors is the color rendering index (R). This is a number which compares a specific light source of interest against a reference source on a 0 to 100 scale. The system is limited and sometimes misunderstood, because a comparison of two sources is meaningful only if the two sources being compared have the same chromaticity. It would not be meaningful to compare the R of an incandescent lamp with that of a cool white fluorescent lamp because the chromaticity of incandescent is 3000 K while cool white fluorescent is 4200 K. A comparison could be made between cool white (R = 66) and deluxe cool white (R = 89) because both have the same chromaticity.

From a design viewpoint, if the appearance of colors is important, one approach might be to select a chromaticity whose warmth or coolness is suitable for a particular application, and then find a source with a high R in that chromaticity.

Often an experienced designer or colorist is called upon to select the color scheme for a commercial area. The lighting engineer must ensure that the color specifications are reasonable for good visual comfort in areas where good seeing is critical, and that the assumptions made for ceiling, wall, and floor reflectances are realistic to ensure a satisfactory lighting design.

Theories of lighting and color perception are continually evolving. One engineer has recently authored a new concept of color perception. While this theory will require time to validate or disprove, a reference is provided for those who are interested [12].

10.3.5 Psychological Factors. There is a great deal that is subjective about how individuals react to a space. Nevertheless, in recent years, studies of the psychology of lighting have provided valuable data as to how statistically significant groups of people react to various kinds of lighting. Criteria have been developed which allow one to use lighting to create impressions of a *public* or a *private* space, for example. These criteria can be extremely helpful in applying lighting in such areas as lobbies, private offices, cafeterias, conference rooms, as well as libraries and general offices [2], [3], [4], [5].

10.4 Illumination Quantity. The Illuminating Engineering Society of North America changed the basis for its recommended levels of illumination in 1979 [8], [9]. The previous system involved single number target values of footcandles (or ESI) for various tasks representing an averaging of assumptions about user eyesight, age, task demand, etc. The new system involves illuminance ranges that correlate with the recommendations in the International Commission on Illumination (CIE) Report No 29. These are summarized in Table 81. This approach can be considered to be an interim step, based primarily on an international consensus. It is intended to replace this system with a scientifically-based method at some future time, pending research yet to be completed.

To determine the nominal design illuminance from the range, Table 82 is consulted and weighting factors assessed. The selection of weighting factor depends on the age of the workers, the reflectance of the task background, and the demand for speed and accuracy of task performance. All these factors are identified by recent research as significant variables affecting task performance.

The individual designer is required to make more specific decisions and to accept more responsibility for the performance of the lighting system with the new IES system than in the past.

There are subtleties and refinements in the new IES system that cannot be covered in detail in this brief summary. One very important one has to do with tasks subject to veiling reflections (which abound in commercial buildings), where use of the visibility metric equivalent sphere illumination (ESI) may be helpful in comparing various lighting systems. It must be cautioned that ESI values, measured or computed, cannot be directly compared with the iilluminance values of Table 81. But an assessment of ESI for several lighting systems of interest can help to determine which are better in creating task visibility.

Equivalent sphere illumination (ESI) is a relatively new metric which involves both quality and quantity aspects of illumination design. It allows the comparison of actual or proposed lighting systems with a reference *sphere lighting* condition using a standardized task and observer.

The sphere lighting condition is a convenient reference condition, not an ideal lighting situation. Imagine a visual task located in the center of a uniformly bright sphere interior. (For flat paper tasks a hemisphere would be satisfactory.) If a small aperture is created in the sphere surface so that the task can be viewed at the reference viewing angle (25°), the contrast of the task can be measured with a visual task photometer (VTP) and the contrast rendition factor (CRF) computed. The VTP can also be used in the field to determine the CRF of the same task under an actual lighting system. When the CRFs are computed for the identical task then allow the effectiveness of the field lighting system in creating task visibility to be compared with that of the sphere reference condition, or with other practical lighting systems.

Table 81
Illuminance Recommended for Use in Selecting
Values for Interior Lighting Design†

Category	Range of Illuminances* in Lux (Footcandles)	Type of Activity
A	20-30-50** (2-3-5)**	Public areas with dark surroundings
B	50-75-100** (5-7.5-10)**	Simple orientation for short temporary visits
C	100-150-200** (10-15-20)**	Working spaces where visual tasks are only occasionally performed
D	200-300-500‡ (20-30-50)‡	Performance of visual tasks of high contrast or large size: for example, reading printed material, typed originals, handwriting in ink and good xerography, rough bench and machine work, ordinary inspection, rough assembly
E	500-750-1000‡ (50-75-100)‡	Performance of visual tasks of medium contrast or small size: for example, reading medium-pencil handwriting, poorly printed or reproduced material, medium bench and machine work, difficult inspection, medium assembly
F	1000-1500-2000‡ (100-150-200)‡	Performance of visual tasks of low contrast or very small size: for example, reading handwriting in hard pencil on poor quality paper and very poorly reproduced material, highly difficult inspection
G	2000-3000-5000§ (200-300-500)§	Performance of visual tasks of low contrast and very small size over a prolonged period: for example, fine assembly, very difficult inspection, fine bench and machine work
H	5000-7500-10000§ (500-750-1000)§	Performance of very prolonged and exacting visual tasks: for example, the most difficult inspection, extra fine bench and machine work, extra fine assembly
I	10000-15000-20000§ (1000-1500-2000)§	Performance of very special visual tasks of extremely low contrast and small size: for example, surgical procedures.

†Adapted from Guide on Interior Lighting, Table 1.2, Pub CIE No 29 (TC4.1) 1975
*Maintained in service
**General lighting throughout room
‡Illuminance on task
§Illuminance on task, obtained by a combination of general and local (supplementary) lighting

Table 82
Weighting Factors to be Considered in Selecting
Specific Illuminance with the Ranges of Values for
Each Category in Table 81*

Task and worker characteristics	Weight		
	−1	0	+1
Workers ages	Under 40	40—55	Over 55
Speed or accuracy or both	Not important	Important	Critical
Reflectance of task background	Greater than 70%	30 to 70%	Less than 30%

*Weighting factors are to be determined based on worker and task information. When the algebraic sum of the weighting factors is −3 or −2, use the lowest value in the illuminance ranges D through I of Table 81; when −1 to +1, use the middle value; and when +2 or +3, use the highest value.

ESI varies greatly in a room from one location to another. It also varies greatly even at a single point depending on viewing direction in the room. For this reason it is desirable to have some method of predicting in advance what the ESI will be at many locations in a space so as to know the best location for tasks subject to veiling reflections. A number of computer service companies have programs available which can provide this information before a particular lighting system is installed.

The design practice committee of the IES now has an approved method for evaluating the ESI in spaces where task locations are not known in advance of occupancy. [10].

10.5 Light Sources. Electric light sources and daylight have a range of characteristics in terms of efficacy (lumen per watt), color, source size (optical implications), lumen maintenance, starting and restarting attributes, and economics.

Table 83 shows the lumen per watt efficacy (not including ballast watts) for all the major general lighting sources. Lamp efficacy, lumen maintenance, life, and optical control are the major factors affecting lighting economics. Economic comparisons of lamp/luminaire combinations as an important basis for selecting an appropriate lighting system. Computer programs are available from computer service companies and from manufacturers.

10.5.1 Incandescent Lamps. Incandescent lamps have tungsten filaments and lamp efficacies generally ranging between 17 lm/w and 24 lm/w. This is the lowest efficacy of any of the light sources used. However, incandescent lamps, due to good optical control may be energy-efficient when used to light a small target from a distance, as with spotlighting in stores or theatrical lighting.

Incandescent lamps are not recom-

414

Table 83
Approximate Initial Efficacies for
(Lumens Per Watt) Range of Commonly Used Lamps

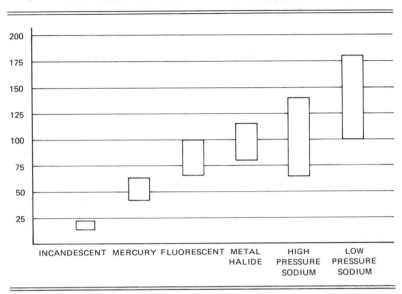

mended for lighting sizable areas having long operating hours. For athletic fields and infrequently accessed storage areas where operating hours are short, incandescent lighting should be considered.

10.5.1.1 Life and Efficacy. Life and efficacy of incandescent lamps are inversely related. This is the only light source for which this is true. The lower the operating temperature of the filament, the lower the rate of tungsten evaporation and the longer the lamp life. However, the lower the filament temperature the lower the lamp efficacy. The factors of life, efficacy, energy cost, and maintenance labor rates have been related in an economic equation to determine optimum lamp life for incandescent lamps. These criteria vary among users, accounting for some of the various life ratings found among incandescent lamps available today.

However, at higher energy costs, the more efficient incandescent lamp should be selected. Some of the extremely long-lived lamps waste considerable energy. For example, it is possible to select the next lower wattage while getting the same amount of light by using standard incandescent lamps. This saves considerable energy and money compared with extremely long-lived incandescent lamps. Difficulty of access and labor cost of lamp repalcement also affect the choice of lamp life. An economic analysis is suggested.

Higher efficiency krypton-filled incandescent lamps are available at wattages slightly below standard values and lives somewhat longer than those of general service incandescent lamps. Their considerably higher initial cost may be paid back with reduced operating cost and longer life.

10.5.1.2 Color. Incandescent lamp color rendition is usually regarded as very good though its spectrum is unbalanced in favor of warm colors and it is comparatively low in the cool colors. A number of colors are available in incandescent lamps by means of filters applied to the lamp bulbs. Many more colors are available using separate color filters as with theatrical and display lighting equipment. These colors are less efficient than the several fluorescent lamp colors available, since considerable light is absorbed in the filter.

10.5.1.3 Tungsten Halogen Lamps. Tungsten halogen lamps are incandescent lamps that make use of the *halogen cycle* to prevent deposits of evaporated tungsten from collecting on the inner bulb surface. Consequently, their lumen output does not drop appreciably during life. Lamp life is about double that of standard service lamps.

10.5.1.4 Dimming. Incandescent lamps can be dimmed simply by reducing the voltage at the lamp socket. Variable auto transformers and solid-state devices are most often used for this effect. Dimming may be desirable for certain special effects where more than one level of illumination is necessary. However, it should be remembered that light output drops much more rapidly than wattage as incandescent lamps are dimmed and lamp efficacy is greatly reduced. Further information on dimming is included in 10.9.3.

10.5.2 Fluorescent Lamp Characteristics. Fluorescent lamps are electric arc discharge sources which depend on a two-step process for generating light. The electric arc discharge through low-pressure mercury vapor generates ultraviolet radiation which, in turn, excites phosphors deposited on the bulb wall of the lamp to generate visible light. The phosphor is vitally important, determining the efficacy, color, and lumen maintenance of the light produced. Phosphor composition also affects the lamp cost.

Fluorescent lamps generate light more efficiently than incandescent or mercury lamps though there is great variation in fluorescent lamp efficacy, depending on color and wattage. Fluorescent lamps also have characteristics of low brightness and diffusion, making them excellent for many lighting applications where high brightness could cause specular reflections in tasks, and where luminaire discomfort glare should be controlled.

10.5.2.1 Types of Fluorescent Lamps. Fluorescent lamps have tubular bulbs and are made in a variety of lengths for a number of operating currents. For example, 430 mA is the typical operating current of rapid start and slimline fluorescent lamps. Ballasts can be selected, however, which will operate these lamps at 200 mA or 300 mA for reduced wattage and similarly reduced light output.

High-output fluorescent lamps are operated at 800 mA and these have about 45% higher light output per unit of length than the 430 mA lamps. There are also extra-high-output fluorescent lamps operated at 1500 mA which generate 60% to 70% more light per unit length than even high-output lamps. The more highly loaded lamps have applications where higher levels of illumination are needed or where higher mounting heights are involved and the number of lamps or fixtures, or both, can be reduced for economic reasons.

10.5.2.2 Reduced-Wattage Fluorescent Lamps. Fluorescent lamps of specific lengths and tube diameters have individual electrical characteristics to which their ballasts must be designed.

416

Consequently, substitutions of different lamp types in sockets designed for a particular lamp can rarely be done satisfactorily, even if the lengths allow a physical fit. However, in recent years due to rising energy costs, lamp manufacturers have made reduced-wattage fluorescent lamps which fit in the existing sockets of particular lamps, and that will operate satisfactorily on the existing ballasts. These lamps reduce the lamp/ballast system power consumption by 10% to 20%, depending on the lamp type and luminaire type. They are available for the 3 ft and 4 ft rapid-start lamps, and for slimline, high-output, and extra-high-output types.

The first-generation of these reduced-wattage lamps reduced light output in about the same proportion as wattage. If less light were acceptable, this was satisfactory. However, if lighting maintenance procedures were improved, such as cleaning luminaires every year or two and replacing all lamps in groups every three or four years, the average lighting level maintained might be equal to, or even greater than, that maintained with the standard, higher-wattage lamps.

A second generation of reduced-wattage fluorescent lamps has now been developed, employing new, more efficient fluorescent phosphors. This has made it possible to provide the same reduction in wattage achieved with the original lamps, yet provide virtually as much light as with standard lamps, the quantity depending on luminaire type. Though the choice of lamp color is limited, this second-generation lamp provides light at lowest cost and least energy use in new lighting installations as well as being suited for retrofit in existing systems where energy reductions are desired but reduced light is not.

All reduced-wattage fluorescent lamps are more sensitive to temperature than standard lamps. Consequently, they are not recommended for use in areas where ambient temperature will be less than 60 °F. (Standard lamps are satisfactory down to 50 °F.) In addition, these lamps are not recommended for use on dimming systems or for use in emergency lighting units. Standard lamps should be used for these applications.

10.5.2.3 Fluorescent Lamp Color. A great variety of fluorescent lamp colors can be obtained by simply changing phosphor components or their relative amounts. Specification of lamp color involves, principally, matters of efficiency and aesthetics.

Much of the early history of fluorescent lamp colors involved trade-offs in lamp efficacy with color rendition. Lamps high in lumens-per-watt efficacy were only fair in color rendition, while those which provided excellent color rendition were substantially lower in lumens-per-watt efficacy.

Currently, however, there are lamp colors available which combine high color rendition with high efficacy, due to improved phosphor technology. Such lamps are considerably higher in price than the high color rendering lamps of lower efficacy, but analyses show they are cost-effective.

For applications of fluorescent lighting in some types of stores, restaurants, and homes, and where good appearance of people, food, merchandise or furnishings is essential, high color rendering fluorescent lamps should be considered.

Several saturated colors such as pink, blue, red, gold, and green are available in fluorescent lamps. These are obtained through the use of fluorescent phosphors in the lamp which generate the color of light desired. In a few cases, a colored filter is employed integrally with the glass tube to increase the color satuation.

10.5.2.4 Fluorescent Lamps and Temperature. The starting and operating characteristics of fluorescent lamps are significantly affected by temperature. See 10.13 for discussion of this subject.

10.5.2.5 Dimming. Equipment for dimming is now available for 30 W and 40 W rapid-start fluorescent lamps. This broadens their possible applications to auditoriums, restaurants, ballrooms, churches, studios, and the theater. It also provides an opportunity to tie in electric lighting systems with daylighting to maintain constant levels of task lighting indoors as daylight varies in quantity.

Various electronic systems have been devised for dimming fluorescent systems. Manufacturers should be consulted for information on characteristics of their equipment such as dimming range, starting reliability at various brightness levels, cost, etc.

Special ballast designs also make it possible to flash high-output fluorescent lamps with good life performance. The ballasts provide somewhat greater cathode heating during operation than with conventional lamp operation. A lead wire from this ballast to the sign contactor provides lamp control during ON-OFF periods. Principal applications are for signs and attention-getting displays where the flashing uses less energy than continuous burning. Some flashing ballasts are designed to permit satisfactory outdoor operation during extremes in weather.

10.5.3 High-Intensity Discharge (HID) Lamps. The mercury, metal halide, and high-pressure sodium lamps make up this family. These are electric arc discharge lamps, requiring ballasts, but distinguished from fluorescent lamps, in general, by higher pressure (a little more than one atmosphere) in their arc tubes and a more intense, shorter discharge path.

10.5.3.1 Mercury Lamps. Mercury lamps have been widely applied for indoor and outdoor applications for many years. However, since the advent of the substantially more efficient metal-halide and high-pressure sodium lamps they are now almost obsolete. Applications for mercury lamps which would presently be justified would lie in the lower wattages, 175 W and lower, where the more efficient metal-halide lamps have no equivalent. For example, in certain stores with lower ceilings and relatively low requirements for illumination level, mercury lamps can be an appropriate choice.

Mercury lamps have very long lives, usually in excess of 24 000 h. However, they are also characterized by poor lumen maintenance, especially on constant wattage (CW) and constant wattage autotransformer (CWA) ballasts. Practical economic life is generally more of the order of 12 000 h to 16 000 h. Used for longer hours they waste much energy with their greatly reduced light output.

The phosphors used in mercury lamps provide reasonable good color rendition. Most of the earlier lamp types with poorer color have been discontinued.

Self-ballasted mercury lamps are characterized by extremely long lives and mean efficacy about the same as incandescent lamps over their rated lives. They use an incandescent tungsten filament in series with the mercury arc tube to replace the magnetic ballast during start up and steady state operation. The filament consumes about 60% of the total lamp power during operation, substantially reducing the overall efficacy of this source. Further, the extremely long life of the lamp works to the disadvantage of the lighting system efficiency due to the fairly rapid degradation of light output from the mercury arc tube. At the end of life, efficacy is lower than that of typical incandescent lamps. Their principal virtue is long life and reduced labor cost for lamp replacement. However, conventional mercury lamps with their separate ballasts would be far more cost-effective.

10.5.3.2 Metal-Halide Lamps. Metal halide-lamps are substantially more efficient than mercury lamps and should be employed where a HID lamp is appropriate and color is important. These lamps are made with clear outer bulbs and other phosphor-coated bulbs.

The phosphor-coated lamp's color rendering is superior to that of the clear lamp, though that of the clear lamp is good. The clear lamp will provide better optical control in a lighting fixture than a phosphor-coated bulb. Therefore, the clear lamp is preferred for flood-lighting equipment where the precise beam control is desired, projecting light over long distances, restricting light to a specific target area with minimum spill, etc.

Electrically, two types of metal-halide lamps are available. One type requires a ballast designed specifically for metal-halide lamps. The other type, available in 325 W, 400 W, and 1000 W ratings, is electrically interchangeable with mercury lamps on most of the commonly used mercury ballasts (about 80% of existing types). The 325 W lamp works on the same 400 W ballasts as the 400 W interchangeable lamp. This makes it possible to upgrade existing mercury lighting systems in retrofit applications simply by changing lamps in existing fixtures without increasing the connected load or energy consumption. In the case of the 325 W lamp, energy use can be reduced and more light provided. If lighting is considered adequate before retrofitting, it may be possible to eliminate some fixtures.

The technology of metal-halide lamps is still evolving rapidly and there appear to be significant opportunities for developing lamps with performance much improved over that currently achieved. At present, however, there are certain limits with some of the lamps as to burning position, or significant differences in lamp performance in one position compared with another. Certain lamps may also have requirements for operation in enclosed luminaires only. Users are advised to consult published manufacturers' data for current information on lamp operating conditions.

10.5.3.2.1 Mercury and Metal-Halide Self-Extinguishing Lamps. Both mercury and metal-halide lamps produce considerable ultra-violet energy in their arc discharge. Some of this energy is very useful with phosphor-coated outer bulbs, reacting with the phosphor to generate light which improves the color rendition of the total light from the lamp. None of this type of ultra-violet gets out of the lamp as the glass outer bulb will not transmit it.

However, in rare instances, the outer bulbs of mercury lamps have been broken during service and the arc tube continues to operate. If people are in the area for an extended period of time, they can experience a temporary reddening of the skin or irritation of the eyes due to the erythemal action of the ultra-violet coming directly from the arc tube.

Consequently, lamp manufacturers are now making a line of lamps having a *disconnecting* feature that will deactivate a lamp within a short time after the outer bulb has been broken. These are available for new or existing installations employing open fixtures where the lamps could be broken. Use of enclosed fixtures which do not permit foreign objects to break the bulbs allow standard lamps to be used.

10.5.3.2.2 Starting Characteristics —Mercury and Metal Halide. Both mercury and metal-halide lamps require five to eight minutes of starting time (warm-up) before they reach full light output. This is because the vapor pressure of the light-generating arc tube gases is quite low at the start and it takes several minutes for the elements to vaporize into the arc stream.

If mercury or metal-halide lamps experience a momentary power interruption during normal operation or a low enough voltage dip, the arc tube will be extinguished and a 5 to 15 min cooling period of *cool down* will be required, because with the lamp off, but the arc tube hot, the vapor pressure in the arc tube is too high for the available voltage from the ballast to restart the lamp. As the lamp cools, the vapor pressure drops and the voltage required to start the arc decreases and, eventually, the available ballast voltage is sufficient. Higher ballast voltage would allow faster restart, but would also increase ballast cost. An incandescent lamp can be furnished in many HID luminaires to provide light during warm-up and for emergency purposes.

Higher voltage ballasts are required for lamps for starting below 50 °F. However, light output is not affected by ambient temperature.

10.5.3.3 High-Pressure Sodium (HPS) Lamps. These lamps are characterized by a relatively high-pressure electric arc discharge (slightly over one atmosphere) in a special ceramic arc tube containing a small amount of sodium in an amalgam form. When the lamp is first started, there is very low pressure in the arc tube and the sodium generates its characteristic monochromatic color.

However, at operating pressure, the spectral output broadens, and all visible light wave-lengths are present, though in different proportions compared with other familiar sources.

Lumen-per-watt efficacy is quite high and since source size is quite small, control of light distribution is good.

High-pressure sodium lamps have three characteristics quite different from other high-intensity discharge lamps: fast warm-up, fast restrike, and much better lumen maintenance. Warm-up to full light output generally occurs within 2 min, and restrike within 1 min. Ballasts for the HPS lamps have a high-voltage, low-current starting circuit generating a pulse of about 2500 V. This provides fast restrike time, usually within less than 1 min after a power interruption. Ballasts are available (at higher cost) which will provide instant restrike for certain wattage HPS lamps should that be necessary. High-pressure sodium lamps provide mean lumen maintenance near 90% over their approximately 24 000 h rated life, substantially better than metal-halide and mercury lamps.

A high-pressure sodium lamp has also been developed providing better color than the standard lamps. This is accomplished by changes in the electrical characteristics of the lamp's arc tube. The improved color involves some sacrifice in lumen-per-watt efficacy and lamp life.

10.5.4 Low-Pressure Sodium (lps) Lamps. These lamps have the highest lumen-per-watt efficacy. Where color recognition is important, this type of lamp may not be acceptable. Most of the spectral energy output is concentrated between 589 nm and 590 nm, and, hence, the light has a highly monochromatic color. The arc discharge takes place in a tube containing vaporized sodium in the free state. Low-pressure sodium lamps are more like fluorescent lamps in physical size than high-intensity discharge lamps.

Low-pressure sodium lamps available have two different characteristics. One has the typical light depreciation wattage relationships of most lamps with light dropping off as operating hours increase. The other type has a wattage that may vary with hours of operation. Any wattage change will affect light output and also the capacity of the power distribution system. Suppliers should be contacted for data on watts throughout life and light output.

If lamp color is not a factor in application, economic comparisons should be made between low-pressure sodium and other lamps such as high-pressure sodium. Such comparisons will involve assumptions for utilization of light with typical luminaires in order to determine how effectively each lamp/fixture combination delivers light to the task(s) and their relative (or absolute) costs. Computer programs are available for such comparisons.

Care should be taken in disposing of these lamps as the free sodium in contact with water can create a fire hazard. For this reason, their use is not permitted in and around coal mines. The manufacturers' specific instructions should be followed in disposing of burned-out lamps.

10.6 Ballasts. All discharge lamps must have ballasts to perform several functions. These include:

(1) To provide the appropriate voltage to start the lamp

(2) To maintain the appropriate voltage and current to the lamp during operation

(3) Usually, provide power factor correction.

Ballasts consume from 3% to 25% of a lighting system's energy. They also have an effect on the life, light output, and lumen maintenance of the lamps in the system. Hence, specification of an appropriate ballast is highly important to satisfactory performance, energy conservation, and economy of a lighting system. Since there are many different types of ballasts, having a variety of characteristics, it is recommended that manufacturers' literature be consulted for specific details and operating data.

10.6.1 Fluorescent Lamp Ballasts. The rapid start ballast is the predominant type in use today. This ballast provides a low-voltage source of heat for the fluorescent lamp cathodes and this allows the lamp(s) to start within 1 or 2 s after voltage is applied.

Rapid-start lamps are available to operate at 430 mA, 800 mA, (high output) and 1500 mA, (extra high output). The most popular and economical type of ballast available operates two rapid-start lamps in series. The power factor is corrected to be in excess of 90%, leading. This slightly leading power factor can help improve the system power factor for a building, since other loads have usually lagging power factor.

10.6.1.1 Grounding. The National Electric Code ANSI/NFPA 70-1981 [13] requires that all fixtures and lighting equipment (including ballasts) must be grounded. Rapid-start ballasts require a starting aid consisting of a grounded metal strip running the full length of the lamp. The metal of the fluorescent fixture housing when grounded normally acts as a starting aid.

10.6.1.2 Slimline Lamp. Slimline lamps are a lamp type that start instantly due to the high open circuit voltage available from the ballast when the switch is closed. They may be operated at any of several currents (for example, 200 mA, 300 mA, 425 mA,) by selection of the appropriate ballast. The most popular and economical ballast available for slimline lamps is the two-lamp series type. However, a lead-lag ballast is also available which operates a pair of lamps in parallel, one at leading and one at lagging power factor so the net result is a high-power factor circuit. The lead-lag ballast is more costly than the series type.

10.6.1.3 Low-Loss Ballasts. Within recent years, manufacturers have developed flourescent lamp ballasts that reduce ballast losses by almost half with respect to conventional ballasts. These ballasts run cooler due to the lower watt loss and last considerably longer. Though somewhat higher in cost than conventional ballasts, they are cost-effective and should be considered in all new lighting and as replacements for ballast failures in existing installations. Energy-conserving ballasts are available both in magnetic core and coil types, and electronic types. The latter operate the lamps at higher frequency, further increasing the system efficacy.

10.6.1.4 Voltage. Ballasts are available for the standard distribution system voltages, and should be operated at no more than 5% higher or 10% lower than the rating. Higher voltage will overheat ballasts and shorten life. Lower voltage will reduce lamp life and lamps may fail to start.

10.6.1.5 Temperature. Extremes of hot or cold can be damaging to ballast life and performance. Ambient temperature in the areas where ballasts are installed will affect the ballast operating temperature. Fixture design will also have an effect. The ballast case hot spot should not exceed 90 °C during operation.

Thermally protected ballasts (Class P) will disconnect themselves from the circuit if their case temperature exceeds 90 °C, or cause the lights to cycle OFF and ON if the factor causing the excessive temperature persists. Ballasts without thermal cutouts will have their lives shortened with operation above 90 °C.

Most ballasts designed for indoor operation of fluorescent lamps provide voltage for satisfactory starting of the lamps at 50 °F (60 °F for the reduced wattage fluorescent lamps). Low-temperature ballasts are available which can provide higher voltage to start lamps in ambients as low as –20 °F.

10.6.1.6 Lamp Burnouts. Ballasts may overheat with lamps flickering near the end of life or when one of a pair of lamps is removed from the lamp holder. Flickering or burned-out lamps should be replaced promptly to prevent possible ballast damage.

10.6.1.7 Lamp Removal. When lamps are removed from fixtures which remain energized, a small amount of energy is consumed by the ballast at very low power factor (except for slimline fixtures having circuit interrupting lamp holders). This is due to the magnetizing current flowing through the ballast primary. A series ballast for a pair of 4 ft rapid-start lamps will consume 6.5 W with the lamps removed. If the fixture will not be relamped within a short period of time, it may be disconnected from its power source by qualified personnel to eliminate this loss.

10.6.1.8 Fusing. It is desirable to use an in-line fuseholder and time-delay fuse with each ballast. This will prevent an entire area from being blacked out due to a failure of one ballast. It also provides a safe means of replacing ballasts without opening the branch circuit breakers.

10.6.1.9 Switching. Circuit breakers

used for frequent switching of fluorescent lamps should be UL listed SWD for this duty.

10.6.1.10 Radio Interference. Radio interference from fluorescent lighting systems may be minimized by the use of appropriate lenses on the luminaire and by the installation of available filters in the line feeding the ballasts.

10.6.2 High-Intensity Discharge (HID) Lamp Ballasts. Most of the previous comments on ballasts for fluorescent lamps also apply to ballasts for high-intensity discharge lamps. Their functions are the same and since most are electromagnetic devices, their characteristics are similar.

Ballasts for high-pressure sodium lamps differ from other high-intensity discharge lamp ballasts in that they have a high-voltage pulse to aid in starting the lamp. This pulse also aids in restarting hot lamps within about 1 min if a momentary voltage dip or power interruption occurs which extinguishes the lamp.

With older high-pressure sodium ballasts, it is necesary to change burned-out lamps promptly, and not keep the ballast energized without a good lamp for an appreciable period of time (several days) without taking the fixture off the line, or the pulsed starting aid would be damaged. At present, some ballasts do not have such limitations.

10.6.2.1 Grounding. High-intensity discharge (HID) ballasts must all be grounded in a manner complying with ANSI/NFPA 70-1981 [13] or local codes where appropriate.

10.6.2.2 Fusing. It may be desirable to use a line fuse with high-intensity discharge ballasts. This will prevent branch circuit breakers from opening in case there is a defective ballast on the line.

10.6.2.3 Radio Interference. A small amount of interference may be detected during lamp starting. There should be no objectional interference during operation.

10.6.2.4 Voltage. Lag- and reactor-type ballasts should have a supply voltage within ±5% of the design voltage. For constant wattage autotransformer (CWA) ballasts, the voltage should be within ±10% of the design voltage.

High-intensity discharge (HID) ballasts are available for a wide variety of utilization voltages.

10.7 Luminaires. Many factors should be considered in selecting luminaires for a space. Some of the principal ones are listed below. Some of these may receive more emphasis than others for a specific installation, but all exert some influence.

10.7.1 General Considerations Affecting Selection of Lighting Systems

(1) Architectural character of the space to be lighted:

    (a) Size and proportions

    (b) Layout of furnishings

    (c) Structural and mechanical features

(2) Designer's concept of how space should appear:

    (a) Lighting patterns which emphasize structure or layout, or are design elements of themselves

    (b) Unobtrusive lighting patterns

(3) Styling of luminaires:

    (a) Simple

    (b) Decorative*

NOTE: Decorative luminaires are often inappropriate for providing illumination. Hence they may be installed primarily for their decorative effect where the general lighting is provided by another system, such as cove or downlights.

(4) Suitability for specific visual tasks or activities (office, store, warehouse, factory):

(a) Light distribution of luminaires

(i) Direct, indirect, and intermediate types

(ii) Diffusion or directional qualities

(iii) Creation of shadows

(iv) Veiling reflections

(v) Uniformity of illumination

(b) Visual comfort

(i) Use of appropriate shielding and diffusing media

(ii) Opaque or luminous-sided luminaires

(iii) Orientation of viewing

(iv) Visual comfort probability

(5) Efficiency:

(a) Utilization of direct, indirect, and intermediate types

(b) Power requirements

(6) Flexibility

(a) Movable office furniture with task lighting fixtures

(b) Movable shelf mounted fixtures

(c) Movable free standing indirect fixtures

(d) Movable plug-in recessed troffers

(7) Maintenance:

(a) Susceptibility to dirt collection

(b) Ease of cleaning

(c) Ease of relamping

(d) Durability

(e) Characteristics of plastics, paints, and metals used

(8) Coordination with mechanical system

(a) Luminaires for air supply

(b) Luminaires for air return and their effect on air changes fan horsepower, etc

(c) Lighting contribution to building heating

(d) Heat redistribution systems

(e) Heat storage/recovery systems

10.7.2 Special Lighting Distributions. Light control techniques are available for luminaires with certain distributions intended to minimize the effects of veiling reflections in visual tasks. These include the use of polarizing materials, batwing and radial batwing distributions, and indirect lighting. It should be pointed out that the geometrical relationship between eye, task, and light source location is the biggest factor, by far, in reducing veiling reflections. If luminaires are not present in the *offending zone* (mirror angle for the eye with respect to task), veiling reflections are minimized. It may not be possible to have this geometry, however, expecially in rooms with multiple occupancy. Hence, the special lighting materials may be helpful.

Comparison of various lighting materials using ESI criteria (see 10.4.1) can be helpful in assessing the relative effectiveness of these materials in creating task visibility in particular lighting situations.

10.7.3 Lighting and Other Building Subsystems. Integration of lighting with other environmental features such as air conditioning, space partitioning, fire protection, and acoustical control is receiving increased attention. There is much opportunity in this area for advancing system design, making lighting an integral part of the building structure and efficiently coordinating all the other control features which are needed in the modern commercial environment. Figure 118 shows a ceiling system that is coordinated with the space module and provides for lighting, air supply, air return, space partitioning, and acoustical treatment.

10.7.4 Visual Comfort Probability. Glare evaulation data have been developed through an empirically-derived formula which assesses all the factors in a room contributing to glare and con-

Fig 118
Ceiling System Providing Illumination, Air Supply, Air Return,
Space Partitioning, and Acoustical Treatment

(The ceiling configuration also improves visual comfort by
shielding most of the luminaires from view.)

firmed by the testing of many people to establish their brightness tolerances on a statistical base. This system is called Visual Comfort Probability (VCP) and has been discussed in 10.3.1. Visual Comfort Probability (VCP) tables for specific luminaires are available from luminaire manufacturers.

10.7.5 Luminaires and Air Conditioning. Lighting imposes a load on cooling systems in the summer and contributes useful building heat in the winter. Certain types of luminaires have characteristics that may be useful in mechanical systems. For example, some luminaires are designed to supply conditioned air to spaces. Others may be used as air returns, while some can do both simultaneously. Neither supply nor return air is essential to utilizing lighting for heating in buildings.

Luminaires designed to supply air have the air supply path separate from the lamp compartment. Since the air temperature ranges from cold to hot during the seasons, this is necessary to ensure good performance of fluorescent lamps which are temperature sensitive. One advantage of air supply luminaires is that they allow for a cleaner, simpler ceiling appearance without an obvious pattern of air diffusers. In addition, they may allow for greater flexibility in modular design by being able to supply air (and return it) within any desired module. Air supply luminaires are appropriate with uniform lighting layouts. They may not be appropriate for non-uniform layouts where luminaires are expected to be relocated during a building's life-cycle.

Air return luminaires provide a path for air to be returned from occupied spaces to the mechanical equipment room by way of the luminaire lamp compartment (for maximum heat transfer) and the ceiling cavity. An alternative is to by-pass the luminaire and return air directly to the ceiling cavity. In the latter case, the following benefits apply, but to a lesser degree than air return through the lamp compartment.

(1) Reduced heat gain in occupied space

(2) Reduced requirement for air exchanges in space due to (1)

(3) Reduced duct size and fan horsepower due to (1)

(4) Reduced luminaire operating temperature and heat radiation thereby improving thermal comfort with higher room air temperature in summer

(5) Reduced fluorescent lamp operating temperature and increased light output by about 10%

(6) Reduced ballast operating temperature and longer life for older type, conventional ballasts. (There is little effect on life of low-watt-loss ballasts which run much cooler than the older type.)

Local codes must be consulted to determine if air-return luminaires or the by-pass technique with air moving through a ceiling cavity may be employed.

10.8 Lighting Application Techniques. The purpose of commercial building lighting should be defined before further describing lighting application techniques.

There are several different purposes of illuminance. The purpose and functional requirements of a lighting system should be clearly defined for each area of a building or its surroundings. Only after the needs are adequately defined can the illuminance system be intelligently planned and executed. Every legitimate need of lighting must be considered. Illuminance systems are purchased, installed, and operated to accomplish specific functions. Lighting is provided for its functional ability to provide for seeing, mood, direction, color, aesthetics, and other considerations of an occupied space.

(1) *Seeing*. It is necessary for objects to be visible in order to perform tasks. One cannot perform a drafting function unless there is appropriate light for the task. Tasks vary quite widely in their illuminance requirements, with respect to both the quantity and quality of the light. Quantity can be calculated and measured easily. Quality is somewhat more difficult to evaluate, though there is much practical guidance in the form of recommended reflectances and brightness ratios for major room and work surfaces. The advent of systems such as Visual Comfort Probability (VCP), Contrast Rendition Factor (CRF) and Equivalent Sphere Illuminance (ESI) have provided further insights into the ability of lighting systems to provide *seeing* and not just quantity of light. These are also useful tools for comparing specific luminaires and lighting systems of interest.

(2) *Mood*. Illuminance may be used to provide a feeling of warmth, comfort, invitation, efficiency, excitement, or urgency. Many lighting installations should be evaluated in terms of the mood to be conveyed. Obvious violations come to mind: the use of cold, glaring fluorescent fixtures in a tavern; or warm, relaxing lighting for a bank teller work area. Generally, it is desirable to provide a business-like environment in the interest of efficiency in offices, schools, and for other daytime activities. A warm, relaxing atmosphere is frequently desired in specialty shops and for evening activities, such as dining or the theater. The warmer colors can help here.

(3) *Direction and Information*. Lighting can be used to give direction. As people pass through areas, they may be subtly, unconsciously drawn to the brighter area, unless the brighter area conveys a feeling of discomfort or danger, such as a glaring spot or floodlight. Exit lights, advertising, and directional signs are examples of lighting intended to convey information. In some cases different color lighting can be used to delineate pathways in large, open-office plans.

(4) *Aesthetic Lighting*. Aesthetic lighting is designated to make objects and people look pleasing to the eye. This is perhaps the most difficult to define and achieve. The lighting designer or architect must be acutely aware of the aesthetic result desired, in order to specify the best lighting systems. Architectural lighting frequently serves other needs, such as security, seeing, identification and mood changes.

There are several lighting techniques that may be evaluated for specific applications. See Table 84. These include:

1. Uniform
2. Non-uniform
3. Task-ambient.

The first two are usually variations of ceiling-installed lighting. Task-ambient is an important sub-classification of non-uniform lighting in which some of the light sources may be integrated into, or mounted on, the furniture; others may be free-standing or movable, or mounted on the ceiling. These system classifications are defined, and their advantages, disadvantages, preferred and non-desirable uses, cautions, and other application techniques are discussed below.

10.8.1 Uniform Lighting. By definition, uniform lighting illuminates equally spaces and areas on and around the immediate work or task area. The use of uniform lighting has been criticised due to the potential for wasted energy from

Table 84
Relative Merit of Uniform, Non-Uniform and Task Ambient Lighting Systems

	Uniform	Non-Uniform	Task Ambient
Description	Even illuminance through-out the area at task lighting level	Most light on the task, with general and non-critical levels reduced	Direct lighting for task up close with ambient lighting from adjacent indirect or ceiling-mounted luminaires
Advantages	Where tasks not defined Where task areas not known Low levels of illuminance Good eye adaptation Uniform appearance	Energy efficient, lower initial cost Provides space interest, emphasizes work areas	Energy efficient Tax benefits Easily movable
Disadvantages	May be least energy efficient Monotonous May have higher initial cost	Must know tasks, task locations Must be moved as task changes	Veiling reflections Space confining Expensive fixtures Wiring and switching may be more difficult
Typical uses	Large homogenous areas Libraries, drafting rooms Clerical offices, supermarkets, cafeterias Gymnasiums, sportsfields Hallways and corridors	Large non-homogenous areas Smaller areas, private offices Low employee density areas Restrooms CRT-microfilm viewing areas Hallways and corridors	Small work areas Furniture systems Open office plans

lighting both task and non-task areas uniformly. Uniform lighting is frequently applied to areas where the task or the task areas is not defined. Typical of these is the 500 lx or 700 lx uniformly applied to speculative office space. Actual tenant use of the space should dictate an area-by-area appraisal of the lighting system in view of its intended use.

The principal application for uniform lighting is for areas where the activity taking place occurs uniformly and continuously throughout the space and where task locations are quite close to-gether, such as in classrooms, or densely-occupied office space (see Fig 119). It should not be installed as a substitute for proper planning when not required. Fixtures may be kept on site but not installed until the specific locations of work stations is known. An alternative approach, considering the 50 to 60 year life cycle of a building during which time tasks may be performed anywhere in the space, is to install luminaires capable of supplying uniform illumination, but with switching controls which would allow a non-uniform lighting result in the space (see Fig 120).

Fig 119
Near-Horizontal Drafting Boards Positioned So Draftsmens'
Views are Parallel with the Luminaires

(If boards are placed between rows of troffers, veiling reflections will be minimized. The near-vertical boards are
positioned at right angles to the rest, but veiling reflections
(and shadows) are minimal for such boards, regardless of
orientation.)

Fig 120
Two U-Shaped Fluorescent Lamps with an Efficient,
Low-Brightness Anodized Aluminum Luminaire, are Employed
in Each Unit to Supply Lighting from the Ceiling

(Flexibility in switching could allow non-uniform lighting distributions in the space, but the owner has the capability of providing task lighting anywhere in the space during the building's life-cycle.)

Disadvantage of uniform illuminance are:

(1) Relatively high energy consumption with the whole space lighted to the same value

(2) A monotonous appearance

(3) Minimum visual stimulus in the area. On the other hand, an even illuminance tends to make small areas look and feel larger.

Typical spaces where uniform illuminance can be used to best advantage include:

(1) Densely-occupied office space

(2) Data-processing centers

(3) Classrooms

(4) Gymnasiums

(5) Mass merchandising stores

(6) Sports fields

In order to promote energy efficiency in uniform lighting installations, consideration should be given to multi-level switching using two-level ballasts, switching one of a pair of ballasts in luminaires, switching of small areas of luminaires, and switching to lower lighting levels near fenestration which can be utilized as a light source during daylight hours.

Coefficients of utilization values published by luminaire manufacturers are used to calculate average illumination levels for uniform lighting. Actual illumination values in a real space will be higher than average in the center of the space and lower near the walls. In small rooms illumination may be 30% higher than average in the center, varying to near average in very large rooms. Consequently, uniform illumination can be reduced if tasks are located near the center of small- and medium-sized rooms. Conversely, work locations near walls should be avoided unless task lighting is provided.

10.8.2 Non-Uniform Lighting. Non-uniform lighting in task areas is achieved by putting more illuminance on the task and less in non-critical and general areas near the task. This concept has become popular, since it has the potential for greater energy efficiency.

Non-uniform illuminance should be avoided in areas where the nonuniformity might cause confusion or misdirection, and thus be a hazard to safety. With reasonable care and design skill, non-uniform illuminance may be applied successfully to most lighting situations. (See Fig 121.) It is usually not feasible for areas such as classrooms, gymnasiums, and spaces which may be used for a variety of tasks at the same or different times. Wall, ceiling, floor, and equipment reflectances are more critical with non-uniform lighting in order to minimize large changes in contrast. Dark walls may impart a feeling of being in a cave, and may cause adaptation problems. The light reflected from walls (luminance) should not present a ratio in excess of 1 to $\frac{1}{5}$ with the visual task as the worker glances up from his task. This requirement can necessitate a wall illuminance equal to that of the task with dark room finishes. Consider a wall reflectance of 20%, task reflectance approaching 100%; then the 5 to 1 illuminance ratio is met with the same illuminance on wall and task. It is therefore important to provide reasonably high wall, floor, and ceiling reflectances. Ideally, the values should not be less than those specified in Table 80. Note that the key work here is *reflected*; the light the eye sees is that which is reflected, and not the incident illuminance. The lighter finishes provide acceptable luminance ratios with non-uniform lighting. This method of energy reduction is far superior to arbitrary methods of illuminance reduction, lamp removal, or other methods which may

Fig 121
2 ft by 2 ft Luminaires Containing U-Shaped Fluorescent Lamps—
Luminaires Mounted on Appropriate Modules of the
"Waffle-Patterned" Ceiling Structure Supply Non-Uniform Lighting

(Units are mounted closer together over work stations, and farther apart in circulation areas.)

detract from the illuminance performance criteria.

Non-uniform illuminance exhibits these advantages:

(1) Energy-efficiency

(2) Low initial cost

(3) No sacrifice in lighting quality with careful environmental design

Application of non-uniform lighting is appropriate where the task and task area are well defined. (See Fig 122.) This assumes a knowledge of what the task is and where it takes place. Greater time, effort, and design capability must be expended to provide an adequate system. System flexibility is required to provide for unforeseen and future contingencies during the life-cycle of a building.

Disadvantages of non-uniform lighting include:

(1) Expenditure of more engineering time spent in design

(2) Need to define task areas and tasks before actual occupancy in order to provide adequate information during design

(3) Potential for confusion in large areas, although the change in patterns can be used to good effect for mood, direction, and information

Spaces where non-uniform illuminance can be used to advantage include:

Offices

Small or private offices

Executive offices

Special purpose areas such as:

Microfilm viewing

Reception areas

General offices where employee density is not great

Restrooms

Schools

Offices

Sewing classes

Cafeteria serving lines

Library checkout counters

CRT and microfilm viewing

Restrooms

Merchandising

Checkout counters

High-profit merchandise areas

Sale merchandise areas

Offices

Carpet sample areas

Advertising

10.8.3 Task/Ambient Lighting. Task/ambient lighting is a particular form of non-uniform illuminance, that is, a combination of task illuminance and ambient (general) illuminance. One potential advantage of task/ambient illuminance is improved energy effectiveness, as in the non-uniform system previously described. (See Fig 123.) There are other advantages.

10.8.3.1 Advantages. Some forms of task/ambient lighting provide illumination that can be readily moved as the task location moves. This has the following advantages:

(1) More light where it is needed, less in other areas

(2) Potential for energy reduction, fewer fixtures and individual control of fixtures

(3) Potential reduced cost of lighting system

(4) Possible tax advantage where lighting is classified as office furniture having accelerated depreciation

(5) Potential for reduced veiling reflections (higher ESI) where geometry between eye, task, and light source is optimized

(6) Ease of cleaning and relamping

(7) Uncluttered ceiling

10.8.3.2 Disadvantages. The use of task/ambient lighting has some disadvantages:

(1) Need for receptacles at all task lighting locations

(2) Care to ensure the contrast ratios

433

Fig 122
Secretarial Area with Uniform Lighting that has Non-Uniform Possibilities.

(Luminaires closest to desks are operated at full light output, while those over file areas are dimmed or switched to half-level. File cabinets, walls, and floors have high reflectance finishes to keep brightnesses balanced with work areas.)

Fig 123
Open-Plan Office Lighted with Task/Ambient Lighting

(Energy efficiency is provided by close-up task lighting from fluorescent luminaires positioned to supply light from directions that minimize veiling reflections. The ambient lighting is distributed uniformly over task and surrounding areas by indirect units accommodating high pressure sodium lamps.)

Fig 124
Task/Ambient Lighting in a Work Station—Two-Thirds of the Light
on the Task is Supplied by Fluorescent Desk-Mounted Luminaires
Located at Each Side of the Desk—This Arrangement Avoids Vieling Reflections

(About one-third of the light supplied on the task (and the total light in surrounding areas) comes from indirect partition-mounted luminaires for high pressure sodium lamps.)

for visual comfort are not exceeded. (This may require wall washing fixtures)

(3) Need for higher ceiling heights or proper indirect luminaire optics to avoid light "puddles" on ceiling. This can be glaring and reduce task visibility

(4) If the *ambient* portion is provided by indirect lighting, the luminaire faces upward and consequently requires more frequent cleaning.

(5) Task/ambient fixtures may be purchased by persons who have the training to select and use them correctly. Installation may be by persons who do or do not understand them.

The *task* component of task/ambient lighting may take two forms: (1) furniture-mounted lighting built into the work station (see Fig 124), or (2) floor-mounted fixtures that can be placed adjacent to a desk. Some fixtures provide both direct task lighting and indirect ambient lighting (see Fig 125). The direct contribution puts more light on the task than in the surrounding area. It is vital that the direct lighting luminaire be positioned so that it will not produce veiling reflections in the task that will reduce visibility. The indirect contribution provides general lighting thereby helping to reduce the contrast between the task and the surrounding areas. Other types of luminaires are also available which supply direct task lighting alone. Their position with respect to workers' eyes and the task is critical to avoid veiling reflections.

The *ambient* lighting component may be supplied in two ways: (1) by conventional luminaires on the ceiling or (2) by indirect fixtures utilizing HID or fluorescent lamps with the output directed to the ceiling and adjacent walls. Some of these units may also provide direct task lighting. Indirect luminaires (see Fig 126) are available in a variety of forms including bollards, shelf or partition-mounted units, and free-standing open units. Some fixtures have both metal-halide and high-pressure sodium lamps for higher efficiency than metal halide alone and improved color compared with high-pressure sodium alone. The basic fixture can be mounted on a floor stand, on shelving, or display fixtures in stores. Applications include libraries as well as stores, schools, and offices. Asymmetrical reflectors are available for providing special light distributions for units that are located adjacent to walls.

Where ceiling mounted troffers are used for ambient lighting, a plug-in system of wiring should be considered so that luminaires can be relocated as the task locations change.

In some applications, totally indirect lighting is being employed. This works best in large rooms with low furniture. Large rooms utilize illumination much better than small rooms because less lighting energy is dissipated on walls and through doors and windows. If the furniture is low, the tasks on desks and drafting boards can be lighted from a large expanse of uniformly bright ceiling which reduces veiling reflections and improves task visibility. However, if furniture, partitions, screens, etc are high, this blocks out much of the ceiling's contribution to lighting the task and visibility suffers. Dark finishes on furniture, screens, etc, compound the problems.

Typical task/ambient applications include the following:

Offices
Open-plan office areas
Reception areas
CRT and microfilm viewing
Inspection areas
Isolated work stations
Schools
Study carrells

Fig 125
Task/Ambient Lighting with a Single Luminaire Providing Both Components

(The potential for veiling reflections with a light source directly in front of the worker is minimized due to geometry, with the relatively high mounting of the lighting unit and the narrowness of the desk from front to back. The top of the luminaire is covered with a parabolic wedge louver, inverted to reduce direct glare.)

(a)

Fig 126
Installation for Luminaires Supplying Ambient Lighting—The Luminaire
is Mounted on Top of a Room-Divider Partition

(It is important that luminaire optics provide for broad distribution of light across the ceiling to make it as uniformly bright as possible.)

(b)

Fig 126
Installation for Luminaires Supplying Ambient Lighting—A Free-
Standing Unit which May be Moved

CRT and microfilm viewing areas
Carpentry and mechanical trades' shops
Food serving line
Merchandising
Sales desks, checkout stations
Merchandise gondolas

The application of task/ambient systems can be successful if the considerations of VCP, contrast rendition, ESI, and good lighting practice are carefully evaluated and compared with equivalent non-uniform and uniform systems. Engineering comparisons should include not only the illuminance parameters, but should include net life cycle cost, tax advantages, energy consumption, and maintenance.

No single lighting system is right for all applications. (See Fig 127). Many different requirements can be found in a single building. (See Fig 128.) Any large project will require that both uniform and non-uniform lighting techniques be applied, and comparisons should be made to choose the best system for the particular task, task area, and illuminance goal.

10.8.4 Special Lighting Considerations for Stores. Lighting is employed as a sales aid in modern merchandising. (See Fig 129.) To realize the potential of lighting for selling requires more than just a general lighting system which provides illumination for the appraisal of merchandise. Lighting can create attraction for specific displays by lighting parts of the display to at least five times the level of the surroundings. This is the purpose of spotlighting and of the lighting units built into display fixtures such as showcase, shelving, wall cases, etc. Effective spotlighting may sometimes be achieved with energy-efficiency by employing 50 W, 12 V PAR spotlamps instead of 150 W R- and PAR-lamps.

Lighting can also be a vital factor in creating *atmosphere* in a store through patterns of brightness and color.

There are many possibilities for lighting patterns to enhance design or provide distinctive appearance. They can vary from small, compact downlights with incandescent or high-intensity discharge lamps which occupy less than 1% of a ceiling area, to complete-ceiling illumination.

The use of high-intensity discharge lamps for lighting stores is increasing. (See Fig 130.) Metal-halide lamps with their good color rendition and high efficiency are replacing some of the unshielded fluorescent strip lighting which has long been the trademark of the mass merchandising store. High-pressure sodium lamps are also being utilized in combination with metal-halide lamps for greater energy saving (See Fig 131.)

10.8.5 Electric Lighting and Daylighting. Daylighting is receiving increased attention in the interest of reducing building energy use. Wherever windows are employed, lighting units at the building perimeter should be switched separately so they can be turned off when daylighting is adequate. Such control may be automatic or manual.

New possibilities for windows and daylighting are being explored in building design with varying degress of success. It must be remembered that windows transmit thermal energy at a far higher rate than a well-insulated wall. Consequently, in order to determine whether windows are truly energy-efficient in buildings, their heat gain in summer and heat loss in winter must be evaluated as well as any saving in electric lighting to determine their net effect on total building energy. The same is true for skylights. If analyses show that windows are effective in the overall net use of energy for a particular building in a particular

Fig 127
Parabolic Wedge Louver

(The parabolic wedge louver provides exceptionally low luminaire brightness in this office. This type of lighting is recommended for rooms where Video Display Terminals (VDTs) are in use to avoid veiling reflections in the terminal faces. Windows must also be covered if daytime brightnesses are likely to be reflected in terminal faces.)

Fig 128
Building Retrofitted with Energy-Conserving Lamp Products Substantially Reduce
Operating Costs for the Building, Yet Not Sacrifice the Environment

1. Ellipsoidal reflector lamps of 75 W replaced 150 W reflector flood lamps in the deep,
baffled downlights in all elevator lobbies, yet provide the same light as before.
2. Reduced wattage fluorescent lamps replaced standard lamps in all office areas.
3. High-pressure sodium lighting replaced mercury lighting in a multi-level parking
garage.

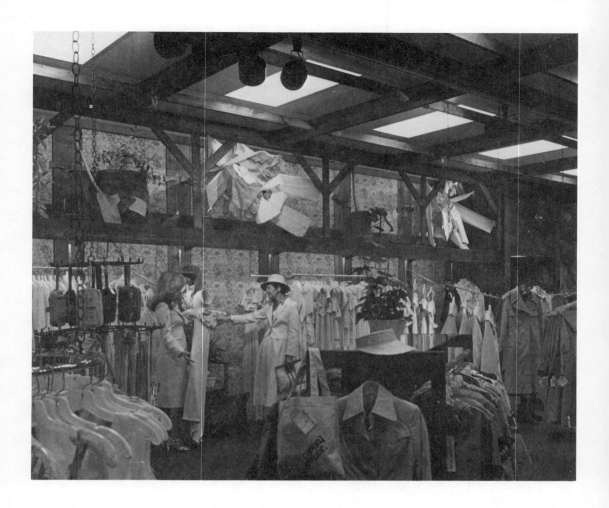

Fig 129
Special Lighting in a Store

The lighting in this store contains three elements to perform specific merchandising functions:

1. A uniform level of general lighting is provided by fluorescent luminaires so merchandise can be appraised anywhere in the space.

2. Fluorescent lighting is concealed in the wall cases to define the perimeter of the store, contribute to a feeling of spaciousness, and enhance the appearance and attractiveness of the merchandise displayed there.

3. Accent lighting is provided on feature displays to create attention. The net effect of the several lighting elements is an interesting, pleasant atmosphere.

Fig 130

Illumination in Foodstore by Metal-Halide Lamps in Recessed Luminaires

(Some owners prefer this type of lighting to bare-strip fluorescent units as there is less glare toward the eye and more sparkle from the merchandise. Economic analyses indicate such lighting can be energy-efficient and economically competitive with unshielded fluorescent lighting.)

Fig 131
Indirect Lighting in a Store
(Luminaires Are Mounted on Top of Display Fixtures—
Each Luminaire Employs One Metal-Halide and
One High-Pressure Sodium Lamp

climate they should be used. For northern climates it may be energy-effective to use large windows with southern exposure and much smaller ones facing north.

10.8.6 Outdoor and Sports Area Lighting. The lighting of parking areas around commercial buildings needs careful design in order to provide for the safety of people and the security of property. To achieve these objectives, adequate amounts of light, properly distributed, are needed throughout the environment, to reveal clearly such hazards as curbs and steps, and to illuminate dark and potentially dangerous areas. To effectively light these areas, luminaires should be selected to meet a specific light level and uniformity, and installed in such a manner as to minimize glare for pedestrians and drivers, and to avoid light spilling onto adjacent properties.

10.8.6.1 Guides for Good Floodlighting Results. Since there are many variables such as pole placement, mounting heights, light level requirements, and the size and shape of the area, floodlights should be selected based on the performance data that is supplied by the fixture manufacturer.

Floodlights are designated by the type and wattage of the lamp they use and by their light distribution or beam spread. Beam spreads can be determined by iso-footcandle diagrams that are supplied for specific floodlights. The beam spread angle defines what are considered to be the outer limits of the luminaire's coverage. Overlapping of beams in multiple floodlighting unit installations is desirable for uniformity and for safety in the event of outages.

To obtain uniform light the distance between poles should not exceed four times the fixture mounting height. A floodlight will effectively light an area out to two mounting heights from the base of its mounting locations. Further separation of poles requires the aiming angle of the floodlight to be raised, resulting in lower utilization of light and increased fixture glare. With proper spacing, sufficient overlap between adjacent floodlights will ensure uniform lighting with minimum shadows.

Outdoor sports lighting is a specialized form of floodlighting. Specific design consideration should be given to each sports lighting application so as to minimize fixture brightness or glare in the eyes of both the players and spectators. Therefore, pole locations, mounting heights, and luminaire aiming should be selected judiciously for each sports lighting system. For example, in aerial sports, the lighting is designed to light the ball in play as well as the players and the playing surface.

10.8.6.2 Light Sources for Floodlighting. Economy of installation and operation is materially influenced by the light source size, wattage, and the amount of light it produces. For these reasons high-intensity discharge (HID) lamp usage has spread for outdoor and sports lighting areas.

The most appropriate HID light sources for outdoor applications are high-pressure sodium and metal-halide lamps.

Almost without exception, the high-pressure sodium (HPS) lamp will be the choice for greatest economy and least use of energy. It also has a life rating as long as mercury lamps and provides reasonable color rendition.

Mercury lamps are inefficient compared with high-pressure sodium and metal-halide lamps and have little merit for application now.

Metal-halide lamps have shorter life ratings than mercury or HPS lamps.

However, these lamps are the preferred choice where the color of landscaping or the appearance of athletic and band uniforms is important; hence, they are often chosen for outdoor sports lighting areas.

10.9 Control of Lighting. In its simplest form electric lighting control is exercised manually by means of a switch located in a luminaire, a wall, or a panel box. This provides an ON-OFF control for a particular luminaire or branch circuit. Often more complex controls are desirable in order to control the level of illumination or provide light at a specific time. Such requirements can be met with either manual or automatic controls.

10.9.1 Switching for 480Y/277 V Distribution Systems. Wall switches approved for 300 V can be employed to switch the lighting fixtures on 277 V branch circuits. According to ANSI/NFPA 70-1981 [13] such wall switches may be employed where the voltage between switches is limited to 300 V, using grounded barriers as necessary where the voltage exceeds 300 V, as when two phases of the 480 V system occupy the same enclosure.

10.9.2 Remote-Control Switching Relays. Low-voltage (usually 24 V) remote-control switching systems can be used for branch circuit and individual luminaire control. This type of control employs a low-voltage switch at the control point to actuate a relay in the branch circuit. Since the branch-circuit wiring goes only to the luminaire and relay and not to the control point, there may be some saving in the wiring. Substantial savings result with remote control in the installations where considerable flexibility is desired and control is employed at several locations.

Lighting contactors are used for con-trolling large blocks of lighting. They are generally available in sizes of 25 A to 1000 A and are mounted in panelboards or separate enclosures. Standard-control voltages are 120 V, 240 V, 277 V, and 480 V. They make it possible to turn blocks of light ON or OFF from convenient locations or from one central location. In addition to convenience of control, installation savings can be realized by using small control wires, thereby reducing power cable runs.

Lighting contactors are actuated electro-magnetically and are held either magnetically or mechanically. Magnetically held lighting contactors are usually controlled by an ON-OFF single-pole single-throw toggle switch and will change contact position upon loss of control voltage.

Mechanically held lighting contactors will not change contact position upon drop or loss of control voltage. The operating coil is only energized during the opening or closing operation, thereby eliminating coil hum and coil power drain. A mechanically held lighting contactor can be controlled from any number of control stations, as shown in Fig 132, or from time switches and photoelectric cell relays.

Auxiliary relays may be used with lighting contactors to accommodate long runs between the lighting contactor and the control switch, for two-wire control, and for control by sensitive contact devices.

Due to energy considerations, the practice of switching large blocks of lighting by contactors is changing in favor of controlling much smaller numbers of luminaires. In some applications energy codes may limit the area where lighting can be on a single switch, or may require that each individual office or work station be switched independently.

IEEE
Std 241-1983

**Fig 132
Parking Area Lighted with "High-Mast" 90 ft Poles**

(The high poles minimize the number of poles and trenching for wiring in the lot; they also provide for more uniform light distribution, less direct glare, and simpler, less confusing patterns of brightness which a multiplicity of low poles sometimes brings. The light sources are 400 W high-pressure sodium lamps.)

Control systems are now available employing microprocessor logic to replace *hard* wiring with *soft* wiring. Coded commands can be multiplexed to control points over a pair of low-voltage wires.

Control points have receiver/switches, the latter component generally a low-voltage relay. Logic functions can be pre-programmed into a control device to turn lighting ON and OFF over a 24 h or a weekly period. Overrides are available, and some systems can be accessed with touchtone telephones. Such systems which control lighting in both time and space save considerable lighting energy compared with past control practices.

10.9.3 Dimming and Flashing of Lamps. Dimming of incandescent lamps has been employed for many years, because changing the voltage at the lamp socket provides a simple method of varying the light output. Several methods of voltage variation may be employed, but the availability of solid-state semiconductor dimmers has made them the most popular type available. Units controlling up to several hundred lamp watts are small and relatively inexpensive. Losses are generally less than 2% of the connected lamp load.

Other possible methods of dimming include:

(1) Variable resistance (rheostat)
(2) Variable autotransformer
(3) Variable reactance
(4) Electronic power tubes.

Dimming devices can be manually controlled by an operator. However, it is often desired to provide the dimming control at locations distant from the load or at several stations. Geared motor drives can be employed for this purpose on low-voltage (typically, 24 V) remote-control systems.

Incandescent lamps can be flashed with a contactor which switches the circuit on and off. Some of the commercial contactors are motor-driven units which can flash lamps simultaneously or in a desired sequence. Small button contactors employing thermal elements can also be used in individual lamp sockets to obtain flashing. Since the lamps are inoperative during part of a flashing cycle, actual service time will be longer than the rated life of the lamps.

Practical dimming and flashing can be done with certain fluorescent lamps. This is described in 10.5.2.5.

10.9.4 Dimming of HID Lamps. The dimming of high-intensity discharge (HID) lamps, through the use of solid-state technology, has not only become practical, but has also materialized as an energy saving method of lighting control. A dimming range from full to about 50% light output can be accomplished without adverse results to the life of the lamp.

A typical dimming system incorporates a centralized control panel with remote dimming controls. Dimming systems can be one or three phase, controlling most standard size HID lamps. Dimming response times, although not instantaneous, are not significantly long, and considering the reduced power consumption that dimming proportionately produces, should not be a major concern.

HID lamp dimming systems can be applied indoors or outdoors. Indoor applications could include schools, hospitals, factories, stores, auditoriums, etc. Outdoor applications could include highways, tunnels, parking lots, shopping malls, etc.

One reason for dimming discharge lamps is to maintain a constant level of illumination on tasks during the life of the lighting system. At the start with new lamps and clean fixtures, the lamps are dimmed well below their maximum output, saving considerable energy. As lamps depreciate and fixtures collect dirt between cleanings, power is increased to keep the task illumination constant.

Interfacing with time clocks, computers, or photo cells can also be considered in a dimming system.

10.10 Lighting Maintenance. Light loss due to dirt, dust, and grime depends upon the type of lighting fixture used, the dirt conditions in the atmosphere, and the time between cleanings. Losses will range from 8 to 10% in a so called *clean* environment to more than 50% under severe conditions. The longer lives of fluorescent and high-intensity discharge lamps reduce frequency of re-lamping and the coincident cleaning of the lighting fixture. The planned effectiveness of a lighting installation can only be achieved by physical maintenance. With proper planning for maintenance during the design period, it is possible to significantly reduce the initial cost, operating cost, and the energy consumption of a lighting system. Energy has always been a major cost component of any lighting system, and when energy cost is reduced, the total life-cycle cost of the lighting will be directly reduced.

Common practice in lighting system design has been to provide excess initial illumination to allow for the reduction in light as system components deteriorate due to dirt and age. The use of light-loss factors in planning installations is a necessary admission that no amount of physical maintenance can keep the output of a system up to its initial level. The value of the light-loss factor used indicates the amount of the uncontrollable depreciation expected, together with the results of effort expended to overcome the controllable factors in depreciation.

The lumen maintenance of most lamps

is published by manufacturers and provides a means of evaluating light output at various points in the life-cycle of a lamp. Some fluorescent lamps at rated life will only produce 80 to 85% of the initial light output. Some mercury lamps produce only 40% of their initial light output at rated life. Obviously, mercury lamps should be replaced in groups well before they reach their rated life, or they will waste much energy and money. Planned lighting maintenance is the most efficient, economical approach to solving lighting system problems. A properly planned re-lamping program will arrest lumen depreciation and avoid many burn-outs, thereby maintaining higher illumination levels without additional energy costs. Reduction of burn-outs gives an added advantage in saving labor, time, and expense otherwise involved in burn-out replacement. A properly planned periodic cleaning program will arrest luminaire dirt depreciation due to dirt accumulation on lamp and luminaire surfaces.

When most lamps in an area are of the same life and operated for the same length of time, the practice of group re-lamping and coordinated cleaning often reduces lighting maintenance costs substantially. This procedure may be utilized advantageously in incandescent, fluorescent, and high-intensity discharge lamp installations. The practice involves replacing all of the lamps in an area at the same time after they have been operated the greater part of their useful life. There are several variables involved, such as the labor cost of individually replacing lamps compared to that of group replacement, and the number, type, and cost of the lamps. If lamps within the same area have different operating hours, group re-lamping may not be practical.

Due to the long life of fluorescent and

high-intensity discharge lamps, these systems should have the lamps and luminaires periodically cleaned several times between re-lampings. Group re-lamping should be scheduled at the same time as cleaning is scheduled.

The timing of re-lamping and cleaning should be in accordance with the plans of the lighting-system designer. If intervals between operations are too long, excessive loss of light results. If intervals are too short, labor, equipment, and lamps are wasted.

10.11 Voltage. The efficiency, light output, life, and power consumption of incandescent lamps are all substantially affected by their operating voltage. For this reason they should be operated at or near their rated voltage to give best value to the user. Incandescent lamps are manufactured and labeled for use with specific voltages at the socket, such as 115 V, 120 V, or 125 V, and many other voltage ratings. See Section 3 for effect of voltage variations on lamp life and efficiency.

The 120 V general-service incandescent lamp is considered to be the standard-voltage incandescent lamp because a large majority of electric utilities provide 120 V service to their customers. Incandescent lamps are also available for operation at higher voltages such as 230 V, 250 V, and 277 V. The latter are less efficient (except for tungsten-halogen types) and not as rugged as 120 V lamps since the high-voltage tungsten filaments are smaller in diameter, longer, and more fragile than those of standard-voltage lamps.

If lower-voltage incandescent lamps are inadvertently inserted in higher supply socket voltages (for example, a 120 V lamp in a 277 V socket) they may shatter with potentially adverse effects on people nearby. Consequently, the application of higher-voltage incandescent lamps is sometimes prohibited by certain users.

Incandescent lamp life is often considered the principal criterion of lamp performance. Actually, lamp efficiency is nearly always far more important. These two factors are inversely related in incandescent lamps. Hence, lamps designed for longer life operate at reduced efficiency; those of high lighting efficiency design (such as photo-flood lamps) have relatively short life. Lamp life design is based on the total cost of light, assuming typical operating conditions and costs which prevail among the great mass of users. In some instances where electric energy rates may be very low or labor cost of lamp replacements high, the economic picture is altered. For such applications lamp manufacturers have a line of special-service incandescent lamps, the life of which is about two and one half times that of general-service lamps. Lamp efficiency is about 15% lower than in the general-service line.

Some incandescent lamps are available with lives of 5000 h to 10 000 h or longer. These are so low in lumens-per-watt efficacy that they are uneconomical for use except where installed in difficult-to-get-at locations, where labor cost to replace is very high or where special equipment may be necessary to change burned-out lamps. A cost analysis is recommended to determine the suitability of their use.

A magnetic-coil fluorescent lamp ballast designed for 120 V primary supply typically can start and operate lamps at +5% or −10% of the design voltage. However, if operated for long periods at the extremes of these voltage limits, the lamps will not operate at their normal photometric, life, and power ratings, and the ballasts may be damaged. Ballast manufacturers suggest a somewhat more narrow voltage tolerance for sustained operating periods. For example, one manufacturer advises that the limits should be 100 V and 125 V for its 120 V ballasts. For a 277 V ballast, the indicated limits are 254 V and 289 V. Manufacturers' data should be utilized in determining recommended voltage limits on specific ballasts.

The life and light-output ratings of fluorescent lamps are based on their use, with ballasts providing proper operating characteristics. Ballasts that do not provide proper electrical values may substantially reduce either lamp life or light output, or both. Ballasts certified as built to the specifications adopted by the Certified Ballast Manufacturers (CBM) do provide values that meet or exceed minimum requirements. This certification assures the user, without individual testing, that lamps will operate at values close to their ratings. Ballasts for high-intensity discharge lamps are often designed with primary voltage taps. Connection should be made to the tap which corresponds most closely to the supply voltage.

Fluorescent and high-intensity discharge lamp ballasts are made for the higher branch circuit voltage (277 V) employed in commercial buildings. There are also 480 V primary ballasts for high-intensity discharge lamps. If such voltage is available for lighting systems, sizeable savings may be realized as a result of reduced wiring and distribution equipment costs. Fluorescent lamps and fixtures will be the same for 277 V lighting as for 120 V. Ballasts are approximately the same size and cost for 120 V or 277 V.

Where the 480Y/277 V power supply is employed in a building distribution system, an effective and economical system is obtainable by connecting fluorescent luminaires line to neutral (see Fig 133). While 277 V panel boards are a little higher in cost, fewer circuits are needed as can be seen in Fig 137 by comparing the two areas similar in size and investigating the quantity of conduit, copper conductors, and branch circuits.

Wye-connected three-phase four-wire supply circuits at 120 V or 277 V line-to-neutral provide a very economical system to supply large general lighting loads of fluorescent or high-intensity discharge lighting. However, the ballasts draw considerable third harmonic current component which flows in the neutral (or fourth) wire. For this reason ANSI/NFPA 70-1981 [13] requires that the neutral conductor be the same size wire as the other three circuit conductors when utilizing three-phase conductors and a common neutral in branch circuits between lighting loads and the serving branch circuit panel board. The neutral conductor connot be reduced in size as is permitted with incandescent and other resistive loads.

Where incandescent lighting is used in certain areas in addition to 277 V fluorescent lighting, 120 V is obtained by dry-type step-down transformers that would serve a 120 V branch circuit panel for both lighting and 120 V receptacles. More complete coverage of this subject is presented in Section 4.

10.12 Power Factor. Incandescent filament lamps operate at 100% power factor. Fluorescent and high-intensity discharge lamps operate with ballasting circuits which generally regulate the current by reactive circuit elements. Because of this, the basic lamp-ballast circuit operates at a power factor generally less than 50%. However, practically

Fig 133
Parking Garage Lighted with 150 W High-Pressure Sodium Lamps

(The luminaires are surface mounted on the concrete slab and the structural beams provide natural shielding. The moderately high reflectance concrete floor and ceiling help provide a reasonable distribution of illuminance through the area with inter-reflected light.)

all ballasts intended for general lighting applications have provisions to improve this power factor to 90% or better.

The most widely used type of fluorescent ballast (the two-lamp, series type) has a slightly leading power factor. This is a desirable attribute since most other building loads tend to have lagging power factor. The lighting designer should specify high-power factor ballasts for his lighting applications.

Certain ballast circuits used in desk lamps, office copy machines, home appliances, etc operate one or two low wattage fluorescent lamps at a low power factor. The total reactive power recorded by these devices depends on the number of such luminaires installed.

Fig 134
High School Football Field Lighted with 1500 W Metal-Halide Lamps

(This high-wattage, high-efficiency source greatly reduces the number of lamps and floodlights required to meet the lighting objectives, compared with other sources of lower output and efficiency. They also reduce transformer capacity and power distribution equipment. Though the life of this high light output source is only 3000 h, this is quite suitable for the relatively low annual operating hours of sports stadiums.)

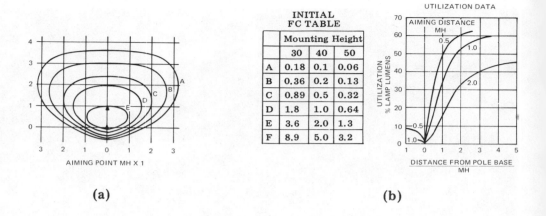

INITIAL FC TABLE			
	Mounting Height		
	30	40	50
A	0.18	0.1	0.06
B	0.36	0.2	0.13
C	0.89	0.5	0.32
D	1.8	1.0	0.64
E	3.6	2.0	1.3
F	8.9	5.0	3.2

(a)   (b)

Fig 135
(a) An Isofootcandle Chart Shows a 400 W High-Pressure
Sodium Lamp Floodlight (b) Specific Footcandles for the Contours

(Manufacturers data should be consulted for specific floodlights of interest.)

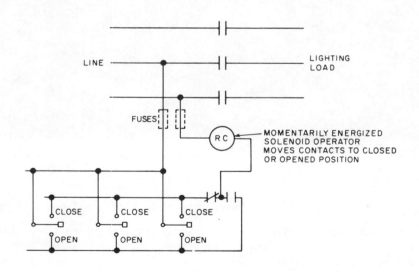

Fig 136
Mechanically Held Electrically Operating Lighting
Contactor Controlled by Multiple Momentary-Toggle-Type
Control Stations

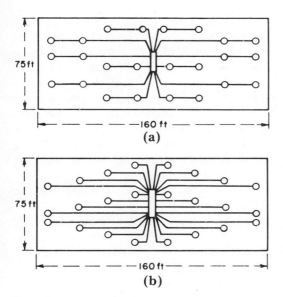

Fig 137
Comparison of Distribution Required
for 277 V and 120 V Lighting Systems
(a) 277 V: 12 Lighting Circuits Requir-
ing 1700 ft of Conduit and 3550 ft of
No 12 Wire (b) 120 V: 24 Lighting Cir-
cuits Requiring 5000 ft of No 12 Wire

10.13 Temperature. The performance of incandescent lamps is relatively unaffected by ambient temperature. Light output and life remain normal in cold or warm weather. Performance is usually satisfactory even in the case of some of the confining luminaires. Where extremely high ambient temperatures are encountered, as in ovens, special lamps should be used which have been manufactured with exhausting and sealing temperatures adequate for the intended service.

The starting characteristics of high-intensity discharge lamps and the starting and operating characteristics of fluorescent lamps are significantly affected by low temperatures. For satisfactory outdoor operation of these lamp types in cold weather, ballasts must be used which supply sufficient voltage to ensure reliable lamp starting. In addition, the ballast must be of a design which will withstand low temperatures if it is mounted outside. In the case of some building-mounted signs or security lighting equipment, it may be practical to locate the ballast remotely in a heated environment.

Rated light output from most fluorescent lamps is achieved with ambient temperatures of 70—80 °F. Above this temperature range light output is reduced about 1% for each two degrees Fahrenheit. In high ambient temperatures, circulating air improves the light output. This is a problem in the design and application of most enclosed or recessed luminaires.

If the ambient temperature around the lamp is reduced below the 70—80 degree range, the loss of light at a rate of 2% per degree Fahrenheit occurs. This is a problem in outdoor applications of fluorescent lighting.

In very cold weather regular fluorescent lamps may not reach their full rated light output, particularly if subjected to air currents. However, if lamps are used in a closed fixture, or shielded from drafts, the ambient temperature around the lamp may build up during operation and light output will increase nearer to rated values. If sufficient voltage is available to start high-intensity discharge lamps in cold weather, they will gradually build up in light output to near normal values, even in open luminaires.

Fluorescent lamps which operate at higher current (800 mA and 1500 mA) will, in properly designed multiple-lamp enclosed luminaires, maintain light output better than lamps of lower current in cold ambient temperatures. In addition, certain fluorescent lamps which operate at 1500 mA have been designed especially for outdoor application. One type is intended for use in open fixtures. In typical outdoor environments the light output of the jacketed lamp is at its maximum in an ambient of –10 °F. Due to the variation in outdoor conditions under which fluorescent lamps may be expected to operate, such as temperature, wind, and equipment, it may be desirable to seek the advice of luminaire manufacturers regarding the best choice of lamps and equipment for a specific climate.

Most ballast designs will start and operate fluorescent lamps satisfactorily down to a temperature of 50 °F. Many will continue to provide reliable lamp starting below 50 °F, but it is recommended that below this value, a ballast that is designed specifically for cold-weather operation be specified. Such ballasts are rated to start lamps reliably down to 0 °F or –20 °F, depending on design. They are available for slimline (430 mA), high-output (800 mA), and extra-high-output (1500 mA) fluorescent lamps.

High temperatures may shorten ballast life or, as Class P ballasts, the thermal protector will open the circuit and turn off the lights. Provision for adequate heat dissipation should be provided for ballasts, both in the design and installation of luminaires. This is especially important for the higher voltampere rated ballasts.

10.14 Ballast Sound. Ballasts for fluorescent and high-intensity discharge lamps produce a very low level of sound if operated in the open on a heavy vibration resistant base. However, when mounted in a luminaire, they induce vibrations into the luminaire. The large radiating surface acts as a sounding board and may radiate audible levels of sound. The sound is a distinctive tonal hum which may be distinguished from other background sounds in a given interior. If it is loud enough to become distinctly audible some occupants may find it objectionable.

Whether or not annoyance is likely to be ascribed to a lighting system depends upon three factors:

(1) The sound level radiated by the lighting equipment

(2) The tonal quality of the particular luminaires in question; that is, the distribution of sound power among the harmonics of 120 Hz which are being radiated

(3) The ambient sound level in the area arising from other sources

In order of importance the factors determining the level and tonal quality of sound are:

(1) The design and construction of luminaire

(2) The design and construction of the ballast

(3) The voltampere rating of the luminaire and the illuminance level. (For a given lighting level, a large number of small luminaires generally yields a lower sound level than a small number of large luminaires.)

The ambient sound level in the area is determined by the activities in the area. For relighting an existing area sound level readings could be made. For new construction the ambient noise levels listed in Table 85 may be used as a guide representing typical experience. It should be pointed out that acoustical treatment has little bearing on ballast hum. This is true because the acoustical treatment will reduce the ambient noise level and the ballast hum by equal amounts. The whole question is whether

Table 85
Fluorescent Ballast Sound Ratings

Application	Ambient Noise Level (Measured with Standard 40-dB, Weighting Network) (dB)
Broadcast studio, church, country residence	20—24
Evening school, city residence, quiet office	25—30
Average residence, public library, study hall	31—36
Classroom, professional office	37—42
Noisy residence, business office	43—48
Store, noisy office, factories	49 and up

the ballast hum becomes noticeably audible above the ambient sound level. Absolute sound level has no practical importance since it is practically never high enough to interfere with speech audibility or create any other objective problems.

NEMA No LE2-1974 specifies a method of rating lighting equipment which makes it quite easy for the lighting designer to determine whether the sound levels will be satisfactory for any particular ambient noise level interior. Certain manufacturers rate their equipment based upon this standard. Other luminaire and ballast manufacturers provide ratings based upon their own proprietary systems. Since the luminaire design is the greatest factor in determining sound levels, rating systems based upon the ballast alone must include large safety factors.

10.15 Lighting Economics. The realm of lighting economics is multifaceted. It can be divided into the following aspects:
(1) First cost
(2) Type and quality of lighting desired
(3) Energy costs
(4) Maintenance costs
(5) Effect on personnel
During the past ten years the order in which these aspects should be considered has been altered. Also, from day to day there is no fixed method to decide which of the above factors are the most important or should be considered first. The aspect of primary importance must be decided on a job to job basis, depending on the user's end needs, type and amount of energy available, energy costs, maintenance, availability, and a number of intangibles such as employee morale, health, comfort, and safety.

10.15.1 First Cost. In today's lighting market, first costs can be very misleading. Not only do first costs vary from fixture type to fixture type, but they also vary among fixtures of a given type. The variance in first costs of fixtures within a category is due to a number of factors, that is, quality of workmanship, durability, attention to detail, reflector type, etc. A fixture may be selected on a first cost basis to serve the client's needs and finances. However, care must be used by the lighting designer in selecting a fixture in this manner. The most expensive fixture is not always the best for the task, and the least expensive fixture is not always the wrong choice. First costs should be weighted in the light of all the other economic aspects.

With respect to first costs, incandescent lighting may appear to be less expensive to install than other more efficient sources if the desired illumination levels are relatively low. However, when the cost of the power distribution system is also considered, the more efficient lighting systems are often lower in first cost. When operating expenses are considered, the incandescent system is far more costly than fluorescent or HID lamp systems.

10.15.2 Type and Quality of Lighting Desired. It is in this aspect of lighting economics that the lighting designer must closely communicate with client or end user of the lighting system. A variety of questions must be answered at this stage of design such as: What are the characteristics of the visual tasks to be performed? What type of luminaire and light source will provide light of the right quality for good task visibility? How important is visual comfort? How important is lighting control? How much light is required for the tasks at hand? What is the client's budget for lighting installation? Once these questions have been answered, one can decide on the type (incandescent, fluorescent, HID) and the quality of lighting that should be installed.

10.15.3 Energy Costs. Energy costs vary from state to state and region to region. Energy efficiency is dominant to most designers today. However, to most clients energy economics is the item of most interest. The lighting designer must weigh the lighting system cost versus the energy economics and make an intelligent choice, with financial guidance from the client.

10.15.4 Maintenance. Lighting designers must obtain maintenance information from the client. How large is the maintenance department, if any, and how skilled are its members? Some end users do not have a maintenance department but call an electrical contractor to service the lighting equipment. For these users a low maintenance lighting system is a wise choice. However, these systems are sometimes more expensive than a system which requires considerable maintenance. In summary, maintenance costs must be weighed versus lighting system costs.

10.15.5 Effect on Personnel. Various types and quantities of illumination will have different psychological effects on the workers. These effects should be discussed with the end user so that the lighting designer can make a responsible decision as to the correct lighting system.

Productivity is usually the key word in most occupational environments. The lighting designer should be aware of the effect illumination has on productivity and call these points to the attention of the client when the illumination budget is being prepared. The trade-off between production levels and illumination quality must be evaluated.

In summary, it is suggested that the lighting designer choose the lighting system on a *life cycle cost* basis. This method of system economics addresses all of the aforementioned aspects so that each can have its proper contributions to the choice of the appropriate lighting system.

10.16 Illuminance Calculations. There are two principal approaches to the calculation of illuminance: one involves situations where uniform distribution of illuminance is desirable, as in densely occupied offices or classrooms; the other is where non-uniform illuminance is desirable as a more energy-efficient way of providing for task performance.

10.16.1 The uniform illuminance method involves the use of coefficients of utilization supplied by luminaire manufacturers, together with maintenance factors applied in a formula to determine the number of luminaires necessary to maintain the desired illuminance in a space. Then it is necessary to arrange the luminaires appropriately so as to provide the desired distribution of illuminance or minimize veiling reflections in tasks, or both. As covered earlier, illuminance values will be substantially higher in the central portions of small- and medium-sized rooms and lower near the walls than the average. This knowledge can save energy, if task areas are confined to central areas of the room, by designing for a lower illuminance for an average room.

10.16.2 The computation of non-uniform illuminance is more complex since it involves a calculation for the direct contribution of each luminaire at a particular point in a room plus the contribution of interreflected light from room surfaces to the point of interest. This process is repeated for each point in the space where illuminance information is desired. This is called the *point-by-point* method of calculating illuminance. Sometimes the computation for interreflected light is omitted to simplify the computations. This may add 5 to 10% to the task light in small rooms, and somewhat more in larger rooms.

Because of the great many individual calculations required, computer programs have been developed to perform them. These are available through various computer services and from manufacturers. Programs are also available for use with programmable hand calculators.

The reader should refer to the IES Lighting Handbook [8], [9], which contains considerable information on the calculation of both uniform and non-uniform illuminance.

10.17 Lighting and Thermal Considerations. Lighting energy in buildings can be used twice during the winter, once for visual purposes (the only reason for its being in the building), and a second time to replace building heat losses when outside temperature is below 65 °F.

Electric lamps are 100% efficient as heat sources. Even the light from the lamps eventually becomes heat. When light or infrared from luminaires is intercepted by people or surfaces in a room, part of it is absorbed and raises the temperature of the surface. That which is reflected is bounced to another surface where another partial absorption takes place. In a brief instant all the light and infrared entering the space from lamps or luminaires is absorbed and is useful in heating if the room needs heat at that particular time.

In the future, when buildings are designed, it will be necessary to evaluate the total impact of each subsystem on energy use, as there may be a mandatory (or voluntary) energy (or power) budget to comply with. Failure to do so could result in buildings that exceed the allotted budget, or, on the other hand, fails to function effectively for the owner.

In order to get the net effect of lighting on building energy use, three things must be checked:

(1) The lighting energy used directly

(2) The load lighting places on a cooling system

(3) The lighting contribution to the heating of a building

In winter, the lighting system keeps the thermostat from turning on the furnace so frequently and using more oil or natural gas, currently the most popular space heating fuels.

In single-story buildings, lighting units installed in a pattern across the ceiling automatically compensate for some of the heat losses through the roof, depending on insulation, lighting BTUs generated, and outside temperature. This compensation begins to take place when the outside temperature falls below 65°, and somewhere between 50 °F and 60 °F all of the lighting energy contributes to replacing heat losses in typical buildings. The luminaires within 10 ft to 15 ft of the walls and windows at the building perimeter also compensate for some of the losses through these surfaces.

Low-rise buildings of one or two stories predominate in existing commercial and industrial building inventory. Over 90% of the existing area of commercial and industrial buildings is low rise.

Lighting is a low-temperature heat source. For example, 4 ft rapid-start fluorescent lamps operate with a bulb wall temperature of only about 105 °F. The 1500 mA fluorescent lamps rise to 140 °F. For example, on a cold morning, the lighting system is not able to overcome a 10 °F overnight temperature setback within 30 min. A higher temperature conventional heating system (such as electric heating elements) should be used for this purpose.

Further, after a fluorescent lighting system has been turned on for some time most of the energy is *stored* in the luminaire. That is, the lighting fixture itself heats up and then the heat is transmitted to nearby surfaces, such as ceiling tiles, the air in ceiling cavities, and the building floor or roof structure. After several hours of operation, some of the building structure as well as the luminaires have temperatures well above that of the air in the occupied space, so they begin to radiate and convect heat into the occupied space. In the middle and late afternoon, heat from the lighting system is entering the room at the same rate it is generated, so the lighting is contributing effectively to the heating.

At 5 pm when many building operations cease, people go home and lights are turned off shortly thereafter. A building setback temperature may go into effect, as well. However, the heat from the lighting system which is stored within the building structure continues to make itself felt, dissipating this energy within the building and delaying the thermostat's first turn-on of the furnace. If heating energy is not used overnight, the stored heat reduces the recovery energy required the next morning since the building will not have cooled down as far by virtue of the storage effect.

In a multi-story building, lighting on the top floor and around the perimeter of lower floors can take care of some of the heat losses through the building boundaries. If the heat in the building interior is to be useful, it must be controlled and redistributed with a system designed for this purpose. Standard mechanical equipment is available to do this.

Sometimes after redistribution of interior zone heat has taken place, some is left over. This energy can be stored if insulated water tanks are available. There is a growing list of buildings employing this concept and saving considerable energy and dollars in so doing. These storage systems can also be used in summer by chilling the water off-peak at night, incurring only an operating expense without a demand charge. Then the chilled water can take the peaks off the cooling requirements during occupied hours.

The control and use of lighting heat in buildings has been treated in some detail because some of its aspects are not well known or understood. It is well recognized that lighting creates a cooling load in buildings in warm weather and an allowance for refrigeration tonnage and some volume of air (or water) must be made in the mechanical system design.

A discussion of the properties of air-return luminaires which provide some advantages in controlling lighting heat in warm weather is included in 10.7.5.

In multi-story buildings, the heat gain from lighting systems in the interior zones of buildings sometimes requires conventionally-designed cooling systems to run in very cold weather. In many buildings this energy is *wasted* to the outside rather than recovered and redistributed to perimeter zones or stored for later use, as described earlier. However, in cold weather *economizer* air systems which use the outside air for cooling the interior zones of buildings can do away with the need for operating refrigeration compressors in winter. Such systems should be used in future buildings to reduce requirement for refrigeration energy to handle lighting loads to the lowest possible level.

Now and in the future, even though more efficient sources will be used for lighting and turned off promptly when not needed, and though less total energy will be used for lighting, as a percentage of the total building energy, it may become greater than at present. Designers and owners should take this into account in thermal design by redistributing, storing, and reusing lighting heat and other internal heat gains of buildings. This can prevent the wasting of building energy, so typical of past practice, and reduce the requirement for energy from new sources to satisfy space heating needs.

10.18 References

[1] ANSI/ASHRAE/IES 90A-1980, Energy Conservation in New Building Design (Sections 1–9).

[2] FLYNN, JOHN E., SPENCER, TERRY J., MARTYNIUK, OSYP, and HENDRICK, CLYDE. Interim Study of Procedures for Investigating the Effect of Light on Impression and Behavior, *Journal of the IES*, Oct 1973, p 87.

[3] FLYNN, J.E., Lighting Design Decisions as Intervention in Human Visual Space (the role of CIE Study Group A). paper presented at Symposium—1974/ CIE Study Group A, Montreal, Canada.

[4] FLYNN, J.E., SPENCER, T.J., MARTYNIUK, O., and HENDRICK, C., The Influence of Spatial Light on Human Judgment, *Compte Rendu*, 18e Session, p 75–03, *CIE Congress*, London, England, 1975, p 39.

[5] FLYNN, J.E., A Study of Subjective Responses to Low-Energy and Non-Uniform Lighting Systems, *Lighting Design and Application*, Feb 1977, vol 7, no 2, p 6.

[6] Federal Energy Conservation, Supplement no 1, *GSA Federal Management Circular FMC 74–1*.

[7] IES/EMS-1, *Recommended Lighting Power Budget Procedure* Published by the Illuminating Engineering Society, 345 East 47th Street, New York, New York 10017.

[8] IES *Lighting Handbook*, 1981 Reference vol. Published by the Illuminating Engineering Society of North America, 345 East 47th Street, New York, New York 10017.

[9] IES *Lighting Handbook* 1981 Application vol. Published by the Illuminating Engineering Society of North America, 345 East 47th Street, New York, New York 10017.

[10] Recommended Practice for the Specification of an ESI Rating in Interior Space when Specific Task Locations are Unknown, prepared by the Design Practice Committee of the IES, *Journal of the IES*, Jan 1977, p 111.

[11] Selection of Illuminance Values for Interior Lighting Design (RQQ Rep no 6), prepared by the Committee on Recommendations for Quality and Quantity of Illumination of the IES

(RQQ), *Journal of the IES*, Apr 1980, p 188.

[12] THORNTON, W.A., Difference in Color Vision, *Lighting Design and Application*, 1980, vol 9, no 2, p 17.

[13] ANSI/NFPA 70-1981, The National Electrical Code.

---

[48] Copies of the National Electrical Code are available from the Sales department of American National Standards Institute, 1430 Broadway, New York, NY 10018.

11. Electric Space Conditioning

11.1 General Discussion. This section deals with heating and cooling aspects of occupied spaces of industrial buildings. The concepts presented here can be applied to areas in which control of ambient conditions is necessary for reasons other than human comfort. This section is meant to serve as a guide for the electrical engineer engaged in working with space-conditioning systems and in selection of equipment and operation of these systems. Discussions of typical process components are presented and means of calculation are given. References in which more detailed information can be found and a glossary of commonly used terms are included.

In former times environmental control considerations did not have importance which is now accepted. There was no energy shortage. In today's world regulations at all levels of government and utility imposed controls affect the engineer's evaluation of problems associated with design of heating and air-conditioning systems for buildings.

Stabilization of energy sources can-

not be expected within the foreseeable future. Indeed, it can be expected that the situation will become more chaotic. The design engineer must make decisions concerning energy availability and economics, environmental impact, and system economics. The engineer will be guided by ethical standards. All of these, with the exception of ethical standards, will change with time and technological advances. It will be necessary to keep abreast of current literature, attend professional meetings, and maintain liaison with representatives of utilities and public agencies.

Table 86 presents a list of states in which energy standards were in effect as of 1980.

All-electric buildings are becoming quite common. These include homes, offices, motels and hotels, schools and colleges, and commercial and industrial buildings. While the advantages associated with the flexibility of electric space-conditioning are obvious, increasing costs of electricity have lessened the attractiveness of all electric space-

Table 86
State Summary Chart

	Laws/Regulations in Effect								Comments
	Standards	Enabling law for standards	Incentives	Energy consumption analysis	Energy consumption disclosure	Solar rights	Life cycle costing	Other	
ALABAMA	•								
ALASKA							•		
ARIZONA		•	•					•	
ARKANSAS			•						
CALIFORNIA	•	•	•					•	
COLORADO	•		•		•				
CONNECTICUT	•	•	•			•			
DELAWARE									
DIST. OF COLUMBIA									
FLORIDA	•	•	•	•			•	•	
GEORGIA		•	•						
GUAM									
HAWAII			•						
IDAHO			•						
ILLINOIS	•		•					•	
INDIANA		•	•						
IOWA	•								
KANSAS		•			•				
KENTUCKY									
LOUISIANA									
MAINE	•						•	•	
MARYLAND		•	•		•				
MASSACHUSETTS	•		•	•					
MICHIGAN	•		•						
MINNESOTA	•							•	
MISSISSIPPI									
MISSOURI									
MONTANA	•	•	•						
NEBRASKA									
NEVADA	•		•					•	
NEW HAMPSHIRE			•						
NEW JERSEY	•								
NEW MEXICO	•	•	•	•		•			
NEW YORK	•		•					•	Standards for equipment efficiency only
NORTH CAROLINA	•		•	•		•			
NORTH DAKOTA	•	•	•			•			
OHIO		•						•	
OKLAHOMA			•						
OREGON	•	•	•		•	•		•	
PENNSYLVANIA								•	Standards for lighting
PUERTO RICO									
RHODE ISLAND	•	•	•	•		•			
SOUTH CAROLINA									
SOUTH DAKOTA			•						
TENNESSEE									
TEXAS	•		•	•				•	
UTAH	•								
VERMONT	•	•	•					•	
VIRGIN ISLANDS									
VIRGINIA		•	•					•	
WASHINGTON	•		•	•		•			
WEST VIRGINIA									
WISCONSIN	•								
WYOMING	•								

conditioning. It is possible that greater use of nuclear power generation may reverse this trend. The very future of nuclear generation however, is now in serious jeopardy. Solar energy may, in time, become a feasible power source but it is not too promising for the near future. Solar energy may supply immediate needs at specific sites.

Electric space conditioning has many advantages but economics is a serious problem. A detailed economic analysis comparing alternative systems, including operating costs, return on investment, and projected availability of power is necessary before meaningful conclusions can be drawn.

Electric space conditioning can be accomplished by centralized or decentralized systems. All-electric buildings may require fewer operating and maintenance personnel, depending on the complexity of the mechanical control systems. A fair economic evaluation of competitive energy systems would include cleanliness, equipment, insurance, taxes, and similar items. Electric space conditioning may affect the aesthetics and, therefore, the architectural approach to a new structure. Alternatively, the presence of a cooling tower might be obtrusive.

Attention to exterior design is important but is only incidental to the basic job of the electrical engineer. His task is the design of a controlled environment which is conducive to the comfort, well being, and efficiency of the occupants.

There are two approaches to the above problem; independent systems and completely integrated systems. There are now alternatives to oil and gas. Solar energy and heat pumps are finding increasing favor. Heat-pump technology is rather well advanced but solar-energy technology is still in the early stages.

This section is meant to serve as a guide and much specific information, not suitable for presentation within the confines of this section, will be found in the references given in 11.6.

11.2 Primary Source of Heat. Space heating and air conditioning are the major components of an electric space-conditioning system. Engineers are using the approach of planned environments more and more.

The primary source of heat for humans is the heat generated within the body. The human body is a heat-generating unit adjusted internally so as to maintain a temperature of 98.6 °F (37 °C) as long as the body is healthy. In a space-conditioning system the object is to regulate the environment so that heat is not dissipated too rapidly for human vitality, well being, efficiency, and comfort. All of the foregoing are closely interrelated. The building system does not warm the body because the body is much warmer than the surrounding air. The rate at which heat is lost from the body is dependent on the air temperature and rate of air movement. Comfort is a function, among others, of the humidity of the air. The surface temperature of the body is about 85 °F (29.4 °C) and the surrounding air is 10 °F (5.6 °C) to 15 °F (8.3 °C) lower. With this differential the heat transfer is from the body outward. The same observations can be made in regard to air conditioning, in which the temperature is controlled so as to maintain the heat flow from within. In this case the humidity assumes great importance. It should be noted here that the basis of regulations regarding quantities of make up air are based primarily on humidity and odor control, not buildup of respiration gases.

11.2.1 Electric Space Heating. Electric space heating is accomplished by resistance heaters or heat pumps. In resistance heaters the heat is generated by passage of electric current through resistance afforded by conducting materials. Heat pumps accomplish the exchange of heat from one medium, such as air or water, to the space to be heated or cooled.

Heating equipment should be located so as to replace most effectively the heat lost to the outdoors and to counteract heat loss or eliminate as far as possible any cold surfaces to which body heat can radiate.

Wall panels and baseboard heaters are, therefore, placed on outside walls, preferably beneath window areas. In this position they can effectively counteract and raise the temperature of the wall and inside glass surfaces. Convection currents set up within the room tend to move downward across the cooler window surfaces, combining with infiltrating air, to produce a constant downdraft across the glass. Heat rising from units beneath the window helps to neutralize this downdraft and prevents the cool air from circulating through the room. Locating the units at the baseboard or lower portion of the walls removes cool air from floors, while the natural rise of heat keeps ceilings warm. Most modern occupied spaces require cooling in interior spaces, even in coldest weather, because lights, people, etc, provide more heat than the system loses. Thermostatically controlled perimeter heaters, which are off during much of the occupied period, do not offset window downdrafts, and a resultant pool of cold air may develop. Separate draft heaters are available to combat the downdraft, providing more heat than is absolutely necessary. Continuous perimeter heaters, having a capacity of 135 Btu to 250 Btu/(h · ft) [(40 W to 73 W)/ft] should be placed under the whole length of the window. These are controlled by thermostats between adjacent heaters, mounted in line with the heaters, but isolated from radiation and conduction. The controls should be independent of the basic system controls, and operate to control downdrafts regardless of whether the basic system is heating, cooling, or off. Within the load-carrying capacity of the control system, one thermostat may control all draft barrier heaters on the same wall of a particular room. Often draft-barrier heaters on several adjacent floors may be controlled by a single thermostat between heaters under windows with identical orientation.

Ceilings heated with embedded cable radiate directly to the floor, thus creaing comfortable temperature levels on both floor and ceiling, as well as objects between. Walls are heated by direct radiation, by reradiation from furniture and other objects and by natural convection of air within the room. Where large window areas are present it may be desirable to provide additional capacity beneath the windows in the form of baseboard heaters or, where slab floors exist, a few runs of floor cable in the slab beneath the windows.

Local control of temperature in each room is a great advantage of resistance-heating equipment and room heat pumps, but particular attention must be given to thermostat location if optimum results are to be obtained. It should be mounted 5 ft or less from the floor and on an inside wall in order to avoid the direct effect of lower temperature on outside walls. It should not receive the direct output of a heater and it should not be in a position to be affected by drafts when doors are opened. Direct hear from lamps or appliances will cause erratic and inefficient operation.

Both line-and low-voltage thermostats are available, the latter operating in conjunction with a relay. Both perform satisfactorily.

A single heater installed as a supplementary source of heat or to serve a specific function, such as heating an entrance way or vestibule, may use a built-in thermostat to sense the temperature in that particular area. It is usually preferable to use a wall mounted thermostat where several heaters are used to serve a room.

Floor heating systems are sometimes provided with floor thermostats for limiting floor temperatures to avoid heat damage to floor finishes or discomfort from excessive floor temperatures. Where this is the only means of control, wide variations are likely to occur as outdoor temperature varies. Operation of room-air thermostats in conjunction with floor-limiting thermostats produce more satisfactory results. More commonly, room thermostats are used exclusively without regard to variations in floor temperature.

11.2.2 Heat Pumps. Heat pumps operate on a reversible-cycle principle. They extract heat from a space during the warm season. During cold weather, they are manually or automatically switched to introduce heat into the space, extracting heat from the air, water, or ground.

Units using water as the heat source are practical where water from a well, stream, lake, or river, substantially above the freezing temperature, is available. Heat drawn from the ground by means of water pumped through embedded pipes is of relatively constant temperature, but the installation is expensive and sufficient ground area is not always available for laying pipes. Also, available heat varies with shifting soil conditions.

Units drawing heat from the air are most widely used since air is always available. The disadvantage here is that large equipment capacities are necessary in areas where the air temperatures drops substantially below 20 °F, since a larger volume of air must be handled in order to extract sufficient heat. Recent developments using compound compressors (two compressors in series) have made operation practical down to −20 °F but the equipment is expensive and it is not yet practical for small heating requirements.

The heat-pump capacity is usually designed to handle the air-conditioning load. When outside temperatures become too low for the installed capacity to handle heating loads, supplementary resistance heaters are usually provided for additional heat. Heat pumps for small rooms developed in the last ten years permit room-by-room control which is not practical with central-heat pump systems.

Although higher in initial cost and maintenance costs than resistance heaters, heat pumps offer the advantage of lower overall operating costs due to the *free heat* extracted from external sources.

11.2.3 Resistance Heaters. Resistance heaters may be classified by type as follows:

(1) Wall and ceiling units
(2) Central furnace or boiler
(3) Radiant-heat cables
(4) Unit heaters
(5) Infrared heaters
(6) Heat-storage equipment.

11.2.3.1 Wall and Ceiling Units. Wall and ceiling units may be surface mounted or recessed, incorporating heating coils, radiant glass, ceramic panels, or finned elements, with or without a built-in fan. Certain types of ceiling units may include built-in fluorescent or incandescent lamps and are used largely in bathroom areas. Infrared lamps are frequently used in these applications as the heat source, and built-in fans are used for circulating the heat.

Baseboard units, as their name implies, are designed to be placed along the outside wall of each room at the location normally occupied by the baseboard. They are constructed in sections of about 2 ft to 12 ft in length and vary from about $3\frac{1}{2}$ inches to 10 inches in height. Sill-line heaters are similar to baseboard heaters but are intended to be mounted with the top of the heater enclosure at window sill-level.

The heating elements are rated from 100 W (341 Btu/h)/ft to 400 W (1365 Btu/h)/ft or more and can be selected to operate at nearly any common branch circuit voltage. A variety of heating elements are installed in the baseboard heaters, including glass panel, metal-alloy strip, ceramic, finned tubing, and metal-sheathed chrome-wire types. One type uses a small *electric boiler*, and hot water circulates through the finned elements.

Provisions are made for wiring the heaters from the back, bottom, and ends, and for connecting two or more sections. The units may be equipped with receptacles for window- or through-the-wall-type room air conditioners. Units may also contain convenience outlets. This permits wall outlets to be located below the heat discharge and avoids portable cord deterioration from heat if the convenience outlets were located above the heater. Baseboard and sill-line heaters normally do not use fans and are of the convection type. Care must be exercised, particularly with the higher wattage per foot units, to assure that the high grill temperature (which can be over 200 °F) will not present a problem.

The heaters can be controlled by either a line- or a low-voltage wall-type thermostat, by built-in thermostats, or by thermostats installed in special baseboard sections designed for this purpose and matched with the baseboard in appearance.

Wall elements are similar to baseboard heaters, but are normally mounted higher than the baseboard type. Many types include fans. They are generally available in larger sizes with higher ratings than baseboard heaters.

Radiant heating panels of glass, ceramic, and metal alloy are designed for recessed or surface wall mounting, similar to resistance-type wall heaters. The heating element may be made up of tempered glass into which is fused a continuous alloy grid. Some units utilize a metallic coating, fired to the back of the glass.

11.2.3.2 Central Furnace or Boiler. The electric furnace is a central heating system and is closest in operation to the fuel-fired furnace, resistance elements replacing the combustion chamber. The heat is distributed by means of blowers and ducts. This equipment has step controls to permit energizing successive sections of the heating element in accordance with the amount of heat needed. The same ducts may be used to distribute cooled air during the summer months from a central air conditioner, or heat pump coils can be employed in the furnace to provide normal heating and cooling. Resistance elements are then used to supplement the heat pump in very cold weather or when quick recovery is desired.

The central boiler is similar in operation to the central furnace, with the duct system being replaced by a piped system. The resistance elements are used to generate hot water, which is pumped through the system to hot-water coils, baseboard heaters, or radiators.

In either type of central system the operating cost is usually higher than that of room resistance heaters. This is due to the loss of heat in the ducts or pipes to unused areas to overshoot of the temperature, and to not normally supplying separate thermostats for each room.

A modification of the central furnace concept distributed unheated air through the ducts, heating being accomplished by duct insert heaters located in the registers of each room or at some point in the ductwork.

11.2.3.3 Radiant-Heat Cables. Electric heating cable may be installed in plastered ceilings, between the layers of gypsum ceiling board, or in concrete floors. The cable has a diameter of approximately $\frac{1}{8}$ inch to $\frac{1}{4}$ inch, exclusive of splices, including plastic insulation. When installed it is laid back and forth on a flat surface with a definite spacing between turns. This spacing depends on the heat loss of the room heated, as does the length of cable.

To determine the length of cable required to heat a room, the heat loss has to be calculated. The loss in Btu/h can be converted directly to watts by dividing by 3.413. The length of cable is determined by the load in watts. Heating cable is available in a variety of sizes. The temperature of each room can be controlled with its own thermostat. Special thermostats or auxiliary relays may be required for heating cables over 4000 W. In operation, the surface temperature of the material in which the cable is embedded is above that of the human body. Cable comprises the most widely distributed heat sources, since it is installed in closely spaced rows over large unobstructed ceiling or floor areas. This type of heating equipment to old construction is rarely adaptable.

11.2.3.4 Unit Heaters. Unit heaters, used chiefly for spot heating of industrial or commercial areas, employ a relatively strong fan or blower to force air through a heating element into the space to be heated. Louvers are usually provided for direction control. A distinctive form of heater is called the schoolroom unit ventilator, although its use has not been restricted to schools. Its flexible operation permits cool outdoor air to be drawn in and directed through heating elements by a blower to the space to be heated. With the heating elements de-energized, the cool outdoor air may be required to ventilate when cooling rather than heating. A return intake permits room air to be mixed with the outdoor air in various proportions before passing through the heating coils.

11.2.3.5 Infrared Heaters. Where very localized heating is needed, infrared radiant heating is finding increased use. Typical applications are religious facilities, bowling alleys, warehouses, manufacturing areas, loading platforms, sidewalk areas adjacent to store windows, drive-in banks, and areas adjacent to concession stands where the infrared heaters provide comfort for the participants, workers, and shoppers. Quartz lamps, having a very high output in the infrared region, are popular as the heat source for such applications. NEMA HE3-1971 (R 1976) [3] [49] provides useful design and applcation information.

11.2.3.6 Heat-Storage Equipment. Heat-storage equipment is available in two major designs, (1) large single units serving a whole structure or (2) small units approximately the size of regular water radiators which can be dispersed about the various rooms in a structure.

---

[49] Numbers in brackets correspond to those of the references listed at the end of this section.

Heat-storage equipment may be useful to take advantage of off-peak utility rates, since it can be charged during the *valley hours*. A heater with a capability of storing 354 kWh is about 5 ft by 3.5 ft by 3 ft high. A heater with a 54 kWh storage capacity is about 3 ft by 2 ft by 1.5 ft deep. Heat is released by convection and radiation and may in some models be assisted by a small fan.

11.3 Load Calculations. Load calculations are required for the purpose of sizing heating and cooling equipment, and they are also useful for estimating energy consumption. The engineer should be aware, however, that with the present trend toward well-insulated structures and lowering of thermostats during unoccupied times, the load calculation may not provide the governing criterion for equipment sizes. In some cases the size of the equipment may be governed by a need for recovery of space temperature control within some reasonable time and, therefore, the equipment will be larger than would be required by load considerations alone.

The rigor of load calculations varies from simple rules-of-thumb calculations of the units/ft^2 type to very sophisticated calculation methods requiring a computer for implementation. The following discussions describe the simplest calculations consistent with meaningful results. Their use should be limited to small buildings having simple heating and cooling systems. In large buildings having complex heating, ventilating, and air-conditioning systems their use should be limited to determining only the general order of magnitude of the load.

The heating load or heat-loss calculation must include conduction through the building envelope, including walls, windows, floor slabs etc. It must also include a ventilation or infiltration load, or both. It is generally safe to ignore infiltration in buildings having ventilation rates of at least 0.15 ft^3/(min $\cdot$ ft^2). In residential construction or in buildings having little or no fixed ventilation an allowance must be made for infiltration.

The cooling load or heat-gain calculation must include all of the foregoing and also must include a solar gain term and an internal gain term. The internal gain term arises from the heat given off by lights, people, and equipment. The solar gain term arises from the transmission of radiant solar energy through windows and from the absorption of radiant energy by opaque exterior surfaces.

The various terms associated with these two heat-flow calculations are discussed in sections 11.3.1, 11.3.2, 11.3.3, and 11.3.4. The calculation of the resistance of building walls to the flow of heat by conduction is discussed in 11.3.1. This section is applicable to both heat-gain and heat-loss calculations. The heating and cooling load associated with the admission of outdoor air to an interior space is discussed in 11.3.2. The heat-gain terms which are associated with the solar load are discussed in 11.3.3. The internal heat gain associated with people, lights, and equipment are described in 11.3.4. These last two calculations are not generally included in heat-loss calculations although, in some cases, common sense will indicate that internal heat gains should be included in heat-loss calculations.

11.3.1 Conduction. The resistance of building materials to the flow of heat is analagous to the resistance of materials to the flow of electricity. Physically, they are related phenomena although the flow of heat does not depend entirely

472

on the motion of electrons as does the flow of electricity. In any case, the parallel is quite useful between voltage and temperature difference and between current flow and heat flow. Just as for electrical circuits, there are parallel and series connections of heat paths.

The basic equation for heat conduction has the form

$$Q = \frac{\Delta T}{R}$$

where

Q = heat energy flow in Btu/(h $\cdot$ ft^2)

ΔT = temperature difference between the two faces of the material in °F

R = resistance to the flow of heat in h $\cdot$ °F/(ft^2 $\cdot$ Btu)

The values of R for various thicknesses of building materials are tabulated in many places. One of the best sources is ASHRAE Handbook of Fundamentals [2]. In some references, such as Marks Handbook of Mechanical Engineering [15] the conductivity rather than the resistance is tabulated. To find the resistance take the reciprocal of the conductivity and multiply by the thickness in feet.

To find the resistance of a composite wall or roof built of several different materials it is only necessary to combine the resistances exactly as would be done in an electrical circuit.

There is one difficulty with this analog. Virtually all heat-flow calculations include the flow of heat between air and solids. The resistance of this part of the circuit is a much more difficult term to deal with as it depends among other things on the orientation of the interface, on the temperature difference, and on the presence or absence of wind or air motion. This interface resistance is generally assumed, however, to have a value of 0.17 for outdoor surfaces and 0.68 for indoor surfaces. The difference arises from the assumption of a 15 mi/h wind outdoors.

Inherent in all of this is the problem of determining the applicable temperature difference. The inside temperature is generally known but the outside temperature must be taken from local weather bureau records or from some standard listing of weather extremes such as is found in the ASHRAE Handbook on Fundamentals [11].

11.3.2 Ventilation and Infiltration. Ventilation is the intentional exchange of outside air for inside air to remove odors and smoke. Infiltration, on the other hand, is the uncontrolled penetration of outside air through cracks, holes, and other openings such as doors, due to a difference in pressure between the inside and the outside. If the quantity of ventilation air is sufficient to maintain a positive pressure inside the building then the infiltration of air may be ignored.

Outside air will, by one means or another, find its way into a building and will therefore impose a load on the heating and cooling system. This load will depend not only on the difference in air temperature but also on the difference in water content, or humidity. The comfort of persons in a building will depend to some extent on the humidity and generally the humidity is controlled to some degree. Recent practice is to control humidity much less closely than was formerly the case. Regardless of any intended control, however, air-conditioning equipment by its very nature will tend to decrease humidity and therefore a dehumidification load should be recognized.

Air and water vapor at atmospheric conditions behave pretty much as ideal gases and, therefore, have the property that their energy content (or enthalpy) is a function only of temperature. Thus, it is possible, knowing the relative quantities of air and water in the mixture and the temperature of the mixture, to calculate the energy content.

Fortunately, there is one measurement which can be used to determine energy content directly without knowing the exact quantity of water vapor present. This quantity is called the wet-bulb temperature and it can be measured with a wet-bulb thermometer. More importantly, the energy content of air can be tabulated on a per pound of dry air basis as a function of wet-bulb temperature only, and the relationship between wet-bulb temperature and relative humidity can be tabulated. A graphical presentation of these two tables is known as a psychrometric chart, which can be found in any air-conditioning manual or thermodynamics text.

In many cases, especially for heat-loss calculations, no load will be associated with differences in humidity. (Unlike an air conditioner, a heater does not necessarily alter the water content of air). In this case the heat load will depend only on the dry-bulb temperatures. The load then, will depend only on the specific heat of the mixture, the quantity of outside air entering, and the temperature difference. The specific heat can be taken as 0.25 Btu/(lb $\cdot$ °F) (air alone has a specific heat of 0.24).

The density of atmospheric air is approximately 0.075 lb/ft^3. With the given assumed specific heat of 0.25 Btu/(lb $\cdot$ °F) we have a specific heat on a volume basis of

$$C = 0.075 \text{ (lb/ft}^3) \cdot 0.25 \text{ (Btu/°F)/lb}$$
$$= 0.088 \text{ (Btu/°F} \cdot \text{ft}^3)$$

since heat-loss calculations are always done on a per hour basis the above number is multiplied by 60.

$$C = 0.0188 \text{ (Btu/°F} \cdot \text{ft}^3) \cdot 60 \text{ min/h}$$
$$= 1.1 \text{ Btu/(min/°F) / (ft}^3\text{/h)}$$

Thus

$$Q = (1.1 \cdot V) \cdot (T_i - T_o) \text{ Btu/hour}$$

where

T_i = inside temperature, °F

T_o = outside temperature, °F

V = ventilation rate, ft^3/min

and

Q = so called sensible heating load/ (Btu/h)

If the humidity difference is considered then the energy content (enthalpy) must be taken from a table or chart. It is tabulated on a per pound basis and therefore these values must be multiplied by the density of air since ventilation is always measured in cubic feet.

Thus

$$Q = 0.075 \text{ (lb/ft}^3) \cdot V \cdot (h_o - h_i)$$

where

V = ventilation per infiltration rate, ft^3/min

h_o = energy content of outside air, Btu/lb

h_i = energy content of inside air, Btu/lb

Q = load on system Btu/min

Since load terms are usually done on a per hour basis, the above equation is multiplied by 60.

and

$$Q = 4.5 \cdot V \cdot (h_o - h_i)$$

The following example will demonstrate the use of these equations.

Given

Outside air temperature = 92° dry bulb
78° wet bulb

Inside air = 75° dry bulb
60% relative humidity

Ventilation rate = 1000 ft³ /min

From a psychrometric chart for 78° wet bulb;

h_o = 41.55

From a chart for 75° dry bulb and 60% relative humidity;

h_i = 30.25

Then

Q = 4.5 · 1000 · (41.55 – 30.25)

= 50860 Btu/h

If the humidity term or latent load is ignored,

then

Q = 1.1 · 1000 · (92 – 75)

= 18700 Btu/h

This indicates the importance of including the humidity effects.

The quantities of air required for ventilation will vary, depending on many factors. However, (5 ft³ /min) per person is generally sufficient for most situations except where smoking is a problem, in which case 10 ft³ or even 15 ft³ /min may be required. Usually a minimum ventilation rate, 0.10 ft³ /min/ft², with 0.15 a more usual figure.

The applicable temperature differences may be taken from weather bureau records but the use of tabulated extremes is recommended as they generally include a frequency of occurrence.

For buildings having no fixed ventilation a reasonable estimate for infiltration is one air change per hour. This corresponds to a ventilation rate sufficient to replace all the air in the space in 1 h.

11.3.3 Solar Effects. The solar load is quite often the single largest term in a heat-gain calculation and it always represents a significant part of the load. Solar energy is initially in the form of radiation but it ultimately arrives in the building by conduction and convection. For this reason it is more difficult to analyze than the other terms in the heat-gain calculation. Fortunately, many emperical tables and charts are available to simplify the calculation.

The solar load is always treated in two parts. First and generally most important is the solar energy which is transmitted directly to the space through the windows. Ordinary double strength window glass, for instance, transmits 85 to 90% of the incident radiation. Second, is the radiant energy absorbed by the opaque exterior walls and ultimately transmitted to the interior.

11.3.3.1 Window Solar Load. The solar energy transmitted through windows, called the solar-heat gain factor, has been tabulated for ordinary window glass as a function of latitude, time of day, month of year, and orientation of window. Such a table can be found in the ASHRAE Handbook on Fundamentals [11] and in several air-conditioning design manuals. The tabular values may be used directly with one important exception. If the windows have shading devices such as a curtain, or blinds, or if they are made of special glass, then a correction factor known as the shading coefficient must be applied. This is a multiplier which ranges from 0.25 for special glass with heavy curtains to 0.9 for double glazed windows with no shade device. The applicable shading coefficient must be taken from the manufacturer's literature or a handbook, or both.

The flow of radiant solar energy through glass and the conduction of heat by the same glass are almost independent processes. For that reason it will always be necessary to add to the solar-heat gain the heat gain or loss due to conduction. Single glazed windows have an R value of about 0.88. The heat flow through windows may be summarized by the following equation.

$$Q = (SC \cdot SHGF) + (T_o - T_i)/R + A$$

where

$$
\begin{aligned}
SC &= \text{shade coefficient} \\
SHGF &= \text{solar heat gain factor} \\
R &= \text{resistance to heat flow,} \\
&\quad \text{h} \cdot {}^\circ\text{F}/(\text{ft}^2 \cdot \text{Btu}) \\
&= 0.88 \text{ for ordinary window} \\
T_o &= \text{outside temperature, } {}^\circ\text{F} \\
T_i &= \text{inside temperature, } {}^\circ\text{F} \\
Q &= \text{total flow of heat into build-} \\
&\quad \text{ing through the window in} \\
&\quad \text{Btu/h} \\
A &= \text{glass area in ft}^2
\end{aligned}
$$

11.3.3.2 Solar Load Through Walls and Roof. The solar load due to solar energy absorbed by exterior walls and the roof is a complex problem in transient heat flow. In simplest terms it involves the absorbtion of solar energy at an exterior surface, the heating of the wall, and the subsequent transfer of part of the energy to the interior, and part of the energy to the exterior. Solutions to this problem have been characterized in terms of a quantity called the Total Equivalent Temperature Difference (TETD).[50]

The Total Equivalent Temperature Difference (TETD) may be used with the wall or roof R value to calculate the heat gain just as if it were a real temperature difference. The TETD will include all heat flow, thus a calculation need not be made with actual outside air temperature. The TETD is a function of the following quantities.

(1) Time of day
(2) Month of year
(3) Mass of wall or roof
(4) Outdoor temperature profile
(5) Indoor temperature
(6) Color of wall
(7) Latitude of building
(8) Orientation of wall

TETD tables are published in almost all air-conditioning design manuals and handbooks. They are usually given for one month of the year, one set of temperature data, and one latitude. Corrections for variations from this set of values will be given with the tables. In using the tables it will be necessary to find the wall construction that most closely matches the wall under consideration, with the weight of the wall being the controlling factor. Some tables are given only in terms of weight of construction.

---

[50] TETD is not to be confused with the Sol Air Temperature, which has no direct application.

11.3.4 Internal Heat Gain. A major factor in calculating the load on an air-conditioning system is the heat sources within the conditioned space such as lights, occupants, and equipment.

This is usually a very simple calculation. An adult occupant generally gives off about 500 Btu/h unless engaged in some vigorous activity in which case 1000 Btu/h is a better estimate. The power dissipated by equipment may be estimated from name plate data.

The only complication that may arise involves the lighting, which is usually the most important item of internal gain. If the lighting system is such that part of the heat produced by the lights is carried away by an air-return system and exhausted outdoors then that part of the lighting heat will not appear as heat gain. In the usual case it is only necessary to sum the total watts for lighting and multiply by 3413 to obtain Btu/h of lighting heat load.

11.4 Energy Conservation. The term *energy conservation* conjures up many varied meanings. In its simplest terms it means minimizing the use of energy in a given application of energy. In the case of electricity, energy conservation often includes the minimizing of electrical demand (kW) as well as energy (kW · h). In this context, the term *energy management* is often used. This should not be confused with the term *load management*, which usually implies control by some party other than the end user.

Energy conservation as an engineering topic may be divided into two broad areas. First is the area of practical energy conservation having to do with the engineering decisions that influence energy consumption in a given building. The second area centers on legislative guidelines and limits for energy consumption based on the weighing of the relative importance of the social versus the economic factors involved in energy resource allocation.

11.4.1 Practical Energy Conservation. Clearly this is a topic which could occupy several books without exhaustion. Many of the energy uses in buildings, while electrical by nature, are frequently the responsibility of the mechanical design team. Heating, cooling, ventilation, heat recovery, controls, etc, are in this group of energy uses. There is, however, considerable interaction between mechanical and electrical disciplines. For example, temperature setback during unoccupied hours could impose an abnormal recovery peak on restoration on an electric heating system.

The following discussion, therefore, is intended only as a perspective on the particular topic of lighting, since lighting is the major use of electrical energy in many buildings. It is not intended to be a discourse on the design of efficient lighting designs, but rather to aid in the evaluation of alternate designs.

The cost of operating a lighting system involves much more than the energy consumed by the light fixtures. The analysis of the energy consumption should include the interaction with all the building systems. In many buildings the lighting system may account for one half the load on the building's air-conditioning system. Furthermore, lighting loads can vary from 5 Btu/ft^2 (1.5 W/ft^2) or less, to 15 Btu/ft^2 (4.5 W/ft^2) or more. Thus, the lighting designer has a major influence in the ultimate size of the air-conditioning system.

Energy consumption may be analyzed, on the basis of total energy consumed, or on the basis of operating costs. An energy basis should include the cost of generation which, for fossil fuels, is approximately 3.3 times the Btu equivalent of the electrical energy consumed. The cost basis will, however, generally have the most immediate and direct application.

The interaction of the lighting system with the other systems arises as a direct result of the heat produced by lighting. In a cooling mode the building system must remove this heat. This is equivalent to an increased air-conditioning cost due to the lighting system of at least 33% for a large well-designed central-air-conditioning system. For smaller, less efficient, or badly-designed systems, this overhead figure might be as much as 60%.

In a heating mode, the heat produced by the lighting system will represent a gain. In some buildings with an inefficient or lightly loaded furnace or boiler, the heating energy from the lighting system may be less expensive than from the furnace or boiler and the illumination would be free. This breakeven point would occur for instance with electricity at $0.06/(kW \cdot h)$ and fuel oil at $1.00/gal and a heating plant efficiency of 40%. This assumption, however, may not hold true in many cases since the heat produced by the lighting system may not be where it is required nor in the proper quantity at a given time.

In the design of lighting systems a number of factors bearing on energy conservation should be taken into consideration. These include:

(1) Use of the most efficient light source and luminaire for the particular application

(2) Provision of adequate local and zone switching

(3) Possible use of time clocks, photocells, and dimming systems

(4) Possible integration of lighting fixtures with air-conditioning systems

(5) Efficient lighting design relating to quality, quantity, and task

11.4.2 Standards and Codes. Most of the standards or guidelines in use today are prescriptive in nature; that is, they are specific design criteria that when applied will minimize the energy demand of a structure. ASHRAE 90A-80 [1] is typical of such standards. They do not usually deal with the performance of the various energy-using components. There is a trend developing towards the use of performance standards for various classes of buildings. Such standards or guidelines will probably set limits on annual energy consumption on some common unit such as per square foot of floor area. The use of performance standards should allow designers greater latitude in design and should, in the final analysis, result in buildings operated in a more energy efficient manner.

The use of any of the numerous computer programs now available is often essential in order for the designer to properly assess the various alternatives available.

11.5 Glossary

air, ambient. The air surrounding or occupying a space or object.

air conditioning. The process of treating air so as to control simultaneously temperature, humidity, and distribution to the conditioned space.

air, recirculated. Return air passed through the conditioner before being supplied again to the conditioned space.

air, return. Air returned from the conditioned space.

air ventilation. The amount of supply air required to maintain the desired quality of air within a designated space.

British thermal unit (Btu). The quantity of heat required to raise one pound of water 1 °F.

calorie. The quantity of heat required to raise one gram of water 1 °C.

capacity, heat. The amount of heat necessary to raise the temperature of a given mass one degree—the mass multiplied by the specific heat.

coefficient of performance (heat pump). Ratio of heating effect produced to the energy supplied.

conductivity, thermal. The time rate of heat flow through a unit area of a homogenous substance under steady conditions when a unit temperature gradient is maintained in the direction normal to the area.

control. Any device for regulation of a system or component.

degree day. A unit based upon temperature difference and time used for estimating fuel consumption and specifying nominal heating loads of buildings during the heating season. Degree days = number of degrees (F) that mean temperature is below 65 °F · days.

dehumidification. Condensation of water vapor from the air by cooling below the dew point or removal of water vapor from air by physical or chemical means.

diffusivity, thermal. Thermal conductivity divided by product of density and specific heat.

heat, specific. The ratio of the quantity of heat required to raise the temperature of a given mass of a substance one degree to the heat required to raise the temperature of an equal amount of water one degree.

heat pump. A refrigerating system employed to transfer heat into a space or substance. The condenser provides the heat while the evaporator is arranged to pick up heat from air, water, etc. By shifting the flow of air or other fluid a heat-pump system may also be used to cool a space.

heating system, radiant. A heating system in which the heat radiated from panels is effective in providing heating requirements. The term radiant heating includes panel and radiant heating.

heating unit, electric. A structure containing one or more heating elements, electrical terminals or leads, electric insulation and a frame or casing, all assembled into one unit.

humidity. Water vapor within a given space.

humidity, relative. The ratio of the mole fraction of water vapor present in the air to the mole fraction of water vapor present in saturated air.

infiltration. Air flowing inward through windows, etc.

insulation, thermal. A material having a high resistance to heat flow and used to retard the flow of heat.

isothermal. A process occurring at constant temperature.

lag. The delay in action of a sensing element of a control element.

load, estimated maximum. The calculated maximum heat transfer which a heating or cooling system will be called upon to provide.

radiator. A heating unit which provides heat transfer to objects within visible range by radiation and by conduction to the surrounding air which is circulated by natural convection.

resistivity, thermal. The reciprocal of thermal conductivity.

solar constant. The solar intensity incident on a surface oriented normal to the sun's rays and located outside the earth's atmosphere at a distance from the sun equal to the mean distance between the earth and the sun. The values for July and January are 415 and 445 Btu/h $\cdot$ ft^2, respectively. The mean value is 430 and the sea level value in July, because of atmospheric attenuation, is close to (300 Btu/h) $\cdot$ ft^2.

temperature, dew point. The temperature at which condensation of water vapor in a space begins. The dew point temperature is a function of pressure and humidity.

temperature, dry bulb. The temperature of a gas, or mixture of gases, indicated by an accurate thermometer after correction for radiation.

temperature, effective. An arbitrary index which combines into a single value the effects of temperature, humidity, and air movement on the sensation of heat or cold felt by the human body. The numerical value is that of the temperature of still, saturated air which would induce an identical sensation.

temperature, wet bulb. The temperature at which liquid or solid water, by evaporating into air, can bring air into saturation adiabatically at the same temperature.

therm. A quantity of heat equal to 100 000 Btu.

thermostat. A device which responds to temperature and, directly or indirectly, controls temperature.

ton of refrigeration. 12 000 Btu/h.

transmittance, thermal (U factor). The time rate of heat flow per unit temperature difference.

velocity, room. The average sustained residual air velocity in the occupied zone of the conditioned space.

11.6 References

[1] ASHRAE 90A-80, Energy Conservation in New Building Design.[51]

[2] IES Lighting Handbooks, Reference and Application Volumes, 1981.[52]

[3] NEMA HE3-1971 (R 1976), Manual for Infrared Comfort Heating.[53]

[4] ASHRAE, *Handbook on Fundamentals*, 1981.

[5] ASHRAE, *Handbook on Equipment*, 1979.

[6] ASHRAE, *Handbook on Applications*, 1978.

[7] ASHRAE, *Handbook on Systems*, 1980.

[8] *Marks Handbook on Mechanical Engineering*, 8th ed, T. BAUMEISTER, Editor, New York, McGraw-Hill, 1978.

[9] NBS COMMITTEE REPORT, Energy Conservation in Public Buildings,

---

[51] ASHRAE publications are available from American Society of Heating, Refrigerating, and Air-Conditioning Engineers, 1791 Tullie Circle, N.E., Atlanta, GA 30329.

[52] IES publications are available from the Illuminating Engineering Society, 345 East 47 Street, New York 10017.

[53] NEMA publications are available from the National Electrical Manufacturers Association (NEMA), 2101 L. Street, NW, Washington, DC 20037.

National Bureau of Standards, General Services Administration, July 1972.[54]

[10] NBS COMMITTEE REPORT, Technical Options for Energy Conservation in Buildings, *National Bureau of Standards, Building Environment Division.*

[11] AMBROSE, E.R. *Heat Pumps and Electric Heating.* John Wiley and Sons, Inc., New York, 1969.

[12] KLAUSS, A.K. Changing Concepts in Ventilation Requirements. ASHRAE *Journal, June* 1970.

[13] ROOTS, W.K. *Fundamentals of Temperature Control.* Academic Press, New York, 1969.

[14] STEIN, E. and McGUINESS, W. *Mechanical and Electrical Equipment in Buildings.* John Wiley and Sons, Inc, New York, 1971.

[15] STROCK, C. and KORAL, R.L. *Handbook of Air Conditioning, Heating and Ventilating.* 2nd ed. Industrial Press, New York, 1965.

[16] *Guide for Calculations of Heating and Cooling Requirements.* Tennessee Valley Authority.

[17] *Heat Pump Improvement Project* (RP 59). Edison Electric Institute Report, May 1971.

---

[54]NBS publications are available from the National Bureau of Standards, General Administration, Washington, DC.

12. Transportation

12.1 General Discussion

12.1.1 Use. Vertical transportation is often mandatory in the multistory structure. Office buildings, hospitals, hotels, department stores, and many other types of buildings must be adequately provided with elevators and escalators. The diversity of the transportation problem requires a variety of equipment to meet various types of service demand. The degree of refinement in control and operation may affect the amount of energy used. However, elevator equipment or systems should never be selected solely on the basis of a reduced energy consumption simply in the interests of economy. Any such saving is insignificant compared with the total elevator operating costs, while a possible deficiency in elevator service may impair the return on the entire building investment.

12.1.2 Design Factors. There is no simple formula for determinating the characteristics of an elevator plant. The number of variables involved does not lend itself to some rule-of-thumb that could be used as a yardstick by those unfamiliar with elevator application and practice. Proper elevator plant design requires a complete knowledge—not only of elevator machinery, controls, and operations—but of mathematical probabilities and human behavior as well. Only the experienced elevator engineer is prepared to advise as to the proper number, size, speed, control, and operation of the elevators that will be required to handle adequately the traffic in any particular building. The engineer is in a position to predict these requirements accurately, even if the building is not as yet erected, because of the wealth of available data from many different types of buildings in many different localities, and through the use of computer simulation techniques. Moreover, he is prepared to see that the elevators have the necessary characteristics for fast and efficient service with particular emphasis on smoothness of acceleration and slowdown, accuracy of stopping, and overall operating simplicity for the personnel of the building and the visiting public. Reliability of equipment and minimum maintenance are also important design factors. ANSI/ASME A17.1-1981 [8] and ANSI A17.2-1979 [3] [55] cover the design, construction, installation, operation, testing, maintenance, alteration, and repair of the subject equipment. Since these standards have been prepared with due consideration for past accident experience and are based on sound engineering principles, they are widely accepted by state and municipal jurisdictions as the standards for safety. Installations should conform to the requirements of this code in order to ensure the safety of those who use, maintain, and inspect such equipment. The designer should be aware of changing requirements in the fire-safety rules and local laws based on them which involve automatic and lobby control for return of elevators to the main floor in the event of fire, and of the required availability for elevator service for the fire department.

12.1.3 Consultations. The elevator manufacturer and the consulting engineer are the logical consultants on matters pertaining to vertical transportation. Consultation must be early. If the building design has already progressed to the point where major structural or equipment changes can no longer be made, then the elevator service may suffer.

12.1.4 Efficiency and Economy. Transportation efficiency and economical operation are mostly a function of equipment and systems alone. Automation has replaced the human element in elevator system control and operation. Acceleration, deceleration, floor leveling, and door operation are automatic. Today, through the development of automatic group-supervisory control systems and

adequate protection devices, large groups of high-speed elevators are operated entirely by the passengers with complete confidence and safety.

12.1.5 Responsibilities. Regardless of the complexities of some elevator systems and the need for expert advice in applying elevator equipment, the customers are certainly entitled to know what they are buying and why they are buying it. Furthermore, they have the prerogative of selection, since it is their money that is being spent. In order that the selection may be a wise one, it is the function of this section to provide a better understanding of elevator systems, and to review briefly the more important units which make up a system with particular emphasis on advantages to be derived and power characteristics to be considered.

The economic success of a commercial building, particularly a busy office building, may very well be determined by the quality of the services rendered to the building occupants. Elevators play a major role in these building services. It is the architect or consulting engineer and finally the building owner and building manager who bear the responsibility of providing an adequate elevator plant and adequate elevator service.

Elevators are available with the traction equipment located in the basement instead of in a penthouse or upper floor. So, basement traction type of underslung traction types of elevators may be used in buildings where owners are considering the addition of future floors, in the retrofit of existing buildings, or where it is desirable to eliminate a penthouse projection above the roofline. Bear in mind, however, that the overhead loadings on the building structure imposed by the elevator installation are doubled with basement applications.

---

[55] The numbers in brackets correspond to those of the reference listed in 12.17 at the end of this section.

12.2 Types of Transportation

12.2.1 Electric Elevators. Almost all modern electric elevators are of the traction type, where the hoisting ropes pass from the elevator car to a counterweight over a grooved driving sheave on the hoisting machine. Motion of an elevator and counterweight in either direction depends on the friction created between the cables and the grooved sheave surfaces by the suspended weights. Elevators are inherently safe; since the loss of rope tension results in the loss of traction and prevents either the car or counterweight from being drawn into the overhead structure. ANSI/ASME A17.1-1981 [8] prohibits the installation of drum-type machines except for limited-travel slow-speed freight elevators without counterweights. Traction machines are of two types: geared machines with suitable gearing (generally the worm-gear type) between the hoisting motor and the driving sheave, and gearless machines where the sheave is mounted directly on the motor shaft. Geared-machine elevators are not widely used for car speeds above 450 ft/min. It is quite common, however, to use gearless machines above 400 ft/min. Gearless-machine elevators have been installed for car speeds as high as 2000 ft/min. Although the design characteristics are such as to classify the geared type as a lower car-speed machine and the gearless type as a higher car-speed machine, it is inevitable that there will be speed ranges where either type of machine is applicable. In making a selection it should be kept in mind that, although the geared machine is somewhat less expensive, the gearless machine provides improved efficiency, smoothness and quietness of operation, and a considerably longer life.

12.2.2 Hydraulic Elevators. ANSI/ASME A17.1-1981 [8] requires that hydraulic elevators have direct-plunger driving machines, where a plunger or piston attached directly to the car operates in a cylinder under hydraulic pressure.

Since the cylinder must project below ground as far as the elevator travels above, the costs entailed in providing this arrangement may make hydraulic elevators uneconomical for buildings above a certain height and may require supplementary plunger support. The car speeds of hydraulic elevators are usually limited to 200 ft min.

Hydraulic elevators are particularly suited to low buildings where overhead space and building load are limited. Hydraulic elevators are powered by a motor-driven pump feeding the cylinder from a supply tank.

The relatively new *holeless* hydraulic elevator has been introduced for two- or three-stop applications. The cylinder is mounted above the pit floor, thus eliminating the need for a costly plumb hole below pit level.

The need for increased motor size and the resulting increased current demand, should be taken into consideration when hydraulic elevators will be located in a heavy-usage environment.

12.2.3 Dumbwaiters. Electric dumbwaiters are used to transport material only. The dumbwaiter requirements of ANSI/ASME A17.1-1981 [8] restrict car enclosure size and capacity.

The machine may be either a traction or drum type. The hoist ropes of the traction type extend from the car over the traction drive sheave on the machine to the counterweight. The hoist ropes of a drum-type machine, extend from the car to the machine where it is wound on a drum; counterweight is not used. Drum-type machine installations are limited to low rises because of the small amount of hoist rope that can be wound on the drum.

Since dumbwaiters are strictly for material handling, they do not have buttons in the cab enclosure and are always operated externally.

Dumbwaiters with only two floors usually come with a *Call and Send* control system where the car can be called to or sent from either landing. With multi-floor dumbwaiters, a *Multi-Button* control system is used where floor-control stations contain a call button and a send button for each landing served in order to send the car to the other landings. For intensive dumbwaiter service, more sophisticated control systems are employed such as *Central Station Dispatching.*

12.2.4 Escalators and Moving Walks.

Escalators and moving walks are applicable where it is desirable to move people continuously. Department stores, shopping malls, office buildings, rail and air terminals, parking facilities, subways, and sports arenas are particularly suited for escalators or moving walks, or both.

Escalators are generally operated at an angle of incline of 30° and are furnished in widths of 32 inches and 48 inches and at speeds of 90 to 120 ft min. Although handling capacities for the 32 in and 48 in escalators are rated at 5000 and 8000 persons per hour respectively, at the 90 ft/min rate, actual observations indicate that 55% to 75% of the rated handling capacity is more accurate.

A recent departure from the conventional escalator is the modular escalator concept in which the drive unit is contained in the truss as opposed to a drive machine at the top end. One drive unit is required for approximately each 20 ft increment of vertical rise. This results in a reduction of motor horsepower rating as well as the space requirement.

The minimum width of the exposed treadway of a moving walk is 16 in. Maximum width is dependent on treadway slope and treadway speed. Maximum width is also governed by code and is as shown in Table 87.

The maximum speed of the moving walk treadway is dependent on both the maximum treadway slope at points of entrance and exit, and on the maximum treadway slope at any other point on the treadway. This speed shall not exceed the lesser of the values determined by Tables 88 and 89.

Moving-walk capacities vary with speed, width, and angle of incline.

Access and egress areas to escalator

Table 87

Maximum Treadway Slope at Any Point	Maximum Moving Walk Treadway Width in Inches		
	90 ft/min Maximum Treadway Speed	Above 90 ft/min to 140 ft/min Treadway Speed	Above 140 ft/min to 180 ft/min Treadway Speed
0 to 5°	Unrestricted	60	40
Above 5 to 8°	40	40	40
Above 8 to 15°	40	40	Not Permitted

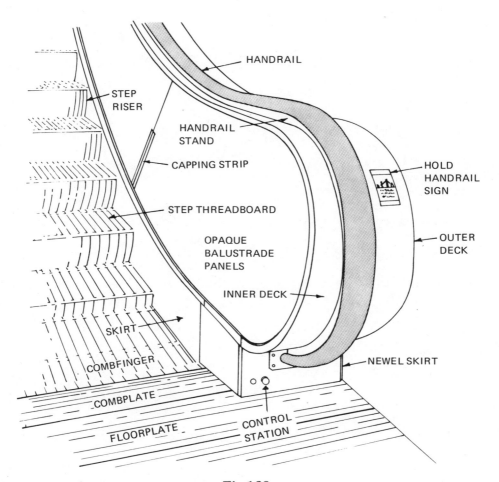

Fig 138
Opaque Balustrade Escalator

Table 88

Maximum Treadway Slope at Point of Entrance or Exit	Maximum Treadway Speed in ft/min
0 to 3°	180
Above 3 to 5°	160
Above 5 to 8°	140
Above 8 to 12°	130
Above 12 to 15°	125

Table 89

Maximum Treadway Slope at Any Point on Treadway	Maximum Treadway Speed in ft/min
0 to 8°	180
Above 8 to 15°	140

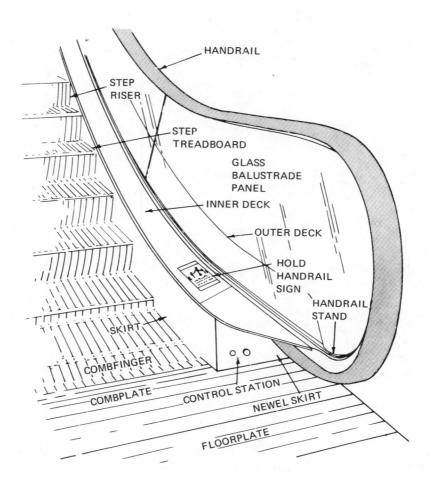

Fig 139
Transparent Balustrade Escalator

and moving-walk installations must be considered in order to avoid hazardous congestion. Distance from the handrail at the newel end to the nearest obstruction (wall, door, etc) should not be less than 10 ft for optimum passenger movement.

12.2.5 Material Lifts. Many vertical materials handling systems are available in addition to elevators for vertical transportation. Elevator-related systems include manually loaded dumbwaiters, dumbwaiters with automated tote-box loading and unloading, and dumbwaiters which automatically load and unload cars. When the code limitation of 9 ft^2 of platform area is exceeded, ANSI/ASME A17.1-1981 [8] requires the material lift to be classified as an elevator rather than a dumbwaiter. The elevator size materials lift can also be equipped for automatic loading and unloading of carts and can be adopted to automatically handle carts or containers on overhead powered conveyor tracks, floor wire or optically guided self-propelled carts, floor or chain vehicles.

Elevator-type material lifts are often equipped with power-operated hoistway doors which are automatically opened to receive or discharge the load and to close when the car is ready to travel. The doors are fire-rated, and important to the safety of the building.

Conveyor-related systems include vertical-selective conveyors, which are continuous chain systems operating the full height of the building. These systems automatically load tote boxes on the up traveling chain and discharge them from the down traveling chain. Such systems can interface with horizontally traveling belts or roller conveyors and are completely automated.

Another conveyor-related system utilized self-propelled electric carriers of tote box size. The box runs in a track on a powered chassis which receives electrical current from a trolley in the track. Switching is provided to direct the carrier to any location. Horizontal travel is by the powered wheels on the carrier and vertical travel by the engagement of a pinion on the carrier in a rack gear in the track.

Elaborate automated storage and vertical conveying systems have been developed to handle anything from finished automobiles to aircraft freight containers. This is done by using the combination of vertical elevator type carriage with material transfer device operating in a rolling tower indexing both horizontally and vertically to store or retrieve the article. Vertical travel energy consumption and power requirements are related to the system used, whether it be the counterweighted elevator type, hydraulic or drum type hoisting. Horizontal travel requirements are similar to those of a railroad, rubber-tired vehicle, chain or belt depending upon the system used. It all totals up to the power needed to start, accelerate, run, decelerate, level or position, and stop including possible consequences of regeneration when the load is decelerated and stopped.

12.2.6 Man Lifts. *Man Lifts* are single-passenger endless belts with attached passenger steps and grasps. These lifts are frequently used in commercial valet-type parking garages in large cities and in other buildings where it is desired to have a high speed of employee travel to remote or high-vertical areas. Man lifts are usually restricted to use by qualified persons only since there are certain hazards associated with their operation.

Man lifts are not included in the scope of ANSI/ASME A17.1-1981 [8] but are covered by ANSI A90.1-1976 [4].

12.2.7 Pneumatic Tubes. Early pneumatic message-tube systems required a person to receive carriers and manually insert them into their destination tube. Each point of origin and dispatch was connected by two tubes.

Today, these systems are computerized, thus permitting single-tube stations, since the carrier can now be dispatched by pushbutton through the computer rather than on the carrier itself. Systems can now accommodate tube sizes up to 8 in round and 4 in · 12 in for special applications.

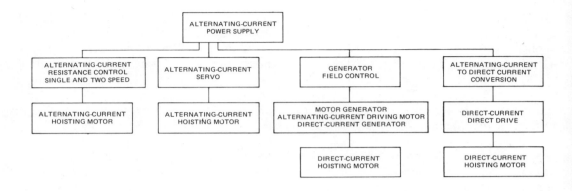

Fig 140
Basic Electric-Elevator Speed-
Control Systems

12.3 Elevator Control, Motors, and Motor-Generators. In general, the principal functions of elevator control are to connect the source of power to the elevator hoisting motor to determine the direction of motor rotation and consequently the direction of car travel, and to dictate the acceleration, deceleration, leveling, and stopping of the elevator at a floor. Such control also provides necessary overcurrent protection for the motor, and includes provisions to handle negative or overhauling loads satisfactorily when the hoisting motor is actually being driven by the elevator. The term *control* should not be confused with the term *operation* as is sometimes the case. Control is the system for regulating the elevator as indicated above. Operation is the method of actuating the control in some predetermined manner so that the elevator, or elevators, respond in accordance with the pattern most suitable to the particular traffic demand.

12.3.1 Control Systems. There are four basic electric-elevator control systems:

(1) Alternating-current resistance control

(2) AC servo control

(3) Generator field control

(4) Direct-current direct drive control

These systems are shown in simple diagrammatic form in Fig 140.

12.3.2 Alternating-Current Resistance Control. Resistance control is used with geared machines for low-speed elevators up to 150 ft/min. Alternating-current controls can be one speed or two speed, with a fast to slow speed ratio of up to 1:6. Motors rated up to about 15 hp are usually started across the line, while larger horsepower motor controllers usually use resistance or wye-delta starting. Alternating-current traction elevators should not be used where maximum smoothness of performance and close accuracy of stopping at a floor are required.

The alternating-current rheostatic control system incorporates a one- or two-speed alternating-current squirrel-cage induction hoisting motor and a type of control which provides, in addition to the principal control functions, phase-reversal and phase-failure protection and reduced-voltage starting through the medium of starting resistances or reactances. The elevator hoisting motor may be started across the line, provided the quality of start is acceptable for the character of service being furnished, and provided that the starting current is within the limits imposed by the electric utility company. Wound-rotor motors are rarely used because of their more elaborate construction, and because they require more complicated control equipment than the squirrel-cage type.

12.3.3 AC Servo. Presently SCR control of an alternating-current motor is generally used with a geared machine and is usually restricted to 400 ft/min.

The control system is a static control, closed-loop system using tachometer feedback to control an alternating-current squirrel-cage induction hoisting motor. The ac motor is controlled during acceleration, running, and deceleration by comparing the tachometer feedback signal to a speed reference signal which provides for smooth operation under all load conditions.

This control system provides accurate floor stopping.

12.3.4 Generator Field Control. The generator field control system may be used for either geared or gearless machines without restriction as to the car speed. The system incorporates a constant-running, constant speed motor-generator set for each elevator to supply variable-voltage direct current to the direct-current, shunt-wound hoisting motor. Control of the hoisting motor is obtained largely through control of the generator field excitation. This gives an inherently smooth performance since the inductance of the generator field reduces the motor armature current fluctuations and consequently eliminates noticeable shocks during the various steps of acceleration and deceleration. In addition to being a smoother and more refined control, it permits the use of smaller and more economical control components, since small field currents, rather than large motor currents, are being handled. Direct-current rheostatic control systems have little application today because of the relatively few power systems which distribute direct current.

12.3.5 DC Direct Drive. Solid-state technology is not only replacing the traditional relay controls, but SCR drives are replacing the motor-generator (MG) sets on geared and gearless installations.

With SCR drives, energy generated by the motor during normal decelerating and regenerative braking conditions is fed back to the line. Static power conversion, eliminating the MG set and its inherent losses, results in more efficient use of power. Motor armature control, using all static, solid-state modules, provides positive, smooth response to all speeds and loads within the rating of the drive.

Solid-state innovation does not detract from safety. ANSI/ASME A17.1-1981 [8] imposes more rigid requirements on static systems.

12.3.6 Hoisting Motors. Elevator hoisting motors in general, and alternating-current motors in particular, are special-purpose motors and must be designed to provide high starting torque with comparatively low starting current. They must have the ability to stand up and perform well under repeated starting and stopping. The motors are rated on the basis of the horsepower required when the car is carrying full load in the up direction at full speed. As this occurs infrequently in actual operation, the time-temperature rating is intermittent (instead of continuous) on a ¼ h, ½ h, or 1 h time basis for a given temperature rise in degrees Celsius, which is based on the type of insulation used. The intermittent type of service permits smaller motors which consume less energy. Alternating-current motors must function at a low noise level to avoid possible complaints by building occupants. Direct-current motors must satisfactorily commutate the peak currents which are experienced when the elevator starts and stops.

12.3.7 Rating of Motor-Generator Sets. Motor-generator sets must have an adequate capacity to handle the peak power demands of the hoisting motor with satisfactory regulation, and should have low free-run losses, since they remain running while the elevator is stopped. They are rated on a continuous basis, usually lower than the hoisting motors with which they are associated, but never less than the root-mean-square hoisting-motor horsepower for the elevator duty cycle. The time-temperature rating is the temperature rise in degrees Celsius, continuously running at nameplate rating, and is based on the type of insulation used. Motor-generator sets and elevator hoisting motors should preferably be matched to obtain the best results from each.

12.3.8 Starting Motor-Generator Sets. Motor-generator sets are usually started through resistance or wye-delta switching. On three-phase alternating-current power supplies, the wye-delta arrangement connects the driving motor in wye during starting and in delta during running. Where the traffic demand is intermittent, and especially with automatic elevator operation, the motor-generator

set is shut down to reduce power consumption after the elevator has been idle for a predetermined time.

12.3.9 Control Design Fundamentals. There are certain fundamentals of control design that invariably produce more acceptable, more reliable, and often more economical equipment. For example, direct-current contactors, switches, and relays are quieter and more reliable in operation than similar equipment of the alternating-current type, and may be timed by capacitors instead of dashpots. The necessary direct-current power for such equipment and elevator machine brakes can readily be obtained from solid-state rectifiers or supplementary excitation generator units. Electronic solid-state devices may be desirable, since they not only reduce the operating currents involved but can usually accomplish the same results more economically than contactors, switches, or relays. In addition, the introduction of static control has provided the means to accomplish more sophisticated operations because of the programming capability.

12.3.10 Rectifiers for Power Conversion. In areas where buildings were originally supplied with dc power, solid-state rectifiers were frequently installed to allow existing dc elevator motors to be retained when the utility converted to ac. Unless there is a positive load imposed on the rectifying units that is sufficient each time to absorb the energy created by negative or overhauling elevator loads, supplementary means must be provided to dissipate the regenerated power. The elevators may overspeed otherwise.

12.4 Elevator Horsepower and Efficiency

12.4.1 Horsepower Calculation. The required hoisting motor horsepower is a function of the rated load for a particular elevator car, the speed of the car in the up direction with this load, and the motor-to-load mechanical efficiency. Since traction elevators are provided with counterweights equal to the dead weight of the car, plus about 40% of the load rating (roughly the average load), the formula for the horsepower may be expressed as follows:

$$hp = \frac{C \cdot V \cdot k}{33\ 000 \cdot e}$$

where

C = rated load of car (including car weight), pounds

V = full-load speed of ascending car, ft/min

k = percent of rated load that is unbalanced by counterweight, usually about 60%

e = motor-to-load efficiency, usually 50—70% for geared elevators and 75—85% for gearless elevators

12.4.2 Ratings. ANSI/ASME A17.1-1981 [8] requires that passenger cars be rated in accordance with the effective area of the car platform in square feet. This is to discourage overloading beyond the capacity of the elevator equipment. It likewise rates freight elevators on a square foot of net platform basis but distinguishes between three separate types of loading: general freight, motor vehicle, and industrial truck. Detailed information on dimensional data in the English and metric units for a wide range of passenger, hospital, and freight elevator requirements for electric and hydraulic system applications is covered in reference [11].

12.4.3 Variations. For practical applications there is a wide variety of loads,

car speeds, and resulting horsepowers. In general it might be said that geared-machine motors vary from 7½ hp at 1200 or 1800 r/min to 100 hp at 500 r/min and gearless motors from 20 hp at 60 r/min to 150 hp at 130 r/min. Motor-generators can have speeds up to 3600 r/min.

12.4.4 Efficiency. The line-to-load efficiency, that is, the ratio of the useful work done in moving the load to the input energy from the supply, varies within considerable limits, depending on the load in the elevator car and whether the car is ascending or descending. It is zero under a balanced load condition at which time the weight of the counterweight exactly balances the weight of the elevator car with the load contained therein.

12.5 Elevator Energy Consumption and Heat Release

12.5.1 Energy Consumption. Elevator energy consumption is measured in kilowatthours per car mile and is a function of the type of equipment used, the load contained in the elevator, and the number of stops made by the car in one mile of travel. Since the latter is a measure of the intensity of elevator service if the elevator is in practically continuous use throughout the normal operating period, the average energy consumption of a particular car in a particular class of building can be estimated. Office-building elevators, for example, usually average 150 stops per car mile for local service, and 75 or less stops per car mile if express runs are involved. By calculating the power consumed in making the corresponding number of average runs in a mile, and adding the power consumed (with generator field control) during the time that the car is stopped, an approximation of the number of kilowatthours per car mile is obtained. Balanced load conditions are usually assumed for the calculation. Typical office-building passenger elevators may consume from 3 kWh to 8 kWh per car mile, depending on the type of equipment and character of the building, and the service demand.

12.5.2 Heat Release. 80—90% of the power taken from the input power lines for the operation of the elevators is dissipated in the elevator machine room in the form of heat. With gearless machines and generator field control, approximately two thirds of the heat released is from the motor-generator set with the remaining one third from the hoisting motor and control equipment. Since performance adjusted for normal temperatures becomes unsatisfactory at excessively high temperatures, it is the responsibility of the architect, consulting engineer, or owner to provide means, if necessary, to hold the ambient temperature of the machine room between 55 °F and 105 °F. Spill air from the building air-conditioning system, separate air-conditioning units, or machine-room ventilating fans are often used for this purpose. The heat released by the elevators can readily be calculated from the power consumption in kilowatt-hours per car mile and the estimated number of car miles per hour averaged on the basis of the number of working hours that the elevators remain in service. Heat release of solid-state motor drives is approximately 40% less than the heat emission of a motor-generator set.

12.6 Elevator Conductors and Diversity Factor

12.6.1 Conductor Size. Elevator conductors must be capable of carrying hoisting motor or motor-generator set driving-motor currents without overheating and, in addition, be of sufficient size to limit the voltage drop to not more than 3%. Conductor sizes are determined by the requirements of ANSI/NFPA 70-1981 [6], Article 620 which states that the ampacity should be based on not less than 140% of the motor-generator set nameplate current, for conductors supplying a single continuously rated motor. Conductors supplying two or more motors shall have an ampacity of not less than 125% of the nameplate rating of the highest rated motor in the group plus the sum of the nameplate current ratings of the remainder of the motors in the group.

NOTE: It is critical to apply the ambient temperature correction factors and to consider voltage drop when selecting conductor insulations and sizes.

12.6.2 Current Ratios. Typical ratios of starting or accelerating current to running current are 1.75 for a high-speed passenger elevator with generator field control, 1.6 or less for a low-speed heavy-duty freight elevator with generator field control, and 2.0—3.0 for alternating-current rheostatic control with reduced-voltage starting. The power factor of the elevator starting peak is better than 90% with generator field control and usually more than 75% with rheostatic control. Passenger elevators in busy buildings may be in motion as much as 60% or more of the time they are in regular service. The time duty for other classes of service ranges down to 25% or even less.

12.6.3 Diversity of Operation. When two or more elevators operate from the same set of feeders, it is improbable that they will have simultaneous current peaks very frequently. Advantage may, therefore, be taken of the diversity of operation of the elevators in calculating the size of the feeder. For a group of similar cars, each with an rms current of I, the line rms current will fall somewhere between nI when the peak coincide and $\sqrt{nI}$ when peaks are in the most random pattern. The diversity factor is the ratio of the most probable current to nI and is a function not only of the number of cars on one set of feeders but also of the type and intensity of the elevator service. The factor represents the percentage of the overall current that may be used in designing the feeder. Typical diversity factors might be 0.97 for a group of two elevators and 0.890 for a group of four elevators under specific operating conditions.

12.6.4 Disconnecting Means. Each elevator is provided with a fused service switch or circuit breaker located in the elevator machine room, which constitutes the means for connecting or disconnecting the particular elevator equipment from the power lines. The fuses or circuit breakers are for the purpose of protecting the branch circuit feeders to the elevator equipment. Circuit breakers inherently interrupt all lines simultaneously and eliminate single phasing which may occur when individual fuses open.

The location of the disconnecting means for elevators shall conform to the requirements of ANSI/NFPA 70-1981 [6]. Article 620-51 requires, that a disconnecting means shall be located in the vicinity of the controller for ac- and rheostatic-controlled elevators. When the machine is not in the vicinity of the controller, an additional manually operated switch shall be provided at the machine connected in the control circuit to prevent starting. On elevators with generator field control, the disconnecting means shall be located within sight of the motor starter for the driver motor of the motor-generator set. When the dis-

connecting means is not within sight of the hoist machine, the control panel, or the motor-generator set, an additional manually operated switch shall be installed adjacent to the remote equipment connected in the control circuit.

ANSI/ASME A17.1-1981 [8], Rule 210.5 specifically requires the power supply disconnecting means for elevators to be a fused disconnect switch or a circuit breaker, regardless of other types of overcurrent protection which may be provided on the power supply line at some other remote location.

12.7 Elevator Operation

12.7.1 Manual and Automatic Operation. Operation is the method of actuating the control. It determines the manner in which the car is operated and its response to different traffic-demand conditions. Operation may be broadly classified under two categories, manual, where the elevator is under the control of an authorized attendant or authorized group of employees, and automatic, where the elevator responds automatically and is therefore suitable for general self-service operation. The modern type of operation is automatic. The term *automatic operation* means a system wherein the elevator starts in response to momentary actuation of devices at the landings, or in the car identified with the landings, or in response to an automatic starting mechanism, and wherein the elevator stops automatically at landings for which stops have been registered.

12.7.2 Single Automatic Push Button. This operation incorporates one button in the elevator car for each floor served and one button at each landing. Momentary pressure of a button automatically initiates and completes car travel to the corresponding floor without interference from other buttons that might be subsequently pushed during the trip.

Single automatic push-button operation was the first type of fully automatic operation. It has little application today in commercial buildings because of its characteristic of responding to one call at a time and ignoring persons at floors passed during the trip who may wish to travel in the same direction.

12.7.3 Selective Collective. This operation extends the usefulness of the automatic elevator by allowing the car to respond to several different calls on the same trip. It incorporates one button in the car for each floor served and up and down buttons at floor landings. Car or landing calls are registered by momentary pressure on these buttons and remain registered until answered by the arrival of the elevator. Calls may be registered in any sequence and at any time. The car stops automatically in response to up landing calls on the up trip, to down landing calls on the down trip, and to furthermost calls irrespective of direction of car travel. Stops are made in the order that the particular floors are reached.

12.7.4 Duplex Collective. This is automatic operation of two selective collective elevators from common landing buttons. The buttons in the cars operate each elevator individually. Landing buttons register calls which will stop either elevator if available for service in the proper direction. The system moves the elevators only on demand, prevents both elevators from starting for the same landing calls, and ensures the utilization of each car in satisfying the overall service requirements.

12.7.5 Group Automatic. This is the automatic operation of several automatic elevators coordinated by a group supervisory control system and applied to modern passenger elevator systems. It provides fast and efficient service and adjusts to the changing traffic needs as dictated by building occupancy. Cars start automatically after closing of the elevator doors and stop automatically in response to calls registered by floor buttons in each car or by up and down landing buttons common to the group. A landing call stops the first available car approaching in the proper direction. Fully loaded cars automatically by-pass landing calls. Doors open and close automatically and are provided with devices both to detect a person in the doorway and discourage passenger interference with the closing operation. Automatic dispatching is included whereby cars selected at designated dispatching points close their doors and depart in a regulated manner.

12.8 Quality of Elevator Performance. A high quality of passenger-elevator performance is essential for most commercial buildings. Not only must the riding public be transported rapidly with a minimum of delay, but the entire performance must be smooth and acceptable to passengers. Floor stops must be accurate to provide safety of passenger transfer.

12.8.1 Passenger Comfort. Fast service, although partly a function of the top speed of the elevator, is limited by the *feelings* of passengers. The rider has no different sensation at a steady-state speed of 1800 ft/min than at lesser speeds. However, the passenger is quite sensitive to acceleration, deceleration, and rates of change of acceleration, which must be consistent with passenger comfort, independent of the top speed of the car.

12.8.2 Leveling. Accurate registration of the elevator platform with the floor landing is important on passenger elevators not only as a time saver but to preclude tripping of passengers upon entering or leaving the car. It is particularly important on hospital elevators where wheeled stretchers or beds with patients are carried. It is likewise essential on freight elevators if small-wheel trucks are used for handling freight. Consequently, automatic leveling is furnished on almost all moderate and high-speed elevators, and to a large extent on slow-speed elevators, especially those in the freight classification.

Leveling systems include one-way automatic leveling which corrects only when the car tends to fall short of the floor landing, and two-way automatic leveling which provides correction for a tendency to either fall short of or overshoot the floor landing. If the type of leveling includes a maintaining feature, it will not only level during the initial stop, but also during loading and unloading if stretching or contracting of the hoisting ropes tend to destroy the original level.

12.8.3 Handicap Requirements. In order to comply with ANSI A117.1-1980 [5] which covers the specifications for making buildings and facilities accessible to, and usable by, physically handicapped people, elevators must have accessible call buttons, special signals, tactile floor indications, minimum door openings and platform sizes, and a myriad of other features. A comprehensive coverage of the elevator requirements can be found in ANSI A117.1-1980 [5] or in reference [13].

12.9 Elevator Doors and Automatic Door Operation

12.9.1 Interlocks. Corridor openings into an elevator hoistway are required to be protected by hoistway doors so interlocked with the elevator control that the car cannot leave a floor landing unless the doors are closed and locked. The car itself must have a door or gate that also requires closure before the elevator can leave (ANSI/ASME A17.1-1981 [8]).

12.9.2 Power-Operated Doors. Passenger-elevator hoistway doors and car doors or gates are invariably power operated. The opening operation is usually automatic and initiated sufficiently in advance of car arrival at a floor to permit entrance or exit of passengers by the time the elevator has stopped level with the landing. The closing operation is automatic on self-service elevators, and is usually dictated by a timing device, light beam, or automatic dispatching equipment. ANSI/ASME A17.1-1981 [8] requires door closing protection by specifying a re-opening device and torque and kinetic energy valves.

Power operation of freight-elevator hoistway doors and car gates is desirable, both from the standpoint of convenience and saving of time.

12.9.3 Passenger-Elevator Doors. Passenger-elevator doors in commercial buildings are usually horizontally sliding and of the side opening or center-opening types. The latter are preferred since entrance or exit of passengers is accomplished more rapidly after the car comes to a stop.

12.9.4 Freight-Elevator Doors. Freight-elevator doors are usually vertically sliding and of the biparting variety. The top of the lower door panel is provided with a truckable sill which will facilitate the movement of freight onto or off of the car.

12.9.5 Fire Rating of Elevator Entrances. Elevator hoistway entrances must have a $1\frac{1}{2}$ h fire rating and must be tested by a recognized testing laboratory in accordance with ANSI/ASTM E152-80 [9].

The interface of the hoistway door jamb and wall, whether it be masonry or gypsum board, is critical and must meet the stringent requirements outlined in ANSI/ASTM E152-80 [9].

12.10 Group Supervisory Control. The typical commercial building, particularly the busy office building, has a coming-to-work traffic peak in the morning, a going-home peak in the late afternoon, and a peak at midday when persons are going to or returning from lunch. These traffic peaks, and the in-between traffic as well, must be handled rapidly and with minimum waiting on the part of the public.

Group supervisory control provides the means for regulation and equalization of service to waiting passengers by automatically distributing the cars in a manner to match the elevator service to the demands and direction of the flow of passenger traffic.

The supervisory control should be capable of operating the elevators with different patterns to match the intensity and mode of passenger traffic; up peak in the morning, down peak in the afternoon, and two-way traffic during the rest of the day.

12.11 Calculation of Elevator Plant

12.11.1 Selection of Elevator. The proper size of a passenger car is a function of the vertical transportation demand during the peak traffic and the type of peak traffic expected whether it be up peak, down peak, or a combination of up peak and down peak. Size is also a function of the passenger load, whether it will be office personnel, hotel guests with luggage, or hospital personnel with accompanying carts and equipment. An additional size consideration is the ability of the building to function with one elevator out of service. The remaining elevators should have sufficient capacity to provide the required handling capacity.

12.11.2 Car Ratings. Elevator cars are rated by their maximum inside area as related to load and as established by ANSI/ASME A17.1-1981 [8]. Capacity load of elevators is based on approximately 1.5 ft^2 per person, whereas nominal passenger capacity is closer to 2.3 ft^2 — 2.5 ft^2 per person.

Passenger-car ratings for commercial buildings range from 2000 lb to 5000 lb, usually in 500 lb increments. A rating of 3000 lb is the minimum recommended for office buildings, and 5000 lbs is the minimum recommended as a vehicular service elevator in a hospital.

Cars for general passenger-use need greater width than depth to facilitate transfer of persons at floors. Hospital cars require a greater depth than width if designed to carry beds, stretchers, etc.

Passenger elevators are often used for freight, especially in high-rise office buildings. They are often referred to as *Service Elevators*, and must be designed in accordance with passenger loading area requirements. They should also be reinforced to withstand the expected loading of hand or powered trucks.

The size of freight elevators varies within wide limits due to the diversity of material that may be carried. Standard ratings may be anywhere from 2500 lb to 20 000 lb. Consideration should be given as to whether the freight is general in character, loaded by industrial truck, or of a specialized nature such as automobiles. Depending on the nature of the load and method of loading, consideration should also be given to the building supporting structure for the elevator guide rails.

12.11.3 Double-Deck Elevators and Sky Lobbies. In buildings with large floor areas and populations that require a large number of elevators or elevators of capacities greater than 5000 lbs, several solutions to reduce the space required for elevators should be considered:

(1) Double-deck elevators—that is, elevators having two decks or compartments mounted one above the other in a single hoistway arranged to serve two consecutive floors simultaneously. To passengers, a double-deck elevator looks and operates similar to a conventional single-deck elevator. The only obvious difference to prospective passengers is that on entering the building they are directed to one deck for service to odd floors and the other deck for service to even floors. This arrangement permits simultaneous loading and unloading of both decks. Once the elevator leaves the main landing (upper and lower lobby) and responds to a hall call it is programmed to serve calls the same as a conventional elevator. This solution, when properly applied, provides equivalent elevator service to a conventional system but requires less space and fewer elevators.

(2) The sky-lobby concept is best suited for structures which can be designed to have two or more separate zones or separate buildings, one above the other. The first floor of the lowest zone is designated as the main lobby. The first floor of the upper zone(s) is called a sky lobby. From this sky lobby all floors in a zone are served by groups of local and express elevators similar to a typical building. Elevator service from the main lobby to each sky lobby is provided by large express shuttle elevators, either single or double deck. Using this concept people desiring service to the upper floors of an upper zone are required to take an express-shuttle elevator to the sky lobby serving that zone and transfer to a local or express elevator to reach their destination. Using this solution minimizes the space required for local and express elevators on the lower floors and should be studied for any building where three or more groups of elevators are required, or where multiple functions are expected in a building such as a hotel above an office building. Conventional elevators may be used for both the shuttle and local/express elevators. Double-deck elevators may be used for the shuttle, or local or express elevators to obtain additional space savings.

12.11.4 Selection of Elevator-Car Speed. Proper elevator speed for passenger cars is a function of the round-trip distance to be traveled and the degree of service desired. In general the taller buildings require the higher speeds based on having sufficient express run to realize the full effect of the rated speed for a good portion of the distance.

The best speed for a passenger elevator is determined by considering the cost of the single elevator unit, the number of units required to satisfy the traffic demand, and the service return expected from the investment.

Freight elevators usually range in speed from 50 ft/min to 200 ft/min and are usually hydraulic in buildings up to five floors.

12.11.5 Quantity of Service. An elevator plant in a commercial building is expected to provide both quantity and quality of service. Quantity of service is measured by the passenger-carrying capacity of the elevator group per unit of time, and becomes important during peak-traffic periods. The riding public measure quality of service in terms of *How long do I wait for an elevator?* and

How long does it take me to complete the ride? It is important at almost any time.

The quantity demand on the elevators is evaluated differently for different classes of buildings. In the case of the office building, it is a function of the potential population who might use the elevators and the expected percentage of this population requiring service during the five minutes of heaviest passenger traffic. For a new building, the former is determined from the net rentable area and the anticipated population density in terms of square feet of area per person. The latter is an experience factor resulting from analysis of many existing buildings of the same type. When an existing building is modernized, the demand data are known and need not be predicted.

12.11.6 Building Classes. Office buildings are classed as single purpose (occupied by one or two large concerns) or diversified tenancy (occupied by a number of small and unrelated concerns). For either type, the normal population density may vary from 90 ft^2 to 150 ft^2 per person. Diversified-tenancy buildings usually have a 5 min arrival rate of 11 to 12% of the population. The 5 min-arrival rate for single-purpose buildings normally varies from 12 to 20% of the population.

12.11.7 Quality of Service. Quality of service is most commonly evaluated by the time interval between succeeding cars, since this reflects passenger waiting time. Intervals of 20 to 30 seconds are considered excellent for office buildings in which time is usually at a premium during the arrival and departure peaks.

12.11.8 Number of Elevators Required. The number of elevators required in a building is basically determined by the relation of the building quantity demand to the unit handling capacity, that is, the handling capacity of a single car of the type selected. The latter is directly proportional to the number of persons that can be accommodated in the car per trip and inversely proportional to the time required to make a round trip. The unit handling capacity for 5 min, the number of required cars, and the resulting time interval between cars may be expressed as follows:

$$HC = \frac{300 \cdot L}{RTT}$$

$$N = \frac{BD}{HC}$$

$$I = \frac{RTT}{N}$$

where

HC = passengers carried by each car in 5 min

BD = building 5 min critical passenger demand

L = passengers loaded per car each round trip

RTT = round-trip time, seconds

N = number of cars required

I = interval between cars, seconds

12.11.9 Round-Trip Time. Round-trip time is a most important factor in elevator plant calculations since it influences both unit handling capacity and interval. It represents the time between successive departures of elevators from a main lobby or the frequency with which elevators pass a point in the building. It is calculated on the basis of proven mathematical probabilities plus empirical data determined through experience.

12.11.10 Interval. Interval is the final check on the elevator calculation. If unsatisfactory, more elevators must be provided to change the round-trip time.

12.11.11 Grouping of Elevators in the Plant. In serving all or part of the floors in a building, one centralized group of elevators should be provided rather than a number of small and widely scattered groups giving parallel service. Elevators in small groups have inherently longer intervals and consequently longer passenger waiting time. Furthermore, it becomes difficult to equalize the demand of each group since prospective passengers will often prefer one or another group if they have a choice of groups serving their particular floor.

Elevators within a group should be located so as to minimize the walking distance required to register a call and board an elevator. The alcove arrangement is preferred for the larger groups with two cars opposite two, three opposite three, etc. No more than four elevators should ever be located in a single line unless such an arrangement is unavoidable.

For a comprehensive coverage of elevator use and traffic analysis for buildings see reference [15].

12.12 Regenerated Energy. When an elevator is operating with an overhauling load such as full-load down, energy is generated by the motor-generator set. Alternating-current drive motors, when operating under normal energy conditions, pump this regenerated energy back into the power company's feeders. If this regenerated energy is pumped back to a standby-power generator and the generator does not have the capacity to absorb this regenerated energy, the generator or the elevator may overspeed.

This regenerated energy may be absorbed in the form of emergency lighting, appliances, or other loads which must always be connected when emergency power is being used in the energy losses of the prime mover of the standby generator or in *dummy* load resistances provided for the purpose. Typically, an engine-generator set can absorb approximately 20% of its rating in kilowatts.

12.12.1. Regenerated power calculations for a variable-voltage gearless machine are based on the following:

(1) Running full-load down = approximately 40% of running full-load up at a 40% negative power factor

(2) Stopping full-load down = approximately 50% of starting full-load up at a 50% negative power factor

12.12.2. A sample calculation is as follows:

(1) Assume that we have a 75 hp motor generator, 460 V, three phase, 60 Hz, with the following characteristics:

Nameplate marking	86 A
Starting (full load)	186 A
Accelerating (full load)	217 A
Running (full load)	118 A

(2) Regenerated current stopping, full-load down:

$0.5 \cdot 186 = 93$ A

(3) Regenerated power stopping, full-load down:

$$93 \cdot 460 \cdot 1.73 \cdot 0.5(PF) = 37 \text{ kW}$$

(4) Hence, a load equal to or greater than 37 kW must be provided to absorb the regenerated power.

12.13 Standby Power Operation of Elevators. Arrangements to permit one or more elevators to be operated when the main power fails are often provided and may be required by local codes. Standby power sources such as turbines or diesel-driven generators are frequently used to supply power for elevators in high-rise

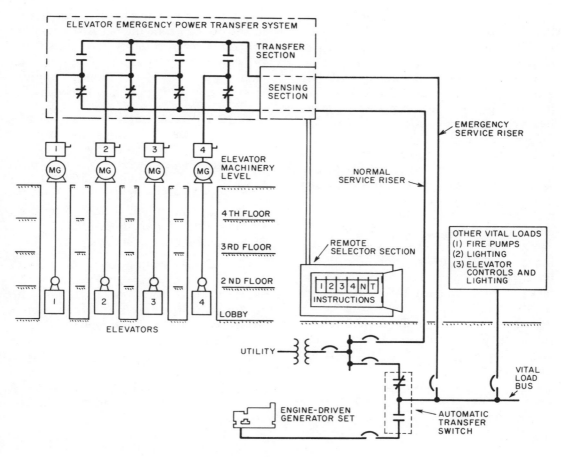

Fig 141
Typical Elevator Emergency-Power
Transfer System

buildings for offices, apartments, etc, and are used for practically all elevator installations in hospitals and health-care facilities. A typical standby power-control system will, upon failure of the main power supply, permit one elevator at a time in each bank to descend to the main floor on power supplied by the standby system.

12.13.1 Standby Power Requirements. The amount of power needed for such arrangements and the cost of the neces- sary equipment vary widely, depending on the number of cars to be used, their rated load, intended running speed, and method of operation. Power demand should be calculated for each individual installation. In addition, provision should be made to dissipate the power regenerated when the elevators are under negative or overhauling loads. This re- generated energy may cause overspeed- ing of the elevators unless a sufficient amount of power is dissipated by the emergency system, by the electric losses in the elevator equipment, or by some other positive load on the standby supply.

There are situations where increased standby power is required for all elevators. Savings in equipment may be made by first supplying power during outages to one-half of all the elevators provided that the traffic can be safely rerouted and the capacity of the operating elevators is adequate to handle the extra traffic. Power should then be transferred to the second bank of elevators shortly thereafter to clear stalled elevators. Power may be left on this bank until power returns to normal.

Where elevator service is critical, it is desirable to have fully-automatic engine starting and power transfer without manual supervision. Operators and maintenance personnel may not be available to handle manual operation in time, especially if the power failure occurs on a weekend or at night.

12.13.2 Elevator Emergency-Power System. Figure 141 depicts one form of an elevator emergency-power transfer system serving a bank of four elevators. The system consists of an automatic transfer switch for each elevator, a sensing and control panel, and a remote selector station. In this system, when the utility service fails, a preselected elevator is powered from the emergency service riser. An operator can select individual elevators from the remote selector station to permit complete evacuation of all elevators. Interlocks in the remote selector allow only one elevator to be connected to the emergency-service riser at a time. The engine-generator set and the emergency riser need be sized for only one elevator, thus minimizing the installation cost.

A more commonly installed system uses one transfer switch to transfer the elevator feeders from normal to standby power at the switchgear room. All of the elevator power feeders are energized, and the limit as to how many elevators operate at one time is established by locking out individual elevator controllers.

12.14 Operation of Elevators Under Fire or Other Emergency Conditions. In the event of fire or other emergency conditions, elevators shall be made available for the use of emergency personnel. The operation is handled in two phases.

(1) Phase 1 requires that all automatic operation elevators serving three or more landings or having a travel of twenty-five feet or more shall have a switch at the main floor, and sensors in each elevator lobby, which, when activated, shall return all elevators to the main or designated floor.

(2) Phase 2 requires that a switch be provided in each elevator which when activated allows emergency personnel to have control of the elevator. For a detailed description of this operation, refer to ANSI/ASME A17.1-1981 [8].

12.15 Emergency Signals and Communication. Elevators that are operated at any time without a designated operator shall be provided with an audible signaling device and a means of two-way conversation between each elevator, and a readily accessible point outside of the hoistway. This may be a telephone, intercom, etc.

More stringent requirements are required in buildings in which an attendant or watchman is not continuously available to take remedial action. ANSI/ASME A17.1-1981 [8] dictates these needs in complete detail.

12.16 Car Lighting. Elevator cars shall be provided with not less than two lamps that will provide a minimum illumination at the car threshold of not less than five footcandles for passenger elevators and two and a half footcandles for freight elevators. In addition, passenger elevators shall be provided with an emergency lighting power source which shall be automatically turned on immediately after normal lighting power fails in the car. Refer to ANSI/ASME A17.1-1981 [8] for complete details.

12.17 References.

[1] A17 Guide, Evacuation of Passengers from Stalled Elevator Cars.[56]

[2] A17 Interpretations.[56]

[3] ANSI A17.2-1979, Practice for the Inspection of Elevators, Escalators, and Moving Walks.[57]

[4] ANSI A90.1-1976, Safety Standard for Manlifts.

[5] ANSI A117.1-1980, Specifications for Making Buildings and Facilities Accessible to and Usable by the Physically Handicapped People.

[6] ANSI/NFPA 70-1981, National Electrical Code.[58]

[7] ANSI/NFPA 101-1981, Life Safety Code.

[8] ANSI/ASME A17.1-1981, Safety Code for Elevators and Escalators.

[9] ANSI/ASTM E152-1980, Methods of Fire Tests of Door Assemblies.

[10] *Basic Building Code.* Building Officials and Code Administrators International, 17926 South Halsted Street, Homewood, IL 60430.

[11] *Elevator Engineering Standard Layouts.* National Elevator Industry, Inc, 600 Third Avenue, New York, NY 10016.

[12] *Standard Building Code.* Southern Building Code Congress International, Inc, 900 Montclair Road, Birmingham, AL 35213.

[13] *Suggested Minimum Passenger Elevator Requirements for the Handicapped.* National Elevator Industry, Inc, 600 Third Avenue, New York, NY 10016.

[14] *Uniform Building Code.* International Conference of Building Officials, 5360 South Workman Mill Road, Whittier, CA 90601.

[15] STRAKOSCH, GEORGE R., *Vertical Transportation, Elevators and Escalators.* John Wiley and Sons, Inc, New York, NY.

[56] This document is available from the American Society of Mechanical Engineers, 345 East 47 Street, New York, NY 10017.

[57] ANSI Standards are available from Sales Department, American National Standard Institute, 1430 Broadway, New York, NY 10018.

[58] The National Electrical Code is published by the National Fire Protection Association, Batterymarch Park, Quincy, MA 02269. Copies are also available from the Sales department of American National Standards Institute, 1430 Broadway, New York, NY 10018.

13. Communication Systems Planning

13.1 General Discussion. This section covers general engineering considerations involved in providing adequate communication systems in commercial buildings. Well planned facilities, for both present and future needs, designed into buildings during initial construction or major alterations will pay dividends throughout the life of the building. All trends point to the continually increasing importance of communication, control, and signaling facilities in commercial buildings.

It is important that the engineer investigate future communication needs while the building is in the planning stages. The communication companies and independent consultants are prepared to offer assistance in planning the needed facilities, and the importance of consulting with them at an early stage cannot be overemphasized.

13.2 Telephone Facilities. A complete system for telephone facilities in a building includes provision for the entrance cables, the main terminal room, the riser cable system, the distribution cable system for each floor, the cable distribution terminals, and the station cabling. Special consideration should be given to buildings that require private branch exchanges, large public telephone installation, electronic switching systems, etc. If an additional private telephone system is planned, early consultation with the equipment supplier and the communications consultants is essential. Early contact with the Building Industry Consulting Service (BICS) at the local telephone company is advantageous to the professional designer on all buildings regardless of the specific tenants or type of telephone equipment being proposed.

13.2.1 Service Entrance. This is the point where the telephone company lines or cables cross the property line and enter the building. The type and location of the service point depends upon telephone company distribution facilities, local practices, and municipal codes. Service will normally be furnished from overhead lines or from underground cables. In most localities the service will be provided by the owner and placed in underground conduits. Buried service may be allowed in some areas.

Again, the owner must provide the necessary trench work. Some areas will require a utility easement for buried service.

The number of entrance conduits required should be based on the ultimate number of telephone lines. For large commercial buildings 4 in conduits are required. Generally one 4 in conduit for every 150 000 ft^2 of usable floor space with a minimum of two conduits is recommended. A No 12 steel wire or equivalent should be provided in each conduit to facilitate drawing in a pulling rope, wire, or cable. A means of bonding, grounding, or insulating cable inside the building is necessary. The openings for cable entrance should always be kept sealed to avoid penetration of water and gas, and to avoid service interruption from flooding or equipment damage from dampness.

Whatever the type of entry, suitable provisions must be made for the appropriate openings in walls, roof, or below grade (with sleeves or conduit) well in advance of the required service date. General recommendations regarding conduit entrances are:

(1) Use corrosion resistant material

(2) There should not be more than two 90° bends

(3) All ends are to be reamed, bushed or capped, or both

(4) Sleeves through foundation walls must reach undisturbed earth to prevent shear

(5) Minimum depth should be 18 in or as required by local codes

(6) Conduit placed on private property must not be terminated in joint use manholes with electrical cables or equipment

Consideration for a dual or alternate entrance provision should be given to the following type of buildings:

(1) Hospitals

(2) Airports

(3) Police stations

(4) Military installations

(5) Power generation or control installations

(6) Radio stations

(7) Television stations

(8) Transmitter sites

(9) Data centers

If the building is served by overhead cable, the cable will normally be supported on poles and enter the building aerially. Proper safety clearance from other utility lines must be provided.

The local telephone company Building Industry Consultant can be of great assistance in designing the service entrance.

13.2.2 Main Terminal Room. The heart of the building's communications system is the main terminal room. It is the main cross-connecting point between the telephone company central office and the riser cable system for the in-building telephones. The entrance cable usually runs directly to this main terminal room after it is brought into the building.

The terminal room should be accessible to communications personnel at all times. It should be located as closely as possible to the center of the riser-cable distribution facilities. The room should be well lighted, ventilated, and properly equipped with electric outlets. Its floor support should be adequate to sustain heavy terminal equipment. This room should be used only for communication equipment, and it should be capable of being locked for security reasons.

For large installations, floor-type main terminals are located in the main terminal room, sometimes called the main frame room. The space allocated to this room must be adequate to meet the original telephone service needs of the occupants and to provide for later installation of additional terminals for growth. Depending upon communication requirements, batteries may have to be installed in the main terminal room.

In buildings having rentable floor area of up to 70 000 ft², wall-type main terminals are commonly used. Each terminal provides connections for the desired number of pairs of cable for voice, data, and EDP equipment.

In some buildings (primarily single-occupant commercial and institutional types), the switching equipment that serves private branch exchange (PBX) or centrex switchboards may be located in the main terminal room. In this event, the room should be dust resistant to protect sensitive relay circuitry.

It is desirable to utilize conduit from the service entrance to the main terminal. Conduit simplifies future additions, minimizes inconvenience, and protects the cable.

13.2.3 Riser Systems. The riser system is the backbone of the commercial building's telephone network. It provides facilities for bringing cables from the main terminal room to the various floors of the building.

The riser cables may be brought up either through riser conduits, sleeves, slots or riser shafts, depending upon the type and size of the building.

Policies of the local telephone company and municipal codes should be checked to determine the responsibility for fire prevention in both used and unused riser facilities.

13.2.4 Riser Conduits and Sleeves. Riser conduits are not usually employed in buildings exceeding 12 stories or covering a large horizontal area as the advantages of having the cables protected by conduits is offset by difficulties of installing and properly supporting the large heavy cables in conduits, and making complicated splices in the splicing cabinets.

The number of riser conduits or sleeves is determined by the usable area in a building. The following table is a guide to determine the proper number of conduits or sleeves:

Total Usable Area in ft² of Floors to be Served (000 omitted)	No of Sleeves or Conduits (Minimum size 4 inches)
0– 50	2
50–100	3
100–200	4
200–400	6

NOTE: For every additional 200 000 ft² or part thereof, add two additional sleeves or conduits

The general design guidelines for riser sleeves and conduit are:

(1) Minimum size 4 inches

(2) Sleeves should be vertically aligned to permit pulling in of cables

(3) Both ends should be reamed and threaded

(4) Initial unused sleeves and conduits should be capped

(5) Sleeves should be stubbed 2 inches above the finished floor

(6) Sleeves should be located immediately adjacent to the wall and to the left of the cable terminating space. They should never be designed to be located close to or in the center of a usable wall

(7) In conduit runs there should not be more than two 90° bends.

13.2.5 Riser Shafts and Slots. Riser shafts are usually employed in larger buildings. The shafts should start from the main terminal room and extend to all floors of the building. If for architectural reasons a shaft cannot be vertically aligned, conduits of adequate size should be provided to connect the bottoms of the higher floor shafts to the tops of the lower floor shafts.

Slots have become a problem for fire prevention so they are generally unacceptable. When used, they should be 4 inches clear minimum to 6 inches clear maximum in width with a minimum 2 inch high curb. All slots should be vertically aligned.

13.2.6 Underfloor Distribution Systems. There are important advantages to providing an underfloor telephone distribution system. The increased efficiency of raceways makes it possible for every desk or each piece of communication equipment to be served from adjacent floor outlets. Equipment changes made necessary by relocation of partitions or movement of personnel or furniture can be undertaken with minimum delay and inconvenience.

With raceways, the need for surface wiring along baseboards or overhead is eliminated, thereby improving the appearance of the building. In addition, changes in service that later require removal or relocation of wiring leave no unsightly areas.

Telephone industry studies indicate that 4 out of 10 business telelphones are relocated every year.

The local telephone company Building Industry Consultant can assist in the design of underfloor distribution systems.

13.2.7 Underfloor Raceways. There are two types of underfloor raceways in general use for communication cable distribution. They are known as underfloor duct and cellular floor.

Underfloor duct is a system of parallel ducts running within the floor slab. A reasonable space between parallel runs or feeds is from 4½ ft to 6 ft. Cross runs and junction boxes should be located every 40 ft or less, depending upon the layout of the area. (This conforms to most office requirements.)

The cellular floor consists of longitudinal cells which actually support the floor slab and which are connected together by means of header ducts. The header ducts should be connected to the cell area at intervals necessary for maximum coverage, generally no greater than 50 ft. Alternating the power and telephone cells can provide a flexible layout. In many cases it becomes desirable to use a third parallel cell for cables serving low-voltage signalling circuits. At least 1 in^2 of area should be provided for each 100 ft^2 of occupied space.

13.2.8 Ceiling Distribution Systems. Placing wires and cables in a false ceiling area close to the telephone-set location is the most commonly used distribution system. Ceiling systems utilize one of two methods to reach the telephone location. One method is to place the wiring in the ceiling space and poke up through the floor structure to the office above. This is commonly referred to as the *"Poke-Thru"* method. The drilling of holes is the responsibility of the owner. The design of *"Poke-Thru"* should be carefully considered by the engineer. Structural damage can be caused to the floors by drilling the holes. The holes, if not properly plugged, can allow the passage of liquids, gases, and dirt to floors above as well as to floors below. Disruption in occupied spaces will be caused due to the noise and debris of core drilling, and by telephone installers working over desks, and on ladders. This method is not usually recommended by the telephone company representative.

The other method is to place the wires and cables in the ceiling space and confine the service to the same floor. This method utilizes partition walls or utility columns to bring wiring down to user locations. The utility column placement is an owner responsibility. Both telephone wires and electrical wires can be placed in utility columns.

If the ceiling area is being used as an air plenum, a metallic-raceway system must be designed for most wires and cables in the ceiling. This is a requirement of ANSI/NFPA 70-1981, The National Electrical Code (NEC), Section 300–322. Most states have adopted this NEC requirement. It is also a part of OSHA Rules and Regulations, Subpart 5, Electrical Section, 1910.308/309. The exception of the use of teflon or aluminum sheathed wire or cable.

Although local practices may vary, most telephone companies require the building owner to remove ceiling tiles prior to the commencement of work in the area. It is not the responsibility of the Telephone Company to replace the tiles. The telephone company should be consulted regarding local practices.

13.2.9 Apparatus Closets. The relay and other equipment needed for key telephone service (KTS) are housed in the apparatus closets. Locating apparatus in closets eliminates unsightly wiring and contributes to improved office appearance. When maintenance or changes are required, communication workers may do their work without inconvenience or distraction to office personnel. Physical conditions within the building frequently dictate the number, location, and type of closets. In general, it is desirable to have one apparatus closet for each 10 000 ft^2 of usable floor area. This can vary for smaller areas and up to a maximum of 20 000 ft^2 of floor area, depending upon access and other layout considerations.

Apparatus closets should be lined with $\frac{3}{4}$ in thick fire-resistant plyboard to a height of 8 ft for mounting equipment. Floor and ceiling should be designed in accordance with the fire codes and proper lighting with switching should be installed. One separately fused 20 A three-wire circuit run is necessary to at least two 120 V duplex receptacles or an electric plug-in strip. A No 6 copper wire or $\frac{1}{2}$ in conduit connected to the building ground is required in all apparatus closets.

There are two types of apparatus closets, the walk-in and the shallow closet. The general specifications for these are given in Table 90. Table 91 gives an idea of the space required for typical centralized key telephone system apparatus closets. Using this table jointly with Table 90, one can determine the closet sizes.

13.2.10 Satellite Closets. Unlike apparatus closets, satellite closets do not contain relay and power equipment. Their primary use is to provide cable terminating facilities, that is, connecting blocks for key telephone system services, station and central office, and PBX or centrex lines.

Table 90
Apparatus Closet Specifications

Specification	Walk-in Closet	Shallow Closet
Minimum depth	3 ft	1½ ft
Maximum depth	None	2½ ft
Minimum width	5 ft	3 ft
Maximum width	None	None
Minimum height of doors	6 ft 8 in	6 ft 8 in*
Minimum width of doors	3 ft	3 ft†

*When shallow closets are used, the center post between double doors should be eliminated, if possible.

†Minimum for single door, 2½ ft for double doors.

Table 91
Centralized Key Telephone
Equipment Space Specifications,
in Linear Feet

Floor Area Served (square feet)	Relay and Power Equipment	Terminal Blocks	Total
Up to 5000	3 ft	5 ft	8*
5001—10 000	3 ft	8 ft	11*
10 001—20 000	6 ft	9 ft	15*

*If more than one wall is utilized, add 1 foot for each turn.
NOTE: This Table is to be used in conjunction with Table 90.

Table 92
Satellite Closet Specifications

Floor Area Served (square feet)	Linear Feet
Up to 2000	2 ft/in
2001—4000	4 ft/in
4001—6000	6 ft

Satellite closets are often necessary because of the inadequate space of a single apparatus closet to meet increasing needs, or the physical layout and shape of the floor. Table 92 gives examples of the dimensions of satellite closets.

13.2.11 Telephone Equipment Rooms. In many buildings involving large, sophisticated installations, additional floor area and adequate floor support are necessary to accommodate the complex switching equipment associated with PBX or centrex telephone services. This is in addition to the space required for appartus closets and satellite closets. Since PBX and centrex services will vary according to the needs of building tenants, each equipment room must be individually designed. Such rooms should be centrally located, well lighted, and ventilated. They should be equipped with power and grounding facilities, be as dustproof as possible, and located in areas not subject to dampness or flood. The local telephone company Building Industry Consultant should be contacted for advice on designing individual equipment rooms.

13.2.12 Public Telephones. Facilities for public telephones are generally needed in the lobbies or arcades of all commercial buildings. While telephone companies will install portable telephone booths at suitable locations, built-in booths with a wide variety of designs, finishes, and colors greatly enhance the appearance of the building. Telephone companies are prepared to work with the building designers toward providing facilities which will harmonize with the interior.

Public telephones should be located where they will be readily accessible to the general public. In many cases, signs harmonizing with the architectural design of the building are desirable to call attention to the telephone locations. Consideration should be given to handicap accessibility when designing public telephone facilities.

There is no simple reliable rule for determining the most suitable number of public telephones that should be provided in the lobby, arcade, or street floor of large office buildings or other types of buildings frequented by the general public. For help with this problem the telephone company's public telephone business representatives should be consulted as early as possible.

13.3 Telegraph Facilities. The requirements for the installation of telegraph facilities in a new building are in general very similar to those for the telephones. Installation details should be discussed with telegraph company representatives in all cases.

Conduits or ducts usually enter the building from a vault in the street. The building end and the vault end of the conduit shall have fittings that will prohibit the entrance of gas, water, and other fluids into the building. Special provisions will be necessary if the cable is brought in from overhead facilities.

Recent technological developments and changes in FCC regulations no longer limit telecommunication services to telephone and telegraph services for building occupants. The entire spectrum of both the commercial and governmental telecommunication facilities are now available ranging from teleprinters to massive data communication centers required to establish, store, switch, decode, and process communication data.

Interconnection to the outside is generally accomplished by utilizing the telephone company or other common carrier cables. The carrier's standard installation specification shall be followed, which will indicate elevation, vault, duct bank, path, and other pertinent requirements. The carrier shall be immediately contacted to ensure the quantity and type of interconnctions that can be provided within the required time frame.

13.3.1 Teleprinter Installations. Teleprinter units come in various sizes and types, the smallest is approximately 18 in deep by 20 in wide and weighs 50 lbs. The unit is located so as to provide the operator with ample work and aisle space, with a lighting level of 70 fc (750 lux), 30 in above the finished floor. The placement of the teleprinter should be considered in light of the effect on adjacent offices. When one unshrouded unit is in operation the noise level is 70 dB(A). The signal connections are accomplished by using 2 pairs of cables, $\frac{3}{16}$ inch in diameter, installed by way of underfloor ducts or methods similar to that of the telephone installation.

Power requirements are 120 V, 60 Hz, 2 A running, 8.2 A starting. Individual circuit breakers are recommended to prevent loss of power to units caused by accidents in other receptacles during normal office operation. Some installations in industry require machines to be on line 24 h per day, 7 days a week, to accept messages during nonattended hours. Power shall be available for this type of operation. Teleprinters come equipped with 5 ft power cord, terminated in a U-ground receptacle or twist-lock receptacle to prevent accidental power disconnection by personnel.

Standard building air conditioning or ventilation is sufficient for the teleprinter's operation. Care should be taken to ensure proper ventilation when a teleprinter is installed in small enclosed quarters or with a sound attenuation enclosure.

13.3.2 Data Communication Installation. Equipment size and installation requirements will vary considerably, depending on the user's communication service needs. It is not unusual to have area requirements as small as 200 ft^2 or as large as tens of thousands of square feet. The specifics should be obtained from the user and his prospective equipment manufacturers, taking into consideration potential growth. Without specific data, the following typical installation requirements should be used to design facilities of any size.

(1) Average live load of communication equipment is approximately 70 lb/ft^2.

(2) Ceiling height above the finished floor is 8 ft 0 in.

(3) Lighting level 70 fc (750 lux) at 30 in above the finished floor.

(4) Aisle space and equipment configuration is dependent on physical restrictions of building, as well as interconnect cable, maintenance, supervision, and manufacturing requirements.

Most systems today have been designed for power and signal cables to enter from below. Hence, raised flooring should be installed over the structural floor, creating a space to be used as a cable raceway. In addition, the raised floor can be used as an air-conditioning plenum, which will provide the necessary conditioned air directly to the equipment air intake louvers. The raised floor shall have an adequate number of floor registers or perforate panels, or both, to provide the required air flow. Cutout trims should be provided to protect against cutting of cables. Ramps should be provided to facilitate movement of equipment and test gear. The height of a raised floor is dependent on available head room, number of planned cables, and pressure restrictions affecting air conditioning. For various requirements regarding the placement of wire/cable in underfloor plenum spaces ANSI/NFPA 70-1981, National Electrical Code and local codes should be consulted.

Consideration should be given to acoustical treatment in equipment operating areas. Noncombustible acoustical ceiling tiles, flame-retardant carpets, wall coverings, curtains, etc, provide for noise reduction and absorption. Proper design and equipment placement can reduce mechanical equipment noise. Duct lining, resilient equipment mounting, location of air handler unit, and size of dampers are some considerations.

Conditioned air should be supplied, which will limit the ambient temperature and humidity to within the limits specified by the equipment manufacturer. Much equipment today can operate without any degradation in performance at temperature levels in excess of 78 °F. For energy conservation purposes, where supercooling is not required, the ambient should be maintained at human comfort levels. Similarly, the economy purposes, an adequately filtered backup ventilation system should be provided to limit the temperature rise 10 °F above the outside ambient in case of main air-conditioning outage. Thermal shock, direct distribution of conditioned air to critical equipment, humidity control, and redundancy all are factors for the designer to consider.

Typical power requirements for a communications facility are 25 W/ft^2 for communication equipment, 2.7 w/ft^2 for lighting, and 15 W/ft^2 for the HVAC system. Service is normally 120/208 V, ac, three phase, four wire, 60 Hz (on some main frames 415 Hz is required) with steady-state voltage requirements for communication equipment of +10%, –8%, and a frequency variation of $\pm\frac{1}{2}$%. The duration of the variation must be taken into account. Generally, a 20% transient, lasting longer than 30 ms, or a voltage outage, lasting longer than 15 ms, will cause errors, loss of information, shutdowns, or equipment damage. Communication equipment tolerances can be exceeded for numerous reasons, including utility or building power system faults, large loads placed on line, utility local utilities network switching, lightning, planned brownouts, or unplanned blackouts. This can occur several hundred times a year. The user has several systems available to protect his facility against voltage variation or outages or both. Constant voltage transformers, the mechanical inertia motor-generator set, and solid-set uninterruptible power supply (UPS) are some types of equipment available to improve voltage regulation.

The system chosen shall be dependent on the type of operation, economics, and consequences that can develop if unregulated voltage is provided. The utilization of a UPS for critical systems is ever increasing. The UPS consists of a rectifier, battery, inverter, static switch, and maintenance bypass switch. The size of the UPS is dependent on the total equipment loads. The usual economic cutoff point is 15 min of battery reserve. This provides sufficient time for orderly shutdown of equipment or to activate an emergency power plant to carry the load. The UPS system shall be located so as to take into account floor loading, to provide sufficient air conditioning or ventilation, or both, and to ensure four air changes per hour in the battery room. A diesel-engine-driven or combustion-turbine-driven generator plant may be warranted to carry the communication equipment beyond the 15 min battery supply. An automatic transfer switch, with necessary monitoring and control features, should be installed to start and stop the generator plant and transfer the load from the utility line to the generator line and back again. Fuel supply should be sufficient for at least 36 h, of full-load operation. The designer should take into account the inefficiency of the inverter, charging currents, etc, when providing for input requirements. Remote status alarms in the communications room for both the UPS and generator plant should be provided. Precautions should be taken to reduce the input current harmonics of the UPS rectifier by using proper line filters or by increasing the size of the generator to deliver twice the kVA rating of the UPS. These harmonics can cause overheating and unstable running of the generator set.

Grounding for communication equipment must be provided. An independent ground riser from the building's earth electrode network shall terminate on an isolated ground bus and be sized on at least 2 kcmil per linear foot of cable run. Single-point grounding or multipoint grounding, or both, may be used depending on the communications equipment's frequency characteristics.

Satellite communication systems are used by certain large data processing users. These systems require special considerations not covered here.

13.4 Interior Communications. A variety of aural and visual communication systems are necessary for the proper functioning of a commercial building. It is helpful in the initial planning stages to anticipate the types of tenants that will generally occupy the building. This will allow for the optimum placement and use of conduit, duct, and terminal locations.

Almost all commercial buildings have certain minimum internal communication systems such as entrance monitoring by way of closed-circuit television, public-address facilities for emergency announcements, etc. Many special purpose buildings such as hotels, hospitals, mass-transit terminals, theaters, multipurpose halls, and arenas have complex communication needs which require thorough analysis and planning. Certain tenants in a building may also require special consideration in the design of their spaces to accommodate unique communication facilities, for example, doctor location and registration systems in medical offices, audio and video production, and recording systems, etc.

In addition to the design of the actual communication system, consideration must be given to power requirements, wiring access, space occupancy, heating and cooling requirements and equipment access for servicing.

Visual recording is commonly done on photographic film or magnetic media. The magnetic recordings may be done on an analog basis (video tape) or a digital basis (storage of computer-generated graphics and characters).

Consultation with the various equipment manufacturers may be sufficient for the design of the communications system. Many equipment distributors provide system design assistance. However, the design of a complex integrated multifunction communications system may require the services of a communication system consultant.

13.4.1 Aural Communications Systems. Aural communications commonly take the forms of audible signals, voice, music, bells and chimes (coded and uncoded), tones, and buzzers. These forms of communications may be used in conjunction with a controlled form of "Pink" noise which is used for the purpose of background noise suppression or masking. Voice intercommunications may be two-way or one-way systems. Two-way systems may be classified as loudspeaking or telephonic types. Direction of voice transmission may be either manually or automatically controlled.

The loudspeaking intercom uses a loudspeaker and an amplifier to provide an audible voice signal in the area near and around the unit. Frequently visual annunciation in the form of indicators or digital displays with memory elements augment the classic intercom, especially with the widespread use of micro-

processor circuitry. Power requirements vary greatly among various intercom systems. Some types are centrally powered from a single supply while other types require power at each unit. Voice amplification and switching of audio paths may be a function of each individual master station or may be located in a central equipment cabinet. Telephone type handsets are commonly used options which provide privacy when desired or where extreme ambient noise is a problem.

Telephonic type intercoms and Private Automatic Exchange (PAX) are switched path intercoms which do not utilize a loudspeaker-amplifier configuration but rather operate as a telephone system within the tenant facility.

Most two-way intercom systems have the ability to make conference calls and access high-level loudspeakers for two-way paging operation. One-way intercom systems include standard loudspeaker paging, inductive loop, and VHF radio pocket pagers.

13.4.2 Sound Systems. Sound systems may be classified in two major categories. The first employs a network of distributed loudspeakers placed throughout the area to be covered. The second employs a single central loudspeaker or loudspeaker array located as needed within the area to be covered.

The proper design of a sound system should take into consideration the following:

(1) Acoustical environment of the area or space (volume, absorption in all of the varying uses anticipated, ambient noise levels, reverberation, etc)

(2) Type of program material or information which is to be reproduced

(3) Transducer characteristics (loudspeaker and microphone directionality,

coverage angles, sensitivity and frequency response)

(4) Underwriters Laboratory's listing for speakers used in emergency signalling systems for building evacuation, fire command, etc

Audio compression and equalization are commonly used to tailor sound systems to fit into tenant locations.

13.4.3 Visual Communication Systems. Visual communications utilize many media that range from a single light to film projection to the cathode ray tube. Single light systems include annunciators, flashing lights and strobes, color coded lights, etc.

Projection systems may be of the front or rear screen variety. Common types of projectors include slide, film strip, motion picture, overhead transparency, overhead opaque, microfilm, microfiche and television.

Closed-circuit television may be monochrome or color. Distribution of the television signal within a building is normally by way of either a video amplification and switching system or a modulated carrier basis on the unused channels of a master antenna system. The necessary bandwidth for the required resolution of the program material generally determines the program distribution method as well as system quality.

Computer controlled visual communication systems involve, in a simple form, an input terminal, the electronic data processing equipment, and a display unit. The Central Processing Unit (CPU) may be unique to the system or part of, or shared with, a larger more complex system. The display units may be alphanumeric readouts, flap-boards, cathode ray rubes, X-Y plotters, etc.

The design of visual communications systems should take into consideration ambient light levels, maximum viewing distance and angle of viewing. Projection systems should also be designed for the required image brightness, evenness of screen illumination and best edge-to-edge focus. Live closed-circuit television origination might require special lighting levels and color temperature as well as the proper studio acoustical condition.

Many visual systems are normally operated in conjunction with a sound system. Using delta-modulation and other forms of digitalized voice techniques, aural messages may accompany the visual display and be processed by the same data processing unit. An example of this is a visual staff register which also gives preprogrammed messages upon access or entry to a facility. Visual communications are used extensively in medical applications where it is imperative to attract attention and generate response to a situation.

13.4.4 Recording Equipment. Recording of audio and visual information is often a part of the overall communications requirements in a building.

The needs of audio recording equipment range from a single magnetic tape recorder or dictation machine to the complete professional recording studio where multiple-channel control consoles, multiple-track magnetic tape recorders, and disc cutting equipment are extensively used.

There are two major types of professional audio and visual recording facilities. One is the production studio where the program is originated. These production facilities require detailed planning to achieve the desired acoustics, lighting, heating, ventilation, and air conditioning. The second type of recording facility is that of duplication of recorded material. This may take the form of high-speed photographic film or magnetic tape duplicating facilities and phonograph disc pressing equipment.

There are many special applications of audio- and visual-recording equipment, for example, continuous recording of a voice communications channel, continu-ous video recording from a television camera for monitoring of security in a restricted area, and photographic record-ing of evidence initiated by an illegal act, etc.

14. Facility Automation

14.1 General Discussion. Because facilities continue to expand in size and their operations grow more complex, the use of automated systems to control and monitor both programmable and non-programmed function becomes an economic necessity.

For the purposes of this section the automation is termed Facility Automation Systems (FAS). The FAS includes all of the hardware, software, wiring, and incidentals that are part of the control and monitoring functions.

14.1.1 Additionally, the FAS includes intangibles such as

(1) Training of operations personnel

(2) Maintenance of the FAS

(3) Engineering drawings, diagrams, and installation test data

(4) Manufacturer data such as instructions, parts lists, guarantees, service availability and recommendations

(5) Owner instructions covering events

14.1.2 In general the FAS will be programmed to control and monitor various facility operations and services, such as

(1) Utilities within buildings and grounds

(2) Energy management

(3) Maintenance

(4) Fire management

(5) Security

(6) Transportation (vertical, horizontal, or both)

(7) Traffic

(8) Communications related to the above

(9) Pollution control

(10) The FAS (itself)

All of the criteria for the controls and monitoring of the above systems and subsystems will be furnished to the FAS designer by the appropriate design disciplines, except item 10, an FAS item. It is possible that the FAS designer may be attached to one of the design groups but specializes in FAS design.

It is entirely possible that the FAS will be specified by one of the engineering disciplines that are involved in the facility design or operations. Although the total FAS design task is a team effort of many engineering disciplines, it is general practice to assign this production and coordination task to the electrical engineering team because of their capabilities in the application, testing, and coordination of commercially available and specially designed electrical and electronic equipment.

14.1.3 The role of the FAS design team is an important item in the success or failure of the system because its design is a most complex task, requiring both technical and managerial skills to ensure coordination, reliability, and cost that is justifiable. For the purposes of this section the FAS coordinator and aides are termed designer.

Whether the facility is new or existing, the FAS designer is faced with many of the same tasks. Essentially these are:

(1) Review the basic principles of the FAS with the owner.

(2) Ascertain the owner's priorities to develop the economics relative to first costs of FAS features versus the operating costs of equipment and systems for review with the owner.

(3) Obtain requirements from the design disciplines, such as HVAC and other mechanical systems, energy management, lighting, security, fire management and architecture.

(4) Obtain the requirements from the operating groups, such as maintenance, security, fire management, traffic, and office machines.

(5) Review these requirements, ascertain which are essential, economically acceptable conveniences or candidates for deletion.

(6) List equipment required by applicable codes and insurance standards.

(7) Select the basic FAS, coordinating with all concerned.

(8) Prepare a preliminary budget and review it with the owner.

(9) Develop the final FAS and a cost estimate.

NOTE: This may require several steps or stages to ensure coordination or compliance with budget limitations, or both.

(10) Furnish to other disciplines both preliminary and final data relative to FAS requirements, such as space, access, HVAC, lighting, security, amenities, and furniture.

(11) Assist in evaluation of bids.

(12) Monitor the FAS manufacturer's installation and tests.

(13) Maintain a project overview so that changes during construction are properly incorporated into the FAS and FAS records.

(14) Ensure that complete documentation is provided including shop drawings, as-built drawings, manufacturer instructions, parts lists, service contracts, guarantees, software details, and other contractually required items.

In summary, FAS design and coordination is a comparatively new and expanding engineering discipline. The hardware, software, wiring, space, and access are expensive. The FAS designer must have an awareness of the economics involved; it may be necessary to offer alternatives in the FAS design so that some features may be added at a later date. The designer will also need the managerial capabilities required to coordinate a vast array of detail within specific time limits, as well as the technical ability to recognize the important facets and consequences of all of the input data and proposed changes from the several sources.

14.2 Equipment and Reliability. The manufacturers of FAS equipment have developed systems and services that are compatible with practically every need, for today and for the foreseeable future. The hardware is generally standardized, but these items and the subsystems are flexible, so that the production models can be coordinated with or tailored to meet, any but the most, unusual project requirements.

It is essential to note that the manufacturer's production-line models have advantages in cost, expediency, parts availability, and test by use over specially designed equipment. Use of the latter should be limited to those instances where a need has been proven or nothing suitable now exists.

Most FAS equipment manufacturers (or their vendors) offer the software as well as the hardware. Software is available as production line items or customized to meet special needs. One of the FAS designer's tasks is the selection of the most suitable of these, alone or in combinations.

The FAS designer must know and understand utility rate structures, such as electric power, natural gas, water, sewage, and steam. The designer must also apply engineering economics to permit sound evaluation of costs, benefits, and payback periods. Additionally, the value or advantages of any selected conveniences may have to be explained or incorporated into reports.

One further point, the FAS must be reliable and readily maintainable. The FAS that is not operable (or is only marginally functional) is far worse than none at all. That is, all of the facilities, operating equipment, and systems depend upon the FAS for proper commands and functions. If either the FAS or any of the facility's equipment, systems or subsystems that are controlled or monitored or both, by the FAS, are in a failure mode, that condition must be made known by an appropriate alarm, readout, or signal. Further, the response(s) of the facility's equipment or systems, or both, to FAS commands must be made known by an appropriate alarm, readout, or signal.

There is little or no margin for error in FAS design, installation, and maintenance. The safety, security, operation, and energy management of the facility are wholly dependent upon the control and monitoring provided by the FAS.

Some form of standby power is essential to ensure continued operation of the FAS during any time that the normal source of electric power is not available.

The FAS designer will need to keep up with the rapid technical progress in microelectronics. The increasing application of the integrated circuit chip in control devices suitable for FAS should be reflected in many ways, including reductions in initial cost and technical advances.

14.2.1 General Description of the FAS and Components. A typical FAS will consist of sensors for input data, remote terminal units (RTU) which include data gathering panels and output control panels, the central processor, and the man-machine interface devices which make up the control console. The power supply to the central processor requires consideration as to reliability to ensure intended operation. It is possible that some form of standby or uninterruptible power may be needed as a matter of choice or to meet applicable codes, laws or standards, or both. Sensors may be analog or digital. An analog sensor provides variable information about the sensed condition such as temperature, pressure, humidity, voltage, and current. A digital sensor (in its simplest form) is like a switch that operates a contact when a certain condition occurs, such as a door opening or closing.

Output control functions can turn fans on and off, open or close dampers, etc, and are typically just a switching action of a relay or the activation/deactivation of an electronic logic element for the purpose of control. The sensors and output devices to be controlled are connected to the RTUs which act as collection points for sensor information and the output control commands.

The transmission link connects the RTU and the central processor, but now the selection of the transmission link travels to and from the processor. Traditionally this consisted of a pair of wires carrying digital information between the RTU and the central processor, but now, the selection of the transmission link requires most careful consideration in the design stage. That is, it must be coordinated with the FAS equipment and be capable of transmitting the data without harmful external influence, such as unwanted electrical *noise*. The transmission link must also be able to handle the data at the required data transmission rate. And lastly, the link will need to meet all of today's needs and those of the reasonably foreseeable future. Reliability and maintainability are inherent design considerations. In high risk areas some form of monitoring or supervision may be advisable to guard against tampering.

Another method that essentially provides two paths is a looped link. In the more sophisticated system two independent loops will be used (ANSI/ NFPA 72 D-1980 [5],[59] Type 1 system).

NOTE: ANSI/NFPA 72 D-1980 [5] limits the number of fire protection devices on multiplex lines and at least one FAS equipment manufacturer uses a coaxial cable for the transmission link.

As technology advances other techniques are being used, such as fiber optic cable, wideband coax, and radio fre-

quency transmission. Each has its advantages and disadvantages that must be considered when selecting a transmission medium for a facility.

The central processor and the man-machine interface devices will be discussed in more detail later in this section. Basically the central processing unit (CPU) is the *brain* of the system, taking information from the sensors and giving instructions to the man-machine interface devices on how and when the information should be displayed. It also sends control action commands to the output controls.

The FAS discussed here is normally used to provide centralized control for facilities that exceed 50 000 ft^2. A facility, from an operating standpoint, may require multiple operating centers. Security may be headquartered in a location other than that of the personnel who run the mechanical systems. A fire command center must be readily accessible to the fire department and is usually on a lower floor, while the mechanical-system control center may be located near the penthouse containing the mechanical systems it controls. That is, the FAS design must often accommodate multiple operating centers for efficient operation of many facilities.

The size and complexity of the FAS may vary greatly. A system may contain only one of the mentioned functional sections (subsystems), be it HVAC, fire management, security, access control, of energy management. For example, for fire management a simple fire alarm system containing only detectors, manual stations, and alarm bells may be all that is required. This usually is accomplished by a stand-alone subsystem. However, fire management systems, which may use the same FAS equipment as the mechanical systems for smoke control are

---

[59] Numbers in brackets correspond to those of the references in 14.12.

usually integrated into a common system for efficiency of operation.

Subsystems described herein may be used alone or combined into a large system. Small facilities can have the subsystems combined just as effectively as the large systems. The prime reasons for combining are efficiency of operation and sharing of equipment, to reduce the initial acquisition and installation costs and to enhance operations and maintenance. In general, combined systems allow the facility operators to perform more functions for less money. However, caution is required when considering the use of one CPU for all facility systems. The advantages of dispersed processing systems which can isolate individual subsystems, or themselves be isolated, require study from technical, reliability, and economic standpoints.

There are nationwide companies that offer remote computer and software services for facility automation, with data links to the user's facilities. These have various advantages and disadvantages that require assessment on a project basis.

14.3 HVAC Control and Monitoring. This system automatically monitors the heating, ventilating, and air-conditioning (HVAC) equipment. It also provides the operating personnel with information on the status of these systems and selected components. Temperature, dewpoint, humidity, pressure, and other key operating parameters are continuously monitored and displayed upon command, or when any abnormal or alarm condition occurs.

This system also provides the means to permit remote control of necessary functions for the operation of the HVAC equipment. From the control console, fans can be turned on or off, fan speed adjusted, dampers positioned, control valves positioned, pump speed controlled, equipment started and stopped, control points adjusted, and all other functions necessary to operate the mechanical equipment of the facility properly and economically monitored.

HVAC facility systems are programmed for several operating modes. Basically these programs are developed by the energy management and the facility operating groups. Their data must be incorporated into the software to ensure HVAC operations that meet occupancy and energy conservation needs. These HVAC programs will account for seasonal needs as well as day, night, and holiday occupancy in each of the individual buildings, or building parts, that comprise the facility. In the case of large areas where occupancy varies greatly over short time periods, the programs may have to include real time or hourly control of HVAC components and possibly a major portion of the lighting.

The types of equipment typically supervised by an HVAC control system are

Air handling equipment
Steam absorption chillers
Boilers
Electrically driven compressor chillers
Air compressors
Condensers
Dampers
Evaporators
Fans
Heat pumps
Heat exchangers
Liquid tanks
Pumps
Refrigerators
Sump equipment
Valves
Control switches (EP/PE)

Reheat devices
Cooling towers
The HVAC conditions and quantities to be monitored or controlled may include
Damper position
Flow
Fuel supply
Gas volume
Humidity/dew point
Real or reactive electric-power consumption
Line current and voltage(s)
Liquid level
Equipment operating time
Leaks
Fan speed
Degree day heating/cooling
On-off-automatic operations
Peak load control
Power failures (main, auxiliary, control)
Pressure
Start-stop operations
Status of miscellaneous equipment and systems
Temperature
Toxic gases
Valve position
Wind direction
Wind velocity
Pump speed
Steam flow
Amount of chilled water generated

14.4 Other Mechanical Systems, Control and Monitoring. The FAS may be called upon to control and monitor several other mechanical systems including such functions and conditions as:
Water pressure
Water level
Water pH level
Pump status

Fire reserve water level
Fire pump status
Fire/smoke detection system status
Fire sprinkler system status
Boiler status
Boiler support apparatus status
Sanitary tank status
Drainage system status
Valve position
Emergency (standby) generator status
Other emergency or standby water supply status
There are many more mechanical systems, subsystems or components, that may require control or monitoring on either a programmed or a manual basis. The FAS designer must incorporate these needs in the concept, budget, and final design.

14.5 Lighting Automation. At present, the customary control of lighting is performed by one or more of the following methods:
Local manual switches
Switches or circuit breakers in panels
Time clock operation of relays
Sunlight sensors operating through relays
The FAS offers the opportunity to provide programmed control of lighting as part of the energy management system. That is, programs are commercially available for the timed control of the various lighting needs, such as
Continuous light in stair, exit and exit pathways
Timed control of indoor, general light
Sensor control of light adjacent to window areas
Timed or light sensor control of light in vehicle parking areas

Continuous night-time lighting for security.

Timed or sensor control of light used for advertising or display

14.6 Miscellaneous Facility Items. There are many other facility operations and conditions that may require control and monitoring, for example

Elevator status

Escalator status

Snow melting facilities

Weather indicators or recorders

Freight area; heat, light, ventilation, occupancy

Standby power systems

Uninterruptible power system status

Emergency lights, pumps, and other related facilities

Whether any of these are to be programmed, manually controlled, sensor controlled, or simply monitored are matters of coordination with the discipline having the design or operating responsibility.

There are obvious precautions, such as:

(1) No remote control over escalators where starting or stopping remotely has inherent possibility of danger to riders

(2) No possibility of remote shut down of an emergency system or safety device

(3) No countermanding of signals or directions by facility operators

(4) A procedure for advising police or fire fighters where to (or not to) enter the facility

14.7 Fire Management. The term fire management as used in this section encompasses fire-alarm systems whose functions are detection of fire (manual or automatic) and sounding of a fire alarm signal for evacuation or other purposes. It may also include smoke control subsystems, more sophisticated occupant notification approaches, as well as other fire safety-related control functions.

14.7.1 Fire-alarm or management systems are required by codes in most commercial facilities. The major objectives of these systems are

(1) Detect a fire as early as possible

(2) Notifying the fire department

(3) Notifying the occupants

(4) Notifying in-house fire wardens (required if system encompasses fire protection signaling systems. See ANSI/NFPA 72 D-1980 [5].

(5) Use HVAC system to contain fire and smoke

(6) Use of HVAC system to create safe havens within the structure for the occupants when evacuation is not practical

(7) Capturing the elevators, according to a preplanned scheme

(8) Provide a fire command station to be used by the fire fighters as a control center during the emergency

(9) Providing an emergency two-way radio or telephone system, or both, for use by fire fighters and rescue oeprations

(10) Providing a voice communication system to direct occupants to safety

(11) Starting fire pumps

The basic elements of a simple fire alarm system are initiating devices, control panel, and indicating devices. The other elements which make up a fire management system are "add-ons" to the basic system.

Initiating devices are the elements which sense the presence of a fire and inform the system. These devices are either manual or automatic, or both. Manual devices are typically manual pull stations which are located strategically throughout a facility and are intended to be operated by an occupant if he discovers a fire. Codes typically require that a manual station be located at each legal means of exit (door or stair per floor), and at intervals along the path of egress.

14.7.2 Automatic devices, of which there are a number of types, sense a property or a result of a fire. The most common automatic detectors are:

(1) Thermal detectors that sense heat

(2) Smoke detectors which sense the visible and invisible particles generated by a fire

(3) Flame detectors which sense the infra-red or ultra-violet radiation from a fire

(4) Rate of rise detectors that signal excess temperature rise in a given time period

(5) Waterflow detectors that sense the flow of water in a sprinkler system.

(6) Tamper devices that signal an unwanted operation

Each of the various detector types has subcategories. It is necessary that an engineer designing a fire management system be fully familiar with the types and subcategories currently available and their correct application.

Indicating devices are used to notify the occupants that a fire condition exists. In the past, bells, gongs or horns, or a combination of the three, have been the primary method used in fire alarm systems for this purpose. Recent system designs can use electronically generated fire signals that are transmitted by way of audio amplifiers and speakers. Systems of this design can also be used to broadcast voice messages giving the occupants specific instructions, which cannot be done with a bell, gong, or horn. Codes in many jurisdictions require these audio systems in high-rise construction. Visual signals are also required by many jurisdictions to signal individuals with impaired hearing. The use of prerecorded messages for directing evacuation or other instructions is controversial. There is the possibility of events that are not predictable and therefore not compatible with prerecordings.

The control panel is the *brain* of the system, taking the alarm information from the sensors, processing it, and activating the indicating and alarm devices. In addition this control panel can also initiate other functions required in a fire management system, such as fire department notification, elevator capture, smoke control, etc. The control panel may include a device for recording the time and the location of any fire or smoke signal (a code requirement in some areas). Alternatively this recorder may be separate and remote from the control panel.

Referring to the Block Diagram (see Fig 142) describing a monitoring system, the central processor performs the function of the control panel. Many systems use a microprocessor or a minicomputer, which are software controlled to initiate the desired output functions.

When an alarm occurs these systems may be programmed to automatically position dampers and operate fans to create areas of positive and negative pressure to reduce the spread of smoke to other areas and to exhaust smoke from the building. The central processor can also initiate the capture of elevators, which sends them to a designated floor for use by the fire fighters and authorized personnel.

Some systems use prerecorded voice messages to automatically direct the occupants instead of sounding bells and horns (subject to the previously noted limitations). These systems often use

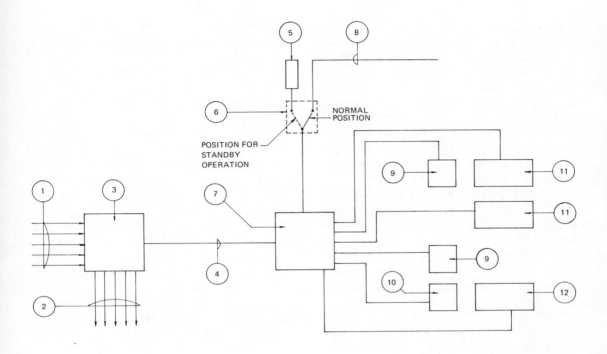

1 Sensors
2 Output Functions
3 Remote Terminal Unit (RTU)
4 Transmission Link
5 Standby Power Source
6 Power Transfer Device
7 Central Processing Unit (CPU)
8 Normal Power Source
9 Printer (prints complete data)
10 Printer (prints selected data only)
11 Operator's Consoles (all functions)
12 Operator's Console (selected functions only)
} Man-Machine Interface Full or Selected
Functions (as needed)

Fig 142
Block Diagram, FAS

several messages that direct people to different areas depending upon the location of the fire. For instance, occupants of the fire area may be directed to go up two floors and those below the area of the fire to go down two floors. Occupants of the floor that will be receiving people will be advised and there may be a general building advisory. Special messages are broadcast to the occupants of captured elevators, advising them of the emergency and instructing them to leave the elevator car when it reaches the designated area. The messages must be coordinated and sequenced properly so that stairwells do not become overcrowded as the people relocate. Usually the messages for the elevator and fire floor are broadcast first.

These systems can also be used to transmit or broadcast special instructions using the public address feature. In this case a person in authority, usually from the fire department, will select the areas he wishes to address from the console and his voice will be broadcast to those selected areas.

Emergency telephone systems are usually provided to give fire fighters a reliable and a private two-way communication between the fire command center and the various floors. The emergency phones may be permanently installed on the floors or phone jacks provided, in which case the fire fighter will carry a handset with a phone plug. If the latter is used a highly visible storage area for a number of handsets should be provided at the fire command center.

For convenience, or for code compliance, the buildings may include a *fire-standpipe telephone system.* Essentially this is a telephone at each standpipe hose or hose cabinet, at selected valves and at external high pressure fire-department hose connections. Where feasible, and in accordance with codes, these telephones may be of the sound-powered type.

Fire management systems must be as reliable as possible because their function is to protect life and property. To ensure that the system meets the desired quality levels and the design has not been changed, approval for the intended purpose as indicated by a UL label is a must when specifying the equipment. Other laboratories or agencies also may be certified to do qualification testing; however be sure that the laboratory or agency is acceptable to the local code-enforcing authority. By specifying the UL listing, you are assured that the equipment is capable of providing the system operation required by ANSI/NFPA 72 series standards (see 14.12). However, the FAS designer is not relieved of responsibility simply by use of laboratory approved or listed products. These are only useful when properly applied and thoroughly coordinated. Further, the owner must be notified that scheduled maintenance and tests are required. Finally, the responses to signals, alarms, and communications require continuing practice, upgrading, and monitoring by qualified personnel. The FAS designer may elect to participate in initiating these procedures.

14.8 Security

14.8.1 Security Systems. Security systems are an essential element in any facility. They have a common characteristic in that they continue to expand in need and techniques. Although every security system must be specifically designed for the individual project, they are all based on the concept of providing safety for the facility occupants (employees and visitors) and the protection of the contents of the facility. The type and depth of the security system varies with the functions at the facility. The security level in any one part of the facility may be entirely different from the security provided in other areas. Banks with generally open or public access will need one type of system. Military installations or offices supporting military organizations have limited access, and thus other security needs. Research facilities, laboratories, and other places where commercial or industrial designs or developments are involved require in-depth security. Stores and normal commercial properties may require limited security precautions. However, location in a known high crime area will tend to increase the depth of the security system.

The concept of security is to make the system fit the need. Excess security is costly to install and maintain. Minimal protection can, of course, be the most expensive simply because the owner believes that all necessary protection has been provided but there is a *weak link*. Security starts with protection at the perimeter of the property. Here a chain-link fence may be used to prevent a casual walk-in by neighborhood youngsters and will identify a specific property line. More protection at this fence can be provided by using outdoor perimeter detection devices which will detect intruders who cross the fence line. Invisible light beams, microwave, E-field, or covered and buried line detectors are just a few of the possibilities. At the perimeter of the building, magnetic switches can be used to monitor perimeter doors and other movable openings. Window foil, traps, pressure mats, and invisible light-beam devices may be used to protect other areas inside the building perimeter. Motion, audio, capacitance, invisible light, and vibration detection devices may be used to protect areas and objects within the facility.

Supervision of the wiring extending to the sensors is important in a security system, as it is in a fire-alarm application. In the case of a fire, a break in a wire is handled as a trouble condition. However, in a security application the loss of supervisory current may be caused by the intruder attempting to compromise (defeat) the alarm system, and it should be treated as an alarm, not as a trouble indication. A ±50% change in line current should indicate an alarm meeting UL Grade A requirements. This is adequate for most general security applications.

Where high security is required, random-digital or high-speed digital interrogation and response are techniques designed to render the circuits most difficult to compromise. Closed-circuit television (CCTV) can be used in manned operations to observe remote areas and extend a single guard to many areas. Cameras can be controlled from the central monitoring location to view a larger area. Cameras are available that operate with very low ambient-light levels and still produce a satisfactory image. It must be remembered that the security office cannot watch a TV screen continuously; thus the CCTV is an adjunct subsystem for specific and intermittent need. Individual applications of CCTV cameras may require certain optional equipment or auxiliaries, such as:

(1) Fixed focus camera
(2) Remotely controlled pan, tilt, and zoom lens functions
(3) Outdoor camera enclosures which may include heaters, windshield wipers, or sun visors. In some instances pan and tilt limiters may be required to avoid direct aiming of the lens into the sun.

Motion detection can be an integral part of a TV system silently viewing an area. When motion is detected in an area viewed by a camera, that camera will be automatically switched on to a monitor for viewing, possibly using light or sound to attract the attention of the security officer.

It is obvious that there are special operational considerations as well as technical design problems in the development of a CCTV installation plan. It is often advisable for the FAS designer to obtain specialized support services to assist in design and coordination with the users of the CCTV system.

14.8.2 Access Control. A security related system has recently been developed which allows access to a facility without the need for locks and keys. In many instances keys and locks are too cumbersome and expensive a method to secure an area or building. Keys are lost or get out of control when people entrusted with them have copies made or fail to return them. Cylinder pins wear and weather also affects conventional locks. A modern and very popular approach has been the card reader. A plastic card, similar to a bank or credit card, is encoded with a hard-to-reproduce cypher. Several different techniques for encoding are in use, such as magnetic stripe or bits, capacitance, photoelectric, and radio frequency. The encoded card is placed in a *fixed* card reader at a point of entry which determines if the card is valid. If the card is valid for that door, time, and day, it will release the door-lock and allow entrance. Systems that require entering a secret number, unique to each card, on a keyboard as well as inserting the card in the reader add another level of security. Unauthorized entry attempts by individuals are processed as an alarm and indicated on the operator's terminal, other signal or display, and alarm printers. The system may also be instructed to record on a printer each entry transaction. The use of card-reader access-control systems is increasing and new innovative ideas are being introduced. The FAS designer is advised to research the available market to be assured he is specifying and obtaining a state-of-the-art access control system.

14.9 Energy Management. The energy management function of the FAS is a major item in the reduction of operating costs and the use of energy. The use of energy and the specifics of control are covered in Section 17. The FAS will provide the mechanics to implement the energy management plan. Here especially, the FAS designer must have close coordination with the energy management engineers. During the preliminary design period many things happen all at once. Fuel and utility costs change, codes change, techniques change, operating methods change, all in short, and sometimes overlapping, time spans.

The initial decision on a single central processor versus dispersed (or individual) processors for HVAC must be made early, and backed up with cost and operational data. Thus, the FAS designer and the energy-management group must develop a basic plan and get it approved quickly. Once that is done the requirements space and other details can be distributed to all concerned. Selection of sensors and methods for responses can then be developed within their own time frames. Similarly, shut down alarms and other features can be developed as the design proceeds. To emphasize, the FAS designer and energy-management team must develop the basic system early.

That is followed by the details of system design, but always as a team effort.

14.10 Central Monitoring Equipment.

The heart of the FAS is the central processor. Refer to the Block Diagram, Fig 142. The central processor may contain a microprocessor or minicomputer, either of which is controlled by software.

All system control is by way of the central processor, which continuously polls (or scans) all the connected equipment looking for changes. On detecting a change, the data is processed to determine if the change is a new alarm, status change, or an operator command requiring service and to select an appropriate response. The central processor then transmits a signal to the appropriate location in the system to initiate the alarm and status changes in a useful form, or to execute other programmed output functions. The central processor also contains programs to detect analog alarm limits, time programs, special action programs triggered by a system event, and other application programs of which there are many.

All operator data is presented on the operator's console or printer, or both. Operator commands and control are all executed by way of the operator's console. Commands should have a positive feedback in the display area of the operator's console indicating that the command has been issued. Some functions may require a signal at the console indicating that there has been proper response to the command.

Commands which a system will typically execute are as follows:

(1) Lock and unlock remote gates and doors

(2) Turn lights on or off

(3) Start, stop, or modulate HVAC components

(4) Start fire pumps

(5) Start, stop, or modulate ventilating fans in response to programmed fire-management needs

(6) Notify police or fire fighting services

(7) Test and reset remote fire alarm and security systems

(8) Test sprinkler systems

(9) Secure access security systems

(10) Start, stop, or modulate remote motors and mechanical equipment

(11) Change status of remote-control systems

(12) Change the control point of remote temperature, pressure, and humidity controllers

(13) Change the position of remote mixing dampers and exhaust dampers

(14) Adjust load shedding set points

(15) Start, stop, load, unload, and test run standby engine generator(s)

From the console, the operator must be able to obtain a display of the status of any of the system inputs. Alarm conditions are displayed automatically and should initiate an audible signal to inform the operator that a condition requiring his response has occurred. He acknowledges his recognition of the alarm condition on the console silencing the audible signal and allowing the system to continue all functions. Fire and security functions also require operator acknowledgement whenever the alarmed point returns to the normal condition.

"Human engineering" considerations should determine the arrangement of signals and manual control devices on the console. That is, the designer should separate (or cluster) HVAC, security, fire management, and energy management items into individual sections on the console, using distinctive colors and shapes for ready recognition of signals and operational parts. Different and readily recognizable sound signals should be used to indicate fire, smoke, security intrusion, system or subsystem failures.

Printers are an important part of the central monitoring section. They provide a hard-copy record of all the activity of the system and should document (1) the point that alarms, (2) when it alarmed, and (3) when it was acknowledged by the operator. Printers may automatically provide the operator with action or advisory instructions so that proper responses are enhanced.

Several printers may be used in a system. A printer may be designated to handle only one kind of traffic, such as alarm only, security only, HVAC only, logs only, or various combinations based upon how the system will be operated.

Printers are electromechanical devices that require maintenance to ensure that they will remain operational. Many systems are designed with a backup printer that will take over automatically when one of the primary printers is not operating. This ensures that important information is not lost. Printers normally assigned to a given function, such as logging, may be programmed as a backup for another machine and print out the failed printer's information as well as its own.

Many systems require control of all or selected functions from locations other than the primary central monitoring location. An example may be the security monitoring, which is normally controlled from the security office, which may be remote from the main control room. To adequately fill this need, an operator's terminal and printer(s) can be located in the security office as well as the control room.

As another example, the building engineer may want to have access to the system from his office. His access to the system should be restricted to only the HVAC or other mechanical systems and perhaps the fire alarm equipment. He should not be able to access or control any of the security equipment.

The use of computers, be they microprocessors or minicomputers, allows flexibility in the programming of the system. The main program routine of scanning and basic control of the machine is contained in an operating system. Software programs which are developed by a manufacturer and used in many of his systems are generally available.

The unique features of each system are accommodated by application programs written specifically for the facility involved. Application software defines the input points, output functions, and the logic of system operation. It is usually entered at the operator's console.

The system should be designed to allow modification of the application software from an operator's terminal. The cost of the terminal is usually justified by the flexibility and convenience of being able to update the software locally, rather than resorting to the manufacturer's facilities.

The modern system in a large facility uses has a color CRT (Cathode Ray Tube) display with a keyboard as an operator's terminal. The keyboard has alpha-numeric capabilities and dedicated function keys for easy operation. This type of operator's terminal allows the system to communicate with the operator in the English language and the nomenclature of the facility itself. An alarm may be displayed as an *Intrusion Alarm at the West Entrance* followed by the action the operator should take, such as *"Call Security on extension 537"* (operator prompting).

The CRT can also generate graphics, such as a floor plan or schematic of an air-handling system to show the operator exactly which equipment or sensor location is involved in a trouble indication or data readout. Current technology has allowed the CRT to display its information in color, further enhancing the system and the functions it can perform.

It should be recognized that color-graphic CRT displays and sophisticated features increase the hardware and software costs and the lead time of a FAS. Complex systems may not be justified for many facilities; however, the continuing advance of electronic technologies will bring costs down and make high-technology systems increasingly easy to justify economically in the future.

Annunciators are often used in control rooms or other locations to provide a continuous visual status indication of selected system parameters These annunciators can be as simple as a grouping of indicator lights with a printed legend for identification. An elaborate annunciator might be a backlit graphic layout of a complete building showing the floor-plan details and lighting the various automatic detection devices on the floor plan when they are in the alarm mode.

Closed-circuit television allows the operator at the central monitoring location to select and view many remote areas for the purpose of security, access control, or monitoring of equipment. Remote cameras can be selected and controlled from the console. CCTV monitors should display date and time data on the screen. Consideration should be given to spare monitors. The console designer should keep the number of monitors which an operator must observe to six. This is the largest number that an individual can effectively watch without becoming confused or losing attention. Combining other detection devices that will cause a camera to be switched to a given monitor and signal the operator is a better alternative.

Systems are available that permit one monitor (CRT) to be automatically sequenced so that it can display information from more than one CCTV camera. In the event of outage of a CCTV that part of the system can be switched to another unit.

14.11 FAS Design and Installation Notes

14.11.1 Documentation. During the initial information gathering period the FAS designer will be meeting with all of the facility design disciplines, the owner, utility company representatives, code enforcing authorities, and vendors of equipment. The information gathered may be written or in some cases, oral. As the project progresses, that information changes, expands, and becomes more detailed. It is therefore essential for the FAS designer to establish and update a written file summarizing the requirements of each group or discipline that furnishes data. Copies, possibly in memorandum form, should be circulated on a regular basis to all concerned. Documentation is a major item in the FAS design criteria. Further, on long term projects, personnel who are new to the project will need that written data.

14.11.2 Contract Documents. The FAS system will be installed by one or more contractors. That installation work will be based upon the contract drawings and specifications that comprise the contract documents.

In general, the contract drawings will show the locations, in plan and elevation, of all the FAS items, from the smallest sensor to the largest CPU. These contract drawings will also include details of construction affecting the FAS installation, or references to other appropriate contract drawings, or both. Interconnection and wiring diagrams will also be provided. In the case where compressed air is used as a control medium, the appropriate details will be shown on the contract drawings. All of these items, except the diagrams, will be developed to a scale that permits estimating of quantities and lengths. Interfaces between the FAS and other systems will also be shown.

Also in general terms, the contract specifications will define the scope, character and time requirements, of the tests required on the FAS. Insurance and beneficial occupancy of parts or the whole FAS system may also be included in the contract specifications.

Installation of FAS systems generally follow these major steps, after the contract drawings and specifications are developed.

(1) Copies of the contract documents are issued to qualified contractors for bids. Issue of the documents is on the same date for all of the bidders. Bid due date and time is the same for all the bidders.

(2) Bids are opened, publicly in the case of governmental agencies, and publicly or privately in cases of private ownership.

(3) The contract is awarded to the lowest bidder, unless there are some extenuating circumstances.

The major point here is that the installation contract probably has been awarded to the lowest bidder(s) on the basis of the information provided in the contract documents. It is therefore important that all relevant information be included. Details are important. For example, assume that the HVAC system includes a motorized valve that is to be controlled and monitored by the FAS. The interface between FAS and HVAC must be clear. Also, the characteristics of the FAS signal must be defined ac or dc, voltage, frequency, current limits, and the like must be given. Then if relays or other devices are needed, they will be included in the HVAC scope of work.

One further task to be performed by the FAS designer is the review of the contract documents produced by the facility design disciplines to ensure compatibility and coordination with the FAS. At the end of the installation there will be acceptance tests. These should be defined in the contract specifications, including what will be tested and any allowable tolerances or variances from specified data. The time of tests and accessibility for witnesses may be specified or noted as an open item to be developed when the installation is complete (or nearly complete).

The following is a generalized list of items to be furnished by the installation contractor(s), at the end of the project.

(1) Certificate(s) of approval by code enforcing authorities

(2) Guarantee(s)

(3) Instruction in the operation and maintenance of the FAS system, and instruction manuals

(4) Special tools for maintenance

(5) Recommended list of spare parts and prices

(6) Service contract (if specified), or in lieu of the service contract, place(s) where parts and service are available

14.12 References

[1] ANSI/NFPA 71-1977, Central Station Signaling Systems.

[2] ANSI/NFPA 72A-1980, Local Protective Signaling Systems.

[3] ANSI/NFPA 72B-1980, Auxiliary Protective Signaling Systems.

[4] ANSI/NFPA 72C-1975, Installation, Maintenance, and Use of Remote Station Protective Signaling Systems.

[5] ANSI/NAPA 72D-1980, Proprietary Protective Signaling Systems.

[6] ANSI/NFPA 72E-1978, Automatic Fire Detectors.

[7] ANSI/NFPA 90A-1981, Air Conditioning and Ventilating Systems.

[8] ANSI/NFPA 101-1981, Life Safety Code.

[9] ANSI/UL 609-1978, Safety Standard for Local Burglar Alarm Units and Systems.

[10] ANSI/UL 611-1978, Safety Standard for Central-Station Burglar-Alarm Units and Systems.

[11] ANSI/UL 1076 1974, Safety Standard for Proprietary Burglar Alarm System Units.

[12] ANSI/UL 864-1980, Safety Standard for Control Units for Fire-Protective Signaling Systems.

[13] ANSI/UL 294-1981, Safety Standard for Access Control System Units.

[14] NEMA SB 4-1971, Training Manual for Local Fire Protective Signaling Systems[60]

---

[60] NEMA publications are available from the National Electrical Manufacturers Association (NEMA), 2101 L. Street, NW, Washington, DC 20037.

15. Expansion, Modernization, and Rehabilitation

15.1 General Discussion. In developing engineering criteria and plans for the expansion or modernization of the electrical power system for an existing building, the electrical engineer or designer encounters problems over and above the electrical design considerations involved in new-building construction.

The point at which electric service is received from the utility can be defined as the service entrance point. In large buildings it usually consists of low-voltage stabs or electrical bus connections extending through a masonry wall into the customer's premises. In smaller buildings it is the point where the utility overhead or underground conductors connect to the building wiring. At this point the customer will install his low-voltage service entrance equipment. For high-voltage service the incoming feed generally consists of one or more sets of high-voltage cables to which the customer is required to connect his high-voltage service equipment.

The service equipment is simply the equipment used for controlling total power to the building. As a minimum this equipment must have provisions for opening and closing the circuit between the customer and the utility and provide protection against electrical faults within the building. It may consist of circuit breakers, fused switches, or similar devices.

For a major building expansion, the designer is faced with the problem of whether to retain the original service equipment and expand it, replace the original service equipment with new equipment in the same location, install new service equipment in a new location, or install an additional new service point to cover the expansion. This section is concerned to a large extent with the problems involved in making this decision and with the related problem of power-system modernization [2].[61] OSHA regulations in effect state that every replacement, modification, repair, or rehabilitation of any part of any electrical installation or utilization equipment

---

[61] The numbers in brackets correspond to those of the references listed in 15.12.

shall be installed or made, and maintained in accordance with the provisions of the 1971 Edition of The National Electrical Code. As a practical matter, most OSHA offices will accept modification made in accordance with the latest NEC, ANSI/NFPA 70-1981 [1] [62].

15.2 Preliminary Study. The most desirable procedure involves an engineering study prior to the development of firm system criteria. All too often a thorough evaluation of the existing system is neglected or is undertaken only when the project is well into the design stage. With a major building modernization or expansion, the preliminary study should be separated completely from the actual design. A preliminary-study final report should contain a full operational assessment of the existing system with emphasis on reliability. Outlines of alternatives of retaining, replacing, rehabilitating, expanding, or supplementing the existing system should be furnished together with preliminary or budget estimates including annual owning and operating costs as well as capital costs. Such a preliminary report may well produce recommendations completely beyond the scope of the conditions which led to undertaking the study.

Some items that warrant investigation in the modification of electrical systems and which are not normally required in new building designs include the following list.

(1) Age and condition of the existing equipment

(2) Maintenance costs and availability of parts for the existing equipment

(3) Automatic operation of the existing equipment as compared to manual control

(4) Voltage, frequency, and number of phases of the existing system as compared to a modern three-phase grounded-neutral 60 Hz system

(a) Availability of additional power from the utility company

(b) Economy of distributing at the existing system voltage

(5) Characteristics of the existing system

(a) Interrupting, fault closing and momentary ratings (including bus bracing)

(b) Load-current rating

(c) Coordination

(d) Susceptibility to voltage disturbances

(6) Compatibility of old and new requirements

(7) Safety of existing system

(8) Code regulations in regard to the existing system and the proposed expansion thereof

(9) Availability of qualified personnel to maintain and operate the old system

(10) Reliability of existing system and outage record including time required for repairs

(11) Energy conservation modifications such as replacement, recircuiting or provision of new controls for lighting systems, installation of more efficient power equipment or provision of automated building control to replace relatively inflexible operational systems.

The study will require a thorough field survey by qualified engineers familiar with field conditions and construction. Even a competent designer, thoroughly familiar with the fundamentals of

---

[62] The National Electrical Code is published by the National Fire Protection Association, Batterymarch Park, Quincy, MA 02269. Copies are also available from the Sales department of American National Standards Institute, 1430 Broadway, New York, NY 10018.

modern electrical design for new buildings may lack sufficient knowledge of old work.

Data regarding the expected remaining life of existing equipment including cables may be important in determining whether to retain such equipment. A history of operational problems and equipment failures over the years is invaluable in predicting the reliability of the existing system and probable rate of future failures.

Coordination of the study with other disciplines is essential. An example of less than good planning is the replacement of multispeed centrifugal blower motors because of motor and controller deterioration and low efficiency when the best decision is to replace the motor-fan combination with a variable-pitch-fan constant-speed motor assembly.

In the analysis of the retention and expansion of the existing system versus providing a new system, the operational and safety characteristics of the equipment should be compared. This includes comparison of certain items such as those listed in Table 93.

An awkward situation which an engineer might well encounter is previous rehabilitation work which was poorly planned. Typical examples follow:

(1) Because of a rash of failures, all wound-rotor motors were rewound or dipped and baked recently. The study now shows a replacement of the motors with new more efficient multispeed squirrel-cage motors with new starters to be the best solution.

(2) All the old knife switch panels were replaced with *Q* type circuit breakers. However, a new service will call for higher interrupting ratings than 5000 A or 7500 A.

(3) Because of the poor condition of thirty-year old type RW cable, particu-

Table 93
Comparison of
Old- and New-System Characteristics

Old System	New System
Open knife switch construction	Dead front
Non symmetrical systems	Systems using grounded neutral
Low-voltage oil circuit breakers	Metal-clad gear
Medium-voltage oil circuit breakers	Air circuit breakers
Fuses with unrated interrupting capability	Current-limiting fuses
Fixed gear	Drawout gear
In-line trip devices	Static relaying
65 °C wire or cable	90 °C wire or cable
Open buswork	Isolated phases or compartmentation
Manual close and trip	Electrically operated close and trip
Local control and operation	Supervisory and automated control
Cascade breaker tripping	Selective breaker tripping

larly in the gutters of switchboards, the building electricians recently completed replacement of all these cable runs. The study now shows that new higher capacity panelboards with larger feeders are required.

(4) All ballasts in the plant were changed several years ago to rapid-start type, eliminating starters. The study now shows that a 480Y/277 V service is justified, and new ballasts will be needed or overall transformation required.

(5) Delamping, removal of unused ballasts, and replacement of remaining ballasts with highly efficient lamp ballast combinations was recently completed as an energy saving measure. A full study shows that recircuiting to provide more

flexible control (half lighting during off hours, perimeter control, control by area) will provide for greater savings, but now much of the completed wiring will have to be redone.

While serious consideration should be given to retaining such previous betterments, the implementation of an overall plan of modernization may justify elimination of, or further modifications to the previous improvement.

Salvage value of existing equipment should not be neglected. Old equipment such as switchboards and conduits is usually not worth much more than junk except for the copper. A recent building inventory of the value of old direct-current feeders indicated a probable scrap value greater than the estimated cost of new feeders for a new alternating-current system. Lead-jacketed cables are additionally valuable because of the easily recoverable lead. The contract should clearly state the disposition and estimated quantities of scrap material. In contracts where the disposition aspects are mute, disputes between the contractor and client may arise concerning ownership of scrap materials.

15.3 Design Considerations. Costs of a rehabilitation project for both engineering and construction may run 50—100% higher than a comparable project involving only new construction. Where field conditions cannot be exactly determined, the estimate of extra work allowances will have to be liberal enough to provide for indeterminable construction difficulties.

There are two considerations that may not have been included in the original design. These have been emphasized because of serious electrical failures that have occurred in high-current electrical systems. The first is the installation of ground-fault protection, now required, which will to a large extent alleviate the problem of sustained low-current arcing ground faults, where the ground fault is insufficient to trip the line-overload elements. The second is to keep the individual main feeds to a relatively low capacity, 1600 A or 2000 A. This is the simplest way of preventing massive power failures. The cost of large circuit breakers rated at 4000 A is also disproportionately greater than the cost of more limited circuit breakers rated at, for example, 1600 A. Even if the service stab is a large one rated at 5000 A, there is no reason why the individual switchboards cannot have much lower ratings. For example, if the service stab is rated at 5000 A, the local codes may require a single service switch, but this switch can serve perhaps three switchboards, building risers, or major feeds, each protected by a main circuit breaker rated at 2000 A. If the local codes permit, it is of course desirable and less expensive to install three individual service switches, each protecting a switchboard.

If existing equipment is to be expanded, it is essential that necessary parts be available. For example, there may be four or five spare spaces in an old panelboard for additional circuit breakers, but such circuit breakers may not be available at acceptable prices or with timely deliveries.

In attempting to expand an old system, the circuitry may become unduly complex. The resulting one-line and riser diagrams should clearly and simply illustrate the system. Operationally this is important. Buildings have been disconnected because the operator or watch engineer could not understand the system and operates a critical switch. Key interlocks, while often necessary, should not be part of very frequently operated equipment. Neither should they be used in such a manner that an emergency operation will be inhibited.

Even though newer wire insulations permit higher cable ampacities, it is doubtful that in a major rehabilitation or expansion replacement of existing cables permitting a 90 °C conductor temperature will be sufficient to supply the much heavier loads associated with such modifications. Note that cables using the 90° rating must have 90° terminations (molded case breakers normally have terminals rated 60° up to 100 A and 75° for higher ratings); a conductor temperature of 90 °C may damage the equipment to which the cables are connected. Since 110 V, 120 V, 220 V, or 240 V systems utilize 600 V cable, upgrading the system by voltage increase may include reuse of the original cables. In reusing main feeders, care should be taken that the neutral is not undersized. The neutral of a system feeding primarily incandescent lighting may have been designed smaller than the line conductors. Replacement with a fluorescent system calls for full capacity neutrals because of third harmonics.

Another important factor is whether the building can be shut down for the rehabilitation project (perhaps after the old tenant moves out and before the new tenant moves in), or whether the building must be kept in operation during the project except for weekends or nights.

15.4 Retaining Old Service Equipment. Although retention of the old service equipment can minimize overall costs, there may be drawbacks to such retention. The original service entrance equipment may have inadequate interrupting capacity when compared to the available fault-current capability of the utility service. It is not the general practice of local utilities to inform a customer when available short-circuit currents from their supply have increased. The customer most often discovers this predicament only when he investigates the need for additional building power.

Older installations will seldom have the expansion capabilities needed for the future, even if immediate power requirements can be satisfied. In trying to expand an old service point, problems will be encountered in matching old equipment with new equipment. It might cost almost as much to purchase a custom-made adaptor or bus-detail section to accommodate new busway to an existing service switch as to replace the switch. Installation of ground-fault protection on an existing circuit breaker may be difficult, and may even be impractical if the circuit breaker does not have an electrical trip; new circuit breakers can be purchased with low-cost ground-fault protection installed integrally. Modifications to older equipment, in the field, are generally much more expensive and may be less satisfactory than comparable factory-installed features.

If the existing service equipment and the main switchboard circuit breakers are otherwise satisfactory, it may be possible to retain this equipment despite insufficient fault-handling capability. The installation of either reactors or current-limiting busway between the utility service point and the service equipment may permit a reduction of the maximum short-circuit currents to values within the rating of the equipment. The reactor has not been too popular because of unfamiliarity in application, possible local code limitations, and because a new service protector or bolted-pressure switch of adequate interrupting rating might be more simple and less expensive.

The operational characteristics of older

circuit breakers may present coordination problems. For example, a service switch instantaneous device is most difficult to coordinate with downstream devices. Older dashpot time-delay features are often erratic and cannot be depended upon for accurate timing. At one time no interrupting rating was designated on older circuit breakers, and these circuit breakers may now be listed at 10 000 A by the manufacturer. If the existing service equipment embodies fuse protection, replacement of the older style of fuses may require certain bus adapter work and the installation of new holders or fuse adapters to either increase the fuse size or interrupting rating or prevent insertion of improper fuses.

If existing equipment is to be retained, the expense of a thorough overhaul should be provided in the overall cost estimate. Buswork bolts must be retightened, and if signs of overheating (blackening) are present, refinishing or the equivalent may be required. Dashpots almost invariably need cleaning and refilling. Circuit breakers should be operated, observing possible points of malfunction, and, if possible, should have tripping characteristics checked by simulating overload and fault currents. Main contact springs on older circuit breakers or fuse spring clips may have lost temper due to overheating.

A factory representative may be required at the work site to perform major or complex field work. The manufacturer will have to agree to any schedule for this work. The manufacturer's quoted living and traveling expenses for time away from the home office by the service representative should be added to the cost for such work. Extended availability of such representatives may be limited and it may be necessary to have the manufacturer train local electricians

or technicians. If contract work is proceeding simultaneously with the overhaul, work-rule regulations may require the involvement of local electricians in a *manufacturer's* overhaul. In one instance, the repair of an open-switchboard circuit breaker by the manufacturer would have cost about one third more than an installed new molded-case circuit breaker.

15.5 Completely New Service Equipment. The installation of a new service point and new service equipment will avoid many disadvantages of trying to retain and expand old service equipment. Outage time can be kept to a minimum if the original service point and service equipment can be kept in operation until the moment of reconnection to the new service equipment. If the new service point is adjacent to the old one, the old service equipment can be removed after the transfer of load and this space allocated for future electrical expansion. Installation of a new service point, allows the engineer an opportunity to increase the service voltage to 480Y/277 V three-phase service that is very much today's standard. In this case it would be necessary to install transformers to step down the voltage to the old parts of the building, if conversion of equipment in the old building to higher voltage is uneconomical. Modern distribution methods may involve less in total cost for a completely new service than the cost of attempting to retain the old service and expand it.

Utility service rules may require the customer to install a new service point with new characteristics when major changes for additions are to be made in the customer's electrical power system. Utilities are often required by the Public Service Commission to continue supplying power to the premises of a customer at the originally installed voltage and frequency. If, however, the existing facilities are inadequate, utility rules may require filing a new application for power which presents the opportunity for the utility to require abandonment of obsolete powers sources. Most utilities will actively resist an attempt to expand an existing 25 Hz, direct-current, or two-phase service. To eliminate an obsolete supply characteristic, the utility has in a few instances been induced to bear part of the customer's conversion costs.

15.6 Additional New Service Point. An alternative to the above is to retain the existing service point and to install a new service point, usually more centrally located (electrically) within the building. The additional service point will generally be located in the new area if a major area expansion of the building is involved. If the additional service point is to provide for an increase in capacity only, the new service point might very well serve only such equipment as motors, while lighting, for example, remains on the old service. Although adding a new service point sounds like a simple solution, utilties may be loath to provide more than one service point without an excess service or other charge, or they may treat the second service as a separate service with a separate bill. The NEC and local codes govern the use of more than one service, and approval of the local inspecting authorities, as well as that of the utility, is required to authorize exceptions. Section 4 describes the various types of multiple metering installations and explains the advantages of each.

In determining whether it is better to undertake the cost of an additional service point, it is necessary to include the cost of the equipment, estimated maintenance costs for old and new equipment, power losses of cables and transformers if significant, and the additional utility power cost for the additional service point if applicable.

15.7 Voltage Transformation. It may be of advantage to install a new service point and to reduce the new and usually higher voltage to a voltage usable by the old parts of the building. If the utility insists on retaining the customer's old lower voltage, it may be advantageous to install transformers to step up to 480Y/277 V for new equipment.

A few hundred feet of busway in the higher current ratings can equal the cost of a transformer. In the typical *compact* building, 480Y/277 V is usually adequate for distribution throughout the building. However, there are some cases where it may be desirable to transform to a higher voltage such as 2400 V or 4160 V. These voltages are very practical for supplying medium-sized motors (in the sizes of 200 hp or above). Such high voltage would be stepped down at the receiving end for uses other than motors.

If medium voltage is transformed up and retransformed down at one sending and one receiving end with no intermediate taps, and no danger of confusion of medium-voltage feeders can exist, the engineer may elect not to install medium-voltage switches. The low-voltage switches can serve as the disconnecting medium at both ends. However, where there are taps emanating from the medium-voltage line fully rated load-break disconnecting switches are usually desirable for operational and safety reasons. The use of medium voltage is recommended only where adequate maintenance and operating personnel are available within the customer's own staff or from the local electrical maintenance agency. Maintenance of medium-voltage equipment by personnel not fully trained therein can be dangerous. ANSI/NFPA 70-1981 [1] requirements for the protection of transformers should also be considered (Article 450).

If multiple service entrances are used with different voltages or if transformation is involved with different voltages, it is undesirable to apply unlike voltages for identical types of usage in the same areas. For example, it is preferable to keep the fluorescent lighting at 208Y/ 120 V for an entire area rather than have a sporadic mixing of different voltages. There is no problem with providing different voltages for lighting in different areas, and there is no problem in providing a different voltage for lighting than for power equipment in the same area, although conspicuous marking of equipment voltage ratings is desirable.

If transformers with a delta primary to wye secondary transformation are used, ground-fault protection may be provided through the neutral of the wye connection of the secondary. Simple computation will show that primary protection only, on such a delta-wye transformer, often cannot be set to properly protect against a ground-fault condition on the secondary. A ground fault on the secondary load conductors will reflect only 57.7% of the current in the primary that an equal-current three-phase secondary fault would develop.

15.8 Distribution of Power to Main Switchboards. A number of methods of distributing power are available, including cable in conduit, busway, cable bus, and cable with continuous rigid cable supports (cable trays). Perhaps the simplest of these methods for low voltage is the busway, particularly the newer types with single-bolt construction. In all cases, the service entrance switch must protect the feeds to the main switchboards.

Where feeds must be concealed, conduit and cable remain in most common use. With one cable per phase, the entrance switch overcurrent device will serve as adequate overload and short-circuit protection. On the other hand, if several paralleled cables per phase are required, protection may not be provided to individual cables by the service-entrance protective device. One way of eliminating this problem is through the use of limiter lugs on each cable. To assure service continuity, it is desirable to place the limiter lugs at both ends of an individual cable so that a fault on one of the cables will operate the limiter lugs at both ends of the cable, leaving the other cables intact. It should be noted that a limiter lug is not in itself a current-limiting device such as a current-limiting fuse. The former is designed solely to protect an electrical system against a catastrophic failure of individual cables. Newer types of limiter lugs with current-limiting features have been developed. Should one of the limiter lugs open, the remaining cables may carry excessive current for extended periods. Therefore, it is desirable to periodically test them either with a clamp-on ammeter or with permanently installed indicating devices.

Although busways and cableways are available for high ampacities (6000 A or more), the original source of power should be divided in such a way that the use of busways over 2000 A or in some cases 3000 A (see ANSI/NFPA 70-1981 [1], Article 300) in normal building installations is avoided. The more massive a bus structure is, the more difficult it is to protect. Cables may be installed in cable trays. If such is the case, ground-fault protection (which is required in all high-current services) is especially necessary for long cable-tray runs. Ground-fault protection should be set low enough to positively detect arcing ground faults which would damage equipment, but high enough to avoid nuisance tripping.

The engineer should consider the effect of failure of a single cable on other cables of the same feeder and even more particularly of other feeders, before using any system involving a large number of conductors in a single enclosure or wireway. Mixing different feeds from different services in the same wireways does not contribute to continuity of service. The local electrical code, for example that of New York City, may be very specific as to the separation of 480Y/277 V multiple services and feeders.

Where more than one set of stabs or takeoffs is provided by the utility for electric service, it is prudent to arrange the building distribution in such a way that outage of a single service will create minimum disruption to building power. For example, if two panels are provided on a floor, one can be from service A, the other from service B. Half the power can be taken from one service and half from the other service. It would not be desirable, for example, to take all the air-conditioning compressor units from service A and all the air-handling units from service B. Some local building regulations do not permit the intermixing of feeds from separate services where such services are widely separated as might be the case when an additional service point is provided.

15.9 Existing Plans. Unless the plans of the existing systems include recent as-built drawings, and reasonable assurance is available that changes have not been made by building staff, the accuracy of such plans is suspect. It is, therefore, necessary to verify on site the accuracy of questionable plans and to record any additions or corrections. Close liaison with the building staff is required. It will be necessary to note building structural changes and to determine acceptable floor loadings for transport and placement conditions of switchgear, transformers, and other major equipment.

The full size of conduits shown on old drawings may not be usable for expansion or modernization. Old conduits may be rusted to the extent that cable pulling problems are accentuated, EMT (thin-walled conduit) may have pulled apart in the ceilings, sections of nonmetallic duct may have been crushed, and tile ducts may contain concrete obstructions. To check empty conduit runs it is desirable that a clearance device be pulled through the conduit.

Where there are operating cables in existing conduits, it is usually impractical to check such conduits prior to the proposed work, except for conditions determinable by visual inspection. In a large project, where data on the conduit condition is unavailable and there are no spare conduits, it may be worthwhile to remove (and replace if necessary) cables in a few typical runs to permit determination of problems which may be encountered in rewiring.

The old system may very well contain code violations by current standards due to code changes. However, if a rehabilitation is undertaken, the code violation may require correction.

Field changes are always expensive when performed as contract extras. If a contractor estimates work where the scope is indeterminate and he is required to give a fixed or lump sum price on such work, it can be assumed that his estimate will be predicated on the worst anticipated conditions.

In a rehabilitation project, knowledge of the exact location of embedded conduits may be essential because of the necessity to cut into concrete slabs for the extension or rerouting of conduits. Such conduits can be located exactly by X-ray methods. However, this location work must be scheduled when personnel will be absent from the vicinity of radioactive hazards in the test area. Where such work is required, it is essential to obtain the services of a firm specializing in this radiographic locating because of the laws, safety requirements, and special techniques pertaining thereto. Metal detectors may help locate conduits.

15.10 Scheduling and Service Continuity. Several well-known methods of scheduling including the Gantt chart, PERT, and critical path (some of which may be computerized) are available to assist the engineer. However, the need for continuous electric service during the construction accentuates the problem compared to new work. Consider some problems which have resulted from poor planning. For example, workers were denied access to an office building at the beginning of the work day because electricians were completing electrical connections. In another case power to a public terminal building failed during rush hours because temporary load reconnections coupled with system transients caused tripping of an old breaker with a defective overload unit.

The schedule should provide extra time for completion of scheduled power outages. For example, such work could

be scheduled to start on a Friday night so that it can be completed over a weekend. An engineering chain of command is especially important during this work. Should critical decisions be required, the availability of engineering assistance or support is essential. The contractor's staff may not be in a position to make project engineering decisions.

Consideration must be given to maintaining a continuous work flow, even though a portion of work is delayed. If, for example, new overload relays for the main circuit breakers are delivered and will not fit (a last-minute determination), the entire project should not be adversely affected.

The design of the job, if staged, should involve the development not only of old and new one-line diagrams together with the applicable plan and detail drawings and specifications, but also detailed phasing schedules. A design engineer must be aware of steps involved in making complex power system additions and alterations. Drawings or sketches and phasing schedules showing the step-by-step procedure must be developed, even though they are of only temporary value. The drawings or sketches should be modified on the construction site as the construction work is performed to show the true or as-built status.

It is often advisable for the designer to specify a new or temporary replacement item rather than resorting to a field modification. Returning to the example of the main circuit breaker with new overload relays—if the overload relays fit perfectly the job can be completed in one night; if not, a large section of the building could be out of service, additional overtime may be required, or the repair may be completed in an incorrect manner. The alternative plans of installing a temporary fuse cubicle or having one available for possible use would have been prudent. In a project involving old construction, contingencies of time are as important as contingencies for costs. Scheduling a modernization project over several years (with perhaps contract escalation clauses for inflation) may be much more effective regarding cost and outage than an accelerated contract. Time permitting, problems can be minimized by keeping the day-to-day scope of work in any area limited.

15.10.1 Example 1: Building Expansion, Medium-Voltage Service. Consider an expansion almost doubling the area of a large terminal building including public, vehicular, office, and tenant areas. The question is whether to expand the medium-voltage power system as a primary-selective system or start from scratch, abandoning the existing system.

In this instance (Fig 143) 13 200 V is provided to the building from four existing feeders. Two of these are standby feeders of the existing primary-selective system. If the primary-selective system is to be retained, two additional normal feeds are required to handle the building expansion. The two standby feeds are adequate for second contingency conditions (one feeder out of service and one subsequent failure).

The installation of a four-feeder network permits the use of the four original feeders with all feeders carrying a normal load, thus eliminating the need for two additional feeders. Where computers are to be installed, the need for an uninterruptible power supply may be eliminated because the primary switching interruptions are done away with.

Note that network primary feeders should, in general, have essentially the same receiving voltages and phase angles to provide proper load distribution between transformers.

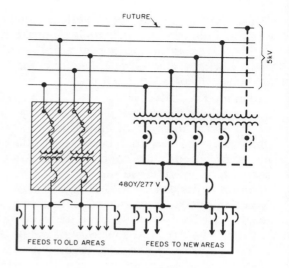

Fig 143
Example 1: Building Extension Medium-Voltage Service. After Connection of the New Substation, the Old Equipment (Shaded) Was Removed

The old equipment is retained in service until all loads have been transferred, one by one, to the new switchboards. Network transformers with high impedance (that is, 7%) are specified to reduce available fault currents, provided increased voltage drop is no problem. Main circuit breakers and buses of switchboards are limited to 2000 A to reduce available fault currents.

15.10.2 Example 2: Building Expansion, Low-Voltage Service. Here the problem is to increase the size of a large building by about half in area. In the preliminary study, perhaps twenty schemes (seven of which are illustrated in Fig 144) are analyzed to obtain the most desirable solutions from economic and operating standpoints. Discussions with the local utility determine which schemes are acceptable. The problem

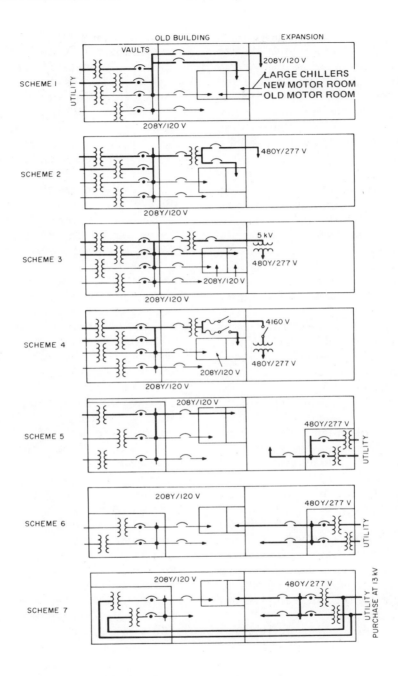

Fig 144
Example 2: Building Extension, Low-Voltage Service, Simplified
One-Line Diagrams of Seven Possible Ways of Handling a Major Expansion
(Heavy Lines Are New Installations)

has been made general and very much simplified for purposes of presentation. The existing building is fed from utility vaults at 208Y/120 V from a common network system. Air conditioning-compressor equipment in the older area is supplemented to provide chilled water for air conditioning to serve the new area as well.

A first analysis involves increasing the capacity of the existing service entrances. This is accomplished by increasing the ratings of the stabs or by adding stabs. The latter is preferable because it minimizes the fuse or circuit-breaker sizes, enhancing fault protection. The old equipment remains in service and intact, provided that fault-handling ratings are adequate.

If transformers and network protectors are owned and operated by the local utility, the utility has to concur in increasing vault sizes. This is probably the utility's first choice since adding network transformers to an existing network installation generally makes most effective use of equipment. With four transformers, two of these should be able to handle the full load under second contingency conditions. During this condition, the remaining two transformers operates at an emergency overload of perhaps 20–30%, so that the maximum rating of the bank is between 60 and 70% of the normal rating of four transformers. Adding one transformer to a bank of four increases the capacity of that bank by more than the rating of the transformer added.

Thought must be given to the physical problems involved, that is, the advisability of carrying exposed heavy-current busway through public areas. The availability of space is important, and the value of the space for rental or other purposes should be a determinant when deciding to locate service entrance equipment in the old or new buildings.

Scheme 1 involves expansion of the original networks by the utility and transmission of power at 208Y/120 V to the new area. With permissible voltage drops the cost may be substantial. (See Fig 144.) The cost of 208 V feeds to the new large chillers would be excessive.

Scheme 2 also involves expanding the 208Y/120 V network, stepping up to 480Y/277 V and supplying this voltage to the new building and to the new compressor equipment in the old building. Overall costs are substantially less than for Scheme 1.

Scheme 3 involves stepping up to 5 kV in the vicinity of the 208Y/120 V expanded vaults and retransforming to 480Y/277 V in the expanded area. Since motor load in the expanded area consists mainly of smaller air-handling units, 2400 V or 4160 V is not a practical utilization voltage. The motor room when expanded in the existing building continues to be fed at 208Y/120 V. As in Scheme 1 the cost of feeding the chillers at 208 V is excessive. There are, however, considerable savings in the feeds to the expansion and in the distribution at 480Y/277 V more than off setting the transformer cost.

Scheme 4, involving 4160 V or 2400 V three-phase distribution, is much the same as Scheme 3. However, each individual load center is each provided with its own transformation from 4160 V or 2400 V to 480Y/277 V and the new air-conditioning compressors are fed at 4160 V or 2400 V.

In Scheme 3 medium-voltage switches are not specified because there are no intermediate taps on the medium-voltage feeder cables. However, in Scheme 4 medium-voltage switches are desirable for isolation purposes, since each feeder serves several loads, and fused switches are required for transformer protection. Where several feeder taps are involved, the provision of such medium-voltage switches enables operation of all but one substation under conditions of a transformer outage, simplifies fault finding, and enhances safety considerations. The first two considerations may not be given the same weight in making the decision if a highly qualified utility-type maintenance crew (not normally available to building maintenance staffs) with special fault-finding equipment is to operate and maintain the system.

Scheme 5 has the main disadvantage of requiring both the installation of a new service for the new building and expanding the old vault. If no major construction is involved in expanding the old vault and the increased air-conditioning load is relatively small, this might be an acceptable scheme. There is a good possibility that the utility would balk at installing at their own cost both the network expansion and new network installation. Approval of the applicable code authorities have to be obtained for a multiple-service installation.

Scheme 6 is perhaps the best choice of all. Vault expansion is not required in the old building, 480Y/277 V is provided for the new motors in the old building without transforming. This scheme can be so desirable that even if conjunctional billing as previously described, is not provided between the old and new service points, the total annual cost may be less than for the other schemes. This does involve the question of multiple services to one building.

Scheme 7 might be the best if sufficient differential exists between high- and low-voltage rates to justify owning the network systems.

In practice, two factors besides cost may militate against the use of lower voltages for the expansion. There simply may not be space to carry heavy conductor-system (busways, or multiple conduits,) through existing areas, and above certain sizes (approximately 1000 hp) chiller motor starters may not be available which can operate on 208 V. If utility considerations strongly favor expansion of the existing vaults and the expansion involves capacity in the order of 2500 kW or more, Scheme 4, medium-voltage distribution to the new equipment, would probably be the first choice.

15.10.3 Example 3: Retention of Existing System and Expansion.
In this instance, the availability of additional power is severely limited in a large thirty-year old building containing offices, commercial areas, freight-handling facilities, and consumer facilities. A series of changes are made which, while not modernizing the existing electrical system in its entirety, provide adequate power for continued expansion (Fig 145).

Power is available at 208Y/120 V, three-phase, from individual network transformers owned by the utility. These transformers are distributed throughout the building in vaults. The existing electrical system is adequate beyond the service switches, except for new heavy load requirements.

The anticipated load is expected to rise steadily over a period of years, perhaps 10% a year, and the distribution of loads on risers and the nature of the additional loads are such that it is entirely practical to retain the existing equipment and existing distribution system for the existing loads and to install new service equipment and

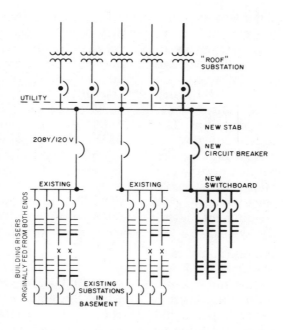

**Fig 145
Example 3: Retention of Existing System and Expansion
(Heavy Lines Are New Installations)**

switchboards for the new loads. At some later date older equipment could be modernized without affecting the new installation.

A second consideration in retaining the old equipment is the need to maintain full service during working hours during the power expansion. By installing new equipment to pick up additional loads the old equipment is maintained intact. After the new equipment is installed, loads are shifted from the old to the new circuit breakers to relieve heavily loaded feeders and provide capacity in the existing feeders for future normal expansion.

For example, a feeder running fifteen floors is cut and fed as two feeders, one from the top and one from the bottom.

New service stabs are provided by the utility, one for each service room.

15.10.4 Example 4: Rehabilitation on a College Campus. A campus is comprised of three buildings, namely, the main building, the library building, and the science center. The center section of the main building was constructed about eighty years ago, while the east and west wings were added about ten to twenty years later. The library and science buildings are modern structures having been built within the past ten to fifteen years (See Fig 146.)

The rehabilitation basically called for the enlarging and renovation of the main building. In the course of investigating the electric service for this expansion and modernization, it was found that each of the buildings on the campus was separately metered and was also supplied at a different utilization voltage.

An economic analysis indicated that it would be prudent to purchase primary power at a single service on the property and feed each of the existing buildings from that point. After discussing the plan with representatives of the utility company, it was decided to purchase power at 13.8 kV and to transform this voltage to 480Y/277 V at an outdoor substation on the property.

This secondary voltage was suitable for the science building and for the planned rehabilitation of the main building. At the library building a 480—280Y/120 V transformer was installed in the vault which previously housed the utility company transformer.

To simplify the switchover and to keep the outage time required for reconnecting the services for the science and library buildings to a minimum, the revenue meters and metering-current transformers were purchased from the utility company, obviating any work in the metering cubicles of either service. The purchase price was considerably less than the cost of having them removed. In addition, it provided the owner with metering facilities for maintaining records of energy consumption at each location for stastical and accounting purposes.

A new switchboard and a new distribution system were installed in the main building. When new loads and all existing loads which were to remain were con-nected to this new service, the original two services were discontinued and the old switchboard was removed.

15.11 Wiring Methods. There are a number of wiring methods which lend themselves to building rehabilitation.

While it is desirable to utilize existing conduits, as a practical matter it may be necessary to abandon all or part of the existing conduit system. If a new architectural finish is being provided, an opportunity exists for running new concealed conduits. However, where physical or cost limitations preclude concealment, one of the surface raceways which are code approved may be used. Where existing boxes are flush mounted, it is possible to install commercially available box extensions which will permit connection of a conduit or a surface raceway. Almost all wiring equipment is available for surface or flush mounting, and where architectural considerations permit, the use of surface-mounted equipment simplifies rehabilitation.

Where a new hung ceiling is being installed or a hung ceiling replaced, an excellent opportunity exists for installing a new lighting system. Where the ceiling is to remain and is structurally adequate, either surface-mounted fixtures or pendant-mounted fixtures may be installed without modifying the ceiling. Caution must be observed in the installation of a surface-mounted lighting fixture to avoid overheating of the fixture or excessive ballast noise when in operation. It should be noted that the cost of relocating a lighting fixture in a hung ceiling might be more than that of installing a new surface or pendant-mounted fixture. Where extensive underfloor wiring has to be provided, particularly for data-processing equipment, the conventional raised-floor approach may be practical. Floor

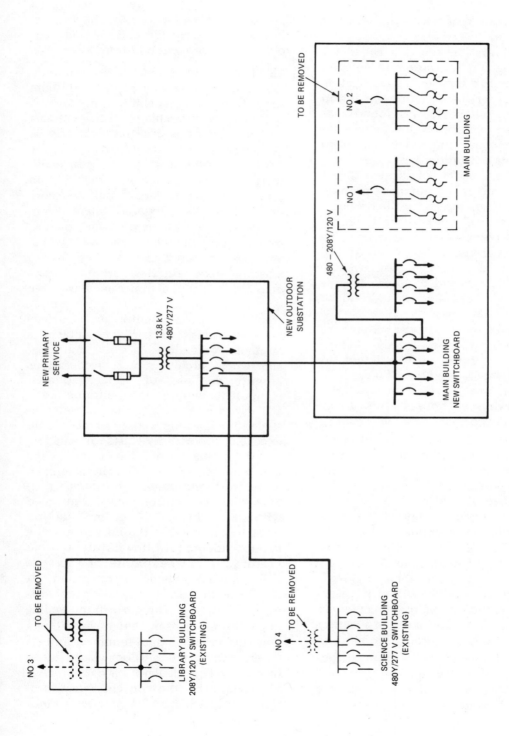

Fig 146
Example 4: Rehabilitation of Electric Service for College Campus
(Heavy Lines are New Installations)

outlets may also be fed from the ceiling below by means of conduit and poke-through boxes; however, only boxes approved for the purpose, and having a suitable fire rating, should be used.

15.12 References

[1] ANSI/NFPA 70-1981, National Electrical Code[63]

[2] GOLDBERG, D. L. Electrical Expansion, Modernization, and Rehabilitation for Buildings. *IEEE Transactions on Industry and General Applications*, vol IGS-6, May/June 1970, pp 219–229.

---

[63] The National Electrical Code is published by the National Fire Protection Association, Batterymarch Park, Quincy, MA 02269. Copies are also available from the Sales department of American National Standards Institute, 1430 Broadway, New York, NY 10018.

16. Special Requirements by Occupancy

16.1 General Discussion. There are many types of commercial facilities. This section covers specific considerations for the most common types.

The preceding sections describe in detail the basic electric systems and components necessary to meet the needs of commercial facilities insofar as modern power supply, electric distribution, transportation, lighting, controls, and communication systems are concerned. There are, however, in each classification of occupancy certain items applicable to that classification and not required in other classifications. There are also certain factors, which if overlooked in certain types of commercial occupancies, may lead to very expensive additions to the system after the facility is completed.

In this section the different classifications of commercial occupancies are defined and the special considerations of some are discussed. Most of the usual requirements such as service entrance, substations, types of distribution, and communication systems are described in the preceding sections.

The designer of commercial facilities may group his design requirements into four basic categories:

(1) Minimum essentials

(2) Dependability of service and provision for contingency

(3) Flexibility for growth and changes

(4) Safety, efficiency, and comfort

Good judgment and selling ability are required by the designer. He should convince the building owner and electric utility company of the need to include in the original installation services for some future loads, and to make provision for facility additions at a later date. Once the facility is finished and all space assigned, the cost of installing additional equipment may be double or triple the cost that would have been incurred had provisions for these future loads and additions been included in a master plan.

Minimum essentials are generally covered by applicable regulatory codes (see Section 1). The design engineer should first determine the enforcement agencies involved with the facility and communicate with them requesting specific information as to the codes, ordinances, and regulations enforced in that area. Copies of these documents should be requested so that they may be reviewed relative to the code editions that have been adopted, and also to check for specific requirements that may not be obtained in communications with the enforcement agency.

Codes, ordinances, and regulations establish the legal minimum requirements for safety, and may not adequately address the needs of specific occupancies. The design engineer must meet the minimum requirements and expand from that point as required by the specific nature of the facility, occupant, and building owner.

In general, whenever static equipment is used, surge protection (including lightning protection) should be considered for power supply circuits and for control circuits which leave the immediate area.

16.2 Apartment Buildings and Condominiums.

A few units or a group of buildings with a thousand units may be involved. The trend in cities is toward more large housing developments with buildings of up to forty or more stories. The risers in such buildings approach the size of those in large office buildings, and the services such as water, fire protection, and heating require heavy equipment and wiring. Air-conditioning loads of the window or through-the-wall types should be anticipated if a central system is not provided. Clotheswashers, dishwashers, and other residential type appliances are important loads. The large number of people in the nation's working force who must do their housework, such as washing and ironing at night, add to peak lighting loads and thus increase maximum demand. The following check list is for apartment buildings and condominiums:

(1) Metering, master or individual; utility, submetering, rent inclusion

(2) Exit and emergency lighting

(3) Telephone system

(4) Annunciator and security systems, closed-circuit television, and intercommunication systems

(5) Central television antenna systems

(6) Individual or central laundry facilities

(7) General lighting, interior and exterior, apartment lighting and receptacles

(8) Air conditioning, heating, and ventilating

(9) Special appliances such as garbage grinders, compactors, and dishwashers

(10) Fire-alarm systems and, if high rise, special life safety systems

(11) Signal systems for handicapped and senior citizens

(12) Parking facilities

(13) Recreational facilities such as swimming pools (ANSI/NFPA 70-1981 [1],[64] Article 680), health facilities, and recreation rooms

(14) Elevators, passenger and service

(15) Restaurant

(16) Stores, shops, and bars

(17) Lightning protection

(18) Snow and ice melting systems for walkways, ramps, and driveways

(19) Cooking and water heating.

16.3 Auditoriums.

An auditorium may be described as a building for large gatherings of people for meetings, entertainment, expositions, or sporting events. Auditoriums may be a part of schools, office buildings, laboratories, churches, or any building where people wor-

---

[64] Numbers in brackets correspond to those of the references listed in 16.38 of this section.

ship, work, or play. Some types, especially in the sporting field, may require special services and some are used for more than one purpose.

16.3.1 School Auditoriums. Most schools have one large room for assemblies, plays, and entertainment. As most of these rooms have a stage, stage lighting should be provided. These provisions, depending on requirements, could range from small systems with light tracks and a few spotlights to electronic dimmer systems with extensive use of spot, flood, and border lights. In all cases stage lights should be provided in the auditorium to illuminate the faces of people on the front portions of the stage. The auditorium lights should be controlled from a central location.

A public-address system is generally required in a modern auditorium and the controls for volume and tone quality may be at the main lighting control location.

Motion-picture, slide, and overhead projection may be required so that service outlets should be provided as well as projection-booth equipment. Convenience outlets for cleaning and general maintenance should be numerous. Building codes dictate the need for an emergency aisle and exit lighting. A dimmer bypass must be provided.

16.3.2 General Auditoriums. These are usually separate buildings which are used as civic centers for entertainment, meetings, exhibitions, and sporting events.

The lighting problem is similar to that described in 16.3.1. There may be requirements for greatly increased lighting intensities on the main floor for illuminating a boxing ring, basketball court, or ice rink. Ice shows may require spotlights scattered throughout the main area. The lighting should be flexible to ac-commodate the different uses of the main floor area.

The type of lighting control and the location of the control booth are important. Communications systems between various lighting centers may be required. Arrangements for television equipment may be provided. Lighting controls may be elaborate for color television which requires higher intensities.

While auditoriums have stages, in many cases they do not require elaborate stage lighting. Many have portable stages, which require portable lighting circuits. Provisions should be made for high-capacity *company switches* to feed portable dimmer boards used by touring shows.

Attention should be given to the problems of glare. High lighting levels can become objectionable unless a great deal of care is used in the design of fixtures, shielding, and ventilation.

The auditorium may be used for exhibitions. This may require a sufficient number of outlets for both 208Y/120 V (and possibly 480Y/277 V) power and 120 V lighting for the various exhibits. Power outlets may require several hundred amperes each. Central distribution centers should be furnished to supply portable distribution outlets. A grid of under-foot raceways should be considered.

The area under the main arena is sometimes used for industrial exhibits or for menagerie, storage, or horse stalls in the case of a rodeo or horse show. Good general lighting should be provided as well as ample convenience outlets for three-phase power and 120 V lighting. This area may have a low ceiling which may require special attention to lighting fixture design to secure the required lighting levels.

Air-conditioning loads, both for the

main auditorium and other smaller rooms, may be major power loads. Ice shows require large refrigerating plants. Continuous ventilation will be required to maintain air quality.

Large auditoriums may have several entrances. Each of these entrances should have an illuminated marquee and signs. The main entrance lobby and foyer may be very ornate and require careful study to provide a satisfactory plan of illumination and decorative lighting. Large signs should be provided so that notices of coming events may be read from a passing automobile. Electric heat for the ticket booths may be required in the colder climates.

Amplifier systems need special attention. Often musical programs need good sound reproductions in an arena which may have poor acoustics. Permanently-installed radio and television pickups should be provided; therefore, local stations should be consulted as to their requirements. Microphone outlets should be positioned at likely points of activities. Telephone and telegraph facilities and press communications may be required. Large numbers of public telephones may be needed, especially in the lobbies and foyers. Electric timers and scoreboards for indoor sports may be widely placed around the arena.

Power services should be of the dual type if possible. A power failure where large crowds of people are assembled in darkened areas could easily lead to disaster. Emergency lighting is usually dictated by local codes, but these are generally minimum requirements. High-intensity discharge lighting is not considered acceptable for emergency lighting due to excessive time to produce acceptable lighting level after a momentary power outage. As codes change due perhaps to some local catastrophe, the requirements

become more rigid so that capacity for increased loads should be the rule.

In auditorium buildings, the following check list should be used as a guide:

(1) General lighting
(2) Stage lighting
(3) Exit and emergency lighting
(4) Public-address systems
(5) Signal and communication systems
(6) Radio and television facilities
(7) Projection and sound equipment
(8) Signs, scoreboards, and timers
(9) Telephone and telegraph facilities
(10) Air conditioning and refrigeration
(11) Ventilation and heating
(12) Special-event lighting and power outlets
(13) Refreshment stands
(14) Elevators and moving stairways
(15) Parking
(16) Floodlights
(17) Snow and ice melting systems for walkways, ramps, and driveways

16.4 Automobile Garages and Showrooms. This classification covers a single room, a sales office, show room and service-shop facilities on one floor, or a multistory building including storage and parking facilities. In the latter type, the so-called automatic parking garages erected on small areas of high-priced real estate in large cities are becoming more numerous as the parking situation becomes more acute. Most parking garages used by tenants, employees, and the general public provide for self-parking with ramps between floors. Passenger elevators may be required for multiple floors.

Rooms used for sales and service generally require higher lighting intensities. Special lighting effects to be used for sales promotion material as well as special features of the new cars may be needed. Turntables showing the complete product in the entire 360 degree range will require motor-driven platforms.

16.4.1 Service Shops. In service shops, lighting intensities of 50—150 fc over the workbenches may be needed. General work area light levels may be satisfactory with 25—50 fc. Rectangular, metal, surface-mounted raceways with convenience outlets every 18 inches running along the back of the benches can be used to distribute the necessary 120 V power for the numerous testing and analyzing equipment. Explosionproof systems are required in certain areas.

Paint-spray booths require lighting intensities up to 500 fc with shadowless distribution. ANSI/NFPA 70-1981 [1] and local codes provide stringent rules covering such application. A 95% explosionproof installation is worthless. As a practical matter it may be best to have all lighting on the outside of the booth shining in through wired glass. Consideration should be given to prefabricated booths.

Particular attention is directed to the NEC provisions relative to hazardous (classified) locations.

Infrared drying, either portable or stationary, is used in many cases. Any such installation require special feeders designed for the particular load. Such a load is best fed from a four-wire three-phase service. This may influence service-entrance and distribution requirements.

Automotive garages have power requirements for elevators, hoists, machine-shop tools, and air compressors. Welding outlets may be required if extensive body work is done. These welders are usually found in ranges up to 45 kW, single phase and three phase. Air compressors should be located near the ser-

vice entrance to reduce circuit costs. Compressed-air lines generally are cheaper to run than conduit and cable. In the machine-shop area, a single length of busway with plug-in overcurrent devices will enable the addition and movement of machine tools at will.

The power supply should be large enough to permit starting a large air-compressor motor, for instance, with a test analyzer on the line without adversely affecting this equipment.

16.4.2 Parking Lots and Garages. Electronic systems may be provided for the collection of tolls, fees, and tickets. Such systems may involve automatic gates, ticket issuing machines, and electronic cash registers. Treadle and loop systems are used to count vehicles for auditing purposes and lot-full inventories. Larger parking lots or structures may require rooms for electronic equipment and perhaps computers. These systems may require isolated power supplies, regulated power, and, in some cases, emergency back-up power from a battery-inverter unit or engine–generator set.

Telephone services should be provided in the waiting rooms for customers as well as for the personnel operating the garage. The following list should be used to check garages in general:

(1) Illuminated directional signs

(2) Remote door controls

(3) Telephone systems

(4) Public-address systems

(5) Signaling and communication systems

(6) Fire alarm

(7) Lighting, general and localized

(8) Power for tools and testing equipment

(9) Elevators and moving stairways

(10) Ventilation systems for enclosed garages

(11) Snow and ice melting systems for ramps and walkways

(12) Electric infrared heating for waiting customers and attendants

(13) Emergency power, especially if handicapped persons are involved

16.4.3 Automobile Washing Facilities. Special loads include pumps, conveyor, fans, washing roll drives, water heating, towel washers and dryers, soap and wax metered dispensers, and an extensive interlock system. Equipment must be separated from the wash area unless it is watertight.

16.5 Banks. The design of bank buildings is apt to involve expressive architecture. Banks today tend to use an open landscape type plan with relocateable partitions. The lighting, both outdoor and indoor, should be in keeping with the architectural design.

Outdoor lighting may include floodlighting the building for decorative purposes and may require provisions for supplying power away from the premises. Short- or long-focus floodlights may be used. The use of electric signs and directional instructions for night banking may require underground cable installations before walks and driveways are constructed. Remote teller drive-in islands for banking directly from automobiles may need lighting, communication, closed-circuit television, and electric heating. Teller exterior walk-up windows may need lighting, communication, and electric infrared heating. Unattended automatic banking machines require power and lighting, communication, and electric heating depending upon whether located on the exterior wall of the bank or in a lobby.

Indoor lighting includes general lighting in the main area. High ceilings permit the use of spotlights and floodlights. Indirect lighting may be used. The tellers' counters should be lighted with care to avoid shadows and glare since visual tasks are demanding and EDP displays may be difficult to read. The directors' rooms and executive suites need the latest in good lighting practices. Decorative lighting as well as numerous convenience outlets should be provided. Provisions for closed-circuit television for banks with multiple branches should not be overlooked. Alarm and communication systems, both visual and audible, demand top priority in the modern bank. Security and alarm systems are normally not a part of the construction contract but close coordination with the suppliers selected by the bank is necessary to ensure that adequate and properly located raceways are included in the building. Facsimile telegraphs or teletype systems tie the bookkeeping and tellers departments for quick checking of accounts. Provision will probably be required for computer terminal equipment at tellers' compartments. Power and data circuits may also be needed.

Closed-circuit television is used both for checking of money and for guard surveillance and protection. Automatic still cameras are often used as well as videotaping.

In a safety-deposit department for the use of the public, banks desire a push button in each booth to call an attendant. Banks also need an intercom system in the vault in the event of people locked in the vault accidentally. A signal system for calling employees should be provided.

The accounting department may have large numbers of business machines. An under-floor duct system or a raised floor (access floor) system at least 6 in high which provides an excellent means for installation of power, data, and communication cables, will permit rearrangement of the office and provide for installation of additional equipment. Emphasis should be placed on the original design. Banks frequently have extensive computer installations (see 16.11).

A central air-conditioning plant may serve the whole building. In smaller buildings, individual units may be used. In central plants, control centers should be used to indicate the status of blowers, pumps, and compressor units, as well as temperatures, pressures, etc.

Some bank buildings have space available for rental. These spaces should be treated as office spaces such as in office buildings (see 16.22).

The check list should include the following items:

(1) General lighting
(2) Interior lighting
(3) Exterior lighting
(4) Security and emergency lighting
(5) Clock system
(6) Fire-alarm and watchman's tour stations
(7) Burglar alarm and holdup systems
(8) Telephone and telegraph
(9) Intercommunication system
(10) Paging annunciators
(11) Electric door locks and controlled access systems
(12) Closed-circuit television
(13) Air conditioning, heating, and ventilating
(14) Business machines, data-processing equipment, and cash machines
(15) Power outlets
(16) Snow and ice melting systems for driveways, walkways, and ramps
(17) Emergency and standby power requirements
(18) Surge protection

16.6 Brokers' Offices. Brokers' offices require information on display, teletype, and special telephone services. These services should be coordinated with the telephone companies in the local area. Brokerage offices and stock, and commodity exchanges have heavy concentration of domestic and foreign communication facilities. Special wireways may be required throughout the building to house these wires and cables. Electronic information displays may be provided at each desk.

Special lighting is needed for the quotation boards. Automatic boards require specialized wiring techniques.

Air conditioning is the same as in general offices: a central system in large offices and individual office units in smaller buildings.

The check list is similar to 16.5.

16.7 Churches and Synagogues. The lighting of churches and synagogues calls for a special study in keeping with the particular type of architectural design. Soft well-diffused lighting is recommended. Such lighting tends to underscore the architectural beauty of the edifice.

With a few exceptions, the modern church building is getting away from the cathedral type of edifice to one of more simple design. The vast majority of churches are being built for service to the congregation and have recreational and educational areas included as well as a main auditorium. In these types of churches, the general conditions for schools and auditoriums may be followed. Gothic churches with their high vaulted ceilings and great stained-glass windows require direct lighting from well-designed pendant lantern-type luminaires. Indirect lighting with brightly lighted ceilings is not recommended for this type of edifice. *Pinhole* down lighting may be effectively used with high ceilings to provide sufficient illumination using spotlights. These spots should be carefully shaded to avoid glare.

The lighting of the altar can be accomplished by spotlights or floodlights concealed in the altar arch or behind ceiling beams. The lighting should be well spread to avoid the *theatrical spot* effect. Dimmers may be used to advantage.

Many churches have beautiful stained-glass windows which can be illuminated so they may be seen at night. High lighting intensities are needed to penetrate the density of the stained glass; therefore, sufficient lighting circuits should be provided for this task. Floodlighting the steeple, dome, or the front of the church may be accomplished with powerful floodlights which, again, calls for the proper size of wiring for this load. Photocells and automatic timers will probably be required to limit such lighting to a few hours each night.

Public-address systems with good coverage throughout the church should be provided. Outlets around the church may be required for the crowds who remain outside for special services, funerals, or weddings. Portable public-address systems and speakers may be needed for those occasions. Radio and television remote pickups may require special outlets and heavy feeders for the large lighting loads required by television cameras. Modern church services require extensive coverage of the altar area with microphones.

Some pipe organs are being replaced by electronic organs. Feeders will be required in the choir loft to handle this load. Tape recorders and record players may also be used. Provision for connecting these to the central public-address system should be made in the original plans.

In the colder climates, the steps, sidewalks, and even the leaders and gutters may need to contain electric heat for melting snow.

Kitchen facilities for the meeting rooms may be needed in today's churches. Electric ranges, hot-water tanks, automatic dishwashers, and large appliances such as toasters and coffee urns have relatively high current ratings.

The check list should, therefore, include the following items:

(1) Special interior lighting
(2) Special exterior lighting
(3) Floodlighting
(4) Emergency lighting
(5) Organ
(6) Public-address system
(7) Tape recorder and record player
(8) Snow melting (where needed)
(9) Kitchen appliance service
(10) Air conditioning, heating, and ventilating

16.8 Clinics.

16.8.1. Dental Clinics. In addition to power for dental chairs, sterilizers, and X-ray equipment, care should be taken to provide sufficient power to lab areas. Many items of equipment now used require 1000 W or more each and may be used simultaneously.

16.8.2 Medical Clinics. X-ray equipment, film developing spaces, and minor surgery lighting require particular attention. Some multi-doctor clinics require rather elaborate call systems. Examination rooms are generally crowded and a detailed layout is required to ensure a suitable and effective location for convenience outlets for such things as power examination table, sterilizer, X-ray viewer, and examination lights. Since patients often look at the ceiling while being examined, overhead lighting should be soft. It is important to consider tamper proof or ground-fault circuit-interrupter receptacles for protection of children in pediatric clinics.

16.9 Atheltic and Social Clubs. The general arrangement of the ground floor of the larger clubs may be similar to a modern hotel. General lighting may be decorative. Outlets for reading and table lamps and service outlets for maintenance equipment should also be provided.

Swimming pools, gymnasium, handball courts, and bowling alleys require special lighting. High-intensity lighting in the gymnasium may require sufficient lighting circuits to take care of special events. The following listing gives the typical wattage per square foot for various sports:

Badminton	2.5
Basketball	3.5
Billiards and pool (general illumination)	3.0
Billiards and pool tables (per table)	450.0
Bowling alleys	2.0
Boxing bouts (over ring)	9.0
Dancing	2.0
Handball and squash courts	2.5
Skating	2.5
Swimming (plus underwater lighting)	2.0
Water polo (plus underwater lighting)	2.5

The upper floors of certain clubs have sleeping quarters for members. The power requirements are the same as for hotels. A centralized antenna for radio and television as well as provisions for air-conditioning needs of each room should be installed. Telephone or intercommunication systems, or both, may also be installed in rooms. Saunas are standard equipment in most clubs. Sauna electric heaters may vary from 5 kW to 15 kW, depending on the room size.

16.10 Colleges and Universities. These facilities often involve a wide divergence of uses. Occupancies such as auditoriums, gymnasiums, hospitals, and clinics, libraries, office buildings, etc, require reference to those classifications elsewhere in this section.

16.10.1 Central Power Plant. The central power plant may be used to supply the heating, cooling, and hot water requirements for the campus buildings. Interest is also growing in including in this plant at least some electrical generating facilities. It is generally a good location for the emergency or standby electric power plant, or both. Cogeneration (tie to utility power) is becoming increasingly acceptable. This building may also serve as the distribution center for the electric system. Underground distribution is usually used between buildings. The designer should take into consideration probable expansion in all phases of power, light, and communications. Underground distribution with spare ducts is recommended for these systems.

16.10.2 Classrooms. Classrooms should be provided with an adequate level of illumination, properly diffused, to eliminate glare, eye strain, and objectionable shadows. Properly engineered fluorescent lighting is considered the best illumination system for general classroom use. Supplemental lighting on chalkboards and bulletin boards may be required.

Public-address systems and clock systems as well as provisions for a closed-circuit television and central antenna system may be needed in classrooms. The convenience outlets should be ample to carry the largest of projection equipment loads for illustrated lectures.

16.10.3 Laboratories. The various sciences as well as other curricula may have laboratories for their particular needs. Flexibiltiy and capacity of power circuits for special apparatus is important. Surge protection is generally warranted.

In electrical laboratories the facilities may become more elaborate. A large range of voltages, frequencies, and direct-current power supplies may be required. The distribution system should be designed so that all these power sources, centrally located, may be transferred into a number of classrooms or test areas. Risers containing all the various feeders may run the full height of the building. On each floor these risers run through a plugboard. By means of jumpers, the power panels in the various rooms on the floor can be plugged into any of the power sources available. Power at frequencies over 60 Hz, such as 400 Hz power, requires specially engineered busway or cable and conduit systems.

Another method is to have plug-in busway in the corridors of the building so that each room, by the proper use of overcurrent device plug-in units, may be fed with whatever power is desired. Circuit breakers on portable cords and plug-in units may be used to protect panelboards, outlet boxes, or test motors and generators. This approach often reduces the need for multiple classes of panelboards in the laboratory space.

The tables and counters should have individual panels which can be interchanged for various needs. By using a module system whereby all the panels are of the same size, it may be possible to interchange with other rooms. Convenience outlets may be required across the back of each table. These outlets should be polarized for the various voltages, currents, and for grounding purposes.

Receptacle slot configurations for each type of service should be standardized throughout the laboratory spaces so that portable equipment may be used in any location without relying on adapters. National Electrical Manufacturers' Association (NEMA) standard configurations are recommended for this purpose.

Special panels and controls are frequently installed in laboratories. It is well to plan with the department heads where these items should be installed in the laboratories.

Power loads are heavy in modern laboratories. The use of wind tunnels for research may require drives of thousands of horsepower. These requirements need to be taken into account in the initial system design. In the mechanical laboratories, large machines for testing and even manufacturing parts for research will produce load densities which will require an industrial type of power distribution system.

Special attention should be directed to any laboratory use which could cause a dangerous condition to occur or result in loss of many months of work if power is lost. Examples are chemical laboratory fume hoods or controlled environment studies where emergency power supply must be considered. Personnel type ground-fault circuit interrupters are advisable where wet conditions exist such as in chemical laboratories.

16.10.4 Dormitories. Dormitories in the modern college have a great need for convenience outlets. The so-called *octopus* plug fed from a single receptacle is used all too frequently in college rooms.

Study lamps have been supplemented by stereos and tape drives, small refrigerators and ranges, clock radios, portable television receivers, and small space heaters for those cold nights in the fall and spring when heating is off. Even though house rules may prohibit such devices, they still are used and the loads should be taken into consideration in the building design.

Laundry facilities should be provided on each floor, and outlets for steam irons and hair dryers should be installed. In the basement, drying equipment is often provided and the use of electric dryers is common.

Public-address, room-annunciator, fire alarm, and clock systems should be provided. Telephones, both in private exchange and pay phones, should be included on each floor. Telephone outlets may be provided in each dormitory room. Cable-television system outlets should also be provided in each dormitory room.

Lounges and cafeterias should be well lighted and may be provided with food warmers and steam tables to keep the food hot as received from a central kitchen. The lounges may have several television sets and may need a central antenna system.

16.10.5 Miscellaneous Requirements.
In addition to the principal types of power requirements the following requirements should be considered:

(1) Outdoor lighting should be provided for the many bikeways, streets, and sidewalks which may be a part of the campus. As this is private property, this type of lighting is usually the responsibility of the college.

(2) Telephone and telegraph systems may be private systems connected to the outside systems through a switching center. Teletype systems may connect with other divisions of schools such as those in a state university organization.

(3) Public-address systems permit rapid and efficient transmission of information between the administrative staff and the instructors.

(4) Signaling. An electric-clock system with a master clock may be used to control secondary clocks throughout the campus; time clocks for recording hours of workers may be controlled from this system. Automatic programming can ring the time signals for the various class periods.

(5) Central monitoring and control of the mechanical, electrical, and security systems on the campus should be considered. An effective central-monitoring system can prove very cost effective for maintenance and for energy conservation.

(6) Exterior and interior fire-alarm and security systems with automatic ties to municipal systems are desirable and may be required.

(7) Snow and ice melting equipment.

16.11 Computer Centers.
Raised floors are mandatory for most computer systems to allow efficient installation and cable concealment. Specifics on computer equipment must be established in order to determine the following requirements:

(1) Temperature and humidity variations acceptable

(2) Voltage and frequency constraints (including transients)

(3) Effect of power interruptions

Generally, computer spaces require air conditioning and humidity systems separate from all other building spaces. Normal voltage fluctuations generally can be tolerated, but transients (or voltage spikes) may upset the computer operations. Surge protection is desirable. Special isolating transformers with electro-static shielding and transient suppressors have proven effective in eliminating this problem. These transformers can also provide voltage regulation if unusual voltage variations can be expected or the equipment is voltage sensitive. Some computer systems are designed with the capability to pick up and continue after a power outage; others may lose the total program. Emergency power should be considered for any computer that is operated continuously. Where interruptions cannot be tolerated an uninterruptible power supply (UPS) may be economical, as determined by a study comparing the costs of computer interruptions with the owning and operating costs of the UPS.

Grounding for equipment and computer power circuits requires detailed consideration. If isolating transformers are used, neutral grounding must be handled according to ANSI/NFPA 70-1981 [1] for a separately derived system. Each computer machine should be grounded to a ground bus and then to an approved ground. Conductors and buses should be insulated so that there is only one ground contact point.

16.12 Convalescent Homes. These may be in the class of smaller hospitals, but there are usually no operating or treatment rooms. The rooms should have ample convenience outlets for handling electric therapy equipment, radio, television, etc. Elevators and movingstairways may be included.

A check list on this classification includes the following points:
(1) Night lighting
(2) Emergency lighting
(3) Convenience outlets in rooms
(4) Laundry facilities

(5) Air conditioning, heating, and ventilation
(6) Kitchen appliances
(7) Elevators and escalators
(8) Nurses'-call and public-address systems
(9) Fire-alarm systems
(10) Snow and ice melting equipment

16.13 Department Stores. In many respects a department store is an oversized small merchandising establishment Fundamentally, the difference lies in the fact that since the whole building is usually occupied by the same owner, expansion or contraction of the space occupied by one department can be effected by reducing or enlarging the operation of others.

16.13.1 Distribution Systems. The distribution systems often resemble those for office buildings. Sufficient flexibility should be provided in the distribution system to accommodate increased needs for power and lighting caused by changes in the use of space. No diversity factor should be used in calculating the lighting loads. Maximum electrical load occurs when the store is open in the evening especially during the evenings of hot-weather when the air conditioning load is added to the power required for lighting.

In a large store, where lighting panelboards are likely to be large, consideration may be given to a duplicate feeder and panelboard system. This reduces the size of the feeder or riser supplying any section of the store, reduces the panelboard size, and may eliminate the necessity of providing a main fuse or circuit breaker in each lighting panel. Another advantage of such a feeder system is the assurance that part of the lighting supplied from any one location remains on in the event of a failure of one of the feeders. Such assurance can be obtained by circuiting adjacent bays or alternating lighting fixtures to different panelboards. Provision should be made for supplying a minimum of 3 W/ft^2 of selling space for general lighting. In addition, a minimum of 1 W/ft^2 should be allowed for spot or accent lighting as well as an additional 1 W/ft^2 for showcase lighting. It is well to allow, in addition to the foregoing, an extra 0.5 W/ft^2 for cash registers and any other special machinery or electric devices used in selling spaces, such as candy cases or pharmacy refrigerators.

In laying out the circuits which supply power for showcase lighting and miscellaneous receptacles, a small panelboard may be placed on each building column. Such an arrangement makes it possible to keep the runs for the required circuits very short. Circuits for cash registers, candy cases, and pharmacy refrigerators should be supplied from the regular, rather than the column, panelboards. By using separate feeders to column panelboards, entire bays of showcase lighting can be switched without affecting the circuits which supply either the cash registers or the refrigeration. Magnetic, mechanically latched contactors for the control of large-area lighting (using keyed control switches) are desirable design features.

The cash-register circuits may be separated to permit checking of cash and accounts after regular store hours, without turning on unnecessary showcase lighting. Computer registers generally require *clean power* and a raceway system should be considered in case central verifying type registers are installed. Surge protection should be included for computer type equipment, for power circuits and for data transmission circuits leaving the equipment area. The refrigeration system circuits for perishable items should be locked-ON or marked DO NOT OPEN to prevent spoilage.

16.13.2 Lighting. Lighting problems in a department store should be considered seriously because the public is involved and so many special flexibilities are required. The designer should consider the following points:

(1) General overall lighting scheme
(2) Accent or spot lighting
(3) Showcase lighting
(4) Exterior show-window lighting
(5) Special areas
(6) Exterior site lighting
(7) Security lighting
(8) Emergency exit and egress lighting code requirements

16.13.2.1 Intensities of 50—70 fc are typically provided in all selling areas; however, energy conservation and energy codes may mandate lower levels. Efficient fixtures and light sources with showcase, accent, or display lighting to bring out a sense of brightness and detail may be used with high-reflectance surfaces to make more effective use of lighting at lower power consumption per square foot. This allows sufficient illumination for appraisal of merchandise when it is taken out of showcases. Close cooperation should be maintained with the interior designer or architect for the store to achieve a pattern of lighting fixtures suspended from or built into the ceiling, which does not distract the attention from merchandise on display, and which is in harmony with the architectural features of the building. Color of light sources is very important in selling and merchandising areas and requires special engineering attention.

The use of cove lighting to supplement the general illumination is common practice. Illuminated coves serve very important functions. The cove-lighted curtain wall provides a terminal or stopping point for the eye when viewing the store generally; hence it gives definition to the space. When direct or down lighting is employed as the principal means of illumination, coves provide the indirect component of light which adds to the softness of the whole interior. Cove lighting usually provides only a small amount of the total light for selling areas.

16.13.2.2 Accent or spot lighting is required by the display department of practically every department store. It is common practice to use the reflector or projector lamps as the source of illumination for this purpose.

Spotlights may be of several types, either concealed in the ceiling with adjustment for direction through the light aperture, or mounted on tracks or outlet boxes and exposed to view. The equipment should be placed in a modular pattern throughout the store to allow for placement of sale merchandise and special mannequin or other displays wherever the store display personnel may desire.

16.13.2.3 Showcase lighting may be provided by means of reflectors placed in the interior of the case to be illuminated. The intensity should be two to three times the intensity of the general lighting in the store. The fluorescent lamp is practically always used as the light source, since its shape lends itself particularly to this application.

16.13.2.4 Show-window lighting in the modern department store should be carefully considered, since it is at this point that the prospective purchaser obtains his first impression of the goods being offered by the merchant. The usual practice is to employ a spotlight every 12 or 18 inches along the length of the window front. These lighting units, when placed as close to the exterior glass of the window as possible, give the flexibility and provide the interest required by the display designer. Color filters may be used to show the merchandise to the best advantage. Recent installations also make use of reflector lamps or border-light strips to illuminate the background of the window. These strips add interest by means of light or color contrast.

16.13.2.5 Certain special areas in a department store require extra electric services. A case in point is the electric-appliance department. Provision should be made, with adequate circuiting, for the demonstration of electric ranges (with special receptacles), toasters, kitchen mixers, refrigerators, and so forth. In case of table- and floor-lamp displays, a large number of receptacles is usually required. Local electric codes may restrict the number of receptacles which may be placed on one circuit and reference should be made to these codes. Receptacles for holiday and special sales displays should be included.

Where radio and television selling space is to be provided, the necessary master antenna receptacles should be installed. Since the type of antenna varies widely with each installation, it is not discussed here. Conduit risers of $1\frac{1}{2}$ in or 2 in from a terminal box in the radio department up through the building to the roof are usually provided.

16.13.2.6 Security lighting should include battery operated lights for main pathways and spotlights for cash register areas.

16.13.3 Communications and Signaling. Telephone and telegraph systems in department stores generally employ a relatively large private branch exchange with telephones in each department as well as in the administrative offices. Order-receiving equipment which involves a heavy concentration of telephone wires may also be required. Communication circuits for credit card verification systems and computerized cash registers are also required.

Department stores also make extensive use of code calling or loudspeaker paging systems to summon personnel to the telephone, signal closing time, and so forth. These systems require signals or loudspeakers located on walls or columns 8 or 10 ft above the floor. Frequently an electric service outlet as well as a communications service outlet is required at these locations.

Adequate provision should be made for public telephones at locations convenient to the customers as stores do not generally handle customers' outgoing calls.

The wire distribution system for these services is quite extensive, although the capacity need not be large except at selected locations.

Paging systems should be considered. In some large department stores local paging systems are installed in various departments. A clock or dismissal system should be considered. In public areas all of these are sometimes combined in a more decorative type of combination unit. Door protection systems are often installed. Where there are many street display windows, portable telephones or portable two-way radios may be provided for inter-communication between the window decorator and an observer on the outside. Merchandise theft control devices are often installed at exits from floors.

16.4 Fire Stations. Fire stations are in two classes, those manned by a permanent force and those manned by volunteers. In the first category, living quarters are generally provided for the firefighters stationed around the clock. Kitchen facil-

ities should be provided and radio and television outlets are needed. The alarm system should be tied into a control headquarters, and special telephone services are usually connected. A cut-out contactor for ranges and other cooking equipment is frequently used to disconnect these items when the station personnel respond to a fire call.

Central headquarters contains the central fire-alarm system with provision for emergency standby power, fire siren, administrative offices, etc. Two-way radio for communications with the trucks and other mobile equipment located here. Fire-alarm maintenance shops call for test and instrument repair equipment. Automatic door openers are installed in all doors for handling fire-fighting apparatus. A tie-in to the traffic signals is provided on many firehouses to flash the red stop signal and halt traffic.

In rural fire stations some automatic devices should be added. Battery charging on all the truck batteries should be on all the time the trucks are in the building. Quick-disconnect plug connectors should be used which disconnect the battery charging leads automatically when the trucks go out on a call. The siren may be operated by remote control to alert members of a volunteer system. There may not be any repair shops on the system, but two-way radio equipment should be on the trucks and a transmitter in the building. The lights should be automatically controlled.

16.15 Gymnasiums. Gymnasiums are primarily used for active sports, and particularly in schools may also serve as multi-purpose areas. Particular attention should be paid to ensure that all electrical lighting and devices have adequate guards to protect against damage from thrown balls. Lighting must be designed to provide minimum shadow and lack of glare to participants. While metal halide or high-pressure sodium (or mix of both) may provide an effective lighting system, restrike time in case of a momentary power outage makes an alternate or emergency light source mandatory for evacuation purposes. Lamps and fixtures are available with standby tungsten halogen or quartz lamps power failure relays included that satisfy the requirements of ANSI/NFPA 70-1981 [1]. Fluorescent luminaires are also included in the system for this purpose. Also, lower wattage high-pressure sodium lamps with an instant restrike feature for emergency lighting are available. Special power requirements include hoist drives for backboards and other gym equipment as well as drives for folding bleachers and dividing walls. If any of these controls use fixed switches the switches should be of the key type. Scoreboard power with convenient control must be considered. Public address equipment requires special attention to ensure appropriate selected coverage for each sport (and other use if applicable) and to prevent feedback in an accoustically *live* space. Floor outlets should be avoided insofar as possible due to potential hazards when such an outlet could be in a playing area for any sport.

16.16 Hospitals

16.16.1 Life Safety. The major consideration unique to this type of facility is the limited capability of patients to protect themselves against electric shock, fire, power failures, temperature extremes, and other hazards which would be much less threatening to the typical occupant of other types of buildings. Patients are normally in some state of debilitation ranging from complete unconsciousness (comatose or anesthetized) to minimal impairment. Although the normal response to a threatening condition in a room or building is the quickest possible evacuation, the hospital staff is faced with the decision as to whether evacuation places the patient in a greater danger than the threatening condition. Measures to be taken for life safety include:

(1) Fire-alarm systems with extensive automatic detection, staff signaling (without patient awareness), automatic smoke removal or building compartmentalization, and extensive system zoning, with widespread zone annunciation for fast staff response

(2) Communication system stations located at the probable staff location during the emergency

(3) Electric power systems which are reliable and zoned to minimize power failure areas and engineered to allow rapid restoration with minimal staff intervention

(4) Lighting systems which are designed to fit the hospital's emergency plans

(5) Special design attention for protection from even very low leakage of electric currents in electrically susceptible patient areas.

16.16.2 Reliability of Critical Systems. Critical systems such as lighting, power, communications, etc, must be designed for reliability of mechanisms and ease of operation based on the skill level of staff utilizing or maintaining the equipment. Although technologically advanced machinery may suit the need, it may be too esoteric for the maintenance staff unless the staff is given training in the operation and maintenance of the equipment. Systems and equipment should be suitable for frequent test under substantially similar conditions to the anticipated failure mode. Redundancy and ease of manual override may add reliability to many systems.

During such operation the following catastrophes should be considered:

(1) Critical equipment should be located above possible flood levels (external or internal)

(2) Normal and backup systems should be electrically and physically located so that a localized catastrophe will not disable both

(3) Staff may be called upon to operate or control a number of diverse systems during adverse situations

Emergency power systems must have the capability to automatically assume the required emergency loads within 10 s after the normal power service fails. In hospitals, special remote supervision of system readiness is legally required. For additional information see ANSI/IEEE Std 446-1980 [8].

16.16.3 Government Regulation. Detailed and extensive codes must be adhered to in all cases where federal financing is involved and in most states. Even though the regulation is extensive, the responsible designer should consider governmental requirements to be the legal minimums and not the standard practice. New electrical systems must normally conform with the following requirements:

(1) USDHEW Pub No (HRA) 79-14500 [11]

These requirements adopt the various reference codes including [9], [1], [2], [6], and [7]

(2) State: Most states adopt the requirements listed in (1) and supplement them with amendments

The review plan or new health-care facilities by state or federal agencies is

standard and is usually done by specialists in health-care facility design.

16.16.4 Designer/Hospital Staff Coordination. Typically hospital staff are not electrically sophisticated and in small hospitals there may be extensive reliance on contractors, service technicians, and consultants. The engineer must match each system to the staff's technological capabilities. Human engineering is an extremely important feature to consider when specifying new equipment or designing systems.

The engineer can improve his comprehension of the human engineering design problem by asking for a written program or by utilizing form-type questionnaires designed to stimulate and assist the user. When discussing electrically esoteric subjects, the engineer should make a special effort to avoid confusing or intimidating the hospital staff. The engineer can obtain a medical dictionary to facilitate his grasp of medical terminology.

For completion of the construction project, the engineer must ensure that sufficient operating and maintenance manuals with manufacturer's shop drawings, engineer's drawings, performance curves, and coordination curves are presented to the staff to ensure proper operation and maintenance. Contract documents should include a formal procedure whereby factory experts, installation technicians, or the engineer performs this function.

16.16.5 Communications Systems. Monitoring, communications, and signaling require systems unique to this type of facility. Consider the following types in discussions with the client.

(1) Telephone (proprietary or customer owned). Portable telephone outlets should be provided in all patient rooms. Public telephones are necessary

in waiting rooms, near solariums, and other public areas

(2) Nurses call (audible/visual). A nurse's call system should be installed for each bed. Indicator (dome) lights outside the door, annunciator panels at the nurses' stations, and system monitoring (duty) stations at nurses' work areas are the usual system components. There is an increasing tendency to use voice communication systems between the patient and the nurses' station. During the night this method has the advantage of allowing the nurse to listen in on the patient from a central point

(3) Paging and public address

(4) Radio paging (pocket pagers)

(5) Television (commercial and closed circuit). Television antenna circuits from a central system to each bed are minimum and special CCTV programming is provided in many hospitals.

(6) Security (intrusion, escape and pharmacy)

(7) Departmental intercoms

(8) Data handling systems (EDP or pneumatic tube)

(9) Dictation systems

(10) Personnel (doctors') register

(11) Emergency services radio

(12) Physiological monitoring (local, hardwired or radio remote)

(13) Emergency assistance signaling (cardiac or nurse assist)

(14) Elapsed time indicators

(15) Central clock correction systems

(16) Remote monitoring of emergency power and HVAC

(17) Medical gas alarm systems

(18) Isolated power ground leakage alarms

(19) Electromagnetic interference (suppression)

With the increasing use of remotely transmitted signals, computers, physiological monitors, and other sensitive electronic equipment, special attention must be paid to electrical noise problems on the power system, and to shielding and grounding. Separate conduit systems or special shielding may be required for highly-sensitive systems.

16.16.6 Isolated Power Systems. These systems consist of a separate primary and secondary wiring transformer with a high impedance, or ungrounded, secondary. A *line isolation monitor* in series with the ground connection measures the ground leakage current and sounds an alarm when it exceeds a certain level (normally 2 m, see ANSI/NFPA 56A-1978 [9]. The 2 m ground leakage current is generally considered to be the maximum permissible for minimization of the risk of spark induced flammable anesthetic explosions or electric shock of patients.

Upon alarm the staff has the option of continuing the procedure with knowledge of the ground fault whereas in a conventional grounded system, power is likely to be interrupted.

These systems are normally provided in operating rooms and are sometimes used in patient areas such as coronary and intensive-care units which are *electrically susceptible*. The engineer should, in conjunction with the hospital and the code authorities, determine the applicability in his project. See [9]

Good practice dictates that the following rules be observed for designing isolated power systems.

(1) Polyvinylchloride (PVC) raceways used with a separate insulated ground wire minimize distribution system losses but may be objectionable for the following reasons:

(a) The code authority may require rigid steel conduit raceways

(b) The designer may not wish to use

PVC due to its combustibility and toxic emissions during a fire

(2) All wiring should have insulation with high impedance characteristics.

(3) Circuits should be kept short to minimize leakage current. Point-to-point runs rather than the usual squared neat circuit layouts should be adopted.

(4) Transformers, switches, circuit breakers and the line isolation monitor should be located either outside the operating room but as close to it as possible, or inside the operating room. The remote alarm indicator should be located inside the operating room. If flammable anesthetics are used, all electrical equipment should be located above the 5 ft level.

Portable X-ray power outlets in operating rooms requiring 208 V or 230 V must be supplied from a separate isolated distribution system. In the past it was common to use a single isolating transformer to supply as many as six operating rooms simultaneously, however, this results in a circuit length not compatible with sensitive line monitors. An isolated distribution panel for portable X-ray equipment can supply up to eight locations by using an electrical interlocked switching system which permits only one branch circuit at a time to be activated, thereby limiting the leakage current. When such a multiple-use distribution panel is used, it should be located as centrally as possible. The limitations in simultaneous use of portable X-ray equipment should be clearly understood by the operating room supervisor before the dicision on its use is finalized.

16.16.7 Equipotential Grounding. The proper grounding of all conducting surfaces in patient-care areas which are *electrically susceptible* is generally considered extremely important. To reduce to a minimum the voltage difference between any two conductive surfaces within reach of the patient, or within reach of staff who might come in contact with the patient, an equipotential grounding system must be provided. The conductive surfaces of portable equipment are grounded through the ground wire in the power cord. Redundant grounding of electrically powered portable equipment and grounding for non-electrified equipment can be facilitated by installation of a plugging system.

An alternative approach to developing a safe environment is to eliminate all exposed conductive surfaces by the use of insulation.

16.16.8 Radiology Departments. Machinery can be used for diagnosis or therapy. In either case good voltage regulation is a prime concern for power circuits for the following reasons:

(1) *Diagnostic machines.* X-ray intensity is voltage dependent and poor regulation degrades the photographic image

(2) *Therapeutic machines.* Variation in intensity creates a problem in regulating the patient exposure

Diagnostic exposure exists for a time period of a few milliseconds. Typically variations in steady-state circuit voltage can be compensated for by manually arranging the machines' input transformer taps. This compensation is of little value however, when the power circuit has insufficient capacity and other loads are causing the voltage to fluctuate.

Diagnostic machines should be fed using independent feeders directly from the main power distribution switchboard if the utility service is at low voltage. In hospitals supplied with primary (medium) voltage, a separate secondary load center is generally used for the x-ray department.

16.16.9 General Requirements. In patient rooms and wards, comfortable lighting for the patient, consistent with the need for performing work in the room, is required. This usually means that several levels of lighting are essential. Silent-type switches should be used, and switches that can be operated by elbows, leaving the operators' hands free, are advantageous.

Night lighting should be installed at a level near the floor in a position which will not disturb the patients while providing *stumble* lighting for hospital personnel. Nightime ceiling lighting should not be located in halls opposite the doorways to the patients' rooms.

Convenience outlets are required as in other types of buildings. Outlets of sufficient capacity should be installed in wards and corridors for floor scrubbing equipment, food carriers, and other portable heavy-load equipment.

16.17 Hotels. The two general classifications are transient and residential. In the transient field there are two types, the commercial or business hotel and the resort hotel. The commercial hotel covered here will be the typical hotel used by the business traveler, rather than the vacationer.

The significant role of electricity in hotels begins with the main power-distribution system which, in large hotels, rivals that found in large industrial plants. Supply power may be at such voltages as 13.8 kV, distributed through load centers of several hundred to several thousand kilovoltamperes. Furthermore these load centers may not all be located in the basement, but may be found on intermediate floors and in penthouses on the roof with medium-voltage feeders running to these locations.

The main ballrooms may have stage-lighting type of control. Restaurants and lobbies will require at least 25 fc—35 fc.

Flexibility in dividable conference and ballroom facilities is very important. Lighting and amplifying systems should have control and equipment interconnect capabilities to permit convenient control of all lights and speakers within the space regardless of the wall configuration.

Many hotels have several stores with street frontages or in arcades. These stores may have separate meters depending upon utility submetering rules and policy. Hotels generally contain a barber shop, newspaper stand/gift shop, restrooms, and possibly a health club. All of these facilities require large numbers of outlets as well as good lighting. Trade association exhibits have heavy power requirements, generally in the ballroom area.

Some of the latest innovations in electric services in hotels include the following:

(1) Annunciator systems for messages, with pilot lights in the guest rooms, which are often integrated with the telephone instruments, to announce automatically when letters, telegrams, or messages are in the mail box for guests

(2) Total-building air conditioning

(3) All-electric kitchens

(4) Moving stairways

(5) Facsimile telegraph systems

(6) Food-control headquarters intercommunication

(7) Dumbwaiters

(8) Individual room control of air conditioning

(9) Central antenna systems

(10) Lighted swimming pool

(11) Electronic security and surveillance systems

(12) Individual control of room security and HVAC through use of the telephone system.

Power distribution may be at 208Y/120 V or 480Y/277 V with 120 V lighting and outlets supplied from dry-type transformers on every two or three floors. The noise level of transformers should be low to avoid disturbing guests.

In the residential type of hotel, the public rooms are usually not as numerous or as decorative as in the transient hotel. Power for individual air-conditioners should be made available if a common system is not provided. A central antenna, radio, and television system may be required. These hotels approach the apartment-house type of building.

A check list on hotels includes the following items:

(1) Medium- or low-voltage service entrance

(2) Primary or secondary unit substations, or both

   (3) General lighting

   (4) Special lighting in public rooms

   (5) Night lighting

   (6) Emergency lighting

   (7) Clock systems

   (8) Fire-alarm systems

   (9) Watchmen's tour system

  (10) Private telephone system

  (11) Public telephones

  (12) Teletype and telegraph facilities

  (13) Intercommunication system

  (14) Public-address system

  (15) Central radio

  (16) Central television

  (17) Television broadcasting

  (18) Bellboy annunciator

  (19) Separate services for stores and shops

  (20) Kitchen appliances

  (21) Elevators and escalators

  (22) Message annunciators

  (23) Lightning protection

  (24) Safe and vault security systems

  (25) Sports lighting systems

  (26) Sign lighting

  (27) Snow and ice melting systems

  (28) Landscape lighting

16.18 Libraries. Libraries need a well-distributed general illumination, since reading and visual work is done throughout the entire area. Lighting should be engineered to provide the required horizontal illumination levels for reading and writing and the required vertical lighting levels on the stacks for identifying book titles.

In the book stacks, which may run several tiers high, it may be necessary to provide outlets for motors on book lifts or dumbwaiters for transporting books and manuscripts to the top shelves. Special outlets are needed at checkout desk and for copying machines.

In large libraries, communications or telephone systems, or both, are needed between library stations. A closing time-signal system is needed. Security systems for checkout control are becoming more important and raceway for such a system should be included in the building structure. This system normally involves screening equipment at exit points and door control of exterior doors.

16.19 Mental Institutions. Requirements for these types of buildings are much the same as for hospitals. Greater emphasis should be placed, however, on greater coverage of emergency lighting. Patients in these institutions are easily disturbed by a sudden *blackout* or noises.

In the rooms for violent patients, the lights should be placed behind heavy glass or clear polycarbonate plates, securely fastened into the ceiling or walls so that the patient cannot reach the lamp. All switching should be from outside the room and wards. General lighting fixtures should be of vandalproof or tamperproof designs. Large institutions may require extensive perimeter and security lighting systems. Security alarm systems, closed-circuit television, and other monitoring systems may be required.

16.20 Museums. Adequacy of lighting circuits and convenient location of outlets are important in museums. The lighting of the exhibits often requires outlets in unusual locations. A high degree of flexibility for supplementary lighting should be provided. Special requirements include close temperature and humidity control. As these buildings are usually made of stone and decorative masonry, these outlets cannot be installed economically after the building is finished.

Some museums exhibit apparatus which may require large power outlets. Special types of power requirements may be served best by a design for flexible power distribution similar to those suggested for laboratories.

An elaborate burglar-alarm and fire-alarm system as well as temperature and humidity variation alarm may be needed especially in large city museums where valuable paintings and other art collections are shown. These systems may tie in with proprietary, central station or municipal police and fire departments. This calls for a close liaison with these agencies.

16.21 Newspaper Buildings. Newspaper buildings are, in a sense, multistoried manufacturing plants, with large motor loads for the presses, conveyors, elevators, and movingstairways. The power-distribution systems are large, with power utilization voltage of 480Y/277 V fed from unit substations which may be supplied by an internal medium-voltage distribution system.

Large systems of telephone, telegraph, teletype, computer facilities, and radio services are required. Television may be added with facilities available for both reception and transmission. Each of the wire services may have connection with not only the whole country but also foreign services. Each service usually has distinct requirements which may call for careful study of specifications as received from the various companies. Radio and television systems will require special shielding to prevent electromagnetic interference. Microwave facilities may be included.

Local telephone requirements are more than in ordinary buildings. Telephone outlets may be provided in floor-duct systems on most of the floors or by use of the newer flat-cable systems laid under carpet tiles. Ducts leading to the switchboard should be oversized to take care of changes in office arrangements.

A check list includes the following requirements:

(1) Large power distribution system at medium voltage or 480Y/277 V

(2) Special lighting in many departments

(3) Emergency lighting

(4) Emergency power

(5) Clock systems

(6) Special requirements for telephone, telegraph, radio, television, computers, and photography

(7) Intercommunication system

(8) Private telephones

(9) Helicopter landing field on the roof or in the yard or short take-off and landing (STOL) airplane facility

(10) Fire-alarm systems

Some newspaper buildings are part of a large office building. If located in the center of large cities, the high value of land dictates tall buildings. The problems then are similar to those of large office buildings.

16.22 Office Buildings. Modern large office buildings may have loads in the order of tens of thousands of kilovolt-amperes. Close coordination with the supplying utility will be required to develop the optimum system. The recommended utilization voltage is usually 480Y/277 V with local transformers supplying 120 V appliance load. Buildings exceeding 20—40 stories in height, depending on the loads, load location, and floor areas, will generally make medium-voltage distribution systems with transformers on upper floors economic. The development of such a system has been dealt with in some detail in earlier sections.

The problem is more acute when these buildings are located on power networks where only 208Y/120 V power is available. In these cases the design engineer may be able to analyze, with the utility company, the economics of supplying such buildings with a medium-voltage service. Such an analysis should include the following items:

(1) Power service
(a) Size, location, and composition of expected loads
(b) Possible transformer location(s)
(c) Primary voltage and feeder arrangement, network or radial
(d) Primary wiring by utility or building owner
(e) Transformers by utility or building owner
(f) Any utility costs to be assessed to building owner
(g) Transformer vault and access requirements
(h) Special metering requirements

(2) Load considerations
(a) Interior: HVAC, lighting, other loads
(b) Exterior: Sign and parking lot lighting, snow melting, electrically operated gates

(3) Special systems
(a) Fire alarm including communication and other systems control required for high-rise buildings
(b) Telephone systems
(c) Remote equipment status and control systems (central monitoring)
(d) Security systems
(e) Lighting and power control systems for conservation of energy

(4) Emergency/standby power
(a) Load requirements
(i) Pathway and exit lighting
(ii) Elevator
(iii) Fire pump and booster pump
(iv) Stairway exhaust and supply fans
(v) Data processing
(vi) Other desired loads such as sump pumps, house pumps, sewage ejector pumps
(b) Location
(i) Type (internal combustion engine or combustion turbine)
(ii) Intake, exhaust and unit silencing
(iii) Fuel supply and storage
(iv) Ambient temperatures and heat rejection
(v) Maintenance
(c) Potential for peak shaving

Office space arrangements are continually changing. Partitions are moved, task areas are shifted, and office equipment is changed or relocated. It is important that the lighting system layout be flexible and that power and convenience outlets can be added or deleted quickly and easily. Frequently these changes must be made with a minimum of interruption to full daily occupancy of the space. Low-voltage switching is often used to change switching patterns. Prefabricated flexible wiring systems (see ANSI/NFPA 70-1981 [1], Article 544) should be considered.

Underfloor raceway systems are often the most practical way of providing for system flexibility. The two most common types are the prefabricated raceways which are cast in the floors, and the cellular system which utilizes the steel form under the concrete slab as part of the raceway construction. In either system, outlets can be provided at the time of construction or with special location and drilling rigs outlets can be installed at any time after construction. These raceways may have dividers to separate power and communication or telephone circuits. The outlets are usually a combination of telephone and power outlets. It should be noted that some codes require fire-proofing around any poke-through assembly. Where heavy load densities are to be expected in both communication and power circuits, the use of conduit on the bottom sides of slabs to serve the floor above usually is a poor choice for new buildings. Utilities run as described above are usually spaced on convenient modular groupings, for example a 6-8 ft spacing. The raceways are usually fed from header-ducts which are fed from conduits or prefabricated assemblies connecting to the electric, telephone, and communication closets. Flat cables for telephone and convenience power outlet branch circuits laid under carpet tiles are now being used (where permitted) instead of underfloor systems or ceiling modular pole systems. Where the density of power and communication and data cables is heavy, consideration should also be given to raised access floor systems.

Office buildings may be occupied by one tenant; however, in most cases, multitenant occupancies are to be expected. Provisions must be made for metering or rent inclusion of electricity in tenants' bills. Tenants may take raw space with only base utilities provided and design their own interiors in which case building standards must govern the tenant construction and plans, and construction should be approved to ensure compliance.

16.23 Parks and Playgrounds. Unless parks and playgrounds are supervised, particular attention must be paid to making the electrical installation as vandalproof as possible. Pathways open to pedestrian, equestrian or bicyclists' use at night must be lighted in such a way that the pathway ahead is always clearly defined. Sports area playgrounds should have lighting as recommended in various publications such as the Illuminating Engineering Society Lighting Handbook [10] and should have key, coin box, or remote control. Swimming pools and fountains are covered elsewhere in this section.

16.24 Piers, Docks, and Boat Marinas. The electrical installations for these facilities must take into account that moisture, particularly salt water, will cause rapid deterioration of metal parts. Rigid hot-dipped galvanized conduit with any cutting or threading galvanize treated has proven generally satisfactory. Plastic coated conduit (with joints plastic covered) has also been good. Mineral insulated copper sheathed cable is also available for this environment as well as metal-clad (interlocked armored) cable with inner and outer extruded protective jackets. Heavy galvanized metal boxes, cast bronze boxes, or nonmetallic boxes (such as molded fiberglass) provide the best life. It is advisable to carry a grounding conductor in all raceways when over water. Thought must be given to possible pier or dock movement and the motion of any floats which are fitted with power. Feeds to floats or to flexible joints in docks can best be handled by the use of type "SO" cable. Transformers exposed to the weather should be of the encapsulated type and in a corrosion resistant enclosure. Service to piers, docks, and boat moorings should have lockable power disconnects at the shore end for use by firemen. Padlocks should be used on switches to be disconnected by firemen.

Lighting should be designed for pedestrian safety as a minimum. Coast Guard obstruction lights may also be required.

Outside lighting is required for night loading and unloading as well as for protection against pilfering. Interior lighting should be designed to accommodate high stacking of commodities.

Power outlets at docks and piers for ship use must be designed for the power requirements of the ships expected to visit the dock or pier. Ships requiring large amounts of power generally have cables with lug connections. Smaller ships, tugs, etc, usually have cables with plugs. Tug companies many times standardize on voltage, current, phasing, and wires for their plug configurations. The engineer must investigate the shore side power requirements for ships expected to dock there regularly. Telephone facilities must also be considered. Piers and docks may also have power requirements for cranes, conveyors, hoists, winches, battery charging for electric truck batteries, pumps, electric doors, moving stairways, and elevator services. Cable tracing (heating) must be provided for piping which would otherwise freeze in winter. Two-way radio and public-address systems may also be required.

Baggage and materials handling facilities will require power outlets, often at 480 V (see 16.37).

16.25 Police Stations. Police stations are general-purpose buildings, but they require some special services not usually found in other commercial buildings.

Communication facilities are a vital part of the services needed. Two-way radio and transmitting equipment is needed to keep the station in touch with radio-equipped patrol cars and foot patrols. This system requires standby power equipment in case of power failure. Fire-alarm connections with local and adjacent county departments should be included, as well as teletype and facsimile tie-ins with other state and federal agencies. In large cities, this system is interconnected with other station houses. In some areas having community cable-television systems, a means is provided for summoning police, fire, and ambulance service. These types of facility connections may be required.

In stations which have confinement cells, a multiple-circuit radio wired to each cell, as well as a public-address system should be provided. A closed-circuit television for cell observation may be required. A projection booth may be used for showing slides and motion pictures. Electric door locks on individual cells and cell blocks may be provided. Adding these to thick walls is a costly item if overlooked in the original design. Standby power and manual backup for electrically operated cells should be provided.

A garage may be required for storage and servicing of patrol cars, emergency vehicles, and repair units. Electric services usually found in garages are required.

Traffic control may be installed in the police station. The latest electronic type of control may be complex and the wire tie lines to the various systems require fairly large wiring installations. Space should be provided for these systems.

16.26 Prisons. Prisons and correctional institutions may include several types of commercial buildings:

(1) Main administration
(2) Cell blocks
(3) Manufacturing buildings
(4) Dining halls
(5) Hospitals
(6) Power plant

All parts of the electrical installation accessible to prisoners must use materials that are designed to prevent them from being damaged. Boxes, fixtures, etc must also be designed to prevent opening by other than authorized personnel.

16.26.1 Administration Building. The administration building is usually outside of the walls or is part of the outer wall surrounding the institution. It contains general offices and the central systems for communication and security. Direct connection to outside law-enforcing agencies is maintained by radio and teletype systems.

Controls for security facilities such as the outer wall lighting, cell block supervisory, and main gate controls are contained in the main building. Particular attention should be paid to the location of these security controls to prevent them from becoming accessible to unauthorized persons. Consideration should be given to the fact that the control room may be seized by prisoners and a backup control means provided in a second secure area. It may well be that recognized modern methods of power distribution in the building will be sacrificed in the cause of security.

16.26.2 Cell Blocks. Cell blocks have radio or intercom systems for each cell controlled by the main set located in the administration building. Electric door locks and an elaborate system in interlocking areas ensure security and safety to the guard force. Lighting in the main

corridors should be of dual service to preclude any outage. All controls must be well away from any possible access by the inmates.

Closed-circuit television may be required to scan the various cell block areas enabling the guards to maintain surveillance over the entire area. Regulations must be checked to ensure that such surveillance is acceptable and does not abridge the inmates' rights. Lighting levels in these areas should be adequate to ensure good television coverage. Camera requirements should also be checked.

16.26.3 Manufacturing Buildings. Manufacturing buildings are a part of many correctional institutions. Power distribution is similar to that of industrial plants.

16.26.4 Dining Halls. Dining halls, if centralized, may be used as auditoriums. Stage lighting, such as that required for a medium-sized high school, may be needed. A public-address system may be required for entertainment, announcements, and orders. Large kitchens will need the usual electric accessories. A projection booth for movies may be required where the dining hall is used as an auditorium.

16.26.5 Hospitals. Hospitals fully equipped to take care of an emergency and any type of operation, are generally required. The power and lighting system is similar to that of a general hospital (see 16.17—. Panelboards are located according to security regulations.

16.26.6 Power Plants. Power plants may be included to generate electric power and to produce the steam, hot water, and heating for the entire institution. Emergency and standby power will be required. Normal electric power may be generated at the power plant but a connection to an outside utility in case of local failure or sabotage is mandatory.

The power plant should be in a secure

location. Power should be distributed underground. The size of the prison determines the distribution voltages required.

Floodlighting of the outer walls with searchlights in the guard towers will constitute a large load. The guard towers may have electric heaters as well as telephones and other intercommunication systems.

16.27 Radio Studios.

Radio studios should be soundproofed; therefore, all ventilation, both heating and cooling, must be of extremely low noise-level design. Air-conditioning loads are generally high in all parts of the building.

These points should be considered:

(1) Air-conditioning loads, centralized or individual units

(2) Shielded wiring for noninterference

(3) Low noise levels of transformers

(4) Quiet fluorescent-lighting ballasts (or remote mounted ballasts)

(5) Clock systems

(6) Paging annunciator

(7) Private intercommunication systems

(8) Special connections to networks

(9) Stage lighting for audience shows

(10) Services for telephone and telegraph news coverage

(11) Special sign lighting

Some stations have the transmitter in or adjacent to the studio. Provision for removing the heat from the large transmitters is required. Power requirements are large, and dual reliable services may be required. Emergency standby equipment with automatic transfer facilities is also needed in cases of emergency.

Transmitter towers generally need obstruction lights for warning aircraft according to Federal Aviation Administration requirements.

16.28 Recreation Centers.

Recreation centers are generally multipurpose facilites. Outdoor requirements can involve parks and playgrounds (see 16.23) with various athletic fields, swimming pools (see 16.34), skating rinks, and parking lots. Indoors, the facility often includes auditoriums (see 16.3) and gymnasiums (see 16.15). In smaller recreation centers a single multipurpose room will serve as a auditorium and gymnasium. Other spaces designed for particular sports, such as handball and racquet ball courts, squash courts, weight training rooms, etc, may also be included. Some recreation centers also include craft shops. Where craft shops are included, attention must be given to outlets for power tools, kilns, etc.

16.29 Restaurants.

There is substantial electrical load concentration in restaurant kitchens. When gas is used for cooking a 50 W/ft^2 connected load is not uncommon. Connected loads of over 100 W/ft^2 have been encountered in all-electric fast-food kitchens (includes ventilation and air conditioning).

Kitchen layouts are frequently made based on a certain brand of equipment. It is important that the engineer check the submittal data of the equipment actually furnished to ensure power-load provisions and that outlet locations are acceptable.

Electrical power in restaurants should fulfill the primary needs of the particular type of building. It is used to preserve and prepare food and provide appropriate lighting.

These points are important:

(1) Adjustable lighting from high to low levels

(2) Air conditioning

(3) Electric cooking and baking

(4) Food warmers

(5) Serving tables

(6) Conveyors and dumbwaiters

(7) Public-address systems

(8) Provisions for background music

(9) Plug-in telephone outlets

(10) Decorative lighting

(11) Parking-lot lighting and lighting for drive-through facilities

(12) Television outlets

(13) Restroom ventilation

(14) Electric hand dryers

(15) Outlets for janitorial equipment

(16) Business machines in main offices

(17) Exit lighting

(18) Emergency lighting and power

(19) Snow and ice melting for ramps, sidewalks, and driveways

16.30 Schools. The general requirements for schools are similar to those for colleges (see 16.10). In the general school building all activities are normally under one roof. The use of individual or connected buildings is common for large schools.

Today greater effort is being directed at more efficient use of electrical energy, except for selected education programs, power requirements have been fairly stable or decreasing in recent years. In lighting, more efficient sources and better lenses have decreased the watts per square foot required for good vision. Special outlets or extra floor outlets are required in home economics, vocational training, business training, computer skills and foreign language spaces. Requirements for auditoriums, gymnasiums, and libraries are covered elsewhere in this section.

16.30.1 General Requirements. Lighting in most areas should involve at least two levels of control. Special systems should include fire detection and alarm, security control, clock system, and program (announcing) system. Both commercial and closed-circuit TV, telephone and central monitoring/control for mechanical systems may be included.

16.30.2 Mechanical Systems. Fans, pumps, chillers, cooling towers, and air conditioners are to be considered. Sometimes electricity is also used for space and water heating. Heat pumps, heat recovery systems, and energy-efficient motors reduce required power use.

16.30.3 Laboratories. The rapid advance in science requires laboratory facilities with semiflexible power distribution features. Convenience outlets and plug-in power outlet raceways may be needed on benches. The electric requirements for the laboratories should be checked in the early stages of design because, frequently, special panels and equipment may have to be manufactured This type of equipment may require long delivery time.

16.30.4 Manual Training. The manual training department may contain large motor-powered machine tools and welding equipment. Courses may be revised and new tools and skills taught, and flexible power distribution systems should be provided. Appropriate shop and tool lighting should be provided. A master cutoff of power for each shop is recommended for instructor convenience and safety. Emergency stop buttons for operating undervoltage release devices disconnecting all shop tool power should be provided.

16.30.5 Kitchen Facilities. School lunch programs have expanded the power requirements of kitchens and cafeterias. Ovens, ranges, mixers, freezers, and exhaust systems require large blocks of power, Heavy-duty power outlets for hot food carts may be required.

16.31 Shopping Centers. A shopping center is a group of stores concentrated in a compact area surrounded by vast parking lots. The majority of these are located on the outskirts of cities and towns, and power is generally supplied at medium voltages. The parking areas should be well lighted. Direct-burial cables are run to the lighting standards, and the feeders are generally controlled by photoelectric cells and time clocks which turn the lights on automatically at night and off after closing hours. The various stores are each wired to the owner's requirements and normally metered separately. See Section 2 for several sets of load figures.

16.32 Supermarkets. The supermarket is a fast growing and changing institution. The small margin of profit and the great volume of business causes the supermarket to attempt new merchandising ideas. These require a flexible electric system. See Section 2 for load figures.

The lighting may be 50 fc—90 fc for general illumination. Auxiliary lighting is also generally used. High-intensity incandescent units over the produce and meat areas have been used to give a better color to the products where color is important. High-intensity areas using valances around the perimeter of the store have also been used. These frequently serve for advertising products as well as for illumination.

The front areas may be highly illuminated for attraction, and the rear of the store may be brilliantly lit to bring the back apparently closer to the front for advertising value and to attract the customer to the rear of the store.

The open freezer and refrigerated cases cause the refrigeration load to be very high. These loads usually have several banks of compressors at several locations in the store, and refrigerant is run to the cases on the sales floor. Three-phase power is usually run to the compressor location, and an additional 120 V circuit is run to the case location for the lighting and small ventilation fans on the case itself.

The freezer cases usually contain electric defrost elements, and the wattage for these elements exceeds that of the refrigeration unit. These defrosting elements may operate on one phase of the motor circuit feeder when the compressor motor is not in operation. The defroster loads are generally considered to have 40—60% diversity when the service entrance is being calculated.

The checkout systems are designed to move the customers out quickly. The capacity usually ranges from three outlets per circuit where only cash registers are used, to one outlet per circuit where an elaborate array of conveyor belts and cash registers is used.

Coffee-grinder loads have increased. These many range from 1—5 hp, three phase or single phase. Choppers and grinders may have two motors running simultaneously, which may be 10–15 hp three-phase motors. Hot-iron sealing machines usually require one circuit per machine. Meat saws range as high as 10 hp three-phase motors. Electric door openers, electric meat tracks, electric rotisseries, and moving displays should be considered in the final survey of loads.

Communication systems range from the simple single pushbutton, and single bell and chime to systems that have speaker boxes located at convenient points in the store where the customer may request location of items. This information is transmitted to the office, and the locations of such items are transmitted back to the customer. This system is usually interlinked with the cash-register locations, enabling the cashiers to communicate with the office when change, stock clerks, baggers, etc are needed.

Refrigeration failure alarms, burglar alarms, closed-circuit television surveillance, and other security systems should be considered.

Signs are an important part of the load and are sometimes elaborate, with running lights and neon lights. Some markets perfer 35—50 ft pylons with large lettering and moving parts.

Special consideration should be given to future electrical loads: conveyors to deliver the groceries from the cashier locations to the customers' automobiles, conveyors to deliver the product from the shelf to cashiers, pushbutton shopping, automatic meat processing, automatic stocking, etc, are all likely items of the future supermarket, and all depend one way or another on electricity.

Provisions should be made for a sound system to provide music and announcements to public areas. Conduits should also be included for computer-type checkout employing stock control. Empty conduits should be installed for regular and pay telephones. Power for vending machines, snow melting systems and other miscellaneous loads should be checked.

16.33 Swimming Pools and Fountains. Where swimming pools or fountains use electric power for motor-driven water pumps, underwater lighting, surface lighting, overhead lighting or outlets for pool cleaning, etc, a serious life-safety hazard can develop. The designer should avoid the use of metallic pipes for water intake and water outlets of pools and fountains, that is, between the pools or fountains and the motor-driven pumps,

regardless of the voltage applied to the motors and electrically operated valves. Underwater lighting should preferably be 6 V, but not higher than 12 V.

The quality of wire, fixtures, etc should be the best obtainable for such wet conditions. The electric pumps and valves should all be isolated from the public.

The grounding system of all electric devices, pump and valve equipment should be of minimum impedance.

Electrical potential should be prevented from developing in any portion of the water. If and when the grounding system or the electric insulation system fails or deteriorates, it should be removed, replaced, and tested. The entire system should be periodically tested to ensure that acceptable insulation and ground resistance values are maintained. The installation shall, as a minimum, comply with ANSI/NFPA 70-1981 [1], Article 680.

The pertinent rules, regulations, and approved equipment lists apply to deck boxes, lighting luminaires and circuiting, bonding of metal structures, grounding connections, ground-fault interrupters, and low-voltage lighting.

The trend to low-voltage lighting, generally 12 V, is gaining increasing attention due to availability of lighting equipment, two-winding transformers, and ground-fault circuit interrupters. Where low-voltage lighting is applied, design details on sizing equipment and wiring are of prime importance. Placing transformers and interrupters in the vicinity of the pool lighting luminaires has proven to be feasible and economical. Lighting fixtures are installed 18 inches below normal water level and mounted in pool walls. Wet niche types are preferred.

If above-water lighting is used, it is imperative that it be designed to reduce surface reflection to ensure that the lifeguard can easily see the entire bottom of the pool.

Effective bonding of all metallic parts of a pool structure, including lifeguard stands, ladders, diving board stand reinforcing steel, skimmers, and other metal parts, can be accomplished with a solid copper conductor not smaller than AWG No 8.

16.34 Telephone Buildings. In addition to conventional offices, portions of these buildings are dedicated to communications switching equipment. Newer facilities use solid-state switching rather than electromechanical; however, the power requirements remain high (per unit area) due to miniaturization of equipment and the resulting higher density. Telephone switching equipment is typically powered from dc power supplies feeding the equipment and a battery bank which *floats* on the line. If a utility service interruption occurs, the batteries assume the load until an on-site ac generator can be brought on line. The chargers are typically oversized so that the battery bank can be recharged at the same time the communications load is served. Battery rooms require special treatment such as venting. Batteries, chargers, and the load should be located in close proximity in consideration of the high currents involved.

Air conditioning is critical and will probably need to be served from the emergency power bus.

Lighting in equipment rooms will need to be arranged to suit the narrow maintenance aisles. Equipment aisle lights are typically a part of the communications equipment racks, which makes the design less difficult. Since only small areas need to be lighted at any particular time each aisle section should have its own switch.

If underground cables enter a building cable vault, special precautions need to be taken against natural gas, or sewer gas leaks entering the phone raceway system and flowing into the vault. Explosionproof lighting, sniffers (gas detectors), etc, may be required.

Maintenance outlets throughout equipment rooms for portable equipment such as soldering irons may be required.

Considering the critical nature of the facility during floods, earthquakes, wind storms, and other natural disasters it should be designed to maximize the facility's potential for survival.

16.35 Television Studios. This type of building is similar to radio studios (see 16.27) except for the lighting loads. General lighting for office buildings should suffice for the offices, corridors, foyers, etc.

Television needs large amounts of concentrated lighting. The lighting designer must take into consideration the requirements of color television. Every show has its own lighting engineer on the production staff. This engineer's preferences for lighting must be considered.

Studio buildings must be designed for color television productions. Transformers, ducts, and cable raceways should be planned and installed during original construction to meet these requirements.

Television studio spotlighting and floodlighting fixtures are mounted on a grid of pipe supports for maximum flexibility. They are wired by means of pigtails, connector strips, and patch panels to a central control console in each studio. Special attention should be given to ventilation and air-conditioning loads produced by the heavy lighting density. Special wiring and shielding to avoid interference with television signals may be required. Telephone and teletype communications may be needed, as well as provisions for network pickup and service. Other considerations include provisions for microwave pickup from offsite trucks, special security systems, videotape editing and reviewing rooms, and emergency power systems.

16.36 Theaters. Theaters are of two general classifications, the motion-picture theater and the legitimate or stage-production theater, but as certain areas in both types are similar they will be covered as a single unit to avoid duplication. Theaters generally have a marquee with illuminated signs and decorations, in addition to a concentration of lighting outlets on the soffit to illuminate the entrance to the theater and to attract the eyes of prospective patrons. Electric signs showing the name of the theater are frequently placed on the front of the building.

The entrance lobby and foyer are generally of a rather decorative nature, and the engineer should study the architectural details so that the lighting will blend properly with architectural treatment and color scheme. The minimum wattage per square foot used in the foyer may be about 2.5 W/ft^2 and in the lobby about 3 W/ft^2. Ticket booths may require a telephone, a signal system, and possibly a special outlet for electrical heater and fan.

The interiors of motion-picture theaters are of a rather plain design, although some of them, where the pictures are combined with a stage show, may more nearly approach the legitimate theater for interior treatment and illumination. The designs are of such a varied character that it is almost impossible to establish any clear formula for an illumination plan. Provision should be made for aisle lights, exit lights, and orchestra and emergency lights.

The projection booth should have provision for a minimum of two projection machines, each supplied by a separate circuit, floodlights, spotlights, a rewinder, exhaust fans, an intercommunicating signal system, and a dc power supply for projection arc lamps. Depending on the size of the theater, provisions may be made at several locations for floodlights or spotlights.

Theaters may have passenger elevators and escalators to serve the patrons using the mezzanines and balconies, and in a number of installations, orchestra lifts, stage lifts, and turntables have been provided.

In the legitimate theaters and the large motion-picture theaters the lighting in the auditorium is controlled by dimmers on the stage switchboard. The border lights, floodlights, and spotlights are controlled individually.

The wiring system may be complicated, depending on the lighting effects required to suit the type of entertainment given. The stage requires stage pockets, several rows of border lights, outlets for special electric effects, motors for operating the fire curtain, contour curtain, roll curtain, heavy drops, ventilators at gridiron, and numerous other items.

16.36.1 General Lighting. General lighting is required for the gridiron, fly galleries, dressing rooms, etc. The power load for air conditioning is a major item which deserves careful consideration in the preparation of the electric system. Convenience outlets should be provided for cleaning and general maintenance work.

16.36.2 Stage Lighting Systems. All lighting fixtures (instruments) involved in a stage presentation are controlled through patch panels and dimmers to achieve desired effects. Extensive wiring is required as each fixture or group of similar fixtures is carried back to the patch panel as an individual two-wire, 120 V circuit. Common neutrals are not used as phase connections vary. If possible the patch panel location should have a view of the stage. Dimmer switch-boards (generally SCR control units) produce heat and require ventilation and should be located in a sound isolated area.

A lighting-control booth with a good view of the stage is necessary and should contain a stage lighting-control panel from which all of the production lighting can be controlled. The control panel controls the many dimmers and contactors to provide each stage lighting scene desired. Each scene will require a different illumination configuration and the sequence of illumination settings is typically repeated for each production. All of this may be preset and sequenced by a microcomputer in the control panel. An emergency *panic switch* is usually incorporated in consideration of the need to immediately restore the theatre egress lighting upon any indication of a threatening condition. The booth also will normally contain follow spots, and slide and motion-picture projectors. A sound-control booth is frequently located near the lighting control booth.

16.37 Transportation Terminals. Transportation terminals consist of passenger terminals for railroads, buses, aircraft interfacing with platforms, boarding ramps, ticket and reservation areas, as well as concessions or stores, theaters, and entertainment areas for the travelers' convenience. Concession areas require general provisions as described under see 16.14, Department Stores.

Provisions should be made for supplying metered or rent-included power to each of the tenants. In some cases, such as airports, the terminal may very well be almost a shell in which most of the areas are finished entirely by tenants, conforming to the owner-established building standards and approved construction methods.

16.37.1 Requirements for All Terminals. The following general requirements apply to all terminals:

(1) Fire-alarm systems, which may include smoke detection systems, are usually interconnected to a central station for transmission of alarms to the fire department. In case of very large complexes, the functions of the central station may be handled in-house. The usual fire-alarm pull-boxes may be supplemented by a number of telephones which may be used for information, porter-call, and police and fire emergencies.

(2) Sprinkler supervisory and alarm systems are tied into the fire-alarm system for transmission of alarms to the fire department.

(3) Public telephones and other telephones served by the local telephone company usually require only conduits. If the system is owned by the phone company they will install equipment and cables. If the building owner purchases the system, the equipment and cables will be installed by the supplier or contractor. A room for telephone services, meeting the standards of the telephone company, must be provided. While the telephone company will install pay stations for various architectural requirements, close consultation with the telephone company is required to establish design standards. Provisions must be made for local private systems

which may be needed by individual transportation companies.

(4) Terminals require public-address systems which may be controlled from more than one location and which may require zoning so that announcements can be provided in different loading, lounge, and baggage areas.

(a) One of the major problems in designing such systems is the use of high-powered speakers, too widely spaced, which results in distorted announcements. The use of lower-powered speakers at close spacing, particularly where noise levels are high, with automatic variable volume control based on ambient sound levels is desirable.

(b) Provision should be made for background music, often transmitted from commercial specialists.

(c) Provisions may be required for automatic departure announcements where established schedules require very frequent departures, such as bus terminals.

(d) Consideration should be given to extending the public-address system into the restrooms, outside loading areas, embarking areas, and perhaps parking lots.

(5) Master-clock systems may be used to ensure that all clocks are maintained at the same time and that hand resetting is not required after temporary power outages or for changes related to daylight saving time. Carrier frequencies (generally 3000 Hz) for master-clock circuits are often superimposed on power distribution systems for clock correction and program signals.

(6) Arrival and departure signs may be displayed on closed-circuit television or variable message signs. Both systems are usually driven in newer installations by microprocessors and may contain elaborate storage information systems involving schedules. Both systems can

include an automatic *roll down* feature wherein messages are kept in a sequence regardless of changes and, as items are removed or added, the proper sequence usually on a time basis is retained.

(7) Security systems in a modern terminal will usually include closed-circuit television surveillance of public areas, taxi loading areas, and other locations where the public must be protected. Closed-circuit television usually includes provision for zooming and scanning (pan and tilt). The special purpose telephones described under fire-alarm systems may also form part of the terminal security system.

(8) Directional signing may consist of back-lit or front-lighted signs with off-on control. Sign colors and designs, usually designed in conjunction with graphic specialists, are essential to provide adequate terminal control. In some cases, advertisement signing is combined with directional signing, the former often being an important source of revenue for the facility. Outlets should be provided for other signs, (that is, Christmas displays) which may be temporarily required.

(9) Outlet systems should be adequate for vending machines, water coolers, temporary displays, and for cleaning and maintenance machines. The latter two items may require 480 V outlets. Codes may require the use of personnel groundfault-protected outlets in certain areas.

(10) Modular system of local wiring, outlets, and metering to accommodate future changes in rentable spaces is desirable.

(11) Electronic dispatching control and surveillance systems may be used to indicate departure time of vehicles or aircraft, to observe such departures, and to direct traffic through the use of spe-

cial signing systems. The location of vehicles in berths may be detected through loop presence detectors, treadles, or similar sensors. Where complex systems are involved, dispatching boards may be used. The dispatching boards may be manually or automatically operated to indicate the status of loading platforms and traffic. Closed-circuit television surveillance of areas where traffic congestion may develop should be provided. Ramps into or out of the terminals usually require close surveillance.

(12) Computer systems are frequently used for control of terminal and tenant systems. Where separate rooms are required for computer installation—as distinct from certain microprocessor or intelligent terminals which may be installed in open areas—raised flooring, separate air-conditioning systems, special fire and smoke detection systems, halogenated flame-extinguishing agent systems, and separate emergency circuits may be required.

(13) Emergency power systems will require the use of engine or turbine–driven alternators. Uninterruptible power supplies are usually required only for special systems such as computers and alarm systems; however, these are often furnished as part of the special equipment installation.

(14) The building supervisory system may be as simple as an annunciator or as complex as a complete computerized building management system. Such a system will include alarm detection and indication points for items such as escalators, elevators, fans, pumps, chillers, cooling towers, temperature detectors and other similar devices. The status of operation such as up or down for escalators, emergency trip-out of equipment, overtemperatures, area lighting status, and similar alarms will be audibly and visually indicated, usually with a reset provision for audible alarms and always with provision for indicating multiple alarms. The building supervisory system may also be used to automatically control heating, ventilating, and air-conditioning equipment and to interface with building electrical power demand equipment to optimize overall power utilization.

(15) Other communciation systems, some of which might be owned by tenants, can include portable radio systems and pagers for larger areas, special building antennas and loops, and automatic printer systems such as teletype, facsimile systems and annunciators. Provisions may also be made to communicate with special staff by the use of *coded* announcements on the public address system. It is usually not deemed wise to announce emergency situations in clear text on the public-address system.

(16) Baggage handling systems generally are fairly simple, involving the use of manual labor for the most part except at airports where systems become quite complex as described below. The major consideration is the providing of telephones at convenient locations for obtaining porter service and to provide some form of alerting signal so that porters will know that they are on call.

(17) It may be desirable to use special spaces such as shafts, utility tunnels, or trenches to provide a practical means of handling cabling at minimum cost. It is often possible to utilize cable trays, particularly for communication wiring. Communication wiring may consist of co-axial and fiber-optic cables, multiconductor telephone cables, data handling, and low-voltage control-system cables.

(18) Since new terminals are usually individualized, highly creative archi-

tectual designs the lighting design must be closely coordinated between the engineer and the architect or lighting consultant. Where high ceilings are involved, HID sources are usually selected as the base source of lighting. Energy conservation and maintenance considerations have created a tendency to move away from incandescent lighting to the more efficient sources regardless of ceiling height.

(19) The following types of systems are often found in terminal buildings:

(a) Snow removal on ramps and sidewalks

(b) Kitchen equipment

(c) Radiant heating

(d) Supplementary air conditioning

(e) Fare collection or change issuing machines

(f) Smoke purge control systems

(g) Emergency escalator shutdown systems

(h) Lightning protection systems

(i) Dispatch booth power, lighting, and control

(j) Infrared heating

16.37.2 Railroad Terminals. Some railroad systems may use an electronic reservation system for trains very similar to that found in airline terminals. Train departures are often displayed on large boards with flip-type discs making up letter matrices and with the *roll-down feature*. Newer terminals often contain extremely large illuminated wall-to-wall advertising displays. Lighting levels are usually kept fairly low, perhaps 8—15 fc, except in ticketing areas, areas where pedestrian safety and security is involved, and in certain limited areas of lounges where patrons may read.

16.37.3 Bus Terminals. Extensive ventilation systems are required to exhaust fumes from bus engines when buses load and unload in closed terminals. Complex

systems of traffic controls are usually required in larger terminals. Information telephones are often distributed throughout the terminal and a large information booth is very common in terminal-type design. Lighting levels in bus terminals may typically range from 10—30 fc and with levels as high as 50 fc in limited areas where highlighting is desired. Security provisions for bus terminals are usually fairly severe, requiring the use of closed-circuit television and emergency telephones. Lighting, used properly, can be a form of security and reassurance to the patrons. Bus terminals today are often installed in urban areas as part of a general area renewal strategy. In such cases, the facade lighting of the bus terminal becomes an important part of the overall improvement in ambiance.

16.37.4 Airports. Airports may range from simple terminal buildings to huge complexes that are associated with major urban areas. A large airport will have most of the facilities common to a small city such as hotels, shopping areas, bus handling systems, extensive traffic and roadway control systems, many of which have been discussed in this section.

Automated baggage-handling systems of the so-called carousel or moving-belt type are supplied from extensive conveyor systems leading to the airport apron areas. For handling aircraft cargo, very extensive electronically-controlled systems, some using linear motors, are utilized. Such cargo-handling systems use industrial material handling technology.

Most ticketing and reservations are handled through computerized systems; however, the main or central computers are often located remotely in an airline building off the airport. Extensive communication wiring systems are required for ticketing and communciations between baggage, apron, ticketing, and gate areas. Extensive signing is required at each gate position as well as signing to indicate the location of gates from which flights will be departing. Provisions have to be made for security control, particularly for the installation of detection equipment and control stations, usually at the entrance to each gate or group of gates for the detection of weapons and explosives.

While terminal area lighting may be held to relatively low levels, perhaps 10—25 fc, much higher levels are usually utilized in the immediate ticketing areas. The control-tower designs which will include provisions for air controllers, ground-traffic controllers, and possibly facility survelliance must be designed in conjunction with the appropriate air-traffic control authorities. Special expertise is required for the designing of these airport systems and of the related runway and taxiway lighting systems. Airport control systems are specified in detail in FAA Advisory Circulars.

At the airline gates, which may be some distance from the ticketing position, 500 kW or more may be required for 400 Hz alternators which provide power to the planes while they are parked. Emergency fuel shutoff and alarm systems are provided where fuel piping systems are used. Provisions are made at the apron for electrical grounding of airplanes. Requirements for plane control at gate positions are usually specified by the airline. Provisions must be made for apron lighting where work will be going on while the aircraft is being loaded, serviced, and refueled.

Hangar lighting is almost always of the HID type supplemented with emergency lighting for power outages. A typical large hangar may require an excess of 1000 kW. Grounding systems have to be provided for the aircraft both in the hangers and on the aprons, and at loading positions.

The power distribution system for larger airports is almost always at medium voltage with energy supplied at the individual buildings either from spot network systems or from medium-voltage primary transfer switches. For reliability consideration at larger terminals, multi-feed selective systems are almost always used.

Extensive parking areas are required for short-term and perhaps long-term parking arrangements. Toll plazas must be established. Newer designs utilize unmanned gate entrances with ticket-issuing machines and manned exits which may contain provisions for automatic fee calculations. The audit systems utilized must provide presence detection of vehicles, time of fee collection, and vehicle counters which are tied in to the auditing device in each toll booth. This may consist of a special cash register or an automatic fee calculation and audit device. The tendency today is to move towards the so-called intelligent terminals which perform all fee calculations and report to a central computer. Parking lot signing utilizing changeable message signs and lighted fixed signs is essential if traffic is directed to different lots.

16.37.5 Rapid Transit Stations. The rapid transit, subway or so-called elevated station is usually an extended platform which provides access to trains. Such stations have public-address systems installed to announce the arrival of trains, safety precautions, and train delays. Lighting levels are relatively low except in the toll collection areas and at points of train boarding. Higher levels of lighting may be used in limited areas where, at night, a feeling of security on the part of the patron is desired. Simple signing systems will indicate which trains may depart first if several trains are parked in a multiplatform station. Where trains are normally parked, warning lights may indicate that trains will depart within a preset time interval.

16.38 References.

[1] ANSI/NFPA 70-1981, The National Electrical Code.[65]

[2] ANSI/NFPA 72A-1980, Local Protective Signaling Systems.

[3] ANSI/NFPA 72B-1980, Auxiliary Protective Signaling Systems.

[4] ANSI/NFPA 72C-1975, Installation, Maintenance, and Use of Remote Station Protective Signaling Systems.

[5] ANSI/NFPA 72D-1980, Proprietary Protective Signaling Systems.

[6] ANSI/NFPA 76A-1977, Essential Electrical Systems for Health Care Facilities.

[7] ANSI/NFPA 101-1981, Life Safety Code.

[8] ANSI/IEEE Std 446-1980, IEEE Recommended Practice for Emergency and Standby Power Systems for Industrial and Commercial Applications.

[9] NFPA 56A-1978, Use of Inhalation Anesthetics.

[10] Illuminating Engineering Society Handbook.[66]

[11] USDHEW Pub No (HRA) 79-14500, Minimum Requirements of Construction and Equipment for Hospital and Medical Facilities. (Stock No 017-022-00643-0).[67]

---

[65] The National Electrical Code is published by the National Fire Protection Association, Batterymarch Park, Quincy, MA 02269. Copies are also available from the Sales department of American National Standards Institute, 1430 Broadway, New York, NY 10018.

[66] This publication is available from The Illuminating Engineering Society, 345 East 47th Street, New York, NY 10017.

[67] This document is available from Superintendent of Documents, US Government Printing Office, Washington, DC 20402.

17. Energy Conservation

17.1 Energy Conservation Requirements. The establishment of a successful electrical energy conservation program is dependent upon the full interest and encouragement of top management, and a formulated company policy committed to saving both energy and the moneys associated with energy and demand savings.

The supervision of the energy conservation program must be delegated to that staff within the organization who will commit the time and resources necessary. The program will not be successful if it is assigned as a part-time duty to staff members whose prime responsibilities lie in other areas. In some instances, sufficient in-house staff may be available to develop a program; however, the management of many commercial buildings will not have access to a staff sufficiently conversant with such a program. As additional services are needed they may be obtained through the use of consultants and some trade organizations whose prime commitment is to the development of energy-conservation programs. In order to ensure the success of a program, periodic reports should be provided to management. Projects showing the energy and cost savings, and payback that have resulted should be included. The report should be reviewed and commented upon by top management. Without the cooperation of the entire staff, and especially the building maintenance staff, the program will likely not be successful.

Since energy conservation deals with equipment and systems which are covered in other sections of this standard, there may be an overlap in the content of this section and some material in other sections.

17.2 Approaches to Finding Economical Energy Recovery Areas. There are two recognized methods:

(1) The audit

(2) *Project shopping list*

17.2.1 The first method is to make a complete energy audit listing major energy-using equipment with nameplate data relating to energy, any efficiency tests, and estimates or measured hours per month of operation. This will include monthly utility data, amounts, and total costs. National Weather Service monthly degree days for heating and cooling must be included. Generally, a one or two year compilation of data is used. The crucial part is an intelligent appraisal of energy usage, what is done with energy as it flows through the processes and facility, and how this compares with accepted known standards. This study includes a prioritied list of projects with a high rate of return.

17.2.2 The second method is used by many. This is to look for specific projects to reduce energy and costs. *Shopping lists* of projects are obtained from associates, newspapers, Department of Energy (DOE) magazines, the local utility, or trade groups. These lists include readjustment of thermostats for heating, cooling and hot water (DOE, July 1979), removing lamps in lighting fixtures, installing storm windows and doors, caulking, additional insulation, etc. An excellent reference is a do-it-yourself guide called *Identifying Retrofit Projects for Federal Buildings* by the Federal Energy Management Program (FEMP) Report 116. Other sources of information are listed in 17.11.

Nationally oriented standards have been issued by the American Society of Heating, Refrigerating, and Air-Conditioning Engineers, Inc (ASHRAE) as part of ASHRAE 90-75R. Today, the latest version of ASHRAE 90-75 issued jointly by ASHRAE and the Illuminating Engineering Society (IES) covers *national* recommendations for Building Energy Performance Standards (BEPS) power budgets for new construction. This standard is changed periodically and other versions such as the one now being considered by the National Bureau of Standards (NBS) with an energy budget may replace existing standards. Another helpful document is *Total Energy Management*, a practical handbook on energy

conservation and management developed jointly by the National Electrical Contractors' Association and National Electrical Manufacturers' Association in cooperation with the US Department of Energy. Energy codes have been published by a number of states, some of which specify materials similar to ASHRAE 90-75R, and others which specify permissible usage on various bases such as watts per square foot, or allowable footcandle levels. Such information is likely to be available from a state energy office (SEO).

17.2.3 An early decision should be made as to whether initial efforts would be to make an energy audit, or to compile a shopping list of projects on which to start work.

17.2.3.1 Nearly all experienced energy consultants, managers, and the federal and state governments recommend starting conservation with an audit of past usage, and a listing of energy-using equipment, nameplate data, and operating times. This is the extent of most governmental supported audits. Audits have been completed by non-technical personnel at minimum cost. Government publications are available for help.

This type of audit lacks historical usage data, efficiency testing of large energy-using equipment, analysis of system operation from known standards, and analysis of alternate system operations, fuels, and equipment. This quality of audit and usage-evaluation information is generally available only from experienced consultants or trained plant engineers. Consulting costs could run $10 per kilowatt of maximum yearly demand, or require $\frac{1}{3}$ man-hour per kilowatt and $10 000 of instrumentation. Many of the best ideas or projects with the best return are discovered only by this method. Such a procedure will often show

how to reduce electrical energy costs 20 to 35% and develop a simple payback of less than three years.

17.2.3.2 A shopping list of projects can give an indication of where to start looking. Many lists are actually a relisting of projects found on other lists, so many of the projects may not pertain to the building being examined. Thus, there is often a search for better project lists that will show the greatest savings.

17.3 Electrical Utility Charges for Energy, Demand, Service, Fuel Costs, and Penalties.

17.3.1 Energy Rates. Historically, electric rates cover the cost to serve the customer, plus 3 to 8% for profit. During the late 1960s and early 1970s, fuel cost increases and inflation turned around the previous annual drop in kilowatthour costs.

17.3.2 Technical Terms Used for Utility Billing

17.3.2.1 Power Factor. Power factor (pf) is a ratio of real power to apparent power. It can be metered and billed when it occurs at a utility peak kilovoltampere period, at the customer's peak kilovoltampere or kilowatt period, or be an average power factor of the three highest peak kilowatt periods, measured as total kilowatthours or kilovoltampere hours for a month, and calculated as an average power factor. It is to the customer's advantage to have power factor calculated at the time of his peak kilowatt usage when he would generally have his highest power factor. Such metering, showing kilovoltamperes at peak kilowatts, generally entails chart or cartridge-type magnetic recorders.

17.3.2.2 Demand. This is the average amount of kilowatthours, or kilovolt-ampere hours metered in a set time interval of 5, 15, 30, or 60 minutes. Demand is measured, recorded, and referred to as kilowatt demand or kilovoltampere demand, as the case may be.

Total utility-system *demand* is extremely important to the utility company because when the system load exceeds the utility-system generating capacity, utilities can only buy the additional energy, if available, and likely will need eventually to build additional generating capacity. Demand charges can frequently amount to 30% or more of the total electrical bill.

17.3.2.3 Rate Methods to Control Utility-System Demand. *Time of day rate.* Due to increasing utility-system demand and its costs, utilities offer cheaper energy at night, weekends, and other off-peak periods. Generally, the energy portion of the bill during off-peak periods is the same as at other times of day, but the demand portion may be reduced or eliminated. This has led to a new concept of energy storage to supply the equipment. In an all-electric building, water can be heated or cooled and stored in large insulated tanks at night, and used to heat or cool the building during the daylight hours. This practice actually requires 3—5% more energy, due to insulation losses, but such a project generally earns a 4 to 6 year simple payback on the investment.

"Sliding Window." Some customers have turned "everything-on" for a few minutes and then dropped-off-loads for a few minutes to control their average kilowatt usage in a preset time period. This has caused system-load oscillations and voltage-control problems on some utility systems. The "Sliding Window" technique was developed to move the demand period to any time within the utility-peak demand period.

17.3.3 Utility Charges

17.3.3.1 **Power Factor (pf).** It is becoming common for utilities to charge for *demand*, based upon the highest kilovoltampere period in a month. To minimize the demand charge, the customer must be operating at 100% power factor during peak demand periods.

Part of kilovoltampere (apparent demand) is kilovar or reactive demand. Kilovar demand does require utility equipment capacity to handle, and it does have wire or resistance losses associated with it, but kilovars have a lower unit cost to the utility than kilowatt hours.

The ratio of kilowatthours to kilovoltampere hours is sometimes called the average power factor. Utility practice of having two kilowatthour demand meters, one with phase changing connections to obtain kilovoltampere hours is rather common. Charges sometimes made for *low power factor* based upon a calculation using the ratio of peak kW divided by peak kVA has nothing to do with average power factor. It is only a mathematical ratio since the peak readings are not likely simultaneous and are often reached on different days. The charge for having a poor power factor has been described in contracts to occur when the power factor is below 85%, 87%, 95%, or 100%. A review of many utilities show a wide variety of practices of handling power factor, including:

(1) No charge, even if average power factor is below contract limits

(2) Charge based upon an estimate not metered except when required by customer

(3) Charges based upon measured peak kVA demand

(4) Assess charge on low average power factor (measured kWh divided by kVAh)

(5) Assess charge on low calculated power factor from metered peak kW noncoincidental and peak Kvar

(6) Charge based upon average of kW and kVA

The recommendation of many utilities is to add capacitors on the customer's system to overcome low average power factor. However, to save the most energy, it may be desirable for the customer to install capacitors in parallel with motors at their terminals and switch with the motor's starter. This will reduce line losses and free-up equipment for additional loads. If capacitors are added at motors, the investment may increase to 400% to 700% of the yearly power factor costs, but the reduction in line loss and the possible elimination of the need for new equipment may make this method the best investment.

The main causes of low power factor are often oversized motors on fans, pumps, and machinery, or motors running for long periods at night or no load. Idle motor waste can be reduced by using *idle machine cycle timers.* They range in price from $35 to $125 and will shut down a motor if it is not cycled, operated, or loaded within a set time (10 s—5 min).

Such a device ($100 to $200 installed) which will reduce energy waste can pay for itself in two weeks to six months in some cases.

A recent innovation developed by a NASA engineer is to add a solid-state control to a motor with a variable load. The control will sense motor current and voltage phase angle which are symptomatic of motor load, and reduce motor voltage accordingly. This improves the motor power factor, reduces noise, heat, and losses. There is virtually no benefit at full load.

17.3.3.2 **Demand Charges.** Demand charges are the costs of the highest average kilowatthour or kilovoltampere hour recorded by the utility using this rate method, during the demand period. These charges were designed to cover the utility line, generator, and other equipment costs required to serve each individual customers' needs.

17.3.4 Billing Segments. The three segments of a billing rate are usually kilowatt or kilovoltampere demand, kilowatthour energy, and a dollar constant charge. With the proper formula and knowledge of the kilowatt or kilovoltampere of demand, the number of kilowatthours used, and with the monthly constant in dollars, the monthly bill can be quickly calculated. Any additional charges, credits, and taxes must be added to the billing. The demand can be included in an energy rate.

17.4 Electric Bill Analysis. The utility assigned representative known as a power sales engineer, customer representative, or rate analyst is the proper logical contact to seek help from in energy analysis and audit work. This individual will likely provide past usage data, rate changes, and contract information along with recommendations on energy-related cost reduction.

17.4.1 Electric utility bills should be analyzed for at least two years on a fiscal or calendar year basis. This is the main source of information relating to airconditioning and heating energy costs.

The first step is to study the electric contract and regulations, and all of the rate changes that may have developed in the two year period.

17.4.2 For each billing period calculate and tabulate total billing, kilowatthours consumed, including tax, actual and billing demand in kilowatts or kilovoltamperes, power factor (if measured), any penalties or surcharges such as fuel clause costs, any credits such as for primary metering or better than normal power factor, number of days between readings, daily average kilowatthours (total kWh/days between readings). Note any rate changes and check monthly cost values with calculations on derived rate, as shown earlier.

17.4.3 Draw curves of above data showing any significant events such as holidays, strikes, outages, season changes, additions or deletions of energy using equipment. Show trend slopes, and occupancy factor, add degree days of heating or cooling, and any other significant feature that may effect energy usage.

17.4.4 With the described curves and data the engineer can usually correlate energy usage with time, weather, wind, occupancy, holidays, particular days of the week, and times of the day. This data will tell a story of how people, weather, events, etc will effect energy usage. Often there are not enough data, or the data are not accurate or precise enough, or significant parts are lacking due to need for more instruments and resulting data.

17.4.5 The above analysis will show the need for greater in-depth analysis of processes, machinery, lighting, insulation, air infiltration, heating equipment, etc. Energy conservation is an economic trade-off. Costs involved to conserve energy and save money are no different than any other investment, although some will accept a lesser return on energy projects for various reasons. A brief cost analysis will indicate the projects to be carried out now and those projects to delay for later appraisal.

The following questions may be of help while the study is progressing:

(1) Is heating the facility more costly

than air conditioning? What is the ratio of cost?

(2) What are the costs of lighting?

(3) How much lighting is used for utility, common areas, outside areas?

(4) How much lighting is used for tenants or business operations? This may be obtained by subtracting item (3) from the billing data.

(5) What are the effects of the heat from the lighting in relation to building heating and air-conditioning costs?

(6) Is it possible that heating and air conditioning occur at the same time?

17.4.6 What is the minimum or least kilowatt load of the facility? This is called the base load and can be obtained with a kilowatt chart recording, or the load current of the main service if allowances are made for the power factor. This is the minimum demand when all except essential equipment is, or should be, turned off.

In the absence of a chart recorder, base load can usually be measured when the facility is closed. For an office building the base load may amount to 30% of the peak load, and can be broken down approximately as follows:

(1) Lighting, night security and outside, 15%

(2) Building heating or cooling, variable, 40%

(3) Cooling towers, pumps, and fans, 30%

(4) Domestic water heating, 10%

(5) Elevators, 5%

Since the demand never drops below the base load, which may be 30% of the maximum yearly peak load, it is important to maximize the efficiency of base-load motors, drives, and lighting systems, and to minimize their usage.

17.5. Become a Better Customer so You Can Purchase Energy with Less Unit Cost.

17.5.1 In most utility systems, the least expensive energy rate is available when the following conditions are met:

(1) Buy at the highest available voltage.

(2) Supply your own primary equipment. It can often be rented from the utility and contractors are available to maintain and repair it. Many utilities use contractor's services.

(3) Keep power factor as high as economically justifiable.

(4) Keep demand as constant as possible, but turn off the quipment when not needed.

17.5.2 In theory, the best customer has a large, high-voltage load of 100% power factor for 8760 h per year. It would be desirable to reduce peak demand to level out consumption.

17.5.3 Work with the utility representative to reduce your costs. Periodically have the representative figure your billing, using all available rates. Change rates when justified.

17.6 Reduce Energy with Minimal Inconvenience and Cost

17.6.1 Studies show that generally 15% of energy can be saved with no awareness by building occupants using typical pre-energy shortage building design. With little inconvenience, up to 35% can be saved.

17.6.2 Reduce the base load (see 17.4.6). This likely consists of heating, ventilating, and air-conditioning (HVAC) loads, which are often oversized with poor-efficiency motors, inadequate controls which permit heating and cooling to operate at the same time. Also included are inefficient or oversized pumps and water chillers. New low-cost solid-state frequency and voltage changing controls for such loads will greatly reduce these energy costs.

17.7 Lighting

17.7.1 General. Lighting is usually the largest single all-year load in the typical commercial building, except for peak periods of electric air-conditioning or electric heating. The heat from the lighting will contribute significantly to the overall air-conditioning load in the summer, and decrease the heat required from heating systems in the winter. In developing an overall energy balance throughout the building, the effect of the *heat-of-light* must be carefully considered. Heat balance studies should be made of large buildings. These can be checked with computer programs for determining the integrated energy building requirements. Air handling systems which are well designed can most effectively handle the lighting heat by recirculating the heat in the air-conditioning system in the winter and by exhausting the heat before it mixes with fresh air in the summer.

The following conservation efforts involving lighting are recommended.

(1) Use of energy-efficient types of lighting in remodeling and in new construction.

(2) Group relamp with lower wattage, high-efficacy lamps and clean lamps and luminaires.

(3) Eliminate unnecessary lighting.

(4) Improve manual and automatic switching and add dimming control.

(5) Lower background ambient lighting levels, improve and use day lighting, and use task lighting.

(6) Reduce glare and veiling reflections.

(7) Reduce lighting levels, depending upon the task requirements.

17.7.2 The Lighting Levels. The quality of lighting is as important as the quantity. Since we see only by contrast, it is more important to work in a glare-free environment, regardless of where glare comes from, than to provide high light levels. Polarized light-fixture panels may be of benefit. The avoidance of shadows in the work area is important. When lighting levels are reduced by turning off lamps, every effort should be made to accomplish this in a way which avoids a spotty lighting installation. Thus, for example, it is preferable to decrease the light level of each luminaire, rather than turn out half the luminaires. This can be accomplished by the use of energy saving lamps, low-wattage dimming ballasts, and the use of dummy tubes.

The level of light required is dependent on the task. For example, if large clear copies of documents are to be read, less light is required than if second or third carbons are to be examined. If, for example, poor quality reading material is limited to one or two individuals, supplementary lighting can be provided without increasing the overall lighting requirements. Sometimes lighting problems in the office can be resolved by requiring a higher quality copy.

17.7.3 Switching. Lighting should be circuited so that half or partial lighting is available during periods of building cleaning, low activity, or even during utility brownouts, so that areas not in use can be separately darkened.

Several techniques in use are listed here.

(1) Luminaires with the capability of having individual lamps or pairs of lamps circuited and switched so that the lighting levels can be varied to match the task, are desirable.

(2) Separate circuits for lighting along the interior perimeter of the building. This permits the outer perimeter lighting to be automatically or manually reduced when sunlight is available.

(3) Schedule cleaning and other special

activities so that the maximum amount of work can be done in a short period of time and then the operation is moved on. This could include cleaning half of one floor while the lights are off in the other half. Some facility managers have found that most cleaning activities can be done during regular business hours.

(4) Photocell switching can be installed to control exterior lighting at little cost per unit. With dimming capabilities on interior lighting, photocell control can dim interior lighting to match changing sunlight from windows, to maintain an acceptable light level. New equipment is appearing that sense the presence of people, and controls individual lamps or groups of lamps to provide lighting when the space is occupied.

(5) The addition of time-clock controls can range from installing complicated master timers to replacing ordinary wall switches with timers. For certain outdoor applications, there are astronomical time switches available which will adjust for the varying length of daylight. Also, timers with an automatic motor-driven spring drive to override electrical outages for up to 16 hours can keep switching times on schedule. With resumption of power, the mechanism will wind-up and reset. There are also many installations that use both clocks and photo switches.

(6) In the larger commercial buildings utilizing computer control, a lighting control package is often provided in the software of the master computer. Typically, lighting can be controlled by floor quadrants on almost an hourly basis using computer systems. Where the system is programmed so that lighting is maintained for absolute minimum hours of operation, it is desirable to have switches which can override the computer in each area. Where lights are turned out automatically, emergency or security lighting should be maintained so that any person who may be in the area is not placed in a hazardous situation due to darkness.

(7) Sophisticated lighting control systems, some of them microprocessor based, are now available to control lighting in small areas with manual override. Low-voltage relays and wireless switching relays with radio-frequency actuators can give precise control while reducing costs.

(8) The question arises whether energy or equipment life is more important in deciding to frequently turn lights off and on. In general, turning lights or equipment off when not in use will save energy, and in times of critical fuel shortage, the general rule should be to turn equipment off. It should be noted that the lifetime energy consumed by a lamp costs many more times than the cost of the lamp, plus the labor cost of lamp replacement. While most fluorescent and all incandescent lamps can be turned off and on without operational problems, high-intensity discharge lights such as mercury, metal halide, or high-pressure sodium require 1 to 10 min to relight.

(9) When reducing lighting by removing fluorescent lights from circuits, both lamps should be removed from a luminaire fed by a two lamp rapid start ballast. The ballast will still consume several watts, but will not be damaged. If the reduction in light is permanent, removing the ballast from the circuit will save an additional three to seven watts per lamp. Lamps can generally be removed from single lamp fluorescent, mercury, and metal-halide ballasts without adverse effects; however, switching or disconnection of the fixture is preferred where practical. Slimline fluorescent luminaires are designed so removal

of the lamp will disconnect the ballast so ballast removal is not necessary. The use of a dummy lamp in a two-lamp fluorescent fixture can reduce lighting by over half, but dummy lamps do increase the cost per lumen hour over a properly designed installation and should only be considered a temporary solution.

17.7.4 Efficiency of Luminaires and Lamps. For comparative purposes, typical luminous efficacies based on only the bare lamp sources are listed in Table 94.

The overall efficiency of the fixture is determined by its design including refractors, reflectors, beam spread, light spread, etc. Architectural requirements may make the fixture into more of a decorative element than a light source. Typical use of a great number of incandescent lamps concentrated for effect should be criticized from a lighting and energy conservation standpoint. The tendency to design special fixtures, which may be quite inefficient compared to those commercially available, well designed units which have been fully tested, is economically questionable. If lighting is to be used for effect, then the areas to be covered should be limited to areas requiring impact. It is the responsibility of the engineer to call to the attention of the architect the lighting design which is not energy efficient. Alternates should be considered. Judicial use of coefficients of utilization and other indices of efficiency can reduce power requirements.

More sophisticated concepts such as Equivalent Sphere Illumination (ESI) and Visual Comfort Probability (VCP) are helpful in developing the most effective and efficient lighting installations.

Computer programs are available for design of indoor and outdoor lighting systems.

17.7.5 Cleaning of Luminaires. The maintained lighting level of any lighting installation can be materially increased by cleaning and relamping frequently. For example, changing the cleaning interval of an office installation from two years to one year can raise the maintained lighting level 10 to 20%. This can save energy in initial designs or by permitting the operation of systems with lamps removed without undue degradation of light levels. Fluorescent lamps linearly drop in light output until the end of life. Where group relamping is undertaken, a simple study will show how the maintained lighting levels can be increased by more frequent cleaning and relamping.

17.8 Operations. An energy conservation program should be developed in commercial buildings. Building designs should incorporate provisions for energy savings. Some of the items that should be considered in developing designs for commercial buildings are as follows:

17.8.1 Metering of Tenants' Areas and Building Operations. Where tenants are to occupy commercial facilities, it is highly desired to have provisions for metering the tenants' area directly. In some states, submetering or the resale of electricity is permitted, while in other states check-metering only is permitted. Where sub-metering cannot legally be used to charge tenants, periodic load surveys may provide for adjustment of the utility billing based on the power consumption, provided that the lease contains such provisions. Tenants who pay directly for energy will be less likely to waste it and will likely monitor their employees in the use of appliances. In many cases, it is desirable to meter sections of buildings or operations so that charges can be allocated to the various areas to empha-

Table 94
Typical Luminous Efficacies Based on Only the Bare Lamp Sources

(Does Not Include Ballast)

Fluorescent	50–80 lm/W	Color rendition — usual. Acceptability for commercial buildings, high quality light, generally acceptable.
Incandescent	15–20 lm/W	High quality light, most architecturally acceptable. Use should be limited to areas requiring special treatment.
HID Lighting		
Mercury, clear color improved	30–60 lm/W	Blue — seldom used, long life, good color, very acceptable for circulation areas. Good color, acceptable for most general
color corrected	. . . .	applications. Excellent color, acceptable for general
deluxe white	. . . .	applications.
Metal additive (Metal halide) mercury	80–125 lm/W	Good color, very controllable, acceptable for all general applications. Medium life.
High pressure sodium	80–190 lm/W	Golden color, very controllable, acceptable for industrial, some commercial, and public areas. Long life.
Low pressure sodium	180 lm/W	Monochromatic, yellow color rendition, satisfactory for parking lots and highway lighting. Seldom used.

size the need for responsibility of energy usage by all groups.

17.8.2 Interlocking or Key Switches. To prevent unauthorized use or tampering with security lighting and power circuits, the use of interlocking or key switches should be considered. On the other hand, in areas of individual offices, use of key switches may prevent turning off lights by the people that use the offices. Placing switches in locked areas or electric closets is equivalent to using key switches and can negate the desire of an energy conscious employee to help save energy.

17.8.3 Load Control. Programs of minimum, maximum, and off-times opera-

tion of major items of equipment should be developed. Plans for the most energy conserving usage of air-conditioning machines and other electrically driven equipment, including air-handling units, should be developed. In some cases, systems are essentially divided into multiples which permit the loss of a single unit without overall loss of the system. In these cases, it might very well be practical as part of a major energy conserving effort to shut down one of these multiple units for extended periods as an enforced energy saving measure. Many areas of buildings are very lightly used or unused during portions of the day. Provisions should be made for automatically or manually in some cases, cutting off power to non-essential areas as part of the normal duties of a watch engineer. Often the control of such systems can be incorporated into the mechanical system control computer that is generally accepted today as a standard in large buildings. These computers are equipped with pre-programmed packages which can be adapted to almost any use within the modern commercial building. For smaller installations and where technical assistance is available, the programmable controller, in conjunction with suitable interfacing and remote supervisor equipment, forms an excellent method for providing very flexible building control. The programmable controller has the advantage of relatively simple development of *software* by untrained personnel for systems up to medium complexity.

17.8.4 Load Shedding. The building staff should prepare for power reductions as required by either the utility or governmental energy agencies. In the event that brownouts become part of the way of life in an effort on the part of the utilities to reduce electrical consumption, arrangements can be made for shedding non-critical loads on a normal operational basis. Load shedding involves dropping a number of loads, perhaps in stages, from the system on a planned basis. A well developed plan of operation for load shedding will make it possible for operating personnel to routinely reduce load requirements without a panic situation developing with normal operating personnel, without serious degradation of service to the public, and without the loss of critical equipment.

17.9 Energy Conservation Equipment. There are a variety of energy conservation devices currently on the market. The designer should be aware that many

of these devices are poorly designed, unrealistic in application, or even fraudulent. A designer should not attempt to assemble complex components, under any circumstances, into a functioning system unless he has specific knowledge of the application of each device. Makeshift application of computers frequently turns out to be disastrous.

One of the more difficult areas of design is to obtain suitable interfaces between the application devices such as motors, heating units, and other utilization devices, and the relatively low-energy control systems. The interfacing of remote-control systems and the application devices requires careful coordination between the mechanical equipment or motor-control center, manufacturer, and the manufacturer of the supervisory control equipment. One major problem is the compatibility of control devices, the controlled equipment, and the signal system.

Where information is transmitted to remote power equipment, it may be wise to check to find out if the device has functioned as required. It is good to know if someone turned the lights on in a local area, or if a piece of equipment was started locally without the computer, or other automatic control intervention. Feedback from the controlled areas or equipment to the control center is thus very desirable.

Listed below are some energy conservation devices and concepts which may be utilized. Some of these are basically energy demand reduction devices and others are more concerned with overall energy conservation.

(1) Load limiters or demand limiters are devices programmed to operate building loads in such a sequence or manner that the billing demand remains at an optimized value. Such devices can

be used to provide alarms when the rate of energy usage exceeds established levels.

(2) Use of automated devices for shutting down or reducing the level of operation of nonessential equipment. Multi- or variable-speed equipment with regulator or feedback control can materially reduce energy requirements. This can be combined with other simple devices as photo-cell control of lighting to be integrated with computerized controls of the heating and ventilating system.

(3) Use of *waste* heat, including that of lighting fixtures, as part of the space-conditioning system.

(4) Efficiency and losses may be specified or used in determining the acceptability of equipment. For example, the cost of rated transformer losses for various loads, calculated for given periods of time such as 10–20 years, can be added on a weighted basis, to the first cost of the transformer in evaluating the low bid.

(5) Energy can be recovered in vertical transportation equipment by utilizing regenerative systems. A descending elevator, for example, can feed back energy into the power system.

(6) Use of high-efficiency motors, drives, belts, and power factor ballast will minimize line and equipment losses. Power factor correcting equipment (capacitors, synchronous motors) and the proper sizing of induction motors all serve to maintain the facility power factor at high values with minimum losses.

17.10 Utilizing On-Site Power. Until recently, much of the thrust of energy management has been directed to shedding of loads. However, the economic advantages to the consumer resulting from the loss of these loads are some-times limited. An alternate or additional approach is to utilize on-site power.

The practice of utilizing on-site generator sets to reduce electric utility charges is fast becoming an economically attractive proposition. Previously, engine generator sets were purchased primarily for their insurance value, with little consideration, or expectation of achieving operational savings. Now more installations of generator sets are being considered strictly on an investment basis.

On-site generator set systems may be less costly to install on a cost per kilowatt basis than are utility central stations. In addition, the existence of emergency systems has increased considerably over the past decade.

Peak-load transfer is one method of improving load factor by selecting loads to be automatically transferred from the utility source to an on-site generator. The generator is electrically isolated from the normal distribution system at all times. The amount of power shaved from the utility peak is dependent on the load or loads connected to the selected transfer switch(es).

Care should be exercised in evaluating the optimum demand setting to ensure adequate peak shaving with minimum loading of the on-site generator.

This transfer method, by its nature, requires no interface or coordination with the utility power source, since there is never a possibility of interconnection.

Figure 147 represents a peak load transfer system providing power to two mechanical loads, each fed by its own automatic transfer switch. The emergency system is comprised of two engine generator sets paralleled onto a common emergency bus. The system also feeds two other transfer switches which are designated as emergency loads. The system may be designed such that the

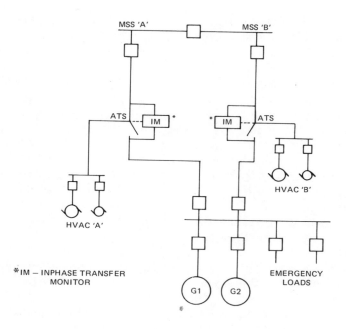

Fig 147
Peak Demand Reduction Using
Multiple On-Site Generators

peak shaving loads are not emergency loads. If this is the case, and a normal failure occurs during a peak demand mode, the peak shaving loads are immediately disconnected from the emergency bus, allowing the emergency loads to be transferred.

17.10.1 Paralleling on-site generators with the utility source is perhaps the most attractive, yet least utilized method of peak demand reduction.

In this system, the generator is directly connected with the normal distribution system. Peaking power is provided directly into the system, with the amount furnished being determined by the engine governor controls. Such systems are usually designed and adjusted so the on-site generator assumes all the load in excess of the limit set point and up to the capacity of the on-site generator. During the peak shaving period, the utility will supply constant power and load fluctuations will be handled by the on-site generators.

Figure 148 depicts a system in which the emergency generators are connected directly into the normal distribution system. Again, note that this is a multiple engine emergency system. Since it is connected directly into the existing distribution, major modifications are generally not required of the system. Furthermore, load selection is not a factor.

Since the on-site power source is to be

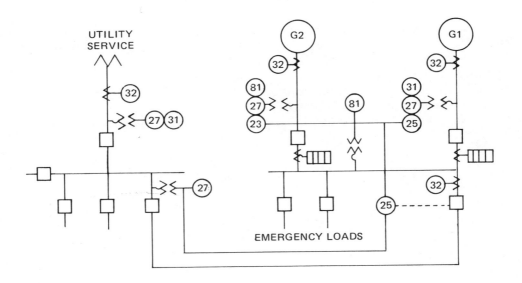

Fig 148
Paralleling Multiple On-Site
Generators with Utility

connected to the utility grid, there are additional considerations. Coordination and cooperation with the electric company are necessary. The utility will require its approval of an adequate protective relaying scheme.

In some instances, the utility company is reluctant to permit paralleling of on-site power with their system. However, this reluctance on the part of utilities seems to be lessening, particularly when it is shown that the purpose of the system is not to infringe on their revenues, but to assist them in handling a difficult situation of dwindling spinning reserves. Some utility companies have even gone so far as to actively promote parallel operation and offer payment for excess power generated.

17.11 Specific Projects. The following documents will be helpful to the reader in locating specific projects to help reduce energy and costs. See 17.2.2.

(1) *Energy User News.* Fairchild Publications, 7 East 12th St, New York, New York 10003. (Weekly.)

(2) *Plant Engineering and Power Engineering.* Technical Publishing, 1301 South Grove Ave, Barrington, Illinois 60010. (Bi-Monthly and Monthly.)

(3) *Power Engineering.* McGraw Hill, 1220 Avenue of Americas, New York, New York 10020. (Monthly.)

(4) *Energy Insider.* Department of Energy (DOE), Office of Public Affairs, Washington, DC. (Weekly.)

(5) *Specifying Engineering.* Chicago, Illinois 60002. (Monthly.)

(6) *Electrical Construction and Maintenance*. McGraw Hill, 1220 Avenue of the Americas, New York, New York, 10020. (Monthly.)

(7) *Lighting Design and Application*. Illuminating Engineering Society, 345 E 47th St, New York, New York 10017. (Monthly.)

(8) *ASHRAE Journal*. American Society of Heating, Refrigerating, and Air-Conditioning Engineers, Atlanta, Georgia. (Monthly.)

(9) *Energy*. Association of Energy Engineers, Atlanta, Georgia. (Quarterly.)

Index